Signals and Communication Technology

More information about this series at http://www.springer.com/series/4748

Anatoly M. Rembovsky · Alexander V. Ashikhmin
Vladimir A. Kozmin · Sergey M. Smolskiy

Radio Monitoring

Automated Systems and Their Components

Springer

Anatoly M. Rembovsky
JSC IRCOS
Moscow
Russia

Vladimir A. Kozmin
JSC IRCOS
Moscow
Russia

Alexander V. Ashikhmin
JSC IRCOS
Moscow
Russia

Sergey M. Smolskiy
Moscow Power Engineering Institute
Moscow
Russia

ISSN 1860-4862 ISSN 1860-4870 (electronic)
Signals and Communication Technology
ISBN 978-3-030-08953-5 ISBN 978-3-319-74277-9 (eBook)
https://doi.org/10.1007/978-3-319-74277-9

Printed on acid-free paper

This Springer imprint is published by the registered company Springer International Publishing AG part of Springer Nature
The registered company address is: Gewerbestrasse 11, 6330 Cham, Switzerland

Preface

At present, we observe the exciting modernization of wireless technologies: the radio communication and radio broadcasting systems, the television systems, and the wireless data transmission. This modernization is directed to increasing of capacity, spectral efficiency, noise immunity, and it is performed on the base of complex modulation types with code and time division, ultra-wideband signals, adaptation to the environment, redistribution and reuse of radio frequencies, etc.

In the civil area, the DVB-T2 digital television systems of the second generation are replacing now the previous first-generation DVB-T digital systems. The second-generation system provides increasing of information transmission rate up to 30% as minimum. Instead of local wireless access systems of the *a*, *b*, *g*, *n* Wi-Fi standards, the systems of the *ac* standard are widely used with the capacity, which value is more in order. In cellular communication networks, the LTE technology has got wide distribution, which has significant advantages compared to the previous UMTS technology.

In the military area, the fast development of wireless communications and control technologies is observed. The conception of modern war assumes the system usage of the military equipment and armaments integrated in the unified stable information network, which should ensure delivering of the reliable and complete information about situation and orders of headquarters to all participants in real time. Systems of wireless telecommunications, which are used at all control levels beginning from the tactical unit and finishing at the strategic level, are concerned as key elements of the military information network.

Methods of radio monitoring should be improved after its object variation. To them, we can relate the combined electromagnetic situation, including the occupancy of radio ranges and frequencies, the acting radio electronic equipment, the high-frequency industrial, medical and scientific installations, the sources of industrial interferences. In order to manage with the growing stream of more complicated tasks without increasing of the employees' quantity, the modern automated control systems should be used. Application of similar systems increases rate and accuracy of radio monitoring tasks fulfillment, releases the operating personnel from routine jobs, and essentially increases the labor productivity.

The Russian IRCOS Company works at Russian and international markets during 25 years. For a quarter of a century, this company made a way from the development and delivery of the radio monitoring equipment with limited functionality on the base of imported radio receivers up to the modern multi-functional high-performance equipment of the own manufacture and the automated radio monitoring systems. This equipment integrates both the equipment of the own production and the equipment of other companies-manufacturers. The automated systems delivered during more than five last years are intended for application in the civil and military agencies, they provide automatic planned and operative modes under the operator control and use the risk-oriented eventual mechanisms of functioning, and they are able to operate effectively at district, regional, and national levels.

This book widens the materials of books “Radio Monitoring: Problems, Methods and Equipment. Springer Lecture Notes in Electrical Engineering 2009” and it is based on the new research results of the automated radio monitoring systems development, obtained by radio and electronic engineers, experts in the software and researchers of the IRCOS Company. That includes constructions and features of radio monitoring equipment, architectures of the software, methods and processing algorithms for radio signals, which purpose are search and detection of radio emissions, measurement of their parameters, identification of radio emission sources and their localization in the open territory and inside the rooms.

In Chap. 1, the architecture of the automated systems is considered, which are purposed to be used at national, regional and district levels by the civil and military agencies. The features of automation units are described, which consist of the radio monitoring equipment, the engineering-technical infrastructure, and the complex of software packets. The examples of the ARMADA and ASU RCHS “Universiada-2013” automated systems constructions are given.

Chapter 2 is devoted to methods of revelation of unauthorized radio emissions in the engineering objects, in the rooms, in the open territory. The sources of unauthorized emissions are considered, the stages of their revelation, the equipment for radio monitoring to be used; the examples of the AREAL automated system construction are given depending on application conditions: along the lengthy boundary, in the restricted part of the open territory, within the limits of the completely closed territory, in rooms, etc. Some typical variants of the system construction are described.

Chapter 3 is devoted to the software architecture, which controls the automated system functioning. The software unifies the geographically distributed equipment of radio monitoring, provides users interaction with equipment on the base of the risk-oriented eventual mechanism and centralized data storage, automatically controls the system status, forms the accounting and report documentation, and interacts with external informational systems.

Chapter 4 contains information on engineering infrastructure of the automated systems. The arrangement of control centers and points, radio monitoring points, application of communication lines between the nodes of the automated system are discussed.

Chapter 5 considers components of the automated systems, which provide the radio monitoring functions fulfillment: digital radio receivers (DRR), automated radio direction-finders, and manpack radio monitoring equipment including their construction and methods for signal processing.

In Chap. 6, measuring methods for radio signal parameters are discussed, which are necessary to control frequency-spatial plans of the radio communications service: the central frequency of analog and digital signals, the width of occupied frequency band, the electromagnetic field strength, the radio interference intensity, as well as the occupancy of the radio frequency spectrum.

In Chap. 7, on the base of the general approach to goniometric, amplitude, temporal, frequency, and time difference of arrival methods of measurements, the algorithms of radio emission sources localization are synthesized. Examples of constructions of time difference of arrival systems are given, and algorithms for synchronization of radio receivers included in the systems are examined, which is necessary for radio emission sources location calculation with high accuracy.

Chapter 8 is devoted to TV and radio broadcasting signals monitoring. The peculiarities of digital TV signals technology, the analyzer of digital TV signals are described as well as the automated complex for multi-channel monitoring of the broadcast content.

In Chap. 9, the reader will be acquainted with problems of construction and functioning of the set of radio signal analyzers for digital communication and data transmission technology including GSM, UMTS, LTE, DMR, APCO P25, Wi-Fi technologies, which use the digital radio receivers of the ARGAMAK-family as the basing equipment, which provide a possibility of radio signal analysis both at the standard and non-standard frequencies.

Moscow, Russia

Anatoly M. Rembovsky
Alexander V. Ashikhmin
Vladimir A. Kozmin
Sergey M. Smolskiy

Acknowledgements

We thank our colleagues from the IRCOS Company for their support and encouragement in writing this book in spite of high everyday workload. Without their enthusiastic help, this book would not have become a reality.

We are grateful to the design engineers Alexander Sergienko, Michael Spazhakin, Alexander Polyakov, Pavel Pershin, and Alexander Litvinov for assistance at describing constructions, functions, and peculiarities of digital radio receivers and radio direction-finders; this information was included in all chapters of the book.

We would also like to show our gratitude to the software developers Dr. Andrey Zhukov, Dr. Andrey Pyatunin, Dr. Sergey Korochin, Anton Novikov, Pavel Alexeev, Dmitry Bocharov, Oleg Bespalov, Alexander Provotorov, the SQA engineers Denis Radchenko, Igor Myakinin, the software analysts Dr. Artem Avdushin, Dmitry Sysoev, Dmitry Alekseev, the system administrators Sergey Barkalov and Maxim Lushin for great support at preparing materials about software functions and interface, architecture and infrastructure of radio monitoring automated systems, and for development of special examples of software operation, which were included in all chapters of the book.

We especially would like to acknowledge the contribution of technical director Alexander Bogdanov who provided information about radio signals modulation analysis to Chap. 2.

We would like to thank the researcher Dr. Anton Tokarev and test engineer Dmitry Gnezdilov for investigations in measurements of radio signals and radio interferences parameters, radio frequency spectrum occupancy. Results of this work were included in Chap. 6 of the book.

A special gratitude goes out to the researchers Dr. Vladimir Manelis, Dr. Igor Kaykov, and Dr. Vladimir Sladkih, whose help enabled us to write Chaps. 8 and 9 dedicated to analysis algorithms of radio communication systems digital signals parameters and data transfer and their realization in the radio monitoring equipment.

We would also like to emphasize the contributions of the researcher Dr. Igor Kryzhko for supporting us with materials about radio signal sources localization, including advanced time difference of arrival method, which were described in Chap. 7, and supporting us with materials about analyzers of the parameters of digital signals, which are included in Chap. 9

We should also mention Sergeeva Evgenia for advising when translating the book and Anna Avdushina for preparing pictures and diagrams for it.

Anatoly M. Rembovsky
Alexander V. Ashikhmin
Vladimir A. Kozmin
Sergey M. Smolskiy

Contents

Acronyms and Abbreviations

(I)BSS	(Independent) Basic Service Set
2G	Second generation
3D	Three dimension
3G	Third generation
4G	Fourth generation
AACH	Access assignment Channel
AC	Alternate current
ADC	Analog–digital converter
ADS-B	Automatic Dependent Surveillance-Broadcast
AES-CCM	Advanced Encryption Standard (AES) in the Counter with Cipher Block Chaining—Message Authentication Code (CBC-MAC) Mode (CCM)
AFC	Amplitude–frequency characteristic
AGC	Automatic gain control
AI	Amplitude imbalance
AM	Amplitude modulation
AMBE	Advanced Multi-band Excitation
APCO P25	Association of Public Safety Communications Officials International Project 25
ARMS	Automated radio monitoring system
AS	Antenna system
ASK	Amplitude-shift keying
ATS	Automatic telephone station
AWS	Automated workstations
BA	Broadcast adjacent cells channels
BB	Baseband
BCC	Base station color code
BCCH	Broadcast control channel
BCH	Bose Chaudhuri Hocquenghem
BER	Bit error rate

BIRT	Business Intelligence and Reporting Tools
BNCH	Broadcast network channel
BPSK	Binary phase-shift keying
BS	Base station
BSCH	Broadcast sync channel
BSIC	Base station identification code
BSSID	Basic Service Set IDentifier
BT	Bandwidth-symbol time
C4FM	Continuous four-position frequency modulation
CA	Cell allocation
CACH	Common Announcement CHannel
CAI	Common application interface
CB	Cell Barred
CCK	Complementary code keying
CDF	Complex digital filter
CDMA	Code division multiple access
CELP	Code-excited linear prediction
CISPR	Comité International Spécial des Perturbations Radioélectriques (in French)
CP	Control physical channel
CPM	Continuous phase modulation
CQPSK	Compatible quadrature phase-shift keying
CRC	Cyclic redundancy check
CSBK	Control signaling blocK
CSBKO	CSBK Opcode
CSMA/CA	Carrier sense multiple access with collision avoidance
CSSI	Console sub-system interface
D8PSK	Differential 8-phase-shift keying
DB	Database
DBMS	Database management system
DBPSK	Differential binary phase-shift keying
DC	Direct current
DECT	Digital European Cordless Telecommunications
DFT	Discrete Fourier transform
DLL	Dynamic link library
DMR	Digital mobile radio
DOM	Document Object Model
DPF	Data packet format
DQPSK	Differential quadrature phase-shift keying
DRR	Digital radio receiver
DSP	Digital signal processor
DVB	Digital Video Broadcasting
DVB-H	Digital Video Broadcasting Handheld
DVB-S	Digital Video Broadcasting Satellite
DVB-T	Digital Video Broadcasting Terrestrial

DVB-T2	Digital Video Broadcasting Terrestrial Second Generation
ECGI	E-UTRAN cell global identifier
EDGE	Enhanced Data Rates for GSM Evolution
EIGRP	Enhanced Interior Gateway Routing Protocol
EJB	Enterprise Java Bean
EMB	Embedded signaling
EMC	Equipment Manufacturer's Code
eNB	E-UTRAN NodeB
ETSI	European Telecommunications Standards Institute
E-UTRAN	Evolved Universal Terrestrial Radio Access Network
EV-DO	Evolution-Data Optimized
EVM	Error vector magnitude
FAIAS	Federal Automated Information-Analytic System
FCCH	Frequency correction channel
FDD	Frequency division duplex
FDMA	Frequency division multiple access
FEC	Forward error correction
FER	Frame erasure rate
FFSK	Fast frequency-shift keying
FFT	Fast Fourier transform
FIR	Finite impulse response
FM	Frequency modulation
FPGA	Field-programmable gate array
FPN	Fixed Part Number
FPU	Floating-point unit
FRMS	Fixed radio monitoring station
FSI	Fixed/Base Station Sub-system Interface
FSK	Frequency-shift keying
FSK-MS	Frequency-shift keying with minimal shift
FSP	Frequency-spatial plan
FTP	File Transfer Protocol
GCS	Generic Continuous Stream
GFPS	Generic Fixed-length Packetized Stream
GLONASS	Global navigation satellite system
GMSK	Gaussian minimum shift keying
GNSS	Global navigation satellite system
GPIO	General purpose input–output
GPRS	General Packet Radio Service
GPS	Global positioning system
GSE	Generic Encapsulated Stream
GSM	Global system for mobile communications
GUI	Graphical user interface
GWT	Google Web Toolkit
GWT-RPC	Google Web Toolkit Remote Procedure Call
HF	High frequency

HPA	High-precision algorithm
HSRP	Hot Standby Router Protocol
HTML	HyperText Markup Language
HTTP	HyperText Transfer Protocol
HWCL	Hardware Component Library
I	In-phase component
I/Q	In-phase and quadrature components
IC	Integrated circuits
ID	IDentifier
IDFT	Inverse discrete Fourier transform
IEC	International Electrotechnical Commission
IEEE	Institute of Electrical and Electronics Engineers
IF	Intermediate frequency
IFFT	Inverse fast Fourier transform
IMBE	Improved Multi-band Excitation
IP	Internet Protocol
IP2	Intercept point second order
IP3	Intercept point third order
IP67	Ingress Protection rating 67
IPSec	IP Security
ISO	International Organization for Standardization
ISSI	Inter-RF Sub-system Interface
ISSY	Input Stream SYnchronization
ITU	International Telecommunication Union
J2EE	Java 2 Enterprise Edition
JMS	Java Message Service
LA	Local area
LAC	Local area code
LAI	Local area identifier
LAN	Local area network
LB	Last block
LC	Link Control
LCH	Linearization channel
LCSS	Link Control Start/Stop
LDPC	Low-density parity check
LLS	Long learning symbols
LML	Labeling measuring laboratories
LNA	Low-noise amplifier
LPF	Low-pass filter
LPT	Line Print Terminal
LSM	Least-squares method
LTE	Long-Term Evolution
LTFS	Local time–frequency scale
MAC	Media access control
MBC	Multiple Block Control packets

MC	Multi–carrier
MCC	Mobile country code
MCCH	Main control channel
MER	Modulation error rate
MIB	Management information base
MIMO	Multiple-input multiple-output
MNC	Mobile network code
MPEG	Moving Picture Experts Group
MRMS	Mobile radio monitoring stations
MS	Mobile station
MS-DOS	Microsoft Disk Operating System
MSK	Minimal shift keying
NAC	Network access code
NCC	Network color code
NID	Network IDentifier
NMEA	National Marine Electronics Association
NMI	Network management interface
NTSC	National Television Standards Committee
OFDM	Orthogonal frequency division multiplexing
OFDMA	Orthogonal frequency division multiple access
ONID	Original Network ID
OOB	Out-of-band signaling
OQPSK	Offset quadrature phase-shift keying
ORMS	Object radio monitoring stations
OS	Operation system
OSM	Open Street Map
OSPF	Open Shortest Path First
OVSF	Orthogonal variable spreading factor
PAL	Phase Alternating Line
PAPR	Peak-to-average power ratio
PBCH	Physical broadcast channel
PCB	Printed circuit board
PCFICH	Physical control format indicator channel
PCIe	Peripheral Component Interconnect Express
PDCCH	Physical downlink control channel
PDSCH	Physical downlink shared channel
PDU	Protocol Data Unit
PER	Packet error rate
PHICH	Physical Hybrid ARQ Indicator channel
PHY	Physical level
PI	Privacy Indicator
PID	Program IDentifier
PJ	Phase jitter
PLCP	Physical Layer Convergence Procedure
PLP	Physical layer pipes

PP	Plot pattern
PPS	Signals Pulse-per-Second
PPTP	Point-to-Point Tunneling Protocol
PRS	Pseudo-random sequence
PSDU	PLCP Service Data Unit
PSK	Phase-shift keying
Q	Quadrature component
QAM	Quadrature amplitude modulation
QE	Quadrature error
QoS	Quality of Service
QPSK	Quadrature phase-shift keying
RB	Radio broadcasting
REE	Radio electronic environment
REF	Reference oscillator
RES	Radio emission source
REST	REpresentational State Transfer
RF	Radio frequency
RFIC	Radio Frequency Integrated Circuit
RFPI	Radio Fixed Part Identity
RIA	Recursive interpolation algorithm
RIP	Routing Information Protocol
RMP	Radio monitoring point
RMS	Root-mean-square
RMSE	Root-mean-square error
RMSG	Radio monitoring and interference search groups
ROM	Read-only memory
RPC	Remote Procedure Call
RPN	Radio Fixed Part Number
SCCH	Specific control channel
SCH	Signaling channel
SCH	Synchronization channel
SCI	Synchronized capsule indicator
SDPSK	Symmetric differential phase-shift keying
SDR	Software defined radio
SDXC	Secure Digital eXtended Capacity
SECAM	Séquentiel Couleur Avec Mémoire (in French)
SFN	Signal frequency network
SFP	Small Formfactor Pluggable
SHF	Super-high frequency
SIB	System information block
SID	Security identifier
SIM	Subscriber Identification Module
SINR	Signal-to-interference ratio
SLS	Short learning symbols
SMS	Short Message Service

SNMP	Simple Network Management Protocol
SNR	Signal-to-noise ratio
SOAP	Simple Object Access Protocol
SoC	System on chip
SOM	Start Of Message
SSB	Single-sideband modulation
SSID	Service set identifier
SSID	Sub-system identifier
STA	Technical analysis
STCH	Stealing channel
STE	System target error
STP	Spanning Tree Protocol
STTD	Space-time transmit diversity
TAC	Tracking area code
TACT	TDMA Access Channel Type
TCP/IP	Transmission Control Protocol/Internet Protocol
TDD	Time division duplex
TDMA	Time division multiple access
TDOA	Time difference of arrivals
TETRA	TErrestrial Trunked RAdio
TETRA PDO	TETRA Packet Data Optimized
TETRA V+D	TETRA Voice plus Data
TS	Transport stream
TSID	Transport stream ID
TSTD	Time-switched transmit diversity
TV	Television
UAB	UDT Appended Blocks
UDP	User Datagram Protocol
UDT	Unified data transport
UHF	Ultra-high frequency
UMTS	Universal Mobile Telecommunications System
UP	Unallocated physical channel
UP	Unified protocol
UP	User packet
UPS	Uninterruptible power supply
USB	Universal Serial Bus
UTC	Universal time coordinates
UUID	Universally Unique IDentifier
VHF	Very high frequency
VLAN	Virtual local area network
VPLS	Virtual Private LAN Service
VPN	Virtual private network
VRRP	Virtual Router Redundancy Protocol
VSAT	Very small aperture terminal
Wi-Fi	Wireless Fidelity

WiMAX	Worldwide Interoperability for Microwave Access
WSDL	Web Services Description Language
XML	eXtensible Markup Language

Abstract

Theoretical and practical problems in construction and application of automated radio monitoring systems for civil and governmental structures including their components: equipment of radio monitoring, the software and engineering infrastructure are considered. The scalable architecture, application of Web technologies, a possibility to use the equipment from different manufacturers, the event mechanism of test fulfillment, and effective self-diagnostics are the typical features of considered systems. The constructive solutions, characteristics of modern equipment for radio monitoring, and their application are given. A great attention is paid to methods of digital signal processing: detection, direction-finding, measurement of radio signal and interference parameters, radio emission source identification and localization. Practical implementations of described methods are presented in applied software packages and in radio monitoring equipment.

Chapter 1
Automated Radio Monitoring Systems

1.1 Introduction

Active development of radio communication services, radio broadcast, television systems and data wireless transmission systems, as well as radar and radio navigation technologies requires improvement of control methods of radio frequency spectrum use. The radio frequency spectrum or (the same) the radio frequency resource is the limited natural resource, and its reasonable usage has the same important sense for the country, as other natural resources, for example, land or water resources. The monitoring of radio frequency spectrum should stimulate development of radio communication equipment, wireless communication systems, and implementation of newest radio electronic technologies. Also, it should develop economics and promote country defense, law and order, life and healthy safety of the citizens [1–4]. All this is impossible without national system of management, which functions consist of planning, regulation and licensing of radio frequency spectrum use, radio electronic equipment usage, international cooperation, and spectrum use methods research.

The radio monitoring is concerned to be one of the most important functions of the control system [1–3]. It is caused by the fact that in reality even authorized use of the radio frequency spectrum does not ensure the planned results due to the complexity of equipment, which is used for information transmission or for industrial technologies and due to complexity of natural processes related to radio wave propagation and interference occurrence.

At present, the most effective radio monitoring systems represent a hierarchy of national, federal, regional, fixed, remote, and mobile radio monitoring stations integrated into the unified computer network, which operates in the real time using complicated software based on the client\server technology [2, 3, 5–7].

In the worldwide technical literature including handbooks and recommendations, published by the International Telecommunication Union (ITU), the term

A. M. Rembovsky et al., *Radio Monitoring*, Signals and Communication Technology, https://doi.org/10.1007/978-3-319-74277-9_1

"spectrum monitoring" is used, which can be replaced by the term "radio monitoring" in Russia.

Each country legislatively defines aims and tasks of radio monitoring, which is based on the established general rules and regulations concerning the international spectrum allocation and its management [8]. Each country—the ITU member—creates the own legislation and relevant rules (taking into account international regulations) to fulfill functions of spectrum administrative management. The regulations include procedures of equipment authorization including specifications and standards for transmitter characteristics.

According to [3], the main charges of the service, which provides the spectrum management at the national level, are:

- Spectrum planning and regulation;
- Financing of the spectrum management, including payment for the spectrum use;
- Distribution and allocation of the frequency bands;
- Frequency assignment and license issue on its use;
- Interaction with other services inside the country;
- International and regional cooperation;
- Standardization and determination of technical requirements on radio equipment;
- Radio monitoring;
- Providing fulfillment of licensee's liabilities in the spectrum use by means of inspections and investigations;
- Fulfillment of spectrum control support, including computers automation, development of spectrum use methods, training.

The radio frequency spectrum control and radio monitoring are closely related since [3, 5]:

- Control establishes the official list of assigned frequencies for emission control;
- Control gives information about frequency bands to be controlled and control tasks;
- Radio monitoring receives specific tasks from the control system: for instance, tasks for search and identification of interferences;
- As a result of monitoring, the occupancy of the frequencies to be assigned is inspected;
- During radio monitoring, parameters are measured, technical correspondence of transmitters to the established norms and issued licensees is verified, and illegal transmitters and transmitters, which parameters do not correspond to norms, are detected and localized.

Radio monitoring is the most important aspect of the spectrum management, and it facilitates to the control process by means of practical monitoring of administrative results. Before decision making on frequency assignment, it is necessary to know their factual state.

Radio monitoring facilitates to spectrum control general operations by means of practical measurement of channel characteristics and band use in order to obtain the channel operation statistics and to estimate spectrum effectiveness. On the base of these data, we can compare the theoretical planning usage and the factual usage. This comparison may be used for correction of the plans.

Radio monitoring is also used to provide responsibilities enforcement in the field of spectrum usage. Having data about factual spectrum occupancy, the agency responsible for responsibilities enforcement can use them to achieve effective spectrum management on the base of correctly formed permissions.

Radio monitoring is tightly connected with functions of inspection and correspondence check, which enable identification and measurement of the interference signals, technical and operation characteristics of radiated signals, as well as detection and identification of illegally operating transmitters.

Radio monitoring information is necessary to be obtained because existing permission does not ensure that the spectrum resource is being used as intended. The equipment complexity, its interaction with other equipment, deliberate malfunction, or misuse may be the reasons for it. These problems are more enhanced due to the facts of fast increasing of the terrestrial and satellite systems' number and utilization of the devices that can cause interferences in these structures, for instance computers, which can create local interference.

Radio monitoring is the component of the national radio frequency spectrum management and the internationally legal protection of radio frequencies or radio frequency channel assignment.

During radio monitoring, the following procedures are executed:

- Measurement and instrumental estimation of emission parameters of radio electronic equipment and high-frequency devices. With the help of these measures, the emission correspondence to parameters established by decisions on radio frequency assignment and emission correspondence to the national standards and technical regulations are revealed;
- Inspection of fulfillment of radio exchange established regulations;
- Inspection of fulfillment of temporal prohibitions and restrictions on radio frequency spectrum usage, introduced at execution of special measures and during emergency situations, by the radio electronic equipment owners;
- Search and localization of radio electronic equipment, which do not use radio frequencies as intended, including the distress radio frequencies of services participating in rescue–recovery operations;
- Search and localization of radio emission sources, which create inadmissible radio interference and also sources which were not allowed to exploitation.

The only possible way to manage with increasing stream of more and more complicated problems without growth of operating personnel amount is usage of the control management modern technologies.

Automated system utilization increases rate and accuracy of measurement task fulfillment, sets free the operators from routine task execution, and increases the labor effectiveness. Besides, the utilization coefficient of measuring equipment is improved owing to possibility of statement and solution of many tasks in the automatic background mode, for instance the task of radio band occupancy monitoring, the search for radio transmitters operating without license, the verification of parameters of registered equipment. Thus, the automated radio monitoring system provides increasing of radio monitoring tasks solving effectiveness at the expense of modern automation technologies on the base of computers.

The automated radio monitoring system should have necessary set of universal functions without which it is impossible to solve tasks of spectrum monitoring. To these universal functions, first of all, the following can be concerned [9]:

- Panoramic spectrum analysis in the real time with maximum rate and resolution;
- Fast search for "new" emissions, measurement of their parameters, identification of their sources;
- Radio channel monitoring, radio signal recording, their technical analysis;
- Measurement of radio emission source field strength, radio signal bandwidth, modulation parameters of permitted and illegal sources;
- Direction-finding and localization of radio emission sources.

Listed functions are necessary both for solution of radio monitoring tasks, performed by the civil services with the aim to regulate radio frequency spectrum, and for tasks of radio electronic surveillance and radio electronic countermeasures, which are the important directions of military service activities.

According to ITU reference, the main tasks of civil services, which regulate the radio frequency spectrum usage, are monitoring of emissions on correspondence to frequency assignment, observation of radio frequency spectrum and measurement of frequency channel occupancy, revelation of radio interference sources, suppression of unauthorized radio emission sources (RES) activity [3].

Systems of radio monitoring and RES localization operating in the interests of military or civil agencies should recognize RES, detect the interferences, and calculate their location and main characteristics.

At present, there are many radio frequency spectrum management systems in the world: SMS4DC (ITU), ICS suite (ATDI), Iris (TES Spectrum Control Systems), SPECTRA (LS telecom AG), and others [2, 10–12]. All these systems are developed on the base of client/server architecture, have a possibility of matching with radio monitoring systems, and permit to solve full spectrum of tasks of the supervision agencies monitoring the radio frequency spectrum.

The most known radio monitoring systems are SCORPIO (TCI) and ARGUS (Rohde and Schwarz) [13, 14]. Both these systems offer the similar set of functions, among which the following are crucial:

- Panoramic spectral analysis;
- Signal demodulation and decoding with possibility of recording;
- Measurement and analysis of signal technical parameters including frequency, level, strength, bandwidth, and estimation of modulation parameters;
- Measurement of spectrum occupancy;

- Direction-finding and localization of radio emission sources;
- Detection of illegal or unknown sources and generation of messages about the detection fact.

Key points of the modern radio monitoring system structure are client/server architecture, centralized database, planning of radio monitoring task execution, interaction with external information systems.

The client/server architecture allows controlling of the geographically distributed equipment from the unified center and supports the multi-user mode. In modern systems, the control of the radio monitoring equipment is performed by servers, and the workstation serves as the client part, which is the control point.

The centralized database provides the radio monitoring system with actual data owing to which we can execute such tasks as source identification and registration of the signal parameters, which values are outside of the authorized bounds.

The planner of tasks optimizes radio monitoring equipment workload and enables its utilization with maximal effectiveness without necessity of operator attraction and without permanent availability of the communication channel with the control point.

Interfaces for integration with external systems enable receiving tasks and reference data from other systems and returning results obtained as well as controlling the radio monitoring equipment. Two types of interfaces are used—exchange files and real-time protocols. The first variant is more universal and less expensive from the implementation point of view. The second variant allows execution of the "seamless" system integration and more effective utilization of radio monitoring equipment possibilities.

As a rule, the most effective radio monitoring systems are developed by manufacturers of the radio monitoring equipment or with their direct participation, since in this case it is possible to use properties and features of produced equipment most completely and to modernize the software quickly during its development and modernization.

The general construction principles of native automated radio monitoring systems will be analyzed below. These principles were developed and delivered by the IRCOS Company intended for application in civil and military agencies, and their components are equipment for radio monitoring, engineering architecture, and software support. To illustrate the described material, we shall consider examples of the real automated systems: ARMADA, AREAL, and ASU RCHS UNIVERSIADA 2013.

1.2 Principles of Automated System Construction

At present, the automated radio monitoring system (ARMS) of various applications and structures intended for utilization in civil, commercial, and military agencies has been developed and delivered by the IRCOS Company. All systems are based on the unified hardware–software platform consisting of radio monitoring equipment and the supplied software. At the same time, the composition, configuration, functions of the specific automated system are defined by purposes and tasks, which are stated for user of this system.

According to a definition taken from the Russian State Standard GOST 34.003-90, the automated system consists of the personnel and the automation equipment complex, which realizes informational technology of the assigned functions fulfillment [15]. In connection with the automated radio monitoring system, the automation equipment necessarily includes the radio monitoring equipment, the software, and engineering-technical infrastructure, which are illustrated in Fig. 1.1.

The radio monitoring equipment produced by the IRCOS Company, including digital radio receiver (DRR), such as the single-channel ARGAMAK-IS, ARGAMAK-M, ARGAMAK-RS, the multi-channel ARC-D11, radio direction-finders ARTIKUL-C, ARTIKUL-M, ARC-RP3 M and others, as well as receiving and measuring antenna systems (AS), various support-rotating devices, acts as the main equipment for the ARMSs. All equipment for radio monitoring is unified in construction, and radio receivers are based on the digital radio receivers of the ARGAMAK-family and have wide bandwidth of the simultaneous analysis, high operation rate, and sufficient dynamic range.

The supplied software consists of the SMO-ARMADA specialized software, which is the main part of the software of the automated system, and the client software packets SMO-PPK, SMO-BS, SMO-DX, and others [16–18]. The software has the open client/server architecture and widely uses Web technologies, and it is able to adjust flexibly to the scale, peculiarities of organization and functions, which the system should perform, and makes possible the radio monitoring equipment operation in the multitask planned, operative, and background operation modes.

The client/server architecture of the SMO-ARMADA software consists of the "thin" client, the application server, and the database server; it enables minimum data transmission between the server and the client. It also provides a small traffic between the application server and the data server and simplifies modernization of functionality and upgrading of the software.

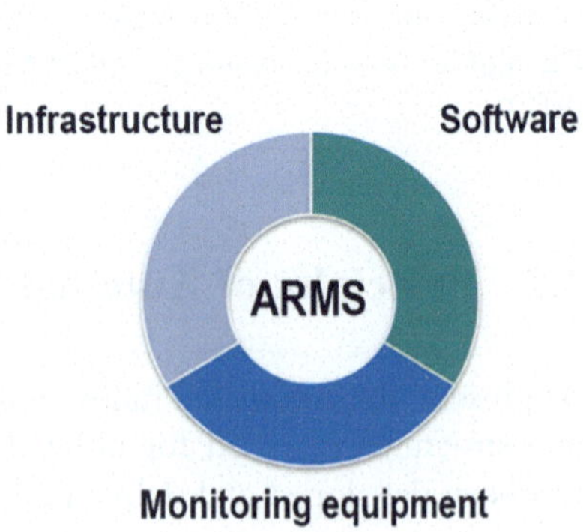

Fig. 1.1 Automation equipment of the radio monitoring system

The engineering-technical infrastructure consists of the data transmission system, the server equipment, the engineering constructions, for instance service rooms for monitoring equipment installation, prepared specific platforms for installation of antenna-mast devices.

Owing to the unified hardware–software platform, all automated systems delivered by the IRCOS Company have general principles of construction and functioning. The following most significant principles can be concerned with them:

- Hierarchy of the architecture and flexible control of equipment;
- Possibility of scaling;
- Usage of geo- and informational technologies;
- Open unified control protocol for equipment;
- Interaction with external informational systems;
- Execution of radio monitoring tasks in the automatic (planned, background) and manual (operative) modes;
- Eventual mechanism of task execution, diagnostics, and self-diagnostics of the equipment.

The automatic system obeys the hierarchy principle, according to which any system lower node operates under the control of the higher node. Control centers and points, radio monitoring stations, separate equipment of radio monitoring, which have control servers with the same type of software, may act as nodes, as it is shown in Fig. 1.2.

At the first hierarchy level, there is the control center (CC); at the second level, there are control points. Control points factually execute the same functions as the control center, but in conformity to the subordinate hierarchy levels. For instance, if the system operates at the national level, then control center provides the system

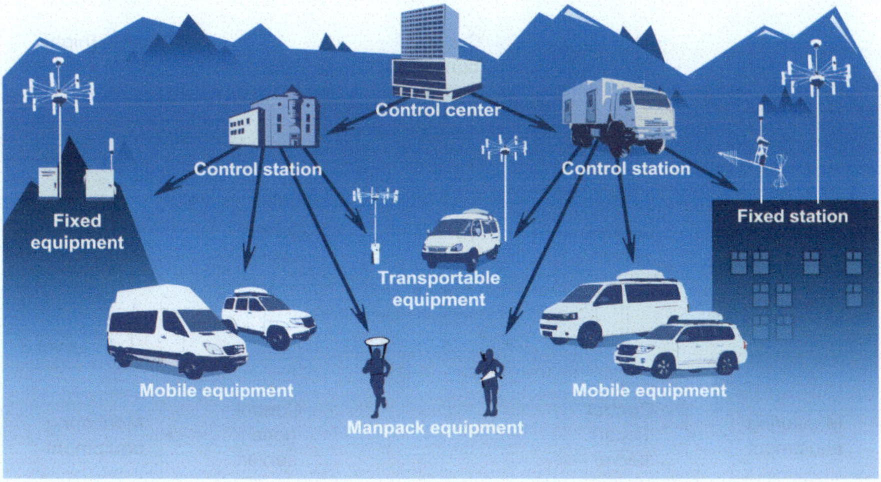

Fig. 1.2 The three-level structure of the automated system

control via the control points, which are situated in regions. The lower hierarchy level is represented by radio monitoring equipment: fixed, mobile, portable, manpack stations, complexes, and equipment. The number of equipment and control levels in the system is defined by utilization scale. It is possible to construct the system with four and more control levels. It is evident that the control structure in the automated system should correspond to the control structure of that administration, in which interest it functions [2, 3].

The control server of the higher-level node obtains information from subordinate (controlled) servers of the lower level, and it also has the urgent access to the radio monitoring equipment, if it is at the lower levels. Interaction of control servers is shown in Fig. 1.3. The task assignment for the node is performed both from the higher levels and directly from the lower level, which enables fast response to local conditions. The control server at lower hierarchy level can control one or several copies of radio monitoring equipment.

The equipment control and data transmission between the nodes is performed via closed channels. In the simplest variant, the OpenVPN technology of the protected network is used. At that, the automatic switching is provided between common, reserve, and emergency communication channels. As a rule, for fixed equipment, the wired or fiber-optical channels act as the common channel, and the wireless channels can be used as reserve and emergency channels. The control by mobile and manpack equipment is performed via the wireless channels.

System users use the automated workstations (AWS) for operation, acting as "thin" clients. At that, there is a possibility of flexible control of equipment. Several variants of similar flexible control of the equipment are shown in Fig. 1.4.

A possibility to control this or that equipment and formulation of the planned or operative tasks for the specific operator working at the AWS is defined by the policy of administration and safety. There is a mechanism of flexible adjustment of

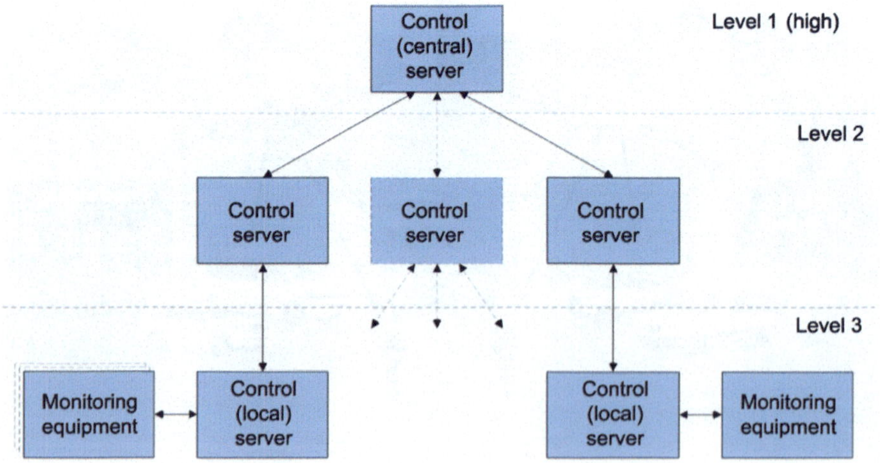

Fig. 1.3 Interaction of control servers in the three-level system

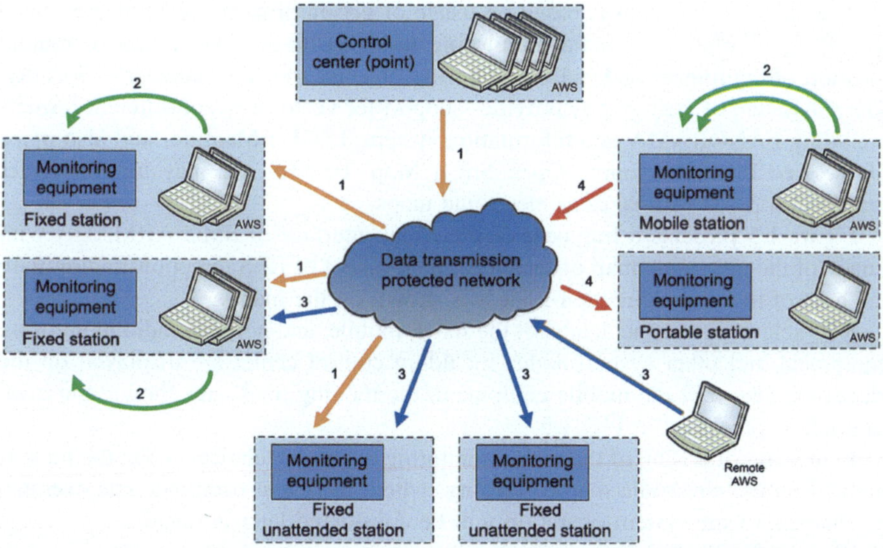

Fig. 1.4 Variants of control of radio monitoring equipment

rights and priorities of users at the access to radio monitoring equipment and to data stored in the databases (DB) of the control servers. At the presence of necessary permissions from any local or remote AWS, the operator can control the arbitrary equipment located in the system.

The automated system is scalable, capable to work at appearance of new nodes or levels. The quantity of levels and nodes is defined by the specificity of application. In cases when the system scale is limited by the separate region or area, we may have no control points and it is enough to have the control center, which formulates tasks and obtains results directly from radio monitoring stations; i.e., the system has a two-level structure. As the extreme variant, for the independently operated equipment, for example, the direction-finder or the measuring receiver, the system has one level and can function at one computer connected to the equipment, and both server and client parts of the ARMS software are installed on this computer.

1. The control center assigns tasks to three direction-finders of the fixed stations—to two maintained stations in which operators are working and one unmanned station without operators.
2. Operators, who work in stations, solve radio monitoring tasks with the help of the equipment located directly in their stations.
3. From the remote AWS located, for instance, in another town, the operator formulates task on direction-finding to three fixed stations.
4. The operator of the mobile station works remotely with the portable station equipment.

The automated system is based on usage of geo-informative technologies since for effective solution of radio monitoring tasks we must have notion of mutual location of equipment and radio monitoring objects, and the character of locality. The subsystem of cartography provides support for vector maps of different formats including PANORAMA geo-information system, INGIT, MapInfo, and also maps of the free Internet-resource Open Street Map. Besides, we may use the raster graphical formats of images as electronic maps.

Figure 1.5 presents a fragment of the client interface of SMO-ARMADA with image of the electronic map of locality. The moment of RES direction-finding with the help of five fixed direction-finders is shown on the map.

RES selected from the database, the fixed, mobile, and portable radio monitoring equipment and other system nodes including control center are displayed on the electronic map. For the mobile equipment, the moving tracks are shown; the track example is presented in Fig. 1.6.

In the case of arising of the radio monitoring event or technical event, the node is marked on the electronic map of locality, where the event occurred. The eventual mechanism of the system operation will be considered later in detail.

The open unified protocol is used in the automated system to control equipment. It is developed with the aim to be used in systems not only of IRCOS Company

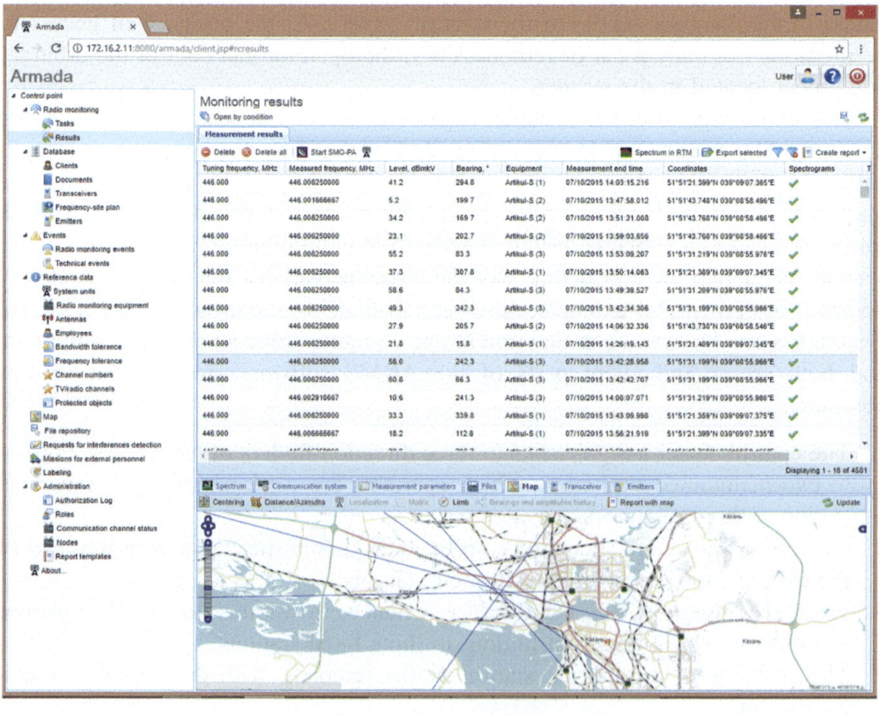

Fig. 1.5 A fragment of the software window with the electronic map

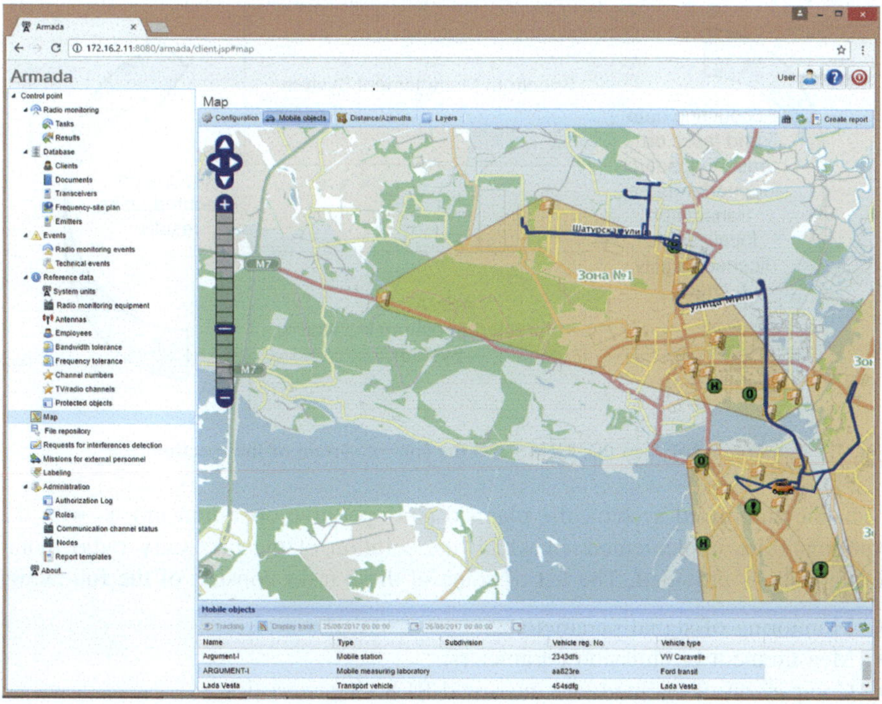

Fig. 1.6 An example of representation of the mobile station moving track

equipment, but of the other manufacturers of the equipment. The protocol is used for control of all types of the radio monitoring equipment: fixed, mobile, portable, and manpack equipment providing the transmission of measurement results and service information. The special software (equipment driver) which transforms the unified commands of the protocol into the command sequence perceived by the specific equipment sample, interacts with the protocol on the monitoring equipment side in order to provide operation of equipment of the outside manufacturer in the system. A possibility of supplementing additional commands without change of already existing operations and the software modules, which processes them, is widely used in the protocol. In other words, if it is necessary to use some useful properties of new equipment, the additional structures are added into the protocol, which are ignored by drivers of the "old" equipment, which already acts in the system.

In the automated system, a possibility of integration with external information systems is anticipated. The exchange is performed on the base of XML-documents using Web services, HTTP, or SOAP. The FTP protocol can be used. Figure 1.7 shows an example of information exchange with the external system of the radio frequency spectrum management.

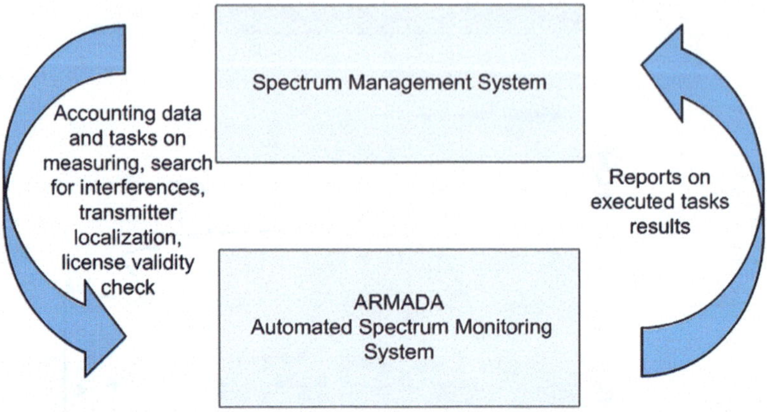

Fig. 1.7 Interaction between the ARMS and the control system of the spectrum use

In the automated system, the planned and operative operation modes with the equipment are implemented, in which the solution of the necessary radio monitoring tasks is provided. The list of some of these tasks consists of the following:

- Monitoring of signal parameters;
- Measurement of emission parameters;
- Comparison of claimed and measured RES characteristics;
- Search for interferences and unauthorized emissions sources;
- RES localization in locality and in rooms;
- Measurement of radio frequency spectrum occupancy;
- Monitoring of analog and digital radio broadcasting and television, as well as cellular and wireless systems of telecommunications.

In the planned mode, equipment located in system nodes functions automatically by the tasks according to the schedule formed in advance. Task assignment is possible from any AWS of the system, and the queue of their execution is defined by priorities of operators who formulated the tasks. Fulfillment results are saved in the DB of the node control server, which fulfilled the task, and are transmitted to the DB of the node, which formulated the task. In the case when channels of communication with the higher node disappear, the execution of the planned task will continue. The local server located at the node controls the task execution. When communication is recovered, results are transmitted to the node, which formulated the task, and to the higher level according to appropriate adjustment.

The planned mode provides a possibility to control a large set of multi-type equipment by one operator from one AWS. The automatic task execution permits to use the radio monitoring equipment maximally fully, including implementation of background tasks of radio monitoring—planned tasks with low priority, for instance, tasks on estimation of electromagnetic environment, search for new emissions, checking of emission correspondence to norms and licenses, measurement of radio frequency occupancy.

The mode of operative radio monitoring is intended for the direct interactive control of one or several equipment of the same type from one AWS. That enables to solve nonstandard frequently rather complex tasks of radio monitoring, which require immediate execution. Access to node equipment is given in accordance with priorities assigned by the operator. Results of operator work in the operative mode can be saved at any moment in the server DB of the control node, from which the control of equipment is carried out, and they are transmitted to the higher node.

The eventual mechanism of radio monitoring and monitoring of the system status is widely used in the ARMS for effective utilization of radio monitoring equipment, increasing of operator's labor productivity, automatic revelation of violations in radio frequency resource use, and occurrence of unauthorized RES. The eventual mechanism enables realization of the risk-oriented principle during radio monitoring, which provides automation of radio monitoring task solution day and night, drawing away the operator attention only in case of obtaining the result, which requires the operator to be involved. In the most of cases, only functions of reaction on the arising events, such as emission parameters, which values are outside of the permissible norms, or detection of illegally acting transmitter are entrusted to the operator. The eventual mechanism is used in monitoring of equipment conditions and the system infrastructure. Events are generated at fault appearance, communication channel disappearance, or the power source fault, at activation of the secure alarm system, etc.

Events generated in the system are divided into three types: radio monitoring events, technical events, and service events. Radio monitoring events are formed during execution of planned radio monitoring tasks, in which the comparison of the current measured parameters of the observed radio emission and an assembly of the formalized beforehand "mask" signs of the normal object conditions are performed. As such signs, levels, and signal waveforms, norms and licenses on modulation parameters are used, including spectrum width and central frequency, RES identifiers, for instance identifiers of legitimate base stations of communication systems, RES coordinates on locality, and other signs. Events are formed in the case of radio emission parameters being outside of the limits of the settled "masks."

Technical events are formed in the case of the status upset of the radio monitoring equipment or the engineering-technical infrastructure, at activation of the alarm signaling system in system nodes, at switching between main and reserved communication channels, at switching to the reserved power source, at finishing of the verification period of measuring equipment, etc. All equipment for system monitoring, network components of the network infrastructure, control servers have the embedded system of state diagnostics and diagnostics of operation parameters of modules and blocks. Information transmission about equipment conditions is executed on the base of the SNMP protocol, and in the case of faults occurrence, the appropriate technical events are generated.

Service events show the moments of start and finish of radio monitoring task execution, and they are used to control equipment in the dialog mode, to transfer control to the external informational systems, to provide monitoring of radio monitoring equipment workload.

Records about events are saved in the DB of system nodes, where they had arisen, and according to the settled adjustments, they are transmitted to the higher nodes to make decisions. Besides messages displaying on the screen and representation of event list in the software interfaces, the eventual information is shown on the electronic map of locality or on the plan of the object under monitoring and is automatically transmitted via e-mail as well as SMS.

The eventual mechanism used in the automated system enables the usage of the risk-oriented approach both to radio monitoring task solution and to providing of the ARMS efficiency.

At present, the most known systems produced by IRCOS Company are ARMADA and AREAL automated radio monitoring systems. We would like to mention here the ASU RCHS UNIVERSIADA 2013, which is the automated system delivered by the IRCOS Company, which fulfilled radio frequency spectrum management during realization of the first-rate international sport action XXVII World-Wide Summer Students Games in Kazan City (Russian Federation).

The automated ARMADA system is used for radio monitoring on locality including detection, direction-finding, localization, and identification of radio emission sources, measurement of emission technical parameters, their comparison with norms and licenses. The system presents information about actual frequency usage and provides technical analysis of new types of radio signals. The automated AREAL system is used for revelation of technical channels of information leakage, search and technical analysis of unauthorized radio emissions, localization of its sources on the guarded territories and in the rooms. The two-level structure of the AREAL system is presented in Fig. 1.8.

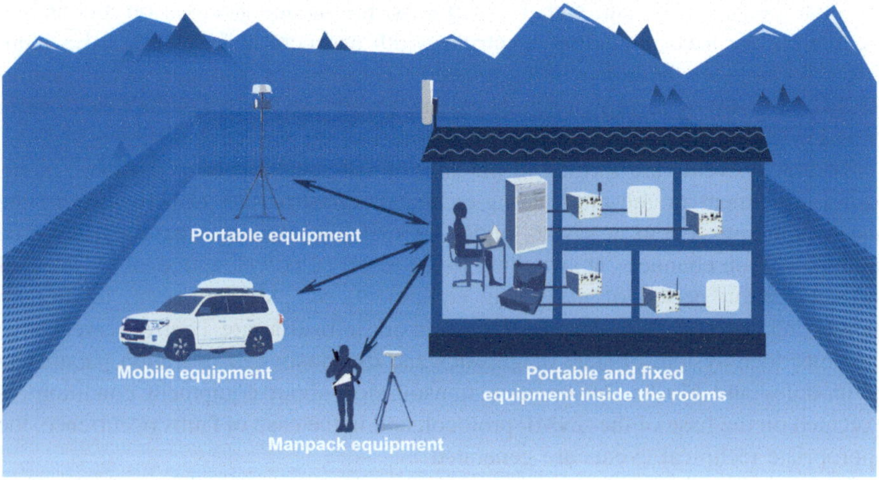

Fig. 1.8 Two-level structure of the AREAL system

The ARMADA ARMS, intended for utilization in the country scale, as a rule, has three or four control levels, while the two-level scheme, in which system nodes, each of which attends to its own object or adjacent territory, is the typical structure of the automated AREAL system. Both ARMADA and AREAL systems are controlled from the unified control point.

Besides general properties, there are some differences in the ARMADA and AREAL systems, which are caused by peculiarities of their application:

- The ARMADA ARMS is intended for operation on locality, its operating zone extends up to the significant area, and covering of the country territory as a whole is possible, while the AREAL ARMS is intended for operation on the limited territories and inside the objects;
- The ARMADA ARMS performs detection, identification, and localization of radio emission sources, which have no licenses on usage, and the sources of intended and unintended interferences, which prevent operation of legal radio equipment. First of all, detection, identification, and localization of unauthorized radio emission sources are related to main functions of the AREAL ARMS;
- Measurement of emission parameters of radio signal sources and determination of their location with the aim of monitoring the correspondence to the issued licenses and to working norms are necessary functions of the ARMADA ARMS at its usage in the interests of the radio frequency service (the national regulator). Usually, the similar tasks are not formulated for the AREAL ARMS.

Further, in this chapter, we shall give the review of radio monitoring equipment applied in the automated systems, and the examples of the ARMADA and the ASU RCHS UNIVERSIADA 2013 automated systems will be considered. Typical usage of variants of the AREAL automated system will be considered in detail in Chap. 2.

1.3 Monitoring Equipment

At present, on the base of the digital panoramic radio receivers of the ARGAMAK-family, the wide set of modern native high-performance multi-functional radio monitoring and direction-finding equipment, which can be remotely controlled and are suitable for effective application in the automated radio monitoring systems practically in any environmental and climate conditions, is supplied. Equipment is supplied in several variants of implementation: fixed unmanned, mobile on the ground and airplane transport bases, transportable and portable, as well as manpack ones [19, 20]. A novelty of engineering solutions, which are realized in equipment and in the control software, is confirmed by patents on methods and devices [19]. Let us make an acquaintance of typical equipment of radio monitoring, which is used in the automated systems.

At first, we start with fixed radio monitoring stations ARCHA-IN and ARCHA-INM. These stations are unmanned; i.e., they do not require personnel presence and are controlled remotely. Usually, these stations are located in densely populated places, large towns, and industrial centers. The necessary quantity of stations is defined by the area, the locality relief, a type of used RESs in the given region. To determine the sources' coordinates, any point of the territory under monitoring should be in the operating zone, at least, of two direction-finding or measuring direction-finding stations.

ARCHA-IN is the fixed measuring unmanned radio monitoring station. It consists of two main devices: the ARTIKUL-C correlation-interferometric direction-finder and the ARGAMAK-IS panoramic measuring digital radio receiver (device to measure electromagnetic field strength). The antenna system of the ARTIKUL-C radio direction-finder and the measuring antenna, and the remote radio frequency (RF) sensor of the ARGAMAK-IS receiver mounted on the roof of the multi-floor building, are shown in Fig. 1.9. The equipment is situated in the special protected secure cabinets (boxes), which enable its arrangement in the open air without usage of service rooms, for instance directly on the building roof. The temperature range of environment should be from −55 to +55 °C.

The integrated radio receiver, which is directly fixed into a basis of the circular antenna array, is the distinctive feature of the ARTIKUL-C radio direction-finder, included in the station structure. Such constructive solution eliminates the antenna effect and provides high sensitivity and high accuracy of direction-finding. Possibility of signal direction-finding in the short-wave range according to the

Fig. 1.9 Antenna system of the ARTIKUL-S radio direction-finder and the RF sensor measuring antenna

surface wave starting from 1.5 MHz is another interesting feature. In the maximal composition, the upper operating frequency of the direction-finder is 8000 MHz

The non-directional active antenna is used as the measuring antenna for the ARGAMAK-IS digital receiver. The directed antenna can also be mounted on the support-rotating device, as shown in Fig. 1.10. The lower operating frequency is 0.009 MHz, and the higher operating frequency is 18,000 MHz at the presence of the ARC-KNV4 converter in the station structure.

The ARCHA-INM is the measuring unmanned fixed radio monitoring station. Its difference from the ARCHA-IN station is that it consists of one receiving block mounted on one mast. The ARGAMAK-IS panoramic measuring receiver with the additional (second) synchronous channel for signal reception is arranged in the receiving block. Such a solution allows integration of the possibilities of measuring receiver and correlation interferometer in one device. The ARCHA-INM radio monitoring station is shown in Fig. 1.11.

Fig. 1.10 Panoramic measuring receiver ARGAMAK-IS with measuring antennas on the rotating device

Fig. 1.11 Radio monitoring station ARCHA-INM

As in the ARTIKUL-C radio direction-finder, the receiver in the ARCHA-INM is integrated into the basis of the antenna system. To measure the electromagnetic field strength, the RF sensor antenna is used, which is mounted over the circular direction-finder antenna array. Transmission of data and control commands via feeder is performed in the digital form via the Ethernet interface (twister pair cable or optical fiber); via this, the following commands and digital data streams are transmitted: time samples of radio signals, spectrums, demodulated signals, etc. Utilization of integrated digital receiver and digital interfaces for data transmission completely eliminates the antenna effect in the feeder, provides high sensitivity and also high accuracy of direction-finding and measuring of signal parameters, keeps the dynamic range, enables utilization of the feeder with length up to 100 m and more, and besides, simplifies the calibration of the measuring paths.

Together with fixed stations, as a rule, the mobile stations with measuring and direction-finding equipment are included in the system structure. Utilization of the mobile stations allows providing necessary coverage area at minimal quantity of fixed stations, which reduces the total system cost.

The mobile stations can be quickly transferred to the region situated beyond the coverage area of fixed equipment. They are effective to fulfill radio monitoring operations, when small power of the transmitter, high directivity of antennas, remoteness of the radio emission sources complicate or make impossible the execution of measurements by fixed equipment. The mobile station performs works on parameter measurements and localization of radio emission sources both in the structure of the automated system and independently.

Different microbuses, for example, Volkswagen Caravelle, Mercedes Sprinter, Ford Transit, automobiles of increased cross-country power, for instance Land Cruiser, UAZ Patriot, can be used as the transport base for the mobile radio monitoring station ARGUMENT-I. The mobile measuring station ARGUMENT-I on the base of the UAZ Patriot is presented in Figs. 1.12 and 1.13 which shows the working places for operators in this station located behind the driver and the front passenger seats. An example of the mobile station on the base of the all-wheel drive microbus Ford Transit is presented in Fig. 1.14.

Fig. 1.12 Mobile measuring station ARGUMENT-I on the base of the UAZ Patriot

Fig. 1.13 Work places for operators in the mobile station on the base of the UAZ Patriot

Fig. 1.14 Station ARGUMENT-I on the base of Ford Transit

The composition of radio monitoring equipment of the mobile station can be varied depending on its purpose. Usually, there are the correlation-interferometric automatic direction-finder ARTIKUL-M and the measuring receiver ARGAMAK-IS in the mobile station. The range of operating frequencies of the mobile direction-finder ARTIKUL-M in the maximal configuration (without changing of the antenna system) is from 1.5 up to 8000 MHz. The mobile station can operate in motion and at parking under control of the automated system or independently. Even at single operation, being in motion, the station provides calculation of location of radio emission sources with high accuracy.

In automated systems, the equipment, which has portable (transportable) implementation, is widely used. Such equipment has rather small weight and dimensions; it enables their transportation by the transport equipment and manually by one or several operators.

The similar equipment is used for fixed and temporal posts, which are not equipped with electric power supply as well as in the open territory. They can be quickly deployed in places, which are inaccessible for transport equipment—in mountains or under conditions of lack of roads, and they are used at temporal posts, for instance on building roofs under urban conditions.

The ARGAMAK-RS S is the compact radio monitoring and technical analysis complex built on the base of the digital measuring receiver—the radio sensor ARGAMAK-RS which has the protected implementation. The omnidirectional measuring RF sensor antenna, the GPS/GLONASS antenna, the power supply unit with emergency accumulator, and the controller are included in this complex as well.

The complex ARGAMAK-RS S is intended for search, detection, identification, recording, demodulation, spectral and technical analysis of radio signals in HF, VHF, SHF, microwave ranges, for measuring of field strength and signal parameters, as well as for localization of radio emission sources when operating in the structure of the distributed system. The overall receiver dimensions are not more than 300 × 300 × 150 mm, and its mass is not more than 7 kg. The external view of the complex is shown in Fig. 1.15.

The ARCHA-IT is the transportable measuring radio monitoring and direction-finding station presented in Fig. 1.16. Its construction is similar to one of the fixed station ARCHA-INMs. The main block is the digital panoramic receiver ARGAMAK-IS, which is used as the measuring receiver. It has additional (second) channel of reception of radio signals, which enables the signal parameter measurement and the correlation-interferometric direction-finding.

As in the station ARCHA-INM, the receiver is mounted in direct closeness to the antenna system. It provides high sensitivity of measurement and direction-finding, and the wide dynamic range. The difference from the ARCHA-INM station consists of utilization of replaceable antennas with low weight and dimensions. The portable antenna system AS-HP5 is used for direction-finding, while the measuring RF sensor antenna must be mounted instead of AS-HP5 for measurement of the electromagnetic field strength. Besides, one can use the set of other measuring antennas mounted on the double-coordinate or triple-coordinate support-rotating device, as shown in Fig. 1.17. To provide the operating frequency range from 1.5 to 8000 MHz, the station is equipped with antenna systems AS-HP-KV and AS-HP2.

ARTIKUL-MT is the transportable automatic correlation-interferometric direction-finder, and it is shown in Fig. 1.18. The ARTIKUL-MT is intended for radio situation monitoring. It uses the double-channel digital radio receiver ARGAMAK, which is not a measurement receiver. Owing to less strict requirements on parameter stability, the box, in which the receiver is accommodated, has

Fig. 1.15 ARGAMAK-RSS

the simplified construction and smaller dimensions compared to the box for the receiver of the ARCHA-IT station. The operating frequency range is from 1.5 to 8000 MHz with the changeable antenna systems AS-HP-KV, AS-HP2, AS-HP5.

The ARTIKUL-H1 is the manpack automatic wide-ranging correlation-interferometric direction-finder, intended for operation in the fixed, mobile, and manpack variants. The changeable antenna systems AS-HP-KV, AS-HP1, AS-HP2, and AS-HP5 provide the operating frequency range for direction-finding from 1.5 to 8000 MHz.

In the fixed variant, the antenna system of the direction-finder is mounted on the mast up to 6 m in height. At operation on the temporal parking, the antenna system is mounted on the folding mast tripod. The external view of direction-finder on the temporal post is shown in Fig. 1.19.

In the mobile variant, the antenna system of the direction-finder is mounted on the automobile roof with the help of the magnetic support, which is illustrated in Fig. 1.20, or is mounted in the luggage box made of radio-transparent material as shown in Fig. 1.21a, b.

Fig. 1.16 Transportable station ARCHA-IT for monitoring and direction-finding with the antenna system AS-HP5 for direction-finding

Fig. 1.17 Measuring station ARCHA-IT with the triple-coordinate support-rotating device ARC-UP3D and the set of antennas

Fig. 1.18 ARTIKUL-MT transportable automatic direction-finder

Fig. 1.19 Arrangement of ARTIKUL-H1 on the temporal post

Fig. 1.20 ARTIKUL-N1 with the antenna system on the magnetic basis

(a) closed luggage box

(b) opened luggage box

Fig. 1.21 ARTIKUL-H1 with the antenna system

For operation in the manpack variant, there is the special lightweight rucksack with shoulder straps and waist belt, to which the changeable antenna system and the double-channel panoramic radio receiver ARGAMAK-2K are attached, as shown in Fig. 1.22. The total mass for operation in the manpack variant does not exceed 16 kg.

The ARTIKUL-P is the portable automatic correlation-interferometric direction-finder intended for operation at the temporal posts, and it is shown in the Fig. 1.23. The antenna system consists of two circular antenna arrays—letterings. The first lettering has the folding construction of the traverses, the plain antenna elements, while the second lettering is realized as removable. As in other equipment, the unified digital radio receivers of the ARGAMAK-family accommodated in the

Fig. 1.22 Utilization of ARTIKUL-H1 in manpack variant

basis of first and second letterings of the antenna system are used in the radio direction-finder. In folded condition, the direction-finder is placed into the special transportable tubule. The total weight of the antenna system together with mast, penal, bracing wires, and anchors does not exceed 24 kg. The operating frequency range for usage of the second lettering is from 25 to 3000 MHz.

The ARC-D11, which is the double-channel panoramic digital radio receiver implemented in the form of the shockproof case shown in Fig. 1.24, finds application as the transported radio receiver. The main purpose of ARC-D11 is solution of tasks on search for unauthorized radio emissions in the engineering objects, buildings, and constructions. The ARC-D11 represents the panoramic digital radio receiver providing the effective search for unauthorized radio emissions using two coherent reception channels on the base of wideband modules of the ARGAMAK-family.

In the basing composition, the ARC-D11 provides connection of one reference and up to four "signal" antennas. To extend the monitoring area, the ARC-BUVM two-channel control module can be connected to it, providing additional connection with eight remote modules, to which the receiving antennas are connected.

The ARC-D11 can be used for radio monitoring execution on locality including direction-finding. In the composition with changeable antenna systems AS-HP-KV, AS-HP1, AS-HP2, and AS-HP5, the operating frequency range is from 9 kHz to 8000 MHz, and with ARC-KNV4 M—up to 18 GHz.

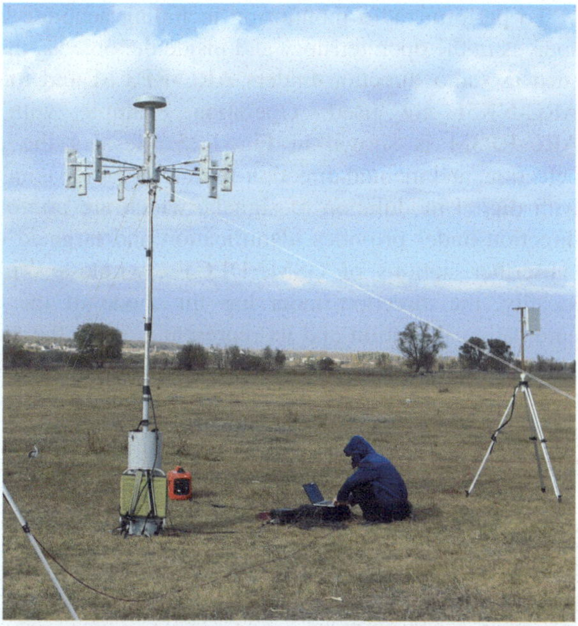

Fig. 1.23 ARTIKUL-P

Fig. 1.24 Double-channel digital radio receiver ARC-D11

To perform radio monitoring including accurate localization of radio emission sources in the open territory and inside the objects, the manpack manual equipment such as radio direction-finders ARC-RP3 M and manpack measuring complexes ARC-NK5I are used. Operation example with the radio direction-finder ARC-RP3M is shown in Fig. 1.25. Besides the panoramic spectral analysis, detection, and manual direction-finding of radio signals with analog modulation and with digital modulation of signals, which are non-overlapping in bandwidth, the direction-finder provides identification and targeted direction-finding of base and subscriber stations of GSM, DECT , wireless Wi-Fi devices. At operation on locality, the direction-finder has the mode of the detected source coordinates' automatic calculation and its representation on the electronic map of locality.

In the AREAL system, the double-channel and single-channel radio receivers such as ARC-D11, ARC-D1T, ARGAMAK-2K with a set of external antennas and remote receiver BUVM modules as equipment for search for the unauthorized sources inside the objects are used.

A presence of additional analog path with the 0.5 MHz bandwidth, which is extremely important for elimination of digital path overloading under conditions of

Fig. 1.25 Handheld direction-finder ARC-RP3 M

complicated electromagnetic situation, is the useful feature of products on the base of DRRs of ARGAMAK-family with bandwidth of simultaneous observation up to 24 MHz, such as ARGAMAK-IS, ARGAMAK-RS, ARGAMAK-CS, ARC-D11, ARC-D1, ARCHA-INM, ARTICUL-C, ARTICUL-M, ARCHA-IT, ARTICUL-MT, ARTICUL-T.

ARGAMAK-IS, ARGAMAK-M, ARC-KNV4, ARGAMAK-RS measuring equipment has the state measurement certificates and approved approaches, which regulate the measurement fulfillment and take into account the ITU-R SM.328, ITU-R SM.377, ITU-R SM.378, ITU-R SM.443, ITU-R SM 1268, ITU-R SM.1880 recommendations.

Radio monitoring equipment based on the ARGAMAK-family DRRs, which are included in the stations and complexes composition, provides measurement fulfillment, panoramic, spectral, and vector signal analysis. Besides, it enables determination of service identifiers and parameters of the telecommunication systems such as APCO P25, DMR, GSM, UMTS, LTE, IS-95, cdma2000, EV-DO, TETRA, DECT, DVB T/T2/H, Wi-Fi etc.

Digital radio receivers ARGAMAK-IS and ARGAMAK-RS have functions of synchronization according to signals of the global navigation satellite system (GNSS) and GLONASS/GPS, which provide reduction of frequency instability of the reference oscillator from 10^{-9} to 10^{-12} and realization of high-accurate measurement of the received signal frequency. It provides measurement of signal parameters of digital TV including the frequency measurement. According to the approved approach, the relative error of the frequency estimation does not exceed $\pm 5 \times 10^{-11}$. At that, the error of signal level measurement does not exceed ± 1 dB, and the error of frequency band measurement does not exceed $\pm 0.1\%$. The values of measurement errors satisfy the State Standards P 55696-2013 and P 5939-2014 and acting norms of the Russian State Committee on Radio Frequencies.

Synchronization from GNSS realized in ARGAMAK-IS and ARGAMAK-RS makes a possibility to determine the signal time of arrival to receivers up to the accuracy of 20–50 ns, which allows usage of the time difference of arrival method for calculation of radio emission source coordinates, which does not require the direction-finder utilization. Localization accuracy depends on the radio receivers' quantity, their mutual location, conditions of the radio signal reception, bandwidth of received radio signals, and other parameters. In the system including the direction-finder and measuring receivers, the hybrid algorithms are used with usage of information about time, levels, and angles of signals arrival, which increases accuracy of RES localization.

1.4 ARMADA Automated Radio Monitoring System

The ARMADA radio monitoring system is intended for the following task solution:

- Radio frequency spectrum monitoring and growth of its usage effectiveness at the expense of the control management automation;

- Execution of planned and operative monitoring of radio electronic equipment;
- Measurement of emission parameters of radio electronic equipment in accordance with norms, licenses, ITU recommendations;
- Automated handling, processing, and storage of radio monitoring results;
- Comparisons of claimed and measured radio system characteristics, administrative and technical information about users of radio spectrum, and radio equipment;
- Search for interference sources and unauthorized emission sources;
- Automation of control processes for equipment with the aim to provide operative radio monitoring of radio frequency situation at large territory;
- Obtaining of generalized information about radio frequency spectrum occupancy for planning of its usage;
- Monitoring of analog and digital TV and cellular and wireless communication systems as well.

The structure of the ARMADA automated radio monitoring system is defined by the specific features of application and can be changed depending on the requirements and destination. Let us consider an example of the radio monitoring system operating at the national level on the small country scale. The system has the following features:

- Hierarchical construction principle, meaning that each node of the system lower level gives an access for nodes of the higher level;
- Usage of the same type software for all-level nodes, in which configuration and function sets are determined by destination of the system node;
- Control of radio monitoring equipment using the open protocol;
- Possibility to use radio monitoring equipment of external manufacturers;
- Open architecture of the software enabling connection of additional software subsystems, blocks, and modules including ones from external manufacturers.

The system functions are as follows:

- Performing of radio monitoring tasks in manual (operative) and automatic (planned, background) modes;
- Assignment of tasks for remotely controlled system nodes;
- Monitoring of the task execution process;
- Storage of tasks and results of their execution in the databases of the node and control center;
- Automated analysis of operation results during the required time interval;
- Formation of reports according to the given templates;
- Export and import operations with the database of frequency assignment;
- Usage of geo-informative technologies for representation of the structure and conditions of system nodes, locations of radio monitoring equipment, results of radio monitoring;

- Storage of organizational regulating documents, measurement approved approaches, reference information;
- Remote diagnostics of the system nodes including radio monitoring equipment;
- Automatic check of verification periods and operation time of radio monitoring equipment.

ARMADA ARMS provides integration with the radio frequency spectrum control system.

To provide radio frequency spectrum monitoring in the country scale, the system has four levels of hierarchy and consists of the following nodes: the national control center, regional and district control points, and radio monitoring equipment. An example of nodes' physical location displayed on the country map is shown in Fig. 1.26. To provide quick reaction on local conditions, besides the control center, there are the regional and district control points, which correspond to the regulations of radio monitoring execution adopted in the country, in the system. Scheme of the ARMADA ARMS at the national level is displayed in Fig. 1.27. Results of radio monitoring task fulfillment, which were initiated at lower hierarchy levels, can be monitored at the higher level.

The lower level is represented by radio monitoring equipment, which is controlled by the higher nodes according to their priority. ARMADA ARMS includes the following equipment of radio monitoring:

- Stationary stations and complexes for radio monitoring;
- Mobile stations;
- Portable (transportable) equipment;
- Manpack equipment.

Fig. 1.26 An example of the ARMADA ARMS control node location

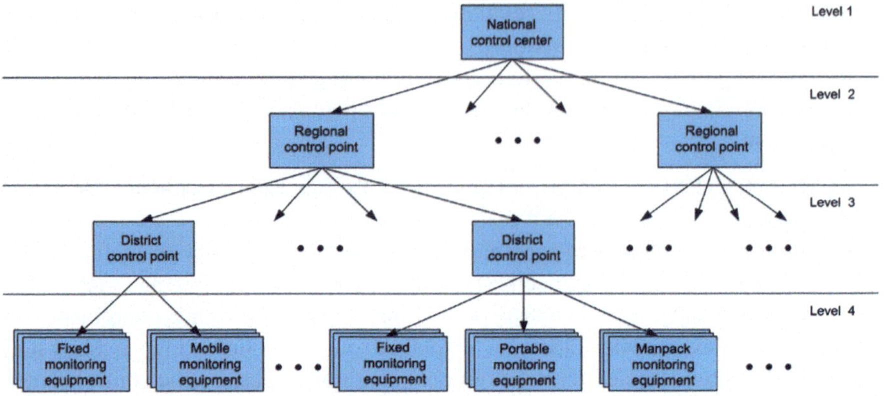

Fig. 1.27 ARMADA ARMS of the national level

The possible compositions of equipment for radio monitoring are given in Table 1.1 for inhabited locality, which differ in area and population. The necessary quantity and composition of radio monitoring equipment are defined by adopted regulations of radio monitoring execution, area of inhabited locality, locality relief, building up character, population quantity, concentration of radio emission sources, and their characteristics.

Table 1.1 Composition variants of radio monitoring equipment

№	Inhabited point	Population, thousands of people	Area, km^2	Monitoring equipment	Quantity
1	Capital	>1000	>500	ARCHA-INB	3
				ARTRIKUL-S	4
				ARGUMENT-I	3
				ARGAMAK-RSS	10
				ARTIKUL-P	1
				ARC-RP3 M	3
2	Town 1	<1000	250–500	ARCHA-IN	1
				ARTIKUL-S	2
				ARGUMENT-I	1
				ARGAMAK-RSS	5
				ARC-RP3M	1
3	Town 2	<500	<250	ARCHA-INM	1
				ARTIKUL-H1	1
				ARTIKUL-MT	1
				ARGAMAK-RSS	5
4	Village	<100	<50	ARC-RP3M	1
				ARGAMAK-RSS	3

To provide maximal coverage and quick reaction on arising problems, the fixed radio monitoring stations are arranged in large towns, which are characterized by extreme occupancy of the frequency range. These stations provide both detection and localization of radio emission sources and measurement of their parameters with the aim to check the correspondence to norms and licenses. Beyond complicated urban conditions, where there are essentially less of radio frequency sources under monitoring, the quantity of fixed equipment of radio monitoring decreases. In order to provide RES coordinates' calculation by the goniometrical method, for instance with the help of fixed stations ARCHA-IN, ARCHA-INM, or ARTIKUL-S N, any monitored point in locality should be inside the zone of electromagnetic availability of the direction-finders, at least, of two stations.

Dimensions of the electromagnetic availability zone of fixed equipment are determined by many factors, including suspension height of the receiving antenna of radio monitoring M, its sensitivity, and RES power. For example, for ARTIKUL-S direction-finders in the frequency ranges 25–100 MHz and 800–2000 MHz and thesuspensionheight of antenna systems of 40 m and 50 m, the radius of the detection zone, which is calculated for RES with 4Wpower obtained on the base of Egli model [21]of the detection zone, is 11.5 and 12.4 km, relatively, as shown in Fig. 1.28,and the area of the possible RES localization is more than 200 sq. km. This can be already quite enough to cover small inhabited point, while for large towns we need to increase a number of fixed radio monitoring equipment: For instance, addition of one more direction-finder leads to increasing of RES localization zone in 1.5 times.

RES localization is possible by not only the goniometric method, but also by time difference of arrival method as well as by combined methods by means of construction of the radio frequency sensor network on the base of measuring compact fixed complexes ARGAMAK-RSS.

In small inhabited points, we can be limited by the presence of one or several sensors, which operating in the automatic mode will provide detection of radio interferences and appearance of radio emission sources, which have no licenses. For

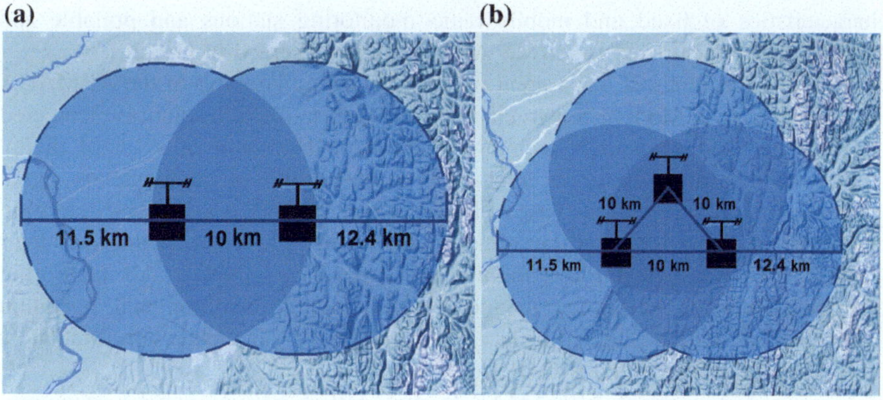

Fig. 1.28 Zones of electromagnetic availability for two and three direction-finders

simultaneous RES localization by the time difference of arrival method, it is necessary to deploy three sensors as minimum.

For inhabited points with small concentration of radio electronic equipment, the organization of two-stage radio monitoring is possible for decreasing of expenses on radio monitoring execution. Usually, the most of violations comprise the radio electronic equipment operation without permissions. Such violations can be revealed by cheap compact equipment, for instance radio sensors, without application of expensive measuring equipment of radio monitoring. At first stage, the detection of the violation occurs, while at the second stage, if necessary, the mobile measuring radio monitoring station moves forward to the region, where violation is registered, with the aim of accurate localization of the violation source, measurement, and registration of violation character.

The system is controlled from the control center, which is accommodated in the capital. Control points are situated in regions and factually fulfill the same functions as the control center, but on the lower hierarchy level, that can be arranged on the fixed or mobile base.

For organization of the direction-finding network, in addition to fixed equipment we can use mobile stations (ARGUMENT-I or mobile complexes on the base of ARTIKUL-H1) and/or portable equipment (ARCHA-IT, ARTIKUL-MT, ARTIKUL-P), which can be quickly transported to the necessary region and form a pair with the any fixed direction-finder to extend the localization area, or to localize RES at all beyond the coverage area of the fixed equipment. Thus, mobile equipment and portable equipment are intended for radio monitoring operation execution in places, where small transmitter power, high directivity of transmitting antennas, remoteness of radio emission sources complicate or make impossible the measurement fulfillment by the fixed equipment. Besides, mobile and portable equipment may operate independently, because they provide RES localization directly during motion or at parking. The manpack equipment, for instance ARC-RP3 M, can be used as a devices of "last mile," which enables RES localization in places that are unavailable for automobiles.

The more detailed information about composition, functions, and technical characteristics of fixed and mobile radio monitoring stations and portable and manpack equipment used in ARMS ARMADA is presented in [19, 20].

Measuring equipment has a great importance for functioning of the ARMADA automated system. Modern constructive, circuitry, and software solutions, tempered manufacture, and adjustment technology allow development of a set of digital panoramic measuring radio receivers with a bandwidth of simultaneous signal analysis up to 22 MHz on the base of DRRs of the ARGAMAK-family. Measuring equipment is controlled locally or remotely via wired or wireless channels, so their usage in the automated radio monitoring systems is enabled. All measuring equipment has the embedded diagnostics systems of condition and operation parameters of modules and blocks. Transmission of diagnostic information is performed according to the SNMP protocol.

Measuring equipment ARGAMAK-IS and ARC-KNB4 M can be accommodated in the temperature-controlled and waterproof secure cabinets (boxes) and mounted directly in the open air without usage of service rooms. On the base of measuring equipment ARGAMAK-IS and ARC-KNV4 M, the fixed and mobile measuring stations and complexes ARCHA-IN, ARCHA-INM, ARGAMAK-ISN, ARGUMENT-I are developed. The portable measuring receiver ARGAMAK-M is used in the structure of the manpack measuring complex ARC-NK5I.

The measuring receiver ARGAMAK-IS is the prototype of measuring radio receiver ARGAMAK-RS in protected implementation, which is intended for measurement of radio signal parameters as well as for search, detection, identification, recording, demodulation, spectral and technical analysis of radio signals in HF, VHF, UHF, and microwave ranges and for localization of radio emission sources at operation in the structure of the distributed system. To eliminate overloading of the digital path of measuring receivers ARGAMAK-IS and ARGAMAK-RS under conditions of complicated electromagnetic situation, the embedded additional analog path with the bandwidth of 500 kHz is used. Comparably, small dimensions of the measuring DRR enable its embedding directly into the antenna system, which provides high sensitivity and measurement accuracy owing to decrease of loss, absence of pickup, and better matching, and at presence of additional coherent reception path, it is possible to construct the correlation-interferometric direction-finder, permitting execution of high-accurate measurements.

Measuring receivers ARGAMAK-IS, ARGAMAK-M, and ARGAMAK-RS provide all necessary types of measurements including measurement of the electromagnetic field strength, frequency, and bandwidth of emissions, decoding, and analysis of service information of digital communication networks and data transmission of GSM, UMTS, IS-95, cdma2000, EV-DO, TETRA, LTE, DECT , Wi-Fi, APCO P25, DMR formats, signals of digital TV DVB-T/H/T2. Receivers ARGAMAK-IS and ARGAMAK-RS have the synchronization function by signals of GPS and GLONASS global navigation satellite systems. It enables execution of high-accurate measurement of the received signal frequency necessary, for instance to measure signal parameters of the DVB-T2 single-frequency networks and to realize the system of localization of radio emission sources, in which the location calculation is performed on the base of information integration: the time of arrival and radio signal levels at points of measurements.

Measuring equipment manufactured by IRCOS Company has Russian state certificates. Measurements correspond to recommendations of ITU-R SM.328, ITU-R SM.377, ITU-R SM.378, ITU-R SM.443, ITU-R SM.1880, and State Standards of Russian Federation P 52536-2006 and P 53373-2009. According to approved approaches, the following parameters are measured:

- Peak, quasi-peak, root mean square (RMS), and mean values of the field strength and power flow density;
- Frequency of radio emissions (non-modulated signals, signals with analog amplitude and frequency modulation, signals with digital types of modulation);
- Bandwidth of radio signals according to X dB and β/2 methods;
- Coefficient of amplitude modulation;

- Frequency deviation of signal with frequency modulation;
- Frequency shift of signals with frequency manipulation;
- Modulation rate of signals with digital types of modulations;
- Subcarrier frequency for stereo broadcasting with FM;
- Parameters of digital TV signals including high-accurate estimation of the central frequency;
- Occupancy of frequency channels.

Approved approaches also regulate measurement execution on the base of unified protocol of radio monitoring equipment control with remote control via network with the TCP/IP protocol.

Besides measuring equipment manufactured by IRCOS Company, equipment of other manufacturers can be used in the ARMADA ARMS; it can be connected to the system with the help of the open unified protocol.

1.5 ASU RCHS UNIVERSIADA 2013 Automated System

From July 6 to 17, 2013, the XXVII World-Wide Summer Students Games was hold in Kazan City (Russian Federation), in which the set of 351 medals was played for 27 kinds of sport among more than 12,000 representatives from 162 countries, which was a record of all Students Games. For conducting this Students Games, 64 sport objects were actively used, and 33 from them were used directly for competitions conduction. More than 120,000 guests arrived to this Students Games. Three Russian and thirteen international TV companies provided direct broadcasting. Each day, more than 30 TV commentators, 200 TV cameras, 15 transported TV stations worked at these games.

Such a large-scale sport event was inevitably accompanied by the sharp growth of radio transmitters' concentration on the restricted territory, and its successful execution was impossible without effective planning and radio frequency assignment, inspection, and licensing of radio electronic equipment and reliable monitoring of assigned frequencies in the real time. It required especially quick and flexible control of frequency usage directly in the place of action to manage with the requests operatively on application of radio electronic equipment, accepted in the last moment. To solve these problems during preparation and execution of XXVII World-Wide Summer Students Games, the automated control system of usage of radio frequency spectrum ASU RCHS UNIVERSIADA 2013 was used, which is the modification of the ARMADA ARMS with additional functional possibilities to control spectrum at big public events [22, 23].

1.5.1 System Composition and Control Structure

Equipment shown in Fig. 1.29 was included in the composition of radio monitoring equipment:

- Unattended fixed radio monitoring station (FRMS);
- Object radio monitoring station (ORMS);
- Mobile radio monitoring station (MRMS);
- Portable radio monitoring equipment, by which radio monitoring and interference search group (RMSG) were equipped;
- Labeling measuring laboratories (LML).

The system provided flexible variants of the equipment control. Task assignment could be provided from the control center, from automatic workstations of maintainable stations or from the radio monitoring complexes, from the arbitrary computer, which has a possibility to be connected to the system, for instance from the AWSs of the interacting agency. To provide data safety, which was circulated in the system, including inside the local networks, the ciphering was used.

The database included in the software structure of all nodes was the most important element of the system software. The database was intended for the account of data about applicants, radio frequencies, and radio electronic equipment, as well as the infrastructure objects, radio monitoring equipment, radio monitoring data. The database provided a possibility of data visualization, report formation, data transmission into other informational systems, etc.

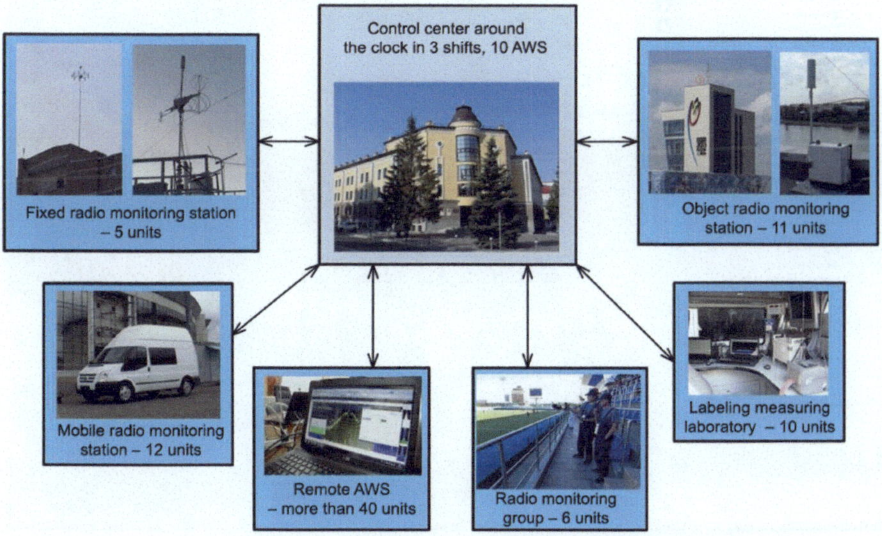

Fig. 1.29 Composition of units for ASU RCHS UNIVERSIADA-2013

The engineering-technical infrastructure included links and nodes of data transmission, the system of service radio communication, the equipment for data transmission, the server equipment, the engineering constructions, etc.

1.5.2 Radio Monitoring Levels in the System

Actions of spectrum control at preparation and conduction of Students Games in Kazan City were divided into three monitoring levels: urban, zonal, and object.

The urban level using the network from five remotely controlled fixed radio monitoring points provided direction-finding, localization, measurement of radio signal parameters.

The zonal level consisting of twelve mobile radio monitoring stations provided direction-finding, localization, and measurement of radio signal parameters including low-power sources. Localization of sport objects of Students Games (orange flags) and boundaries of three radio monitoring zones (the fourth zone included shooting ground located beyond the town) are shown in Fig. 1.30. In each zone, there were simultaneously up to two teams of mobile complexes as well as

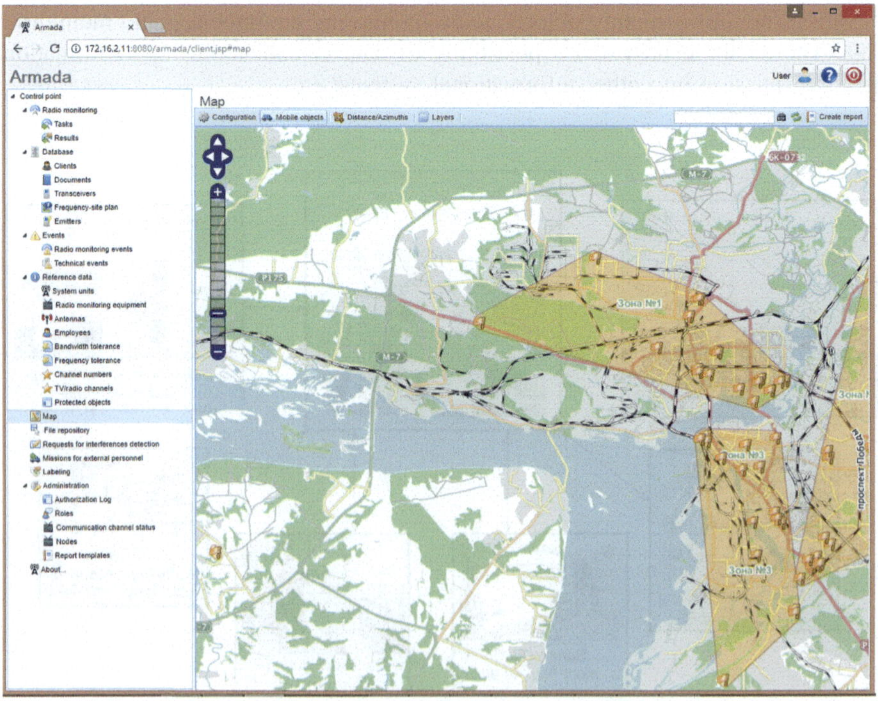

Fig. 1.30 Objects of Students Games and radio monitoring zones

radio monitoring and interference search groups equipped with manpack equipment.

To provide object (local) radio monitoring level, eleven ORMSs and six radio monitoring groups equipped with manpack equipment were attracted, which enabled search and localization of interference sources in places very difficult to be accessed. The main operation on frequency ranges monitoring at the object level was fulfilled in the automatic mode using the eventual mechanism of radio monitoring. ORMS performed round-the-clock scanning of the most important frequency ranges. If the radio emission sources, not registered in the account database of ASU RCHS, began to broadcast, the notification about the presence of the radio monitoring event on this or that object appeared at operator's AWS, after which, according to decision of the operator on duty, the search and localization were executed for unauthorized emission sources.

The personnel of ASU RCHS was integrated into the operative control center, which was divided into the personnel of the control center and external personnel—crews of LML, mobile radio monitoring stations, and RMSG.

1.5.3 Organization of the Control Center

In the control center, ten automated workstations for operators were deployed, with the help of which the control was fulfilled for fixed and object radio monitoring points, mobile stations and radio monitoring groups, special transport, and the system of service radio communication. The work of the situation center operators on duty is shown in Fig. 1.31.

Fig. 1.31 Work of operators on duty in the control center

In the subsystem structure of control center, the set of the server equipment of the central database, the AWSs of the personnel, the video wall, the equipment for conference communication, and video conferences were included. The server equipment included three servers, two of which were integrated into a cluster. The third server was intended to store the reserve copies of the system data. The control center server equipment is shown in Fig. 1.32.

The subsystem of communication and data transmission provided data exchange in the control center, and with external nodes—system control objects. The network equipment enabled operation from two Internet providers (one of which provided the main channel, while another—the reserve channel for data transmission), and automatic switching at disappearance and appearance of the main communication channel was included. The server to control operation of radio communication service network deployed on the base of the digital communication platform MOTOTRBO was also included into the communication subsystem. The network of service radio communication had in its structure three repeaters performing radio

Fig. 1.32 Control center server equipment

Fig. 1.33 Antennas of service radio communication repeaters on the telecommunication tower

communication in all parts of the city and forty-eight subscriber stations. Antenna systems of the one repeaters as well as its equipment are presented in Figs. 1.33 and 1.34.

Beyond the control center, in the mobile radio monitoring stations, in labeling measuring laboratories, in RMSGs, in direction of Students Games, as well as in the power agencies, more than 40 units of remote AWSs were deployed. With the help of remote AWSs, the full-scale operation was provided in the system. At that, reservation of the control channels was performed. In the case of impossibility to use the wired channels, the automatic switching onto wireless 3G channels was executed.

1.5.4 Monitoring Equipment

During Students Games, two types of fixed radio monitoring points, FRMS, in which antenna systems were mounted of the high building roofs, and ORMS installed directly on the objects of Students Games, were used. The MRMS and

Fig. 1.34 Repeater equipment of service radio communication

manpack equipment, by which the RMSGs were equipped, were also used. Peculiarities of radio monitoring equipment utilization are illustrated in Fig. 1.35.

Figure 1.36 shows location of fixed radio monitoring equipment during preparation and conduction of Students Games. When Students Games were finished, the ORMS concentration within the limits of the city became redundant. Therefore, the most of ORMSs was transferred to the other inhabited localities to be used as measuring stations.

The remotely controlled measuring unmanned fixed stations ARCHA-IN were the main equipment of FRMS, in the structure of which the fixed direction-finder ARTIKUL-C (1.5–3000 MHz) and measuring panoramic device for field strength measurement (0.009–8000 MHz) were included. The equipment of ARCHA-IN in Kazan is presented in Figs. 1.37, 1.38, and 1.39.

The object radio monitoring points were also based on the measuring receiver ARGAMAK-IS, which had temperature-controlled and waterproof secure cabinets (boxes) equipped with sensors of alarm signaling, signals from which passed to ASU RCHS. ORMSs were settled directly in the most important sport objects and provided round-the-clock monitoring of radio electronic equipment of small radius

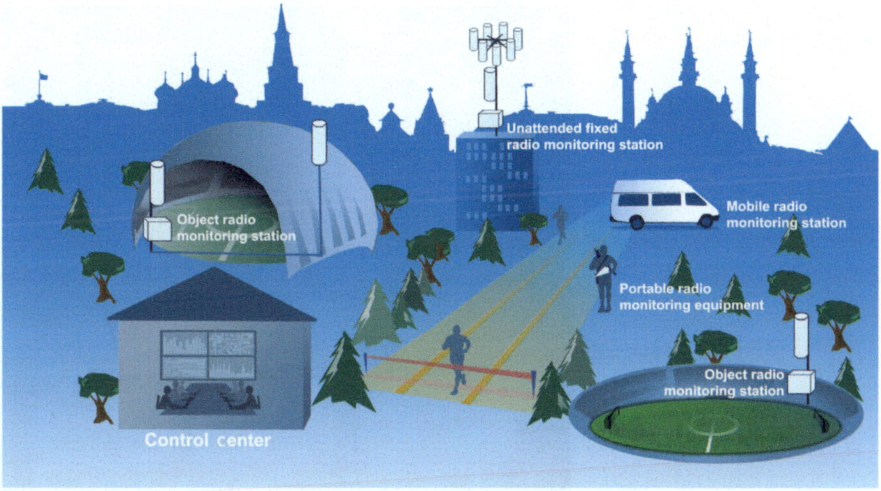

Fig. 1.35 Peculiarities of radio monitoring equipment utilization

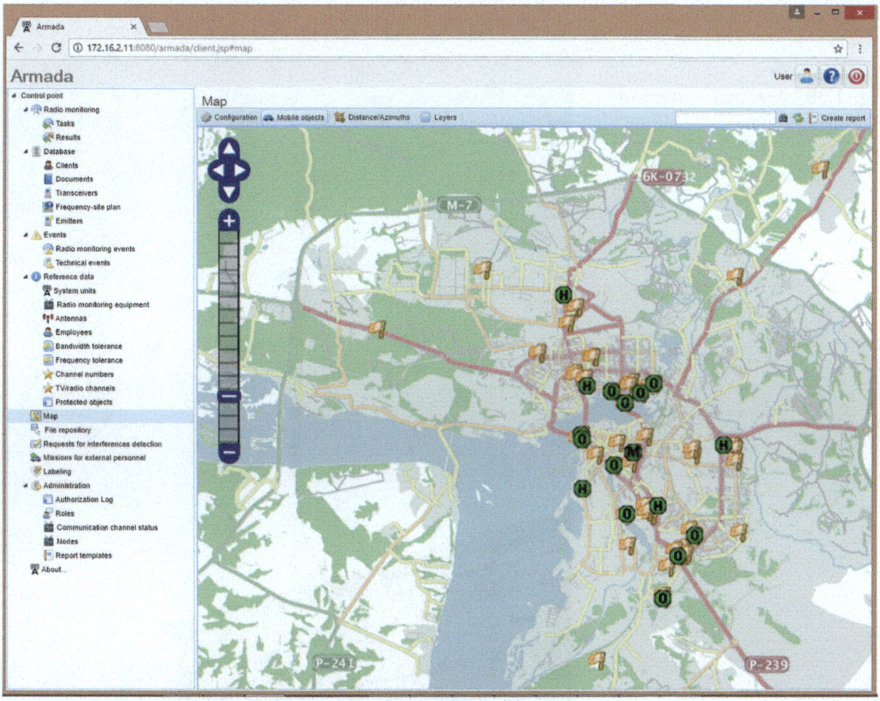

Fig. 1.36 Location of fixed radio monitoring equipment

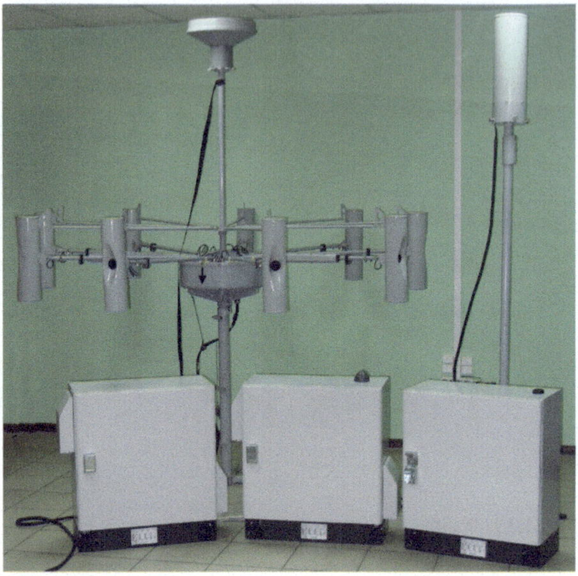

Fig. 1.37 Equipment of ARCHA-IN station

Fig. 1.38 Antenna system of measuring device

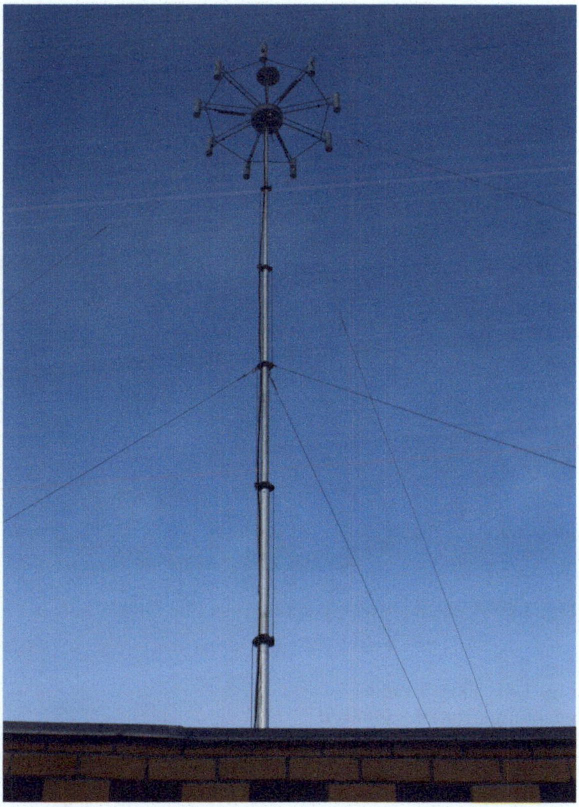

Fig. 1.39 Antenna system of the direction-finder

of action, and they performed identification and measurement of radio signal parameters in the range from HF and to lower part of UHF range including TV video cameras with the wireless channel, the analysis of service identifiers and parameters of cellular communication systems, and the wireless access. The higher operating frequency of ORMS was 8 GHz. Features of several ORMSs accommodation are explained in Figs. 1.40, 1.41, 1.42, and 1.43.

FRMS and ORMS were remotely controlled from the control center and, if necessary, from the mobile stations or from radio monitoring groups. The control was provided via the wired channel, which was reserved by the wireless 3G channel as well as by the radio channel for transmission of emergency messages on the base of deployed network of service radio communication MOTOTRBO.

Mobile equipment was used to execute radio monitoring operations, when small power of transmitters, high directivity of transmitting antennas, remoteness of the radio emissions complicated or made impossible the fulfillment or measurements and localization of sources by fixed equipment.

The mobile stations ARGUMENT-I and stations of other type were used as MRMS during the Student Games. All MRMSs had possibility of connection to ASU RCHS; however, the direct control of radio monitoring equipment from the

Fig. 1.40 Center of rowing sport types (the ORMS antenna is marked by orange circle)

Fig. 1.41 Object radio monitoring station, arranged on the roof of the rowing sport type center

Fig. 1.42 Object radio monitoring point located inside the stadium "Kazan-Arena" (the antenna system of the emergency channel MOTOTRBO is marked by orange circle)

system was realized only for equipment of ARGUMENT-I stations. These stations provided measurement of radio signal parameters up to 43 GHz and automatic direction-finding from 1.5 to 8000 GHz. To expand range of monitoring and amplitude direction-finding up to 43 GHz, as well as for station operation as LML, the spectrum analyzer integrated into ASU RCHS and the measuring antenna P6–69 (18,000–40,000 MHz) were contained in the station. The external view of the mobile station ARGUMENT-I and the workstation for its operator are shown in Figs. 1.44 and 1.45.

Data exchange between mobile station and ASU RCHS was provided with the help of the wireless channel of 3G modems. Also, all main competition objects in the preparatory period were equipped with special places for wired connection of the mobile station to the Internet; therefore, during parking near such objects, the wired connection was used with the help of the Ethernet cable.

The manpack equipment was used by radio monitoring groups in places difficult to be accessed: on roofs of higher buildings, inside the premises, including the sport objects of the Students Games. The handheld direction-finders ARC-RP3 M and also the portable receivers with a set of directed antennas were used as the manpack equipment. Work of radio monitoring group is shown in Figs. 1.46 and 1.47.

Fig. 1.43 Accommodation of ORMS antenna inside the stadium “Kazan-Arena”

Fig. 1.44 Mobile station ARGUMENT-I during on-duty period near the sport object

Fig. 1.45 Workstation for the mobile station operator

Fig. 1.46 Radio monitoring group during search for interference on the stadium

1.5.5 Subsystems to Provide International Activity

To solve the radio monitoring tasks during execution of big public events, the ARMS ARMADA included the following additional functional subsystems:

- The requesting service;
- Testing and marking of radio transmitter (radio frequency source).

Fig. 1.47 Monitoring of electromagnetic situation by the radio monitoring group on the stadium. The manual direction-finder ARC-RP3 M is used as the radio monitoring equipment

The requesting service was intended for automatic consideration of request on utilization of radio transmitter. In the informative portal of the Students Games, there was a service enabling portal users to apply requests on equipment utilization. Requests were automatically transmitted to the account database of ASU RCHS, where they got the status "accepted." Specialists of the Students Games Direction, using the ASU RCHS AWSs, which was located territorially in the Students Games Village, performed preliminary processing of obtained requests and made a decision: to refuse the request informing the applicant about it or to continue its processing. In the case of positive decision, the request got the status "on consideration." The stages of request passage are schematically shown in Fig. 1.48.

All "on consideration" requests had passed processing by specialists of radio frequency department using AWSs, located in control center. According to the request processing results, requests were refused with immediate informing of Direction employees, or were included in database as "planning." All this was accompanied by works on allocation of the frequency–time resource to the radio transmitter equipment on the requested sport objects. For these requests, after its estimation by Students Games Direction specialists and possible additional processing, the electronic document "Recommendation about utilization conditions for radio transmitter equipment" was formed. It gave a right to applicants to test their equipment and to execute marking procedures.

The subsystem of testing and equipment marking was used for technical inspection of parameters correspondence to "Recommendations about utilization conditions for radio transmitter equipment" and marking of equipment by the colored label. Testing assumed the check of correspondence of the real technical

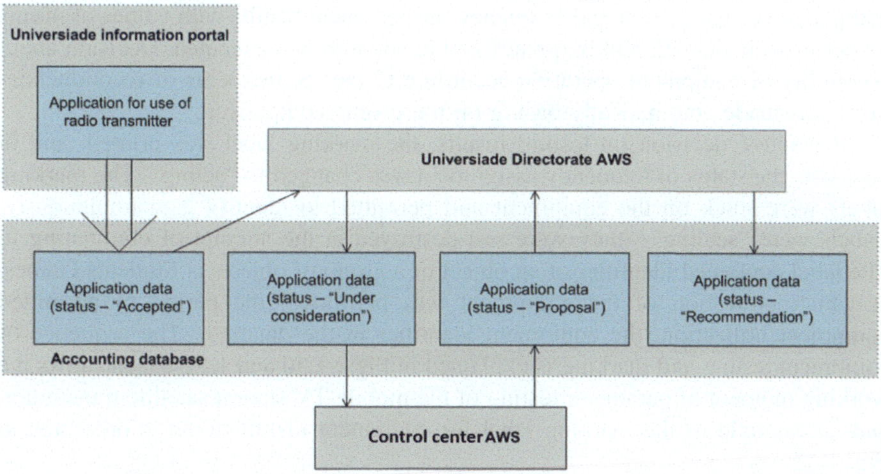

Fig. 1.48 Passage of request on equipment utilization

characteristics of radio equipment emission (frequency, bandwidth, level) to issued recommendations. By the measurement results, the decision was automatically made about possibility of equipment marking. Testing and marking were fulfilled in labeling (marking) measuring laboratories, which were deployed fixed and on the base of mobile station. The local database of LML was automatically synchronized with the central database of ASU RCHS via the networks for data transmission. At that, the LML operation was provided both at the presence and at disappearance of the communication channels. Interaction of LMLs and the central database of control center is shown in Fig. 1.49.

Before beginning of testing, the equipment user reported an identifier of the approved recommendation. According to this number, the list of frequencies to be checked appeared at AWS of the LML officer. During testing, the automatic

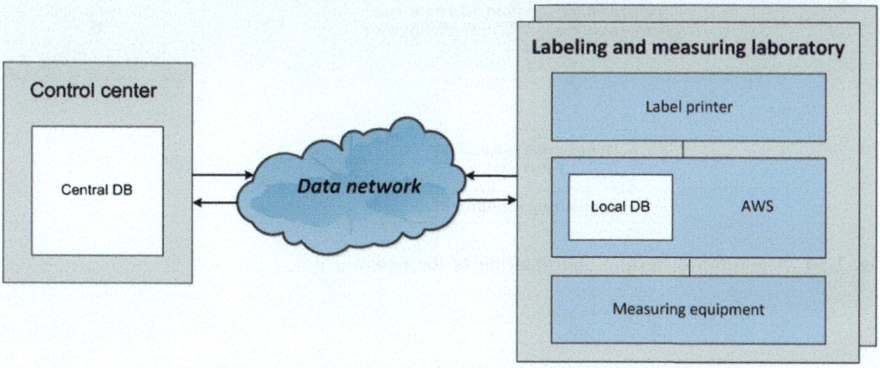

Fig. 1.49 Interaction of LMLs and database of control center

comparison of bandwidth and frequency measurement results with values of norms on permissible deviation of frequency and bandwidth was executed. Decision about possibility of equipment operation according to the requirements of recommendations was made automatically basing on the results comparison.

At positive decision on testing results, the marking label was printed, and in database, the status of frequency assignment was changed to “acting.” The marking labels were stuck on the equipment and permitted to identify it unambiguously. Labels were “sealing”—they were self-destroyed at the attempt of eliminating it. The label contained identifier of an object or a group of objects at Students Games, at which utilization of the equipment was permitted, the period of permitted equipment utilization, the equipment identifier in the database. The sequence of equipment testing and marking is explained in Figs. 1.50 and 1.51 which shows the working moment of parameter testing of the mobile TV station satellite transmitter, and an example of the marking label for equipment identification is presented in Fig. 1.52.

The subsystem of testing and marking permitted to plot on the marking labels the bar codes containing all necessary information in the coded form. According to bar code reading results by scanners on terminals of entrance check, it was possible to track the equipment, which was brought into the sport objects. However, in connection with the large quantity of the Students Games objects, and accordingly, the large number of terminals of entrance check, the necessity of training of large number of employees, this functionality of ASU RCHS was not used in Kazan City.

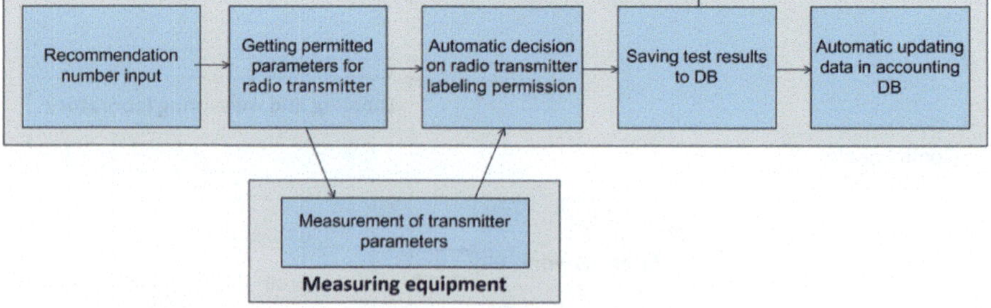

Fig. 1.50 Algorithm of testing and marking of the equipment

Fig. 1.51 Checking of parameters of satellite transmitter of the mobile TV station

Fig. 1.52 An example of the marking label

1.5.6 Execution of Radio Monitoring

The subsystem of the planned mode provided the automatic fulfillment of radio monitoring tasks according to the given plan (schedule) including measurement of signal parameters, localization of radio emission sources in locality, detection of new sources, monitoring of emission parameters of registered sources and their comparison with norms, determination of frequencies and frequency ranges occupancies. The special significance at operation in the automatic mode had application of the flexible system of radio monitoring events, which used spectral and time masks. This enabled radio monitoring equipment operation in the automatic mode for the detection of interference and for search for emission parameter deviations of the radio frequency equipment.

During preparation and conduction of the Students Games on the base of missions with the formation of radio monitoring events, the monitoring of "protected" ranges was executed, i.e., of those equipment ranges for users permitted to the object. These ranges were adopted by the Students Games Direction, as well as equipment ranges of emergence and socially important services.

A possibility of simultaneous control of large quantity of equipment and also autonomous execution of tasks by the equipment at loss of communication channel can be attributed to features of the subsystem of automatic task fulfillment. At autonomous operation, results of task execution were saved to the local control servers, which were deployed on the controllers of the equipment. Transmission of task execution results to the central database was recommenced automatically at communication channel recovery.

Figures 1.53 and 1.54 show the variants of representation of mission's execution results in the planned mode in the interface of ASU RCHS.

The operative mode was used when it was required to make necessary decisions in the complicated cases of the search for the interference sources and to fulfill the immediate localization of radio emission sources in locality. In the operative mode, the operator directly, in the real time, controlled the necessary group of the radio monitoring equipment. In the operative mode, the simultaneous synchronous or asynchronous control by several direction-finders and by radio receivers, the access to account data was made.

Factually, during conduction of the Students Games, all fixed radio monitoring equipment fulfilled the automatic missions with usage of radio monitoring events. At arising of the event, for instance the signal appearance, which operated outside

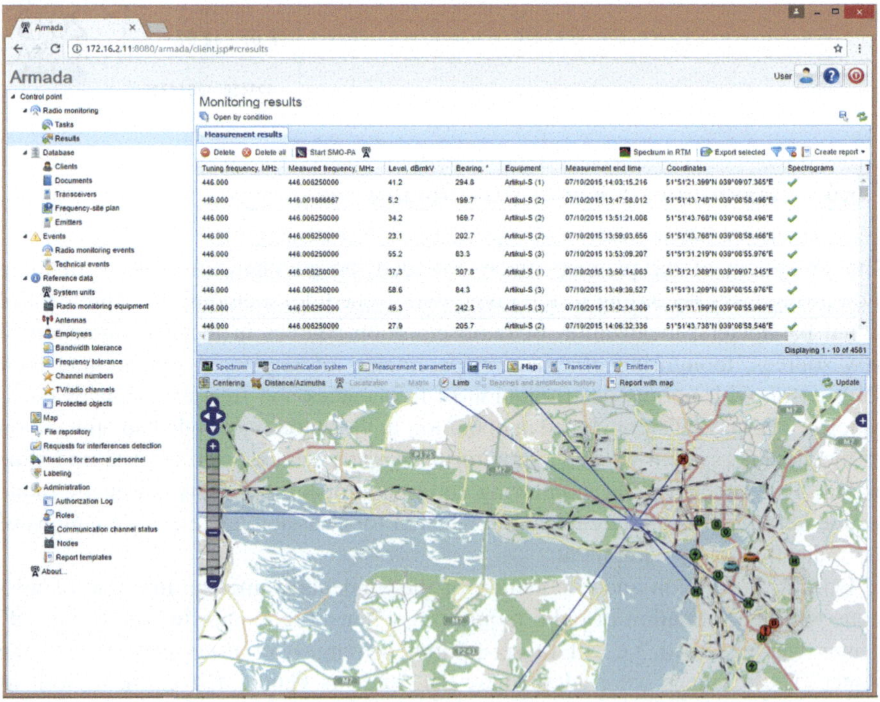

Fig. 1.53 Interpretation of direction-finding results on the map

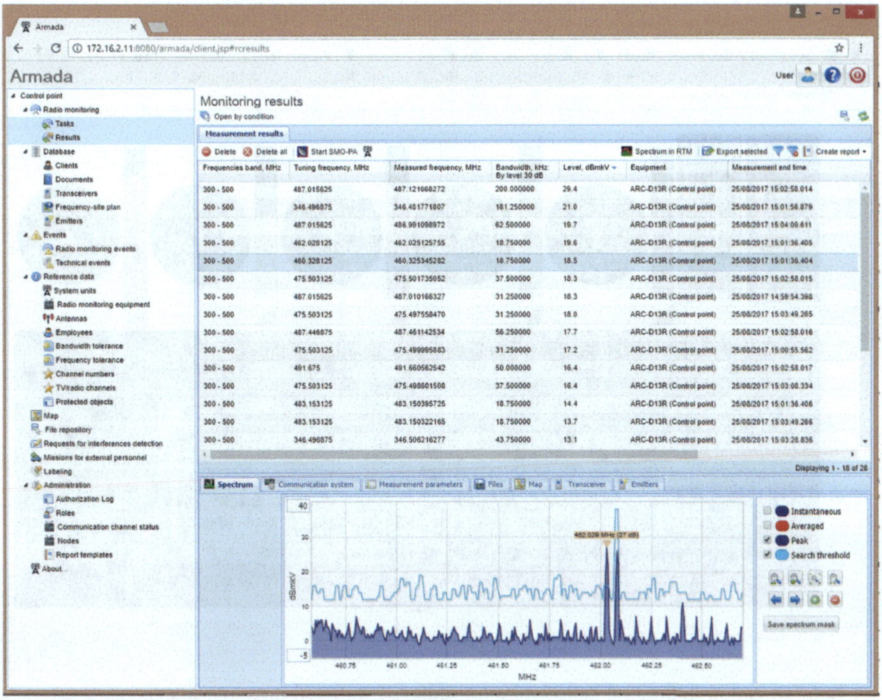

Fig. 1.54 Signal detection on the radio monitoring event (signal outside the mask)

of the spectral mask, the message appeared for control center operator, who started working in the operative mode for detailed analysis with the aid to reveal danger degree of this event and to make the necessary decision.

Examples of the operative mode windows are presented in Figs. 1.55, 1.56, 1.57, and 1.58.

1.5.7 Sequence of Radio Monitoring System Deployment

Works on deployment of ASU RCHS UNIVERSIADA 2013 were based on the plan of actions to provide radio frequency spectrum management during preparation and conduction of XXVII World-Wide Summer Students Games-2013 in the Kazan City, which was adopted by the Head of Roskomnadzor of Russian Federation in September 2010. On the base of this plan, the radio frequency service of Privolzhsky Federal district had developed the technical task and organized the competition, at which the IRCOS Company was chosen as the developer, provider, and integrating organization of the ASU RCHS system.

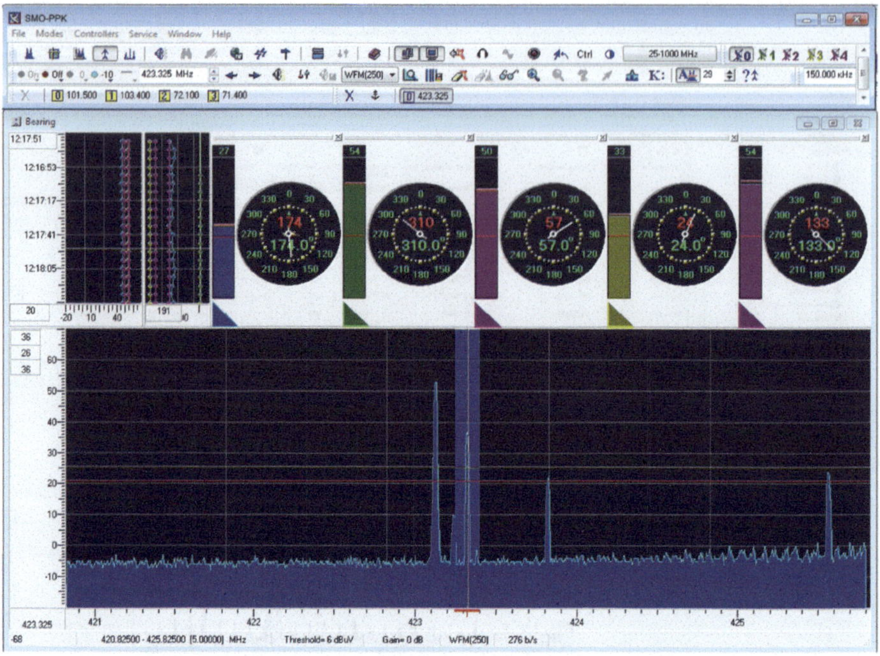

Fig. 1.55 Direction-finding of radio emission source. Five fixed direction-finders are used

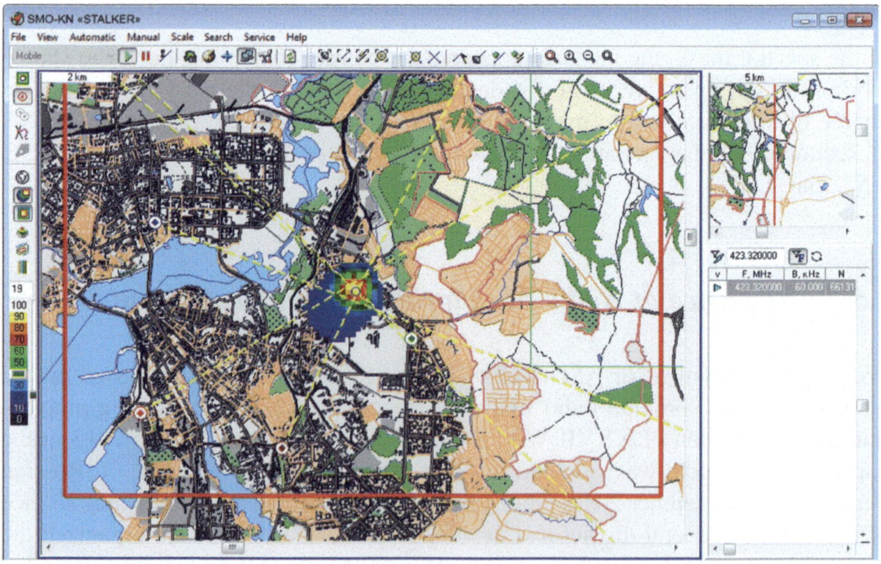

Fig. 1.56 Localization of radio emission source on the map

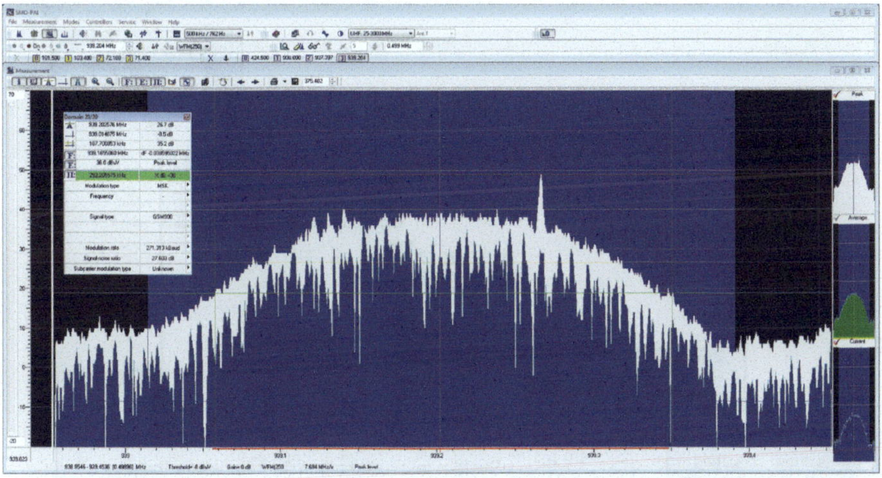

Fig. 1.57 Measurement of the occupied bandwidth and the central frequency of the signal

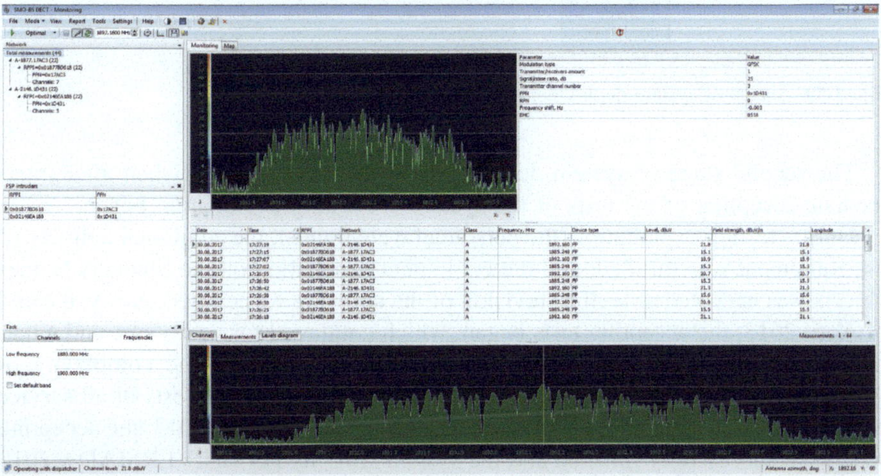

Fig. 1.58 Measurement of the Digital European Cordless Telecommunications (DECT) base station parameters

Implementation of ASU RCHS was executed in several stages as shown in Fig. 1.59. At first stage, in April 2012, the delivery of pilot-party of equipment was provided consisting of two ORMSs and one FRMS, the beta-version of system software, and engineering preparation of the situation center and objects for accommodation of the radio monitoring equipment. During the pilot-work of the system, the correctness of solutions was checked, necessary revisions of the software were made, and revealed errors were corrected.

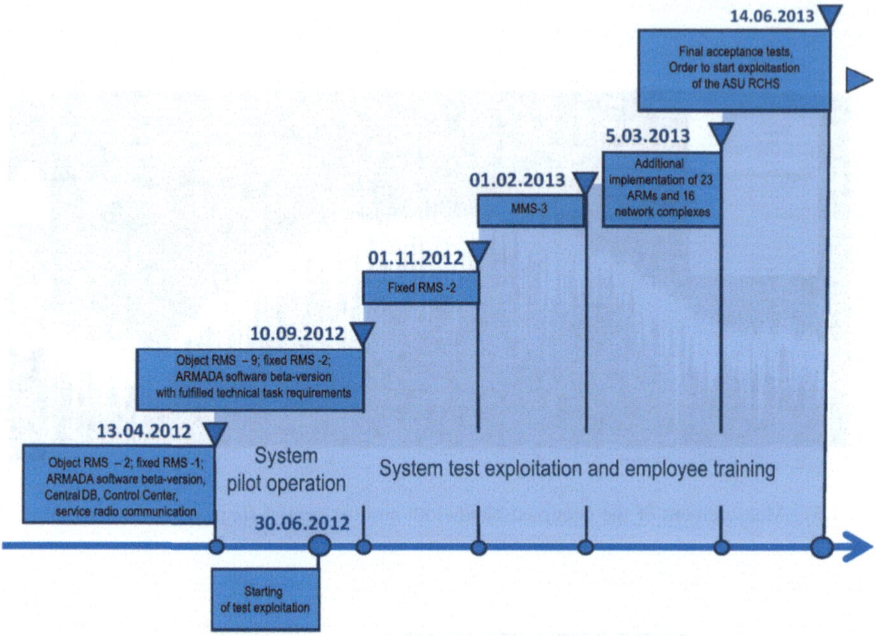

Fig. 1.59 Stages of system deployment

The second stage of system deployment began at July 2012, when after intermediate accepting of the tests of the system, its testing exploitation began.

During testing exploitation, the personnel were trained, the additional delivery of the equipment was fulfilled, and the equipment installation on the Students Games objects was executed as well as revision of the software was carried out. At the end of April 2013, the accepting tests began, which continued for nearly one and a half months. Such a long period of tests could be explained by the complexity of ASU RCHS, its multi-functionality, necessity of participation in tests of all service personnel, and leaders of the higher level. In the middle of June 2013, the accepting tests were successfully finished and the system ASU RCHS UNIVERSIADA 2013 was accepted to exploitation.

ASU RCHS UNIVERSIADA 2013 during preparation and conduction of the Summer Students Games in Kazan City provided the remote control of territory-distributed fixed, mobile, and manpack radio monitoring equipment, testing and marking of radio frequency transmitters, interaction with external informational structures. The system provided effective personnel control, coordinated task formulations, monitoring of its execution and making necessary decisions in the real time.

With the help of the subsystem of request service in the period of preparation and conduction of Students Games, 285 requests on radio frequency equipment were received, 39 of which were rejected. For equipment testing, ten LMLs were deployed including two fixed and eight mobile stations. Totally, monitoring and marking of 8 368 radio frequency equipment was executed including 6 714 for land mobile service, 1 364 devices of small operation radius, 20 equipment of fixed satellite services, 266 units of fixed service, and 4 units of radar service.

During 12 days of Students Games execution, 207 violations of frequency spectrum usage were revealed by the radio frequency service employees.

The experience of exploitation of ASU RCHS UNIVERSIADA 2013 was used by the radio frequency service at XXII Winter Olympic Games of 2014 in Sochi.

1.6 Summary

All considered automated radio monitoring systems manufactured and delivered by the IRCOS Company are based on the unified hardware–software platform consisting of equipment of radio monitoring and the SMO-ARMADA software. The client/server architecture of automated systems consisting of the "thin" client, application, and data servers has the following general properties, which do not depend on the scale and destination of the system:

- Hierarchical scaling architecture, use of Web, and geo-informational technologies;
- Possibility of utilization of equipment of different manufacturers on the base of the open control protocol;
- Interaction with external informative systems;
- Execution of radio monitoring tasks in automatic (planned, background) and manual (operative) modes;
- Eventual mechanism of task execution, diagnostics, and self-diagnostic of the equipment.

At the same time, the composition, configuration, functions of the specific automated system are defined by purposes and tasks stated before users of this system.

The automation units of the radio monitoring system consist of three main parts: the equipment of radio monitoring, the engineering-technical infrastructure, and the software. They provide solution of all radio monitoring tasks including detection, identification, and localization in locality and in premises of radio emission sources and measurement of radio signal parameters.

All equipment of radio monitoring on the base of DRR of the ARGAMAK-family, included in the structure of stations and complexes of the automated system, provides execution of high-accurate measurements and panoramic, spectral, and vector signal analysis. Besides, they enable determination of service identifiers and parameters of telecommunication systems such as APCO P25, DMR, GSM, UMTS, LTE, IS-95, cdma2000, EV-DO, TETRA, DECT , DVB T/T2/H, Wi-Fi.

Synchronization from GNSS realized in DRRs ARGAMAK-IS and ARGAMAK-RS enables usage of the time difference of arrival method for coordinates' determination of radio emission sources, which does not require the direction-finder utilization. In the system including the direction-finder and measuring receivers, the hybrid algorithms are used with usage of information on time, levels, and angles of signals arrival, which increases the accuracy of RES location calculation.

ARMADA and ASU RCHS UNIVERSIADA 2013 systems presented as the examples have the hierarchic structure, at the head of which there is the control center. When ARMADA system is used on the country scale, here are additional control points in it, which decrease the response time of the radio frequency service on the local conditions, for instance appearance of spurious emissions of unauthorized radio frequency equipment and interference arising.

All systems have the mutual central database of radio monitoring, in which radio monitoring data are collected from the local databases of measuring and direction-finding points. In turn, the accounting and reference data and planned tasks on radio monitoring are transmitted to the radio monitoring points. The control of radio monitoring equipment is executed from the any automated workstation if the operator has the appropriate access rights.

Automated systems enable operation in operative and planned modes. At operation in the operative mode, the remote control of radio monitoring equipment is executed by the operator in the real time. At operation in the operative mode, the set of measurements is given, the time period, in which these measurements should be fulfilled, and also the list of equipments, which should execute the required measurements.

In the automated system composition, there are the subsystems of cartography providing the interpretation of radio monitoring equipment, their current condition (condition of the equipment, availability via the network, etc.), accommodation of registered radio frequency equipment and their brief characteristic, tracks of mobile complexes during the defined time interval, plans of buildings and constructions, localization of radio emission sources, etc. The data transmission is performed via the wired communication channels (Ethernet, fiber-optic) and the wireless communication channels (3G/4G), and automatic switching is fulfilled at communication loss in one of the channels.

In the automated ARMADA system, the fixed, mobile, portable, and manpack radio monitoring equipment finds their application. Their composition and quantity depend on many factors: locality character, type of building up, concentration and types of acting radio frequency equipment, etc. Permissible system cost is also of no small importance factor. To save money, the organization of two-stage radio monitoring, when facts of violation are revealed by comparatively cheap equipment, and violator registration, formation of protocols is executed on the base of measurements made by high-accurate measuring equipment. Metrological characteristics of the used equipment satisfy to norms and regulations accepted in Russian Federation and ITU recommendations.

The automated system ASU RCHS UNIVERSIADA 2013 is a variety of the ARMADA system; it is intended for radio monitoring execution at large international sport actions. The peculiarity of the system is a presence of additional functions, including management of request passage for utilization of radio frequency equipment, bringing in by delegations, their testing and marking. ASU RCHS UNIVERSIADA 2013 during preparation and conduction of summer Students Games in Kazan City provided remote control of territory-distributed fixed, mobile, and manpack radio monitoring equipment, testing and marking of radio frequency equipment, interaction with external informational structures. The system enabled effective control of the personnel, coordinated task formulation, monitoring of their execution and making necessary decisions in the real time.

References

1. About communication. Federal Law of the Russian Federation (as amended on 6 July 2016)
2. Handbook on computer-aided techniques for spectrum management (CAT) (2015). ITU-R, Geneva, 174 p
3. Handbook on national spectrum management (2015) ITU-R, Geneva, 333 p
4. Handbook on spectrum monitoring (2011) ITU-R, Geneva, 659 p
5. Recommendation ITU-R SM.1537. Automation and integration of spectrum monitoring systems with automated spectrum management
6. Recommendation ITU-R SM.1139. International monitoring system
7. Recommendation ITU-R SM.1370. Design guidelines for developing advanced automated spectrum management systems
8. Radio Regulations (2012) ITU
9. Rembovsky A, Ashikhmin A, Kozmin V, Smolskiy S (2009) Radio monitoring. In: Problems, methods and equipment. Lecture Notes in Electrical Engineering. Springer, 507 p
10. Bykhovsky, MA (ed) (2012) Fundamentals of management of the use of the radio-frequency spectrum. Frequency planning of television and radio broadcasting networks and mobile communications. Automation of radio frequency spectrum management, vol 3. KRASAND, Moscow, 368 p
11. Intelligent Radio Monitoring System for Efficient Spectrum Management HYUN-SEOK YIM, YUN-HO LEE, KYUNG-SEOK KIM. Department of Radio Engineering Chungbuk. National University 12 Gashin-dong Heungduk-gu Cheongju Chungbuk, pp 361–763 SOUTH KOREA
12. Spectrum management system for developing countries (2015) Version 5.0. ITU
13. Monitoring Software R&S ARGUS. Technical Information. ROHDE & SCHWARZ GmbH & Co. KG Mühldorfstraße 15. 81671 München. Germany. 25-June-2012, 8SPM-Kr, Version 5.4.6
14. SCORPIO Spectrum Monitoring System TCI INTERNATIONAL, INC. 3541 Gateway Blvd. Fremont, CA 94538-6585 USA
15. GOST (Russian State Standard) 34.003-90. Information technology. Set of standards for automated systems. Automated systems. Terms and definitions
16. SMO-ARMADA software package of automated spectrum monitoring. http://www.ircos.ru/en/sw_armada.html. Accessed 28 Nov 2017
17. SMO-PA/PAI/PPK panoramic analysis, measuring and direction finding software package. http://www.ircos.ru/en/sw_pa.html. Accessed 28 Nov 2017

18. SMO-BS Software packages for wireless data communication system analysis. http://www.ircos.ru/en/sw_bs.html. Accessed 28 Nov 2017
19. The Catalogue 2017 of IRCOS JSC. http://www.ircos.ru/zip/cat2017en.pdf. Accessed 28 Nov 2017
20. Radiomonitoring: problems, methods, means (2015, in Russian). In: Rembosky, AM (ed) 4th edn. Hot Line-Telecom Publ., Moscow, 640 p
21. Egli JJ (1957) Radio propagation above 40MC over irregular terrain. Proc IRE 45:1383–1391
22. Alekseev DA, Ashikhmin AV, Kobelev SG, Kozmin VA, Rembovsky AM, Sysoev DS, Tsarev LS (2014, in Russian) Features of automated spectrum management system at 27th Summer Universiade in Kazan, Electrosvyaz (4):34–41
23. Report ITU-R SM.2257-2 (June 2014). Spectrum management and monitoring during major events. SM Series. Spectrum Management, 64 p

Chapter 2
Unauthorized Radio Emission Revelation

2.1 Introduction

Monitoring of the radio communication regulation maintenance and established order of radio frequency equipment utilization, defense of governmental and commercial secrets and inter-corporative information, confrontation to industrial espionage, ensuring of the confidentiality of negotiations and meetings were and are the problems of the current importance for radio communication and security services. Due to the sharply increased amount of radio frequency systems used in the territory and inside the premises, their assortment growth, transfer to new technologies of information transmission, we can manage with increasing flow of more complicated tasks on revealing unauthorized radio emissions without growth of the operating personnel in number by means of modern automated radio monitoring systems utilization only.

Therefore, the high-priority applications of the automated radio monitoring systems and the included radio monitoring equipment are a revelation of unauthorized radio emissions of any nature, localization of unauthorized radio emission sources in the restricted territory and in premises of buildings, in the transport vehicles, and other objects. Main consumers of such systems are military, monitoring, guarding and department services [1–3].

We can distinguish several areas of radio monitoring systems application, which determine their difference in structure and engineering units:

- In restricted territory without fixed radio monitoring boundary in the local events region, for instance, counter-terrorism operations, training exercises, and sport events;
- In district areas within the limits of the guarded restricted territory;
- In premises or in zones of one or several multi-floor buildings (in one or several inhabited localities);
- In separate room or transport vehicle.

A. M. Rembovsky et al., *Radio Monitoring*, Signals and Communication Technology, https://doi.org/10.1007/978-3-319-74277-9_2

The goal of the present chapter is to describe tasks and structure of the AREAL automated radio monitoring system on the base of the typical variants. Main methods of unauthorized RES revelation will be also discussed, ARMS operation in the restricted territory, in buildings, and in separate objects will be analyzed, and features of search and localization of unauthorized RES in engineering objects, buildings, and constructions in the adjacent territory will be described.

Main functions and peculiarities of radio monitoring equipment included in the ARMS will be considered:

- Stationary and mobile radio monitoring stations ARCHA-IN, ARCHA-INM, ARGUMENT;
- Portable direction-finders ARTIKUL-H1 and ARC-RP3M;
- Stationary system ARC-D15R for radio monitoring in premises (zones) of the controlled objects;
- Portable complex ARC-D11 for radio monitoring of the separate premise or transport vehicle as well as radio monitoring in the district.

Functions of the software, which controls the ARMS, will be also considered. Moreover, special focus will be on the procedures of technical analysis of radio signals necessary at the determination of unauthorized RES origin, parameters, and character of the transmitted information.

2.2 AREAL Automated Systems for Radio Monitoring in the Restricted Territory and Objects

The modern level of information systems development provides centralized control of several equipment at once from one automated workstation, thus decreasing the required personnel and workstations number, and automatic solution of typical tasks, reserving the final decision making to the operator and processing of supernumerary situations. Integration of equipment into unified informational system enables the centralized remote control of equipment, the inspection of its performance capacity, the administration of access to information of different user groups, and the storage and processing of information obtained. Such an approach enables to create unified account database of the RESs, to reduce the cost of deployment, implementation and system support. At that, it is expedient to realize the risk-oriented methods, at which the system solves the stated tasks in an automatic mode and gives notifications to the personnel about arising of supernumerary situations (events). The event content may be very different: Information on the revelation of the signal, which presents a potential risk, or information about equipment failure can act as eventual information. Notification of the personnel about events arising should be produced in the mostly short periods using all available notification channels: the urgent presentation on the display, notification via mobile phone, etc.

AREAL ARMS, as ARMADA ARMS, consists of three main parts: the radio monitoring equipment, the necessary network, the service and engineering infrastructure, and the customized software as shown in Fig. 2.1.

At the development of the hardware–software platform, we took into account experience, advantages, and drawbacks of native and foreign automated radio monitoring systems, recommendations of International Telecommunication Union [4]. This platform has the following peculiarities, which simplify its application: an open architecture with the possibility to expand functionality, the flexible system of radio monitoring tasks content generation, and the possibility of integration with other information systems.

Equipment of the ARMS is based on the digital receivers of the ARGAMAK-family [5, 6]. The engineering-technical infrastructure includes cable lines and nodes for data transmission, server equipment, engineering structures including rooms for the hardware. The unified software SMO-ARMADA, which is based on the client/server architecture and the Web technology, is used in the ARMS [7].

Radio monitoring, which is provided by the AREAL ARMS, may be continuous, periodic, or single, and accordingly to it, the components of the system may change. The AREAL ARMS can be used in the autonomous mode as well as the subsystem (a cluster) of the ARMADA ARMS.

The main task of the AREAL ARMS is revelation and localization of unauthorized RES, and it can be solved by means of search for "new" radio emission, estimation of risks, determination of its localization and displaying on the objects electronic plans and territory maps together with saving information in the radio electronic environment databases. Thus, the AREAL ARMS ensures:

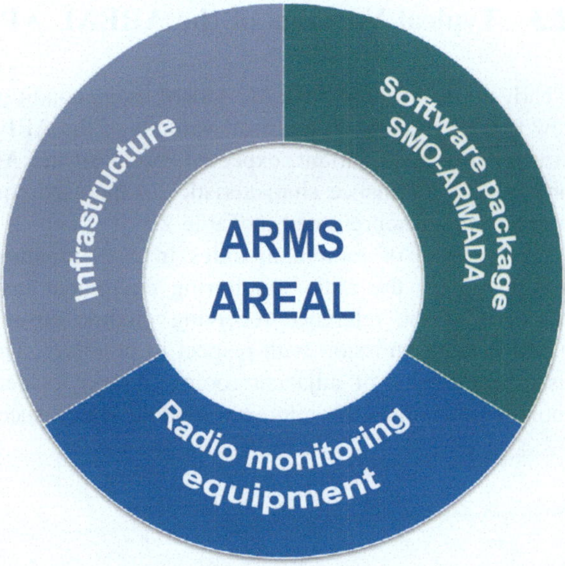

Fig. 2.1 Components of the AREAL ARMS

- Monitoring of frequency range occupancy in the deployed territory and obtaining information about factual frequency usage;
- Revelation of radio electronic environment variations and detection of new radio emissions;
- Search and localization of radio emission sources in the district;
- Classification of RESs being detected, estimation of their value or risk;
- Support of the database on RESs in the monitored district part, in the monitored zone, in rooms, and in transport vehicles;
- Processing and systematization of radio monitoring results, database complement;
- Displaying of the system structure and system nodes condition, the location of RES, and the radio monitoring results on the map and the object plan;
- Revelation of radio networks and radio directions in relation to radio emission sources;
- Monitoring of fulfillment of radio monitoring tasks on radio emission sources operation;
- Automated analysis of accumulated results of radio monitoring over the required time period;
- Export and import operations with reference databases as well as with the results database;
- Automatic monitoring of equipment operation time and its technical condition;
- Reports generation according to operation results.

The tasks related to obtaining of total information access to the radio emission sources are not considered in this book.

2.3 Typical Variants of the AREAL ARMS Utilization

Nearly all manifold of AREAL ASRM usage conditions can be narrowed down to a few typical variants. The typical variants of the AREAL ARMS usage, the ASRM structure for each variant, expected results of the ARMS utilization, composition and basic performance characteristics of the used equipment, and content of radio monitoring are represented in Table 2.1.

A presence of remote modules in each monitored room (zone) is the most acceptable for the radio monitoring equipment inside buildings as well as the presence of the reference receiving module ensuring the priority reception of external radio emission with respect to premises. The practice shows that we can use the modules of adjacent rooms (zones) located at some distance from the monitored room as the reference module with a success.

Table 2.1 Typical variants of the AREAL ARMS

Variant	Conditions of tasks solving in this variant	Monitoring duration	Monitoring periodicity	Expected result
AREAL-1	Along lengthy boundary of monitoring and in the adjacent territory	Continuous	–	Coordinates and parameters of new RES, permanently upgraded database of RESs
AREAL-2	In the restricted area in the region of local activities (counter-terrorism operation, training exercises, sport events, etc.)	Period of action conducting	As required	Coordinates and parameters of new RES, upgraded database of RESs
AREAL-3	In the restricted area, in two stages: first stage—primary radio monitoring and determination of RES location	Depending on radio electronic environment estimation results	As required	Coordinates of RES and places for transportable equipment location
	Second stage—enhanced study of radio electronic environment	Depending on the radio exchange intensity	Single, cycled, or as required	Upgraded database on RESs and RES coordinates for the specific district part
AREAL-4	Within the limits of monitored, fully restricted closed territory	Continuous	–	Parameters and coordinates of unauthorized RES
	Outside monitoring of radio emission levels from the object rooms, RES localization displaying on the object plan	As required	Cycled or as required	Suggestions on elimination of revealed violations
AREAL-5	Remote radio monitoring in the premises (zones) of objects	Continuous	–	Parameters and coordinates of unauthorized RES
	Radio monitoring in the separate room or in the transport vehicle	Depending on radio electronic environment	Cycled or as required	Suggestions on rejection of revealed information leakage technical channels

The real zone of electromagnetic availability of the AREAL ARMS is defined by a number of equipped equipment for tasks solution, their sensitivity to electromagnetic field, and by the height of its antenna systems. Precision of unauthorized RES localization depends on the instrumental precision of the used direction-finding equipment and on a number of equipment units, in operation areas of which the RES are detected simultaneously.

2.4 Monitoring Equipment for Unauthorized RES Revelation

Depending on the conditions of tasks solution for unauthorized RES revelation within the limits of the AREAL ARMS, we can use fixed, mobile, or portable radio monitoring equipment. Features of fixed and mobile equipment:

- All radio monitoring equipment of the ARCHA, ARGUMENT measuring stations has the technical and metrological parameters, which fully correspond to recommendations of International Telecommunication Union (ITU);
- Presence of several independent autonomously controlled posts, wide functional possibilities of stations, and performance characteristics make possible their utilization in the interests of both radio communication services and military structures.
- Stationary stations are supplied in maintainable and unmanned versions (maximal range of operation temperatures is from −55 to +55 °C, comparably small dimensions).
- All equipment in the unmanned version has sensors of alarm signaling (sensors of hit, tilt, openness/closeness).
- Automatic direction-finders ARTIKUL-C and ARTIKUL-M on the base of digital receiver ARGAMAK have the bandwidth of simultaneous analysis up to 24 MHz, the operating frequency range is 1.5–8000 MHz (without change of antenna systems), ensure single-channel direction-finding with rate more than 100 bearings/s, multi-channel direction-finding with rate up to 2500 MHz/s.
- Digital receiver of the radio direction-finder ARTIKUL-S is embedded into the antenna system. It increases precision, sensitivity, dynamic range and provides an absence of antenna effect. The mass of the direction-finder's antenna system is about 30 kg.
- Measuring DRR ARGAMAK-IS of ARCHA and ARGUMENT stations has the operating frequency range from 9 kHz to 8000 MHz, with the frequency converter up to 18 (40) GHz. The bandwidth of simultaneous analysis is up to 22 MHz, which allows processing of input signals in the real time. The rate of the panoramic spectral analysis in the real time is from 6 kHz/s to 10 GHz/s.
- DRR ARGAMAK-IS provides all necessary types of radio signal measurements including electromagnetic field strength, frequency, and bandwidth of radio emissions as well as many other parameters. Decoding and service information analysis of digital communication networks of GSM, UMTS, IS-95, cdma2000, EV-DO, TETRA, LTE, DECT , DVB-T1/H/T2, DMR, and APCO P25 standards are ensured together with the information transmission.
- High-quality stable reference generator on the base of the rubidium frequency standard used in DRR ARGAMAK-IS has a possibility to correct frequency with resolution of 10^{-13} order by the software. This makes possible the synchronization of the receivers in the system on long time intervals with the draft of mutual signal delays, which do not exceed ±30 ns. It allows creation of time difference of arrival (TDOA) measuring system for radio emission source localization.

- DRR ARGAMAK-IS ensures the simultaneous connection of two remote radio frequency sensor equipment antennas (up to 3 GHz and to 8 GHz) and up to five additional antennas with possibility of their commutation and remote orientation change in azimuth and polarization.

Features of transportable and portable equipment:

- Transportable equipment is intended to operation at temporal parking, to which they are delivered by the mobile carrier, while the portable equipment can be delivered by the operators themselves.
- Antenna system of the most direction-finders (ARTIKUL-T, ARCHA-IT, ARTIKUL-MT, ARTIKUL-P) has the embedded double-channel receiver, which is controlled via Ethernet. This allows application of the feeder up to 100 m in length without degradation of the sensitivity, dynamic range and without arising of the antenna effect.
- DRRs with two physical reception channels are used in all above-mentioned transportable systems for direction-finding, which ensures usage of the correlation-interferometric method for direction-finding in range from 1.5 to 8000 MHz.
- The transportable ARCHA-IT station, the ARTIKUL-T direction-finder, and the ARTIKUL-MT direction-finding complex on the base of ARC-D11 have a bandwidth of simultaneous analysis up to 24 MHz, the rate of the panoramic spectral analysis up to 20 GHz/s with spectrum resolution 6 kHz, and the rate of multi-channel direction-finding up to 2.5 GHz/s, while single-channel direction-finding at more than 100 bearings/s. These products have advanced functional possibilities; ensure a fast search for new radio signals, their recording and technical analysis, automatic direction-finding, determination of unauthorized RESs location together with their displaying on the cartographic background, the multi-channel reception of wide spectrum of transmissions in the bandwidth of simultaneous analysis, etc.
- Analysis of service information is ensured for the following digital communication and information transmission networks: GSM, UMTS, IS-95, cdma2000, EV-DO, TETRA, LTE, DECT, DVB-T1/H/T2, DMR, and APCO P25, WiMAX.
- ARCHA-IT, ARTIKUL-MT equipment and complexes on the base of the ARTIKUL-H1, and ARC-D11 equipment provide possibility to operate both at parking with antenna located on the mast and at moving with antenna system located in the baggage box. Unauthorized RES are localized using mutual operation of direction-finding pair at parking, and using autonomous operation of direction-finder at movement.
- It is possible to install equipment onto the unprepared platforms with minimal deployment time (within the 30 min interval) and folding time for all station posts.
- Possibility to exploit antenna systems and receivers under complicated climatic and natural conditions (including mounting climate, heavy winds, and precipitations) without the usage of additional protection measures.

2.5 AREAL-1 Automated Radio Monitoring System

AREAL-1 ARMS provides constant monitoring along the lengthy boundary and in the adjacent territory. Distinguish feature of the AREAL-1 variant (see Fig. 2.2) is an absence of clear boundary over all perimeter of the monitored district part. Size of the monitored territory is determined by the electromagnetic availability zone of located monitoring equipment, and it may be expanded with the help of mobile and manpack units.

The following equipment is accommodated in the monitored territory:

- Radio channel monitoring, radio signals recording, their technical analysis;
- Unmanned stations ARCHA-IN and ARCHA-INM for radio monitoring and direction-finding;
- Direction-finders ARTIKUL-ST in the fixed unmanned implementation;
- Mobile stations for radio monitoring and direction-finding ARGUMENT;
- Portable direction-finding complexes on the base of ARTUKUL-M, ARCHA-IT, and ARTIKUL-MT;
- Communication and data transmission equipment;
- Control centers to which data from radio monitoring equipment are transmitted in the real time or periodically.

Radio monitoring results are sent from monitoring equipment to the control center or are accumulated for further automated processing; at that, the control center may be combined with mobile station. Taking into consideration the permanent radio monitoring character, the used equipment must have, as far as possible, the best performance characteristics concerning the operating frequency range, performance, and functionality.

The AREAL-1 ARMS, in the full structure, ensures the automated radio monitoring tasks fulfillment including:

- Search for new emissions in the range from 0.009 to 40,000 MHz with the rate of panoramic analysis up to 20 GHz/s;

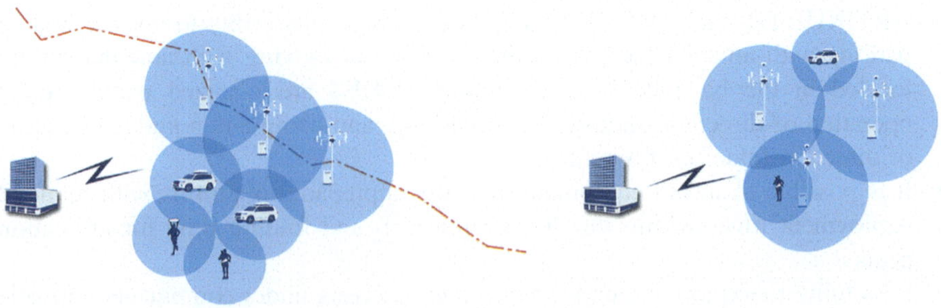

(a) Along fixed boundary (b) In the restricted area

Fig. 2.2 Structure of the AREAL-1 ARMS

- Automated direction-finding and determination of unauthorized RES locations in the district with their displaying on the cartographic background in the frequency range from 1.5 up to 8000 MHz;
- Automated direction-finding and determination of unauthorized RES locations in the district with their displaying on the cartographic background in the frequency range from 8 to 40 GHz;
- Manual direction-finding and determination of unauthorized RES locations in the district with their displaying on the cartographic background in the frequency range from 0.3 to 18,000 MHz;
- Recording, technical analysis, and automated radio signal processing of radio signals in the frequency range up to 40 MHz.

The main performance characteristics of fixed (ARCHA-IN, ARCHA-INM) and mobile (ARGUMENT-I) stations are presented in Tables 2.2 and 2.3. The typical values of sensitivity by electromagnetic field and instrumental precision (RMS) are given in Table 2.4.

ARCHA-IN—two posts, which function independently, namely: direction-finding ARTIKUL-SN on the base of the direction-finder ARTIKUL-S and the measuring ARGAMAK-ISN on the base of the DRR ARGAMAK-IS. They can be controlled by the only one operator and solve tasks both mutually and separately. ARCHA-INM—equipment of posts are combined, tasks of direction-finding and measurement are solved in turn, and measuring equipment is developed on the base of the modified DRR ARGAMAK-IS. ARGUMENT-I—mobile measuring station for radio monitoring and direction-finding on the base of the direction-finder ARTULUL-M1 and DRR ARGAMAK-IS.

Table 2.2 Main performance characteristics of the fixed stations

Parameter	ARCHA-IN	ARCHA-INM
	Value	
Single-channel and multi-channel direction-finding, panoramic analysis		
Operating frequency range (basing structure/maximal) (MHz)	20–3.000/1.5–8.000	20–3.000/1.5–3000
Rate of panoramic analysis (GHz/s)	Up to 20	
Rate of direction-finding (MHz/s)	Up to 2.500	
Sensitivity on field and precision (RMS)	See Table 2.4 (AS-PP4, AS-PP5)	
Antenna system height	Up to 100 m	
Field strength and radio signal measurement, panoramic analysis		
Operating frequency range (basing structure/maximal) (MHz)	20–3000/0.009–18.000	20–3.000/20–8.000
Maximal bandwidth of simultaneous analysis (MHz)	Up to 40	
Radio signal recording in bandwidth (MHz)	Up to 40	

Table 2.3 Main performance characteristics of the mobile station ARGUMENT-I

Parameter	Value
Single-channel and multi-channel direction-finding, panoramic analysis	
Operating frequency range (basing structure/maximal) (MHz)	20–3000/1.5–8000
Rate of panoramic spectral analysis with 6 kHz spectrum resolution (GHz/s)	Up to 20
Rate of direction-finding (GHz/s)	Up to 2,5
Sensitivity on field and precision (RMS)	See Table 2.4 (AS- MP)
Measurement of the field strength and radio signal parameters, panoramic analysis	
Operating frequency range (basing structure/maximal) (MHz)	20–3000/0.009–40,000
Maximal bandwidth of simultaneous analysis (MHz)	Up to 40
Radio signal recording in bandwidth (MHz)	Up to 40
Antenna system height, located on the mast	Up to 100 m

Table 2.4 Typical values of sensitivity to the electromagnetic field and instrumental direction-finding precision (RMS)

Antenna system	RMS dependence (deg) on frequency (MHz)	Dependence of sensitivity by field strength (μ V/m) on frequency (MHz)
AS-PP4/ AS-PP5	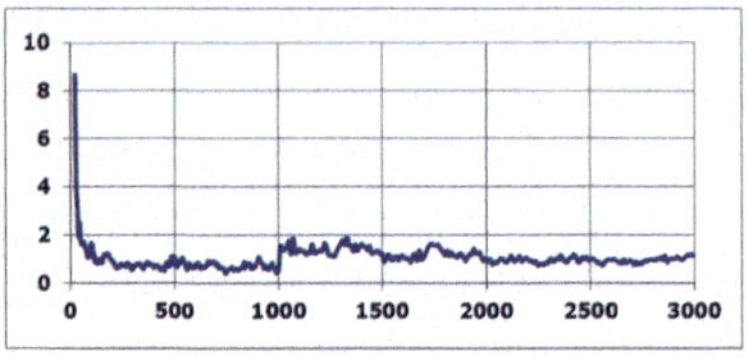	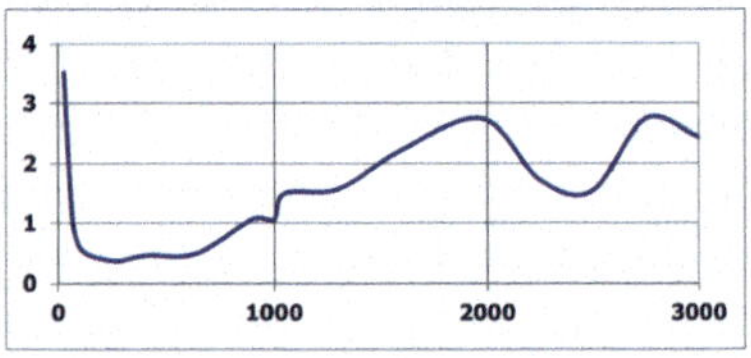
AS- MP	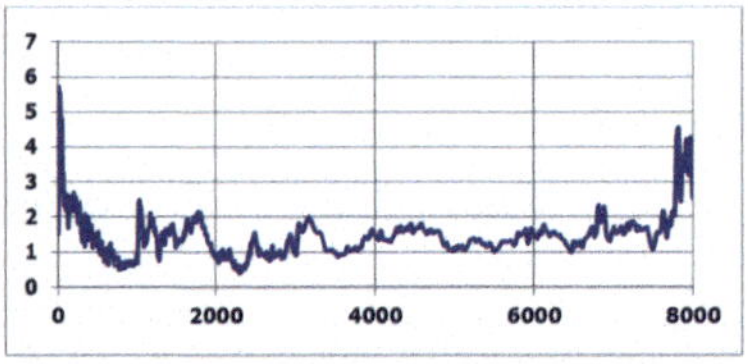	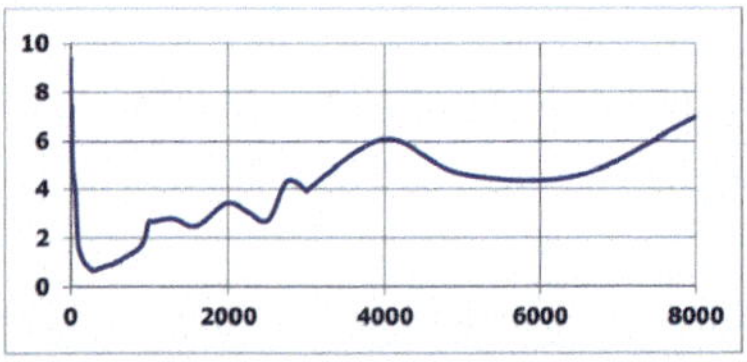
AS- PP17	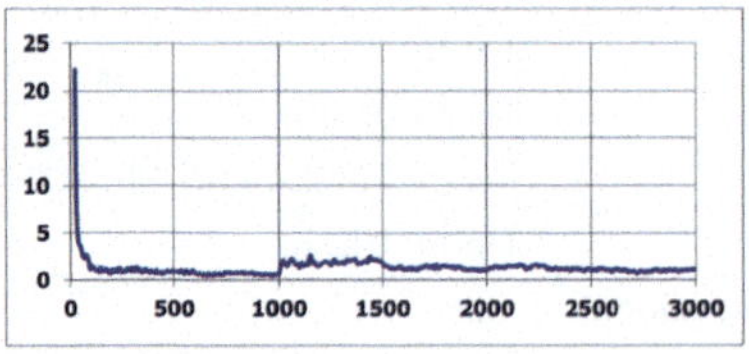	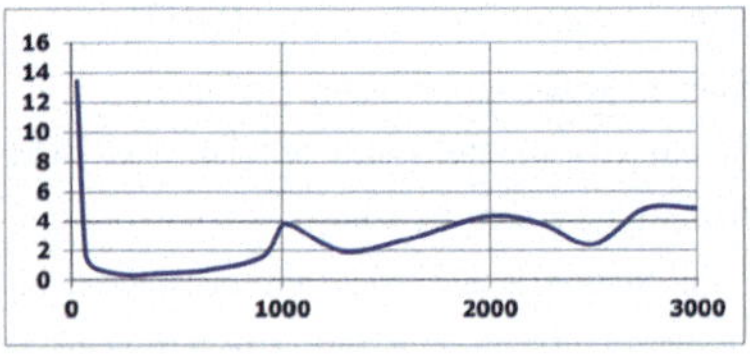
AS- HP-KV	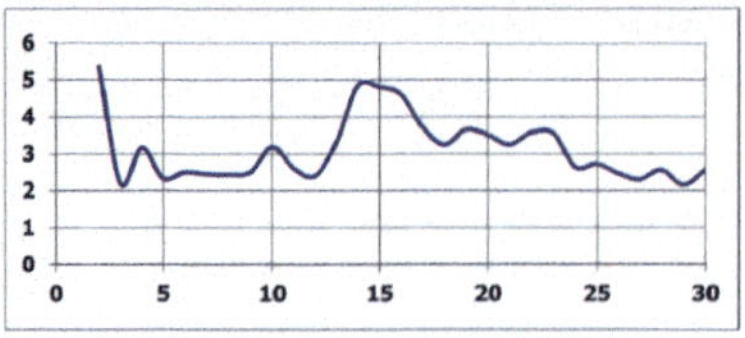	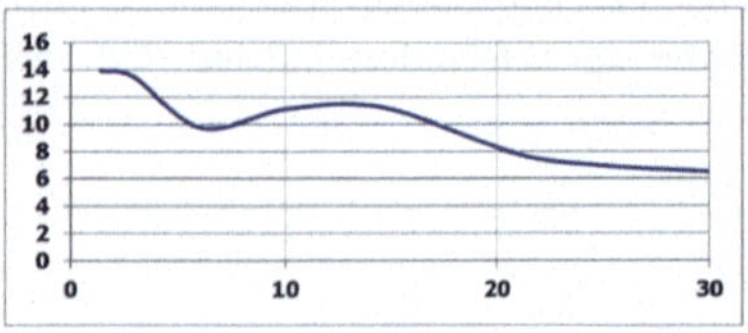

(continued)

Table 2.4 (continued)

Antenna system	RMS dependence (deg) on frequency (MHz)	Dependence of sensitivity by field strength (μ V/m) on frequency (MHz)
AS- HP5	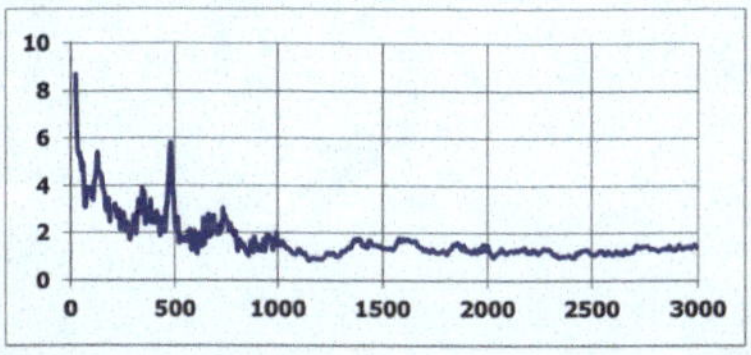	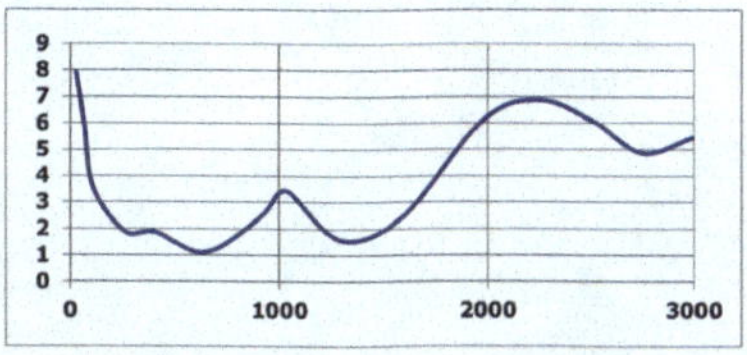
AS- HP2	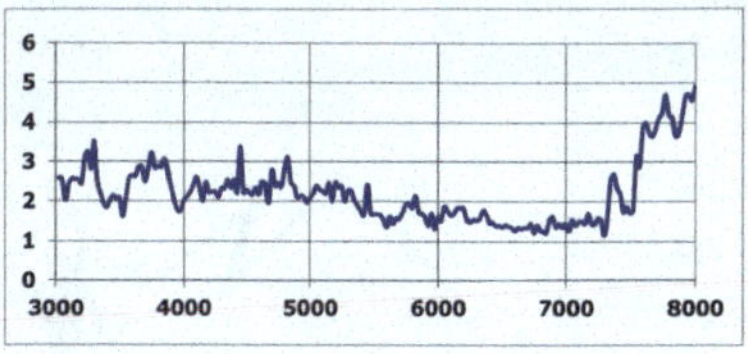	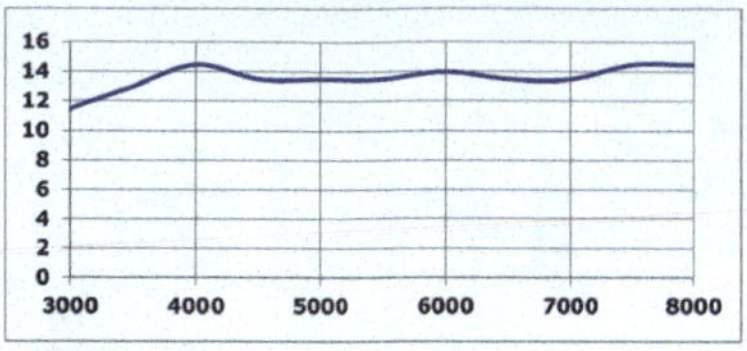

2.6 AREAL-2 Automated Radio Monitoring System

AREAL-2 ARMS (see Fig. 2.3) provides single or periodical monitoring in the restricted territory, in which REE is insufficiently explored. As in the AREAL-1 case, this system variant is characterized by the absence of the clear boundary over all perimeter of the monitored distract part. A size of the territory under monitoring is defined by the electromagnetic availability zone of located (permanently or temporarily) equipment and it can be expanded with the help of mobile and manpack equipment.

Reasonable limitation of tasks for this variant, for instance the absence of enhanced study of REE, relates to time limitation and interest in minimizing of the content and overall mass indices of the equipment. The problem of enhanced study of REE is related to the AREAL-3 variant.

In the AREAL-2 ARMS full composition, the following equipment can be used:

- Mobile stations of radio monitoring and direction-finding ARGUMENT;
- Portable equipment based on ARCHA-IT, ARTIKUL-T, ARTIKUL-MT, ARTIKUL-P, ARTIKUL-H1, or ARC-D11;
- Manpack direction-finders ARC-RP3M;
- Mobile radio monitoring and direction-finding stations ARGUMENT-I or mobile direction-finding complexes on the base of equipment ARTIKUL-M, ARCHA-IT, ARTIKUL-MT, or ARC-D11;
- Communication and data transmission equipment;
- Remote or regional control center, which can be combined with mobile station.

The AREAL-2 provides the automated radio monitoring tasks solution, including:

Fig. 2.3 Structure of the AREAL-2

- Search for new emissions in the frequency range from 0.3 MHz to 18 GHz with the panoramic analysis rate up to 20 GHz/sec;
- Automated direction-finding and determination of RES locations in the district with their displaying on the cartographic background:
  - Mutual operation of direction-finders in the range from 1.5 to 8000 MHz for ARGUMENT-I, ARTIKUL-H1, ARC-D11 with AS and from 25 to 3000 MHz for other equipment in the variant of their arrangement in the temporarily posts, or arrangement of one of them on the car or any other transport base [5];
  - Autonomous operation of direction-finders in the range from 1.5 to 8000 MHz for ARGUMENT-I, ARTIKUL-H1, ARC-D11 with AS and from 25 to 3000 MHz for other equipment at arrangement of the automated direction-finder on the car; in the range from 110 to 8000 MHz at arrangement of the automated direction-finder ARTIKUL-H1 on the operator;
- Handle direction-finding and radio emission sources localization in the district with their displaying on the cartographic background in the range from 0.3 to 18,000 MHz at autonomous utilization of the manpack direction-finder ARC-RP3 M;
- Recording, technical analysis, and automated radio signal processing of radio signals in the frequency range up to 40 MHz.

The main performance characteristics of the equipment are listed in Table 2.5.

Table 2.5 Main performance characteristics of transportable and portable equipment

	Transportable monitoring equipment			Portable monitoring equipment		
Characteristic	ARCHA-IT (AS-HP5)	ARTIKUL-MT (AS-HP5)	ARTIKUL-T (AS-PP4)	ARC-D11 (AS-HP-KV, AS-HP5, AS-HP2)	ARTIKUL-H1 (AS-HP-KV, AS-HP5, AS-HP2)	ARTIKUL-P (AS-PP17)
Operation at parking, AS on the mast	Yes	Yes	Yes	Yes	Yes	Yes
Operation in motion, AS on the car	Yes	Yes	No	Yes	Yes	No
Operation in motion, AS on the operator	No	No	No	No	Yes	No
Admissible AS exportation distance from operator	To 80 m	To 80 m	To 80 m	To 6 m	To 6 m	To 80 m
Operating frequency range in direction-finding and reception modes	20–3000 MHz	On the mast without AS change: • basic 20…3.000 MHz, • full 20…8000 MHz. On the car with changeable AS: 20…3000 MHz or 3…8 GHz	On the mast without AS change: • basic 20–3.000 MHz • full 1.5–8.000 MHz	With AS changeable on support or in the car: • basic 20–3000 MHz • full 1.5–8000 MHz	With AS changeable on support or in the car: • basic 20–3000 MHz • full 1.5–8000 MHz	25–3000 MHz
Performance of direction-finding	To 2.500 MHz/sec				To 150 MHz/sec	
Maximal bandwidth of simultaneous analysis	1 MHz (1.5–30 MHz) 10 MHz (20–110 MHz) 20 MHz (110–220 MHz) 40 MHz (220–3000 MHz)				5 MHz	

(continued)

Table 2.5 (continued)

	Transportable monitoring equipment			Portable monitoring equipment		
Characteristic	ARCHA-IT (AS-HP5)	ARTIKUL-MT (AS-HP5)	ARTIKUL-T (AS-PP4)	ARC-D11 (AS-HP-KV, AS-HP5, AS-HP2)	ARTIKUL-H1 (AS-HP-KV, AS-HP5, AS-HP2)	ARTIKUL-P (AS-PP17)
Dynamic range by third- and second-order intermodulation	Not less 75 dB					
Dynamic range with attenuator/without attenuator	≥ 140 dB/ ≥ 80 dB					
Rejection of reception spurious channels	Not less 80 dB	Not lass 75 dB				
Relative frequency instability	10^{-9}	5×10^{-7}				
Panoramic analysis rate with 6 kHz spectrum resolution	Up to 20 GHz/sec				Up to 3.2 GHz/sec	
Tuning precision	1 Hz					
Demodulation in real-time mode	AM, WFM, NFM, LSB, USB, CW					
Demodulation in deferred processing mode	AM, FM, FSK, FFSK, PSK, QAM, DQPSK, π/4 DQPSK, MSK					
Multi-channel reception for estimation of the level and demodulation of the signal	Up to 128 channels of 25 or 50 kHz in width					
Control	1 Gb LAN				USB. Control using TCP/IP protocol	
GPS/GLONASS	Embedded				Embedded	
Operative temperature: AS and reception unit	From −40 to +50 °C					
Equipment in the car cabin	From −10 to +45 °C					
Maximal attitude of installation over the sea level	4.500 m					

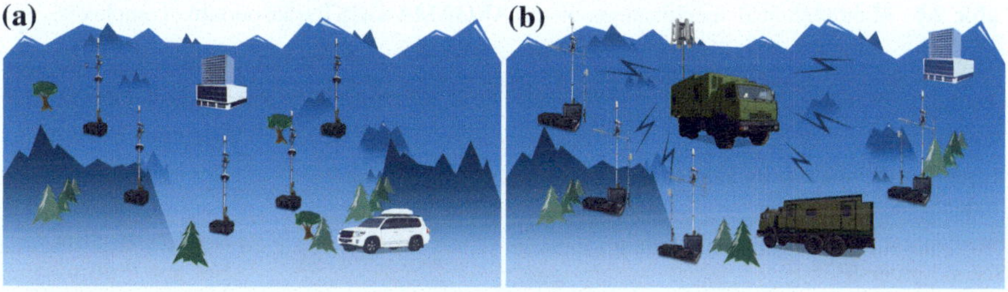

Fig. 2.4 Structure of the AREAL-3 ARMS

2.7 AREAL-3 Automated Radio Monitoring System

AREAL-3 ARMS (see Fig. 2.4) provides enhanced study of REE in the restricted and adjacent territories. The distinguish feature of this variant is also an absence of clear boundary over the perimeter of the district part under monitoring. In the full composition of the AREAL-3 ARMS, the following equipment can be used:

- Mobile stations of radio monitoring and direction-finding ARGUMENT;
- Portable direction-finding complexes based on ARCHA-IT, ARTIKUL-MT, ARTIKUL-H1, or ARC-D11;
- Transportable complexes of recording and technical analysis ARC-CST;
- Communication and data transmission equipment;
- Remote control center of obtaining and postprocessing of the accumulated information.

In this variant, tasks solution is carried out on the base of the direction-finding pair, which ensures direction-finding of new RESs, and the network of unmanned transportable complexes for recording and technical analysis ARGAMAK-CST, which are remotely controlled, and which data in transmitted to the remote control center or is accumulated and later is transmitted and analyzed in the control center (Fig. 2.4a, b).

Radio monitoring tasks solution in the AREAL-3 ARMS is performed in two stages. At the first stage, the mobile station moves forward to the monitored district to specify relief data specification, to determine the unauthorized RES location, and to select places for transportable complex ARGAMAK-CST accommodation. The obtained information is used at the second stage, when transportable complexes ARGAMAK-CST are deployed in the district, which, if necessary, are connected by the wideband communication network. When communication channels between stations are absent, the data exchange (assignment of tasks and results obtaining) can be performed with the help of removable information data storages.

At the remote-control center, the information obtained from the mobile stations and transportable complexes is processed. Data transmission to the remote-control center can be performed via wired or wireless communication channels as well as with the help of removable information data storages.

Table 2.6 Main technical specifications of the ARGAMAK-CST transportable complex

Parameter	Value
Operating frequency range	9 kHz–8 GHz
Sensitivity	Not more 1 μV
Reception attenuation by side channels at image and intermediate frequencies	Not less 80 dB
Interception point on intermodulation: Second-order IP2 Third-order IP3	Not less 30 dBm Not less 0 dBm
Noise factor in the range 20–8000 MHz	Not less 12 dB
Noise factor in other ranges	Not less 15 dB
Dynamic range on intermodulation of third and second order	Not less 75 dB
Maximal bandwidth of simultaneous analysis	Not less 24 MHz
AFC irregularity in the bandwidth of simultaneous analysis	Not more ±1 dB
Relative tuning frequency error:	
With internal reference generator	Not more 5×10^{-7}
With ARC-OG1reference generator	Not more 1×10^{-9}
Limits of input radio signal level measurement	−10 … +130 dB μV
Continuous radio signal recording in band	Up to 24 MHz
Control interface	Ethernet
DC supply voltage	From 21 to 30 V
Consuming power	Not more 150 W
Overall dimensions, not more:	
Remote sensor unit	800 × 260 × 180 mm
Receiving and processing unit	600 × 600 × 300 mm
Mass without cable	Not more 20 kg
The equipment may operate under temperatures: For equipment in the premise For equipment located outdoors (at field strength measurement)	+5 …+40 °C −20 …+55 °C

Performance characteristics of the mobile station were presented earlier in Table 2.3. The main technical specifications of the ARGAMAK-CST transportable complex are presented in Table 2.6.

Fig. 2.5 Structure of the AREAL-4 ARMS

2.8 AREAL-4 Automated Radio Monitoring System

AREAL-4 ARMS (see Fig. 2.5) is purposed for:

- Search for unauthorized RES in the fully controlled restricted territory;
- Outward monitoring of the signal levels of the sources emitting in the objects rooms;
- Localization and displaying of RES on the cartographic plan of the object.

The distinguish feature of this variant is a presence of the perimeter fixed boundary of the district part under monitoring and restricted access for unauthorized persons. In this territory, the industrial objects and buildings with institutions can be situated.

This variant implies the uninterrupted radio monitoring activity in the closed district parts with deployed military and other services, industrial and R&D institutions. Area of electromagnetic accessibility of the system is defined by the area of electromagnetic availability of equipment, which is installed temporally or permanently. It can be expanded with the help of mobile and manpack equipment, and it should maximally coincide with real sizes of the monitored territory.

The following tasks can be solved with the help of the AREAL-4 ARMS:

- Permanent radio monitoring within the limits of the fully restricted closed territory, which goal is the search and identification of unauthorized radio emissions, determination of sources location, and formation of databases with radio electronic environment objects parameters within the limits of the total electromagnetic availability zone of all equipment involved in the ARMS;
- Outside (relative to buildings) radio monitoring of emissions, measurement of their level, and comparison with approved norms as well as localization of building parts from which emissions arise.

To execute radio monitoring in the territory, we can use fixed (ARCHA-INM) and mobile (ARGUMENT) stations for radio monitoring and direction-finding, fixed (ARTIKUL-S N), mobile (ARTIKUL-M) and transportable (ARTIKUL-MT) direction-finders, portable automatic (ARTIKUL-H1) and manpack (ARC-RP3 M) direction-finders (see Fig. 2.5).

The tasks of measurement of radio emission level at the monitored objects boundaries, determination of unauthorized RES location and their displaying on the object cartographic plan at outside (relative to building) monitoring can be solved with the help of the mobile ARGUMENT-I station with a set of measuring antennas in combination with the video observation equipment and the SMO-SECTOR software package [8, 9].

2.9 AREAL-5 Automated Radio Monitoring System

AREAL-5 ARMS (see Fig. 2.6) provides radio monitoring inside the buildings, one room or transport unit with purpose to reveal unauthorized RES, their identification, estimation of its risk, and localization in the controlled objects zones.

Stationary distributed systems of remote radio monitoring ARC-D13R, ARC-D15R can be used inside objects for radio monitoring. Remote modules of these complexes provide receiving of radio signals in operating frequency band; they are located in the rooms (premises) of the building. One or a few remote modules are placed outdoors, so it is possible to compare signal levels inside and outside of the building. Signals from remote modules process to the local control server, which analyzes REE in the building, reveals new signals, and determines their sources location. Distributed systems of radio monitoring complexes are deployed in each building, which REE should be monitored uninterruptedly. Control of the system is performed from the control center, which can be situated in the monitored territory or at some distance.

In cases, when REE is monitored periodically in one room or automobile, or emissions are monitored in premises where constant deployment of equipment is impossible, the ARC-D11 equipment can be used for search for the unauthorized radio emissions. As the DRMS itself, the equipment in this case can be controlled

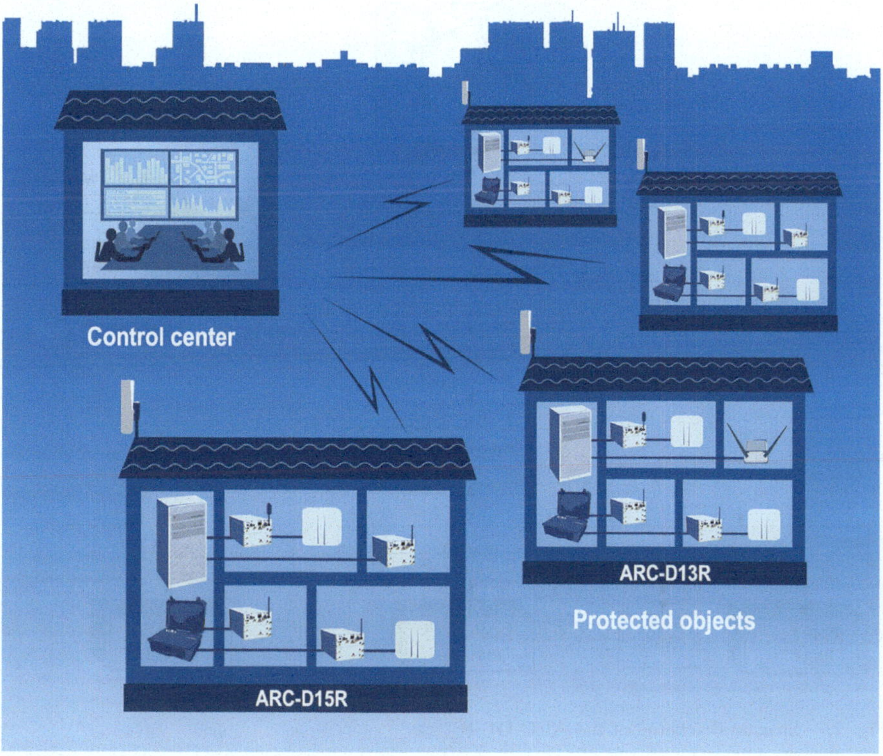

Fig. 2.6 An example of the AREAL-5 system deployment within the limits of one inhabited locality

remotely. In both cases (DRMS and ARC-D11 utilization), the ARC-RP3 M manpack direction-finder should be used to adjust the RES location.

Stationary system of remote radio monitoring in premises ARC-D13R supports monitoring of the rooms inside the objects. Its structural scheme is presented in Fig. 2.7.

The system is based on panoramic DRRs, in which analog part is embedded into the remote modules (RM) and digital part in double-channel control module of digital-analog processing and commutation BUVM-R. External view of the BUVM-R and remote module to be located in rooms are presented in Fig. 2.8.

The remote modules RM4P, which are purposed for obtaining information about radio electronic environment at the outer perimeter of the guarded object, have thermal proof and waterproof features (see Fig. 2.9).

System monitoring can be executed from the operator's AWSs, which are connected to the servers controlling the BUVM-R via data transmission equipment.

To enlarge unauthorized RES detection functions, which use signals of Wi-Fi standard to transmit data, the ARC-ABC analyzers (see Fig. 2.10) of Wi-Fi networks can be connected to the system.

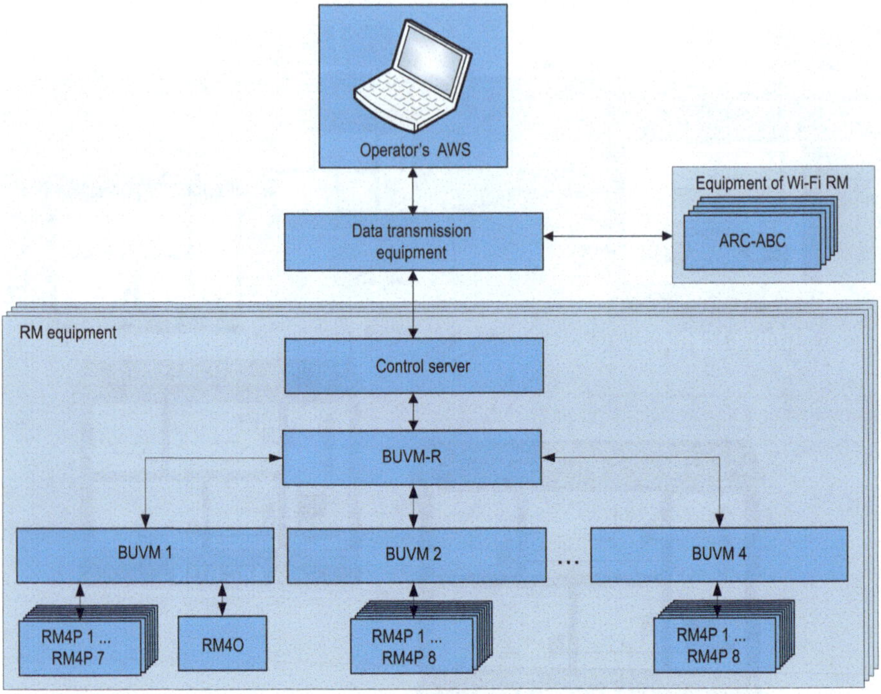

Fig. 2.7 Structural scheme of the ARC-D13R

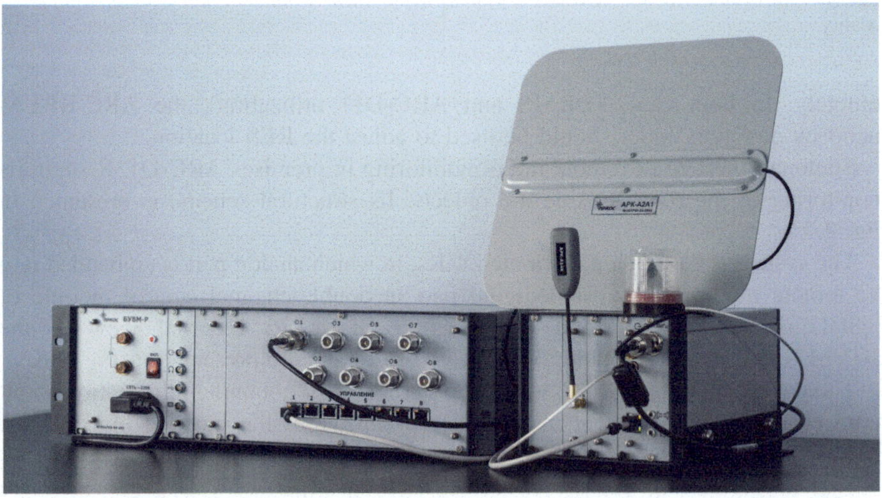

Fig. 2.8 RM4P module for premises

Fig. 2.9 RM4P module

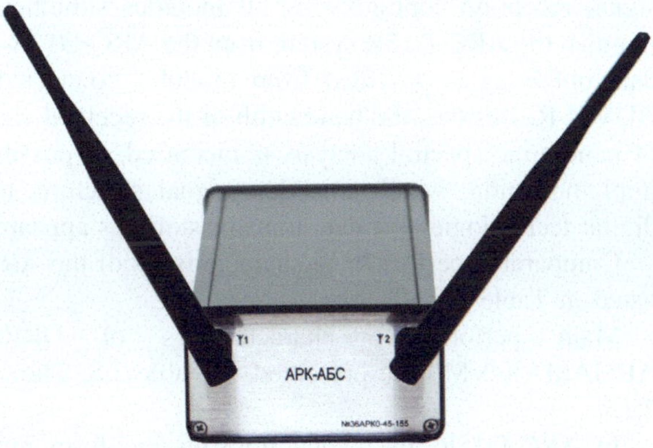

Fig. 2.10 ARC-ABC analyzer of Wi-Fi networks

The next step in development of equipment of stationery radio monitoring in premises is the system of remote radio monitoring ARC-D15R, in which structural diagram is shown in Fig. 2.11.

The system provides detection and localization of continuous, periodical, and short radio emissions, including operation in the same frequency band.

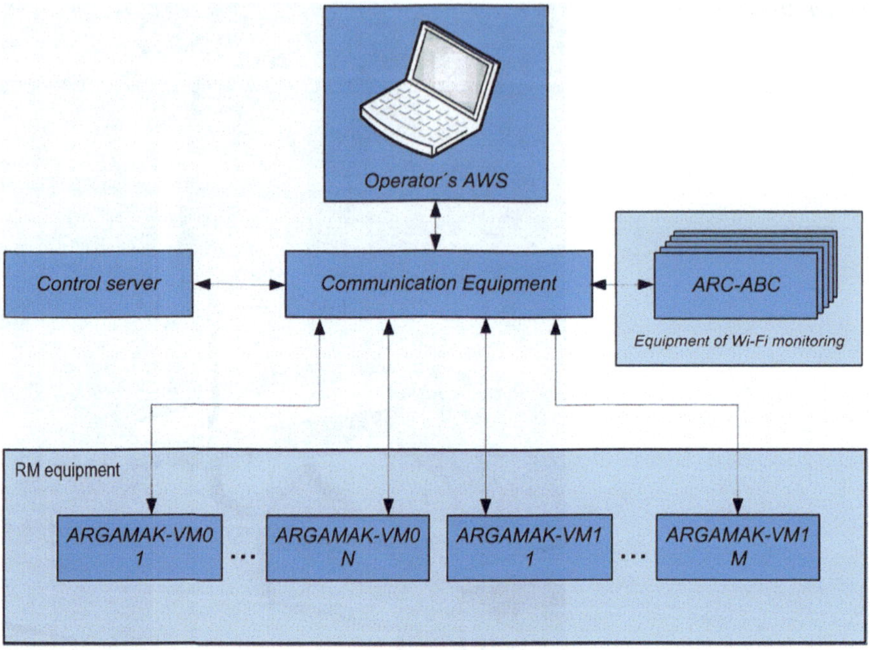

Fig. 2.11 Structural diagram of the ARC-D15R

Application of the wide-range radio receivers as the remote modules provides signal reception continuity by all modules simultaneously, which favorably distinguish the ARC-D15R system from the ARC-D13R one, in which the consecutive data obtaining is provided from modules connected to the signal inputs of the BUVM-R. Besides, the bandwidth of the received signals is expanded and the rate of panoramic spectral analysis is increased. A possibility of "address" processing implementation, which considers signal structure, to be used for devices using digital technologies for data transmission has appeared.

Comparable performance characteristics of the ARC-D13R and ARC-D15R are listed in Table 2.7.

Main performance characteristics of DRRs ARGAMAK-VM1 and ARGAMAK-VM0 are presented in Table 2.8. They are shown in Figs. 2.12 and 2.13.

In ARC-D15R, the signal transmission from remote modules to the control server is provided in the digital form, thanks to which the pickup influences on the cable transmission lines are eliminated, restrictions on the distance between remote modules and the control server are removed, and also, a possibility of technical calibration of HF cables between BUVM-R and remote modules disappears due to its different length during the system deployment.

To search for unauthorized radio emissions in the separate monitored room or automobile, the ARC-D11 and ARC-RP3 M are used.

Table 2.7 Comparable performance characteristics of the ARC-D13R and ARC-D15R

Parameter	ARC-D13R	ARC-D15R
	Value	
Operating frequency range in complete configuration (MHz)	0.009–18,000	
System integral sensitivity (transmitter power in the premise of 8 m × 8 m area, which is detected with probability 0.99)	50 μW	
Synchronization of all remote modules with timing to the unified timescale	No	Yes
Synchronous (in time and frequency) radio signal reception by all remote modules	Pair-wise only	Yes
Application of hybrid algorithms of unauthorized RES localization according to estimations of amplitude and signals time of arrival	Amplitude estimation only	Yes
Rate of panoramic spectral analysis with 6 kHz spectrum resolution, MHz/s	Up to 2,500	Up to 10,000
Number of monitored rooms	Up to 100	Not limited
Remote modules for rooms and in the protected implementation	Frequency converters ARC-VM1, ARC-VM0	DRR ARGAMAK-VM1, ARGAMAK-VM0
Output data	If output 41.6 MHz	Digital output IQ
Interface for control and data transmission	RS485	Ethernet
Analysis of parameters of wireless communication and data transmission systems	GSM, IS-95, cdma2000, EV-DO, TETRA, Wi-Fi, WiMAX, UMTS, DECT, LTE, DMR, APCO P25	GSM, IS-95, cdma2000, EV-DO, TETRA, Wi-Fi, WiMAX, UMTS, DECT, LTE, DMR, APCO P25
Distance from operator to reception modules	Up to 80 m in length of HF cable	Up to 150 m in length of the control cable

Double-channel panoramic radio receiving device ARC-D11 (see variants of utilization in Fig. 2.14) is purposed to solve wide variety of tasks concerning revelation of unauthorized RES at the monitored objects and in the territory.

The following functions can be concerned as main ones for ARC-D11:

- Double-channel synchronous radio monitoring in the real time;
- Revelation of unauthorized radio emissions and localization of its sources in objects under monitoring and in the district;

Table 2.8 Main performance characteristics of ARGAMAK-VM1 and ARGAMAK-VM0 digital receivers

Parameter	Value
Operating frequency range in the basing/maximal configuration	20–8000/0.009–18,000 MHz
Maximal input signal level	+23 dBm
Measurement limits of the input radio signal level	0–120 dB μV
Range of attenuator values	0 dB…40 dB by 1 dB step
Measurement error of the sine signal level	Not more than ±3 dB
Relative error of measurement of non-modulated signal frequency and AM signal, not more:	
At operation from embedded reference generator	5×10^{-8}
At synchronization from GNSS	1×10^{-10}
Single-signal dynamic range for 12.5 kHz spectrum resolution (without taking attenuator into account), not less:	
In the range 0.009–3000 MHz	85 dB
In the range 3000–8000 MHz	80 dB
Intercept point on second-order intermodulation IP2, not less:	
In the range 20–3000 MHz	40 dBm
In the range 3000–8000 MHz	30 dBm
Intercept point on third-order intermodulation IP3 in the range 20–8000 MHz, not less	0 dBm
Selectivity on side reception channels, not less:	
In the range 0.009–3000 MHz	80 dB
In the range 3000–8000 MHz	70 dB
Rate of panoramic spectral analysis for 6 kHz spectrum resolution	Up to 10 GHz/s
Maximal bandwidth of simultaneous analysis	22 MHz
Speculative recording mode (cycle buffer recording of signal until trigger fires)	Yes
Interface for control and data transmission	Ethernet 1000 Base-T
Range of operating temperatures (for external remote module ARGAMAK-VM0)	−40 …+55 °C
Degree of protection from external influences (for external remote module ARGAMAK-VM0)	IP67, IK10
Consuming power	Up to 35 W

- Automatic radio signal direction-finding in the frequency range from 1.5 MHz to 8 GHz, automated direction-finding in the range 8…40 GHz;
- Continuous recording of radio signals in the range up to 40 MHz;
- Multi-channel (up to 64 channels) radio signal recording with the spectrum bandwidth 12.5/25 kHz in each of two independent reception channels;
- Technical analysis, modulation type determination and estimation of radio signal parameters of the wireless communication and data transmission systems;

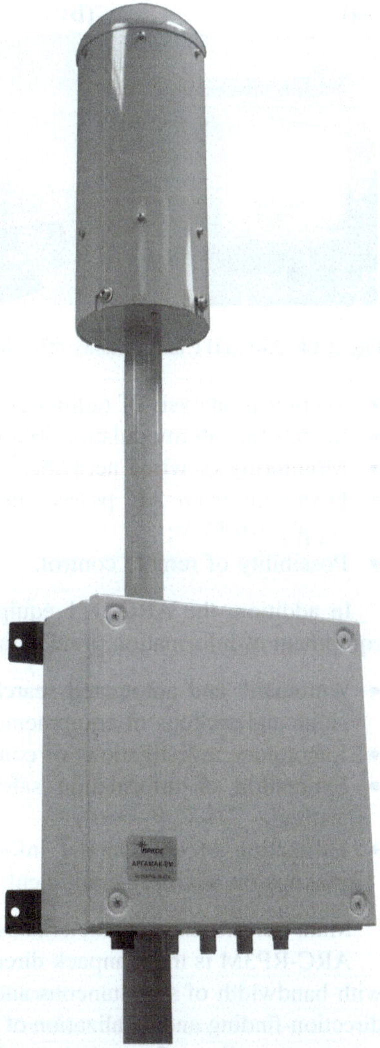

Fig. 2.12 ARGAMAK-VM0 remote module in the protected implementation together with the omni-directional antenna

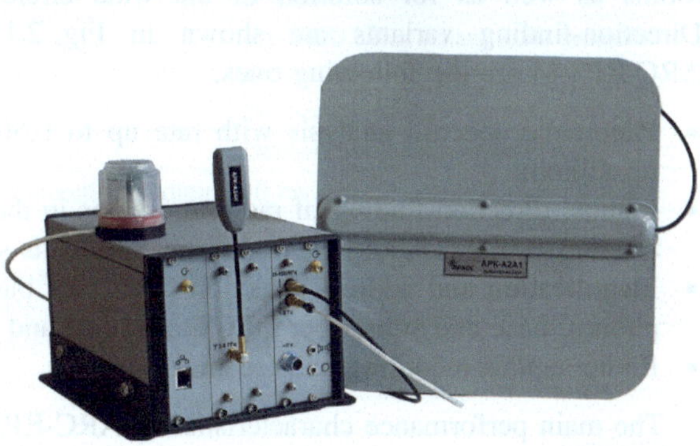

Fig. 2.13 ARGAMAK-VM1 remote module with the antenna set for operation in premises

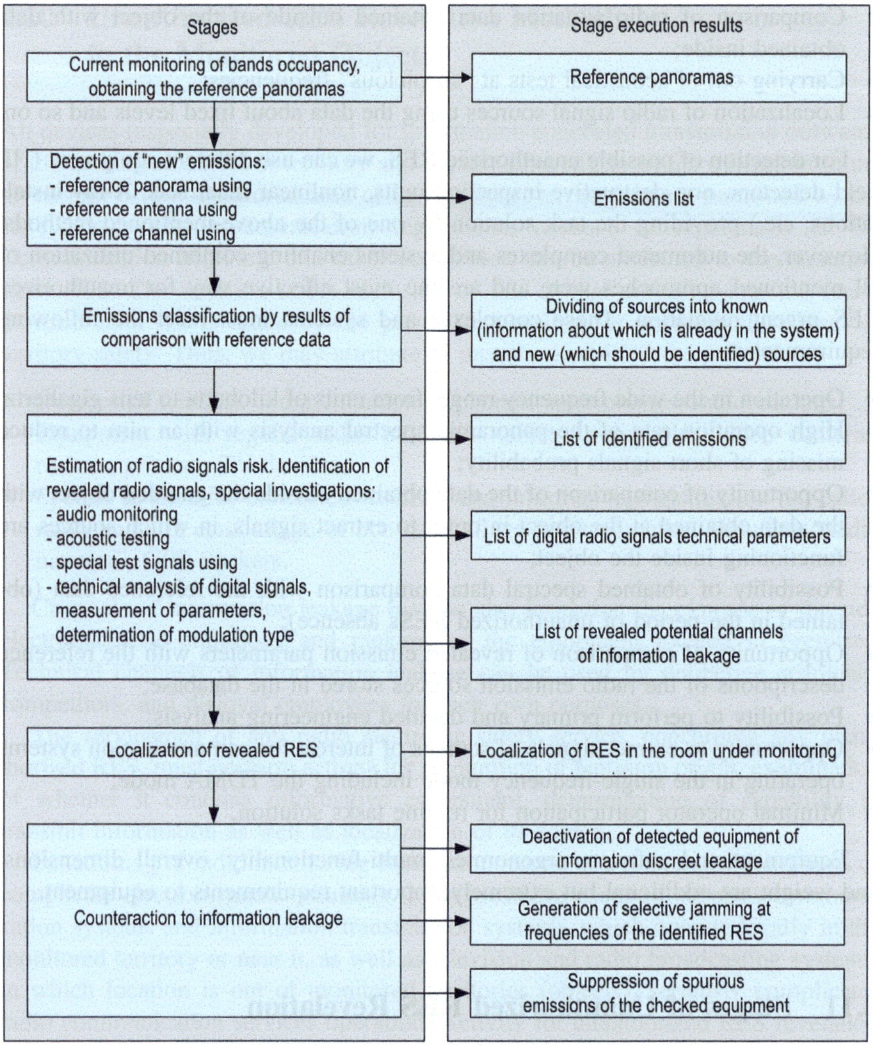

Fig. 2.16 Stages to reveal technical channels of information leakage

We understand the collection of spectral data accumulated during large time interval and obtained at monitoring realization as the "reference" panorama. At that, it is assumed that there were no suspicious signals during the period of "reference" panorama accumulation.

The aim of the second stage is the detection of "new" (not included into the "reference" panoramas) emissions, in which sources are situated inside the perimeter under monitoring and therefore potentially risky from the point of view of information leakage possibility. To solve this problem, we recommend using double-channel or multi-channel receivers. The "reference" antenna, which is situated outside of the signal reception zone under monitoring, but providing the reception of external signals, is connected to the one channel. Signals from reception antennas, which are situated in the zone under monitoring, are applied to the other inputs.

If the number of antennas exceeds the number of receiver inputs, then processing of these antennas signals can be arranged at the expense of the antenna commutator, which switches different antennas to the receiver inputs in turn. The practice shows that for premise monitoring with the area up to 100 m^2, it is usually enough to have from two to four antennas. We can recommend using antennas with ulterior polarization and quasi-isotropic directional pattern.

When using the single-channel receiver, the measurement of the signal power level from different antennas can be performed time-consequently only. Due to the modulation usage, and taking into account that the signal bandwidth is also unknown and during measurement, we take its estimation, an error arises in the power estimation, which leads to growth of probability of false detection or signal missing. Besides, a drawback of single-channel equipment application is the long duration of signal processing.

The double-channel method also ensures a higher probability of short and pulse signals detection under equal conditions. Application of the double-channel receiver, in which channels are synchronously tuned to the same analysis bandwidth, reduces the measurement duration and increases the precision of the signal power estimation. Accuracy growth is achieved at the expense of the fact that taking the time sample is made synchronously in two channels, and estimation of the power level is provided under the same conditions. Since at decision making, the comparison of relative values occurs, the errors related to the inaccuracy of power estimation algorithm do not practically influence on the result.

In the near-field zone, at distance from 1 to 10 m, the emission level from low-power unauthorized RES significantly exceeds the level of powerful (but remote) sources. Comparison of radio signals obtained from the "reference" and "signal" channels allows the revelation of radio emissions generated inside the guarded zone.

The third stage assumes a separation of detected (at the second stage) signals into known ones (information about which was obtained earlier and included into the database) and new ones, which represent the objects for investigation in detail and that is fulfilled at the fourth stage. This stage can be performed in the automatic mode.

The aim of the fourth stage is checking of the revealed "new" radio emissions risk from the position of the possible information leakage. The automated testing methods can be used for monitoring; however, they do not always guarantee the unambiguous identification of all risky emissions. In this connection, the additional "expert" checking methods can be required, which are oriented on operator participation, whose experience (with account of in-depth technical analysis possibility) can help to make a correct decision in complicated cases.

Acoustic identification is one of the methods for radio emission identification for the analogous types of frequency modulation (FM). For this method, the acoustic testing signals of different types formed by testing loudspeakers are used for identification of revealed radio emissions after receiver tuning to the frequency of one of the "new" unauthorized RES. The received signal is demodulated, after which its correlation processing with the modulating signal is performed in temporal and spectral domains. According to the correlation factor value, we can make a decision about belonging of the received signal source to the radio microphone class located in the premise.

To keep efficiency of the suggested approach in the case of such hiding methods as the spectrum inversion and the frequency mosaic, we can use the two-frequency signal or the chirp-signal as the testing signal, and signal processing is performed in the spectral domain. A decision about radio microphone presence can be made at the presence of spectral components of the testing signal over the time interval, which equals to testing signal sounding duration, and its absence outside this interval.

The uniqueness of the acoustic signal transformation lies in the base of the radio microphone identification algorithm with static scrambling in the frequency domain, i.e. a possibility to unambiguous recover of the initial signal according to the transformed one. It means that at multiple repetitions of any acoustic signal, the emitted radio signal will also be repeated. The similar recurrence allows determination of radio microphone presence.

Application of radio microphone identification methods on the base of utilization of the specific acoustic tests can be mated with a possible violation of the search security, because testing signals have sufficiently specific character and, therefore, it is more preferable to use typical sounds (fast rhythmic music, continuous speech). In addition, the acoustic identification method is inapplicable for complex modulation types and when information transmission is implemented in the digital form. In this case, the radio signal fragments recorded in the pass bandwidth of radio monitoring equipment can be used and signals can be processed within the limits of the signal technical analysis (STA) task. Traditionally, STA can be divided into the primary and detailed (in-depth) analysis.

The primary TA is executed at posts in the volume, which is sufficient for fine tuning of the equipment, determination of emission presence on the interference background, and identification of the emission source. The aim of the primary TA is the identification of unauthorized RES on the base of its radio emission parameters estimation. At the primary STA, the uncomplicated operations with radio signals are fulfilled with the aim to measure radio emission characteristics, to determine modulation type and some radio emission parameters after demodulation by the universal analogous demodulators at the carrier or sub-carrier frequencies.

The aim of the detailed (in-depth) technical analysis is not only identification of unauthorized RES on the base of its parameters estimation, but estimation of the informational component. At the in-depth technical analysis, the operator has a right to perform much more mathematical transformations of radio signals in the various forms of its representations and to get practically complete information on the detected signal including the revelation of conditions and reasons of its arising and the origin of the unauthorized radio signals. For instance, it is possible to determine the principles of radio signal generation fulfilling the reverse transforms, to obtain a short fragment of initial message and then to analyze it on correspondence to acoustical testing signals (speech or music), which were transmitted in the room under monitoring, i.e., to perform the emission identification.

The technical analysis of radio signals is performed in the deferred processing mode according to earlier recorded fragments of radio signals. Recording of radio signal fragments is executed in the automatic mode during unauthorized RES detection or in the manual mode by the operator command.

Decision about unauthorized RES risk can be made both at the investigation of the radio signal itself and after its demodulation at the stage of the bit-flow investigation from the demodulator output. As a result of the bit-flow analysis of demodulated unauthorized RES, we can obtain information about the transmitted information character (for instance, the acoustic background in the monitored room, the spurious emissions of the automatic phone station in the office, monitors, or power supply units) and the type of transmitted data (e.g., vocal communication, transmission of the facsimile messages, inter-computer exchange including transmission of pictures, documents, archives).

The collection of features and technical parameters of the detected unauthorized RES, which enables to make a conclusion about radio signal origin and presence of informational component in it, to reveal unauthorized RES and to promote its localization in the room under monitoring or in the guarded territory, is the result of the technical analysis.

The fifth stage is the localization of unauthorized RES can be performed by different approaches: by means of acoustic test implementation, by the amplitude method, or with the help of radio direction-finders.

Localization of radio microphones with simple modulation types is performed on the base of delay measurement for signal, which is emitted by the radio microphone with regard to the testing acoustic signal (so-called acoustical location). In the simplest form, algorithm can be realized with the usage of two acoustic loudspeakers.

In the cases, when unauthorized RES functions continuously or periodically, one can use the equipment, which is simple and easy in exploitation, for localization and for manual direction-finding (indicators or field detectors) together with visual inspection of the room.

At the presence of radio monitoring equipment, which is distributed over the guarded object area, we can narrow the search region at the expense of preliminary source localization by the amplitude method on the base of the data about registered signal levels.

The sixth stage related to counteraction to information leakage is performed depending on the results of five previous stages.

2.12 Unauthorized RES Revelation in the Adjacent Territory

Search for unauthorized RES inside the guarded object is not always possible with the application of equipment located in the object itself. The operation restrictions, the absence of access to the building to be monitored, requirements to the special secrecy of monitoring execution of unauthorized RES search, and other reasons can be an obstacle for this search. At such circumstances, the problem of unauthorized RES revelation in the object under monitoring with the help of radio monitoring equipment, which is located outside of the object (on boundaries of the monitored territory), is really relevant.

Such a method gives a possibility to reveal most of radio emissions, which sources are located inside the object under monitoring, when operating outside. This method precision is from several tens to units of meters, which enables unauthorized RES localization within the limits of a floor, a part of the building, and a group of rooms. The further search for the unauthorized RES source can be performed with the help of the manpack direction-finder ARC-RP3 M.

Technical and operative implementation of this method is possible with the help of the ARC-D11 equipment as a unit with radio directing-finding antennas AS-HP5 (20–3000 MHz), AS-HP2 (3–8 GHs), and the ARC-IG1 equipment [5], installed on the transport vehicle in combination with the software packages SMO-PPK and SMO-SECTOR [8, 11].

Special software enables to obtain bearing panoramas by azimuth and elevation from radio emission sources with narrowband and wideband modulation types, high-frequency house-hold and medical installations, and other sources of the electromagnetic field, which are situated inside the object under monitoring. If the car, on which the radio direction-finder is located, can be equipped with the ARC-IG1 device consisting of the one or several video cameras, then the timing of the radio emission source position to the specific area of the object under monitoring becomes possible. For purposes of outside observations of radio emission sources in the lengthy objects, we can apply software packages SMO-PA and SMO-SECTOR [8, 11, 12].

To increase the authenticity of source revelation besides direction-finding, we may perform an estimation of signal levels, which consists in the comparison of signal spectrum amplitudes obtained in the near-field area of the object under monitoring and at sufficient distance from it. If the source is located inside the object under monitoring, the signal amplitude, as a rule, at distance of several tens meters, essentially exceeds the signal amplitude at removal for more than several hundreds of meters from the object. At that, signal levels from the spurious sources do not practically change.

The direction-finding of the object is performed from several positions located at the different sides of the object. For each position, with the help of a video system, the object angular position, with regard to a mobile station, is saved. Then, we calculate bearings for all sources, which are also saved. On the base of bearing values obtained from different positions, the sources frequencies, which signal angle of arrival coincides with the angular position of the object for the given position, are determined. As a result, the frequency list is formed, which may belong to sources located in the object.

Specification of search results is provided in the single-channel direction-finding mode. Obtained bearing values are displayed over the object image, which is obtained from a video camera. At that, the direction-finder determines not only the azimuth but the elevation as well. This enables the visual estimation of source location in the object under monitoring with precision within the floor or the part of the building. The single-channel direction-finding is conducted from several positions of the mobile system in order to reduce the possible interference influence.

As an example, Fig. 2.17 shows the main window of the SMO-SECTOR software. The possible location of the radio emission source is indicated on photograph of the object under monitoring.

The ARC-D11 in combination with the additional direction-finding antenna system provides an implementation of the correlation-interferometric method for direction-finding. For instance, with a set of direction-finding antenna systems AS-HP-KV, AS-HP5, and AS-HP2, we may perform detection and automatic direction-finding of signal sources in the frequency range from 1.5 to 8000 MHz. The equipment can be used in the mobile version or in the temporary posts. In the last case, the automatic direction-finding from 8 to 40 GHz can be realized with the help of the converter with directional antenna located on the support-rotating device controlled remotely in azimuth and elevation. Rotating device mounting on the portable mast allows increasing of the direction-finding precision and widening of the activity zone of the radio direction-finder.

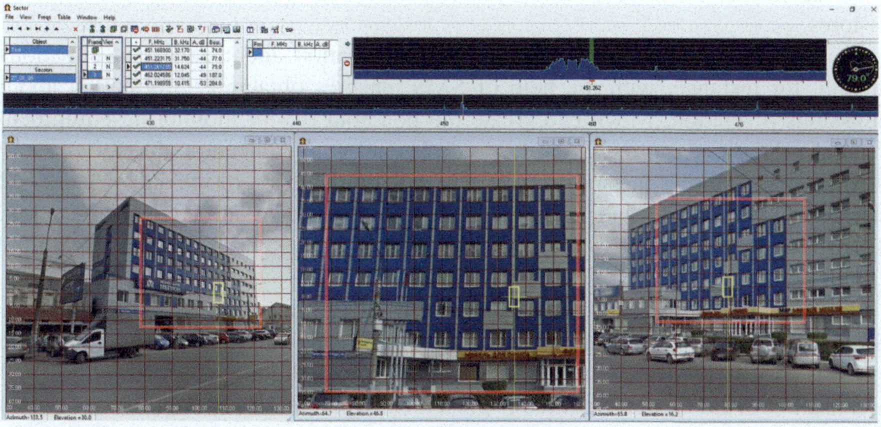

Fig. 2.17 An interface of the SMO-SECTOR software

2.13 Software Support

The software packages SMO-ARMADA, SMO-DX, SMO-PA, SMO-STA, and other special software packages are used as the basing software support for radio monitoring equipment [7, 8, 11, 13, 14].

The SMO-ARMADA software package provides the automated radio monitoring system operation. It has the client/server architecture, which is capable to adjust to the scale, arrangement peculiarities and functions, which must be ensured by the system of unauthorized RES revelation. The SMO-ARMADA system software enables round-the-clock service mode (including without operator participation) of solution of various radio monitoring tasks and search for sources of concealed obtaining information, with employee notification about the revelation of potentially risky sources or equipment failures via notifications by e-mail or SMS.

During radio monitoring, the revealed radio emissions can be compared to the reference descriptions stored in the database. According to the results of such a comparison, the signal division is executed into the new ones, which information is going to be manually processed by the operator, and known ones, which need not be processed.

In this system, the solutions on reservation of data transmission channels are applied, while at disappearing of the main channel, the ARMS automatically switches operation to the reserve channel.

Peculiarities of distributed systems architecture, which assumes the installation of SMO-ARMADA local applications on the equipment controllers, ensure a possibility of long-term autonomous equipment functioning in the automatic mode in accordance with the given schedule even at communication channel disappearance. In this case, all results obtained are stored in the local database and will be automatically transmitted to the system central database when communication restores.

A presence of alternative channels for notification about abnormal situations arising, for instance, with the help of SMS, ensures obtaining information about emissions revelation, which can create the potential risk or monitoring equipment state changing even in the case of wired channel communication absence.

Figure 2.18 shows the spectrogram example of detection of the narrowband FM signal, which coincides in spectrum with a signal of the broadcasting radio station. This source was revealed as a result of comparison of the current spectrum with the "reference" one.

Figure 2.19 shows an example of detection results displaying in the SMO-ARMADA interface. In the lower part of the screen, we see the signal, which exceeds the energy detection threshold, while in the additional window the result of localization of the detected source is presented. Quadratic icons indicate the places for receiving antennas localization, and the icon color shows the registered signal level. The most probable unauthorized RES location is marked by cross-point.

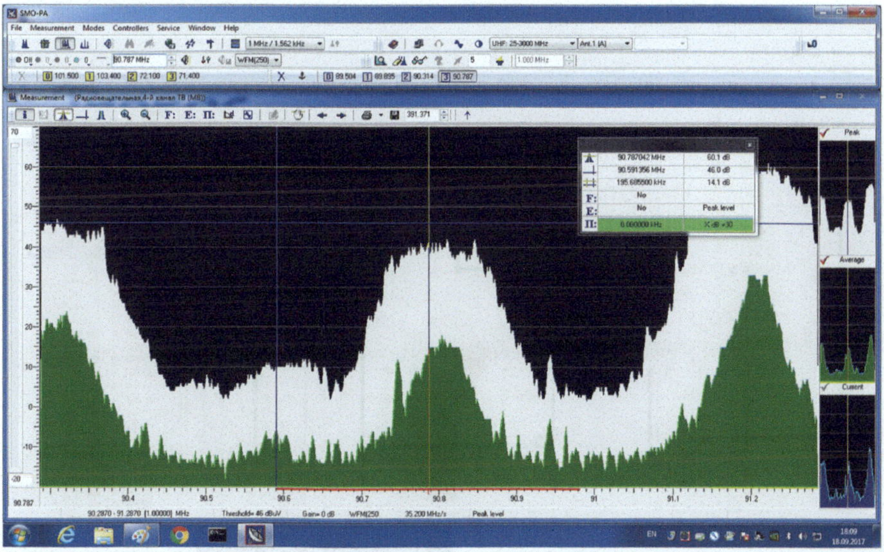

Fig. 2.18 Unauthorized RES with narrowband FM, whose emission is masked by the powerful radio signal of the broadcasting station

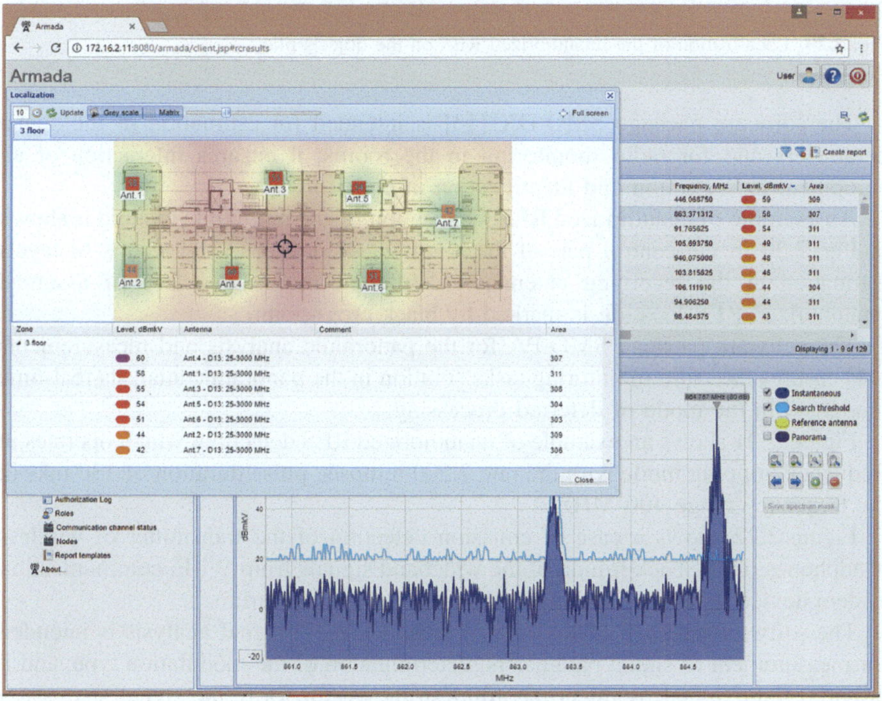

Fig. 2.19 Unauthorized RES localization in the room under monitoring

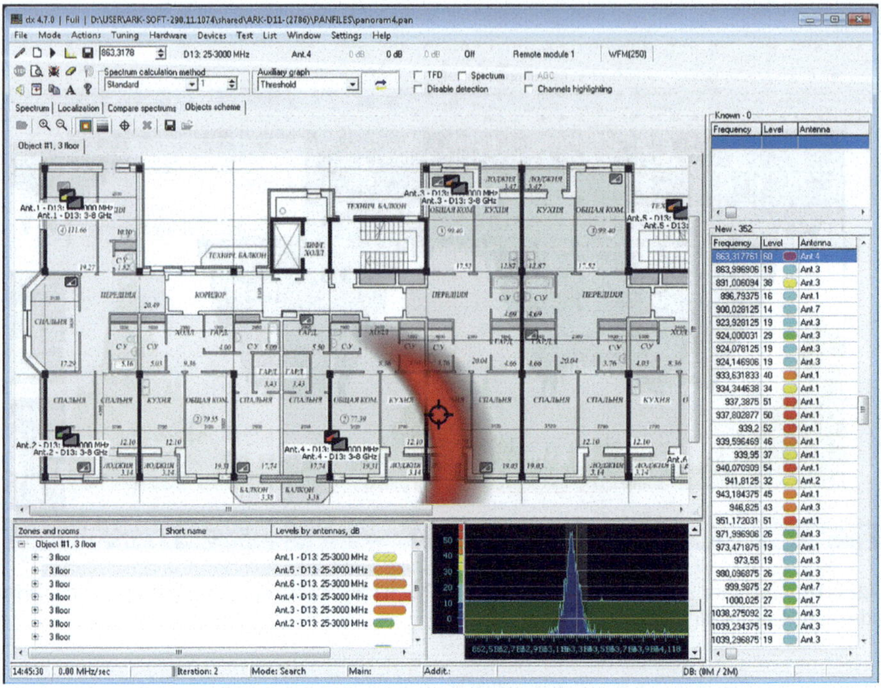

Fig. 2.20 Localization of the unauthorized RES on the objects plan

The software package SMO-DX [13] is intended for estimation of the signal parameters and for radio monitoring in the rooms; it ensures interaction of all equipment for detection and identification of emissions.

An example of unauthorized RES localization by the amplitude method is shown in Fig. 2.20 in the central part of the window, according to the results of levels estimation at the receiving antennas in different rooms. A place of assumed unauthorized RES location is marked by black cross-point.

The software package SMO-PA for the panoramic analysis and measurements [11] ensures the radio monitoring tasks solution in the panoramic analysis real-time mode and in the mode of deferred processing.

Figure 2.21 shows an example of unauthorized RES detection, which operates in frequency hopping mode with the rate 7.500 jumps/s, pulse duration of 130 mks in the frequency range 400 MHz.

Figure 2.22 shows a case of emission detection of the transmitter of wireless headphones on the background of the wideband signals from Wi-Fi communication system device. For detection, we use the probability spectrum.

The software package SMO-STA of technical radio signal analysis is intended for measurement of signal parameters, determination of the modulation type, and it provides radio signal fragment recording in the vector form, the signal analysis at the radio frequency and detected signals and signals at the sub-carriers, the analysis

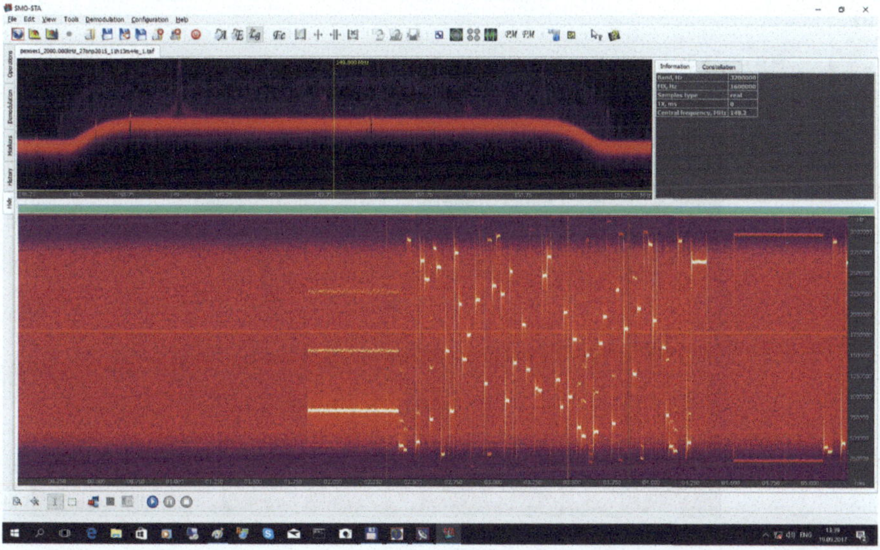

Fig. 2.21 Emission of the data transmission system with frequency hopping

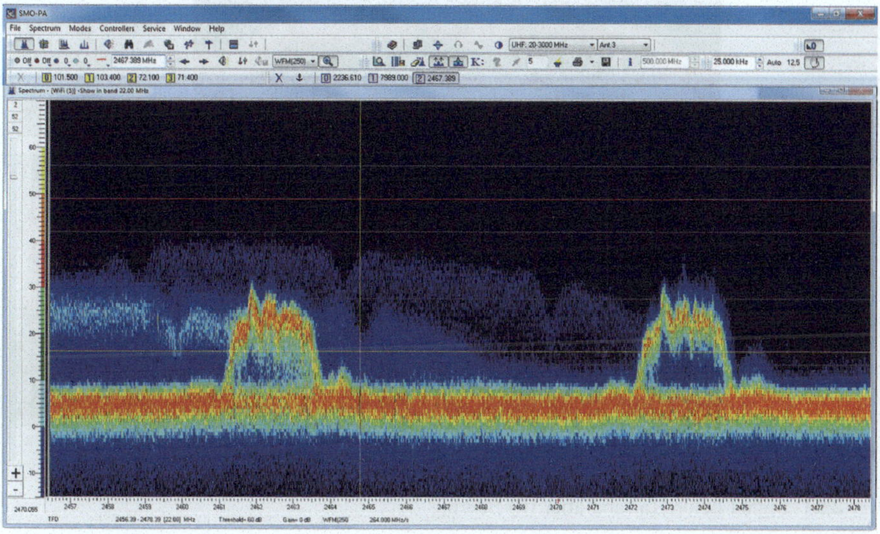

Fig. 2.22 Emission of the wireless headphones transmitter on the background of Wi-Fi communication system device operation

of bit flows with the aim of determination of signal generation principles and decoding in the wide list of protocols and standards [14]. The signal analysis can be performed both in the real time and using deferred processing. Application includes

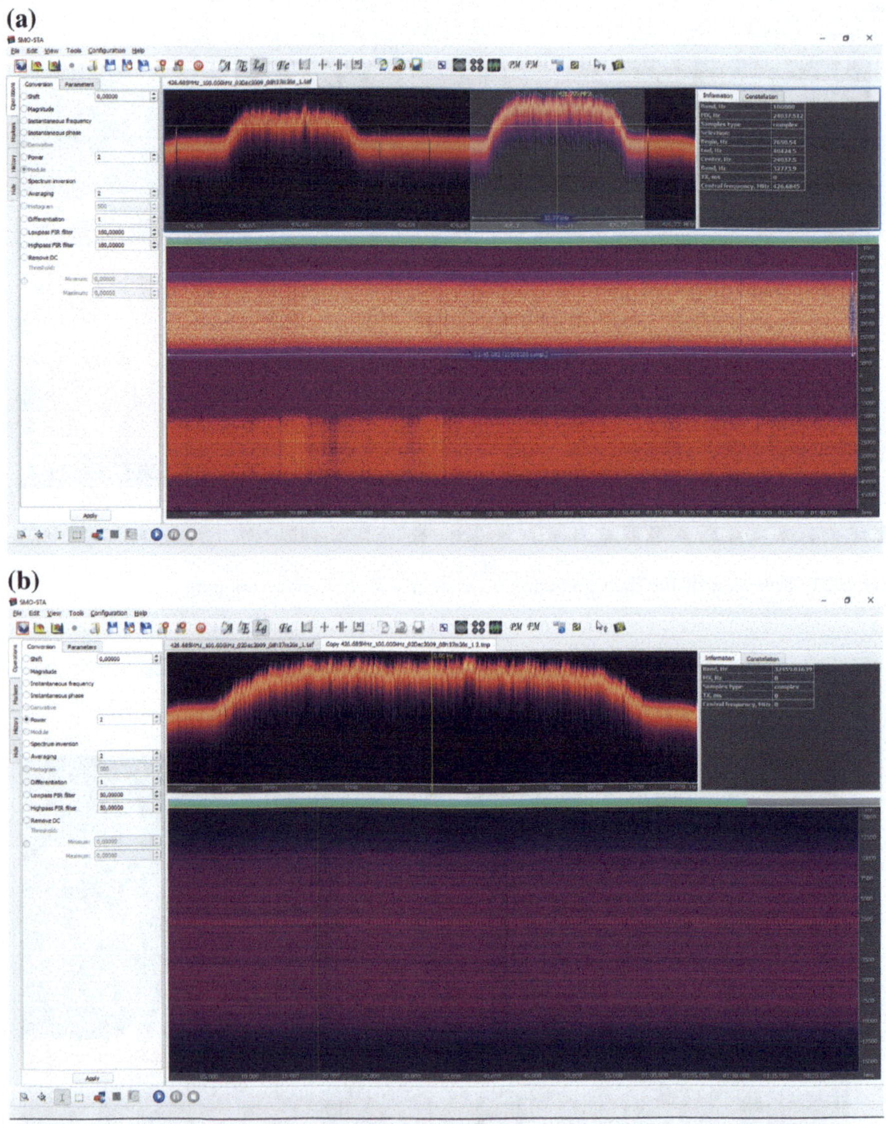

Fig. 2.23 Signal before (**a**) and after (**b**) filtering

functions of scaling, filtering (Fig. 2.23), fulfillment of various necessary signal transformations, for instance the rejection of the frequency and time signal fragments and involution (Fig. 2.24), derivation, the analysis at sub-carrier, the analysis of bit flows (Fig. 2.25), as well as solution of series of other tasks necessary for decision making about the emission character.

(a)

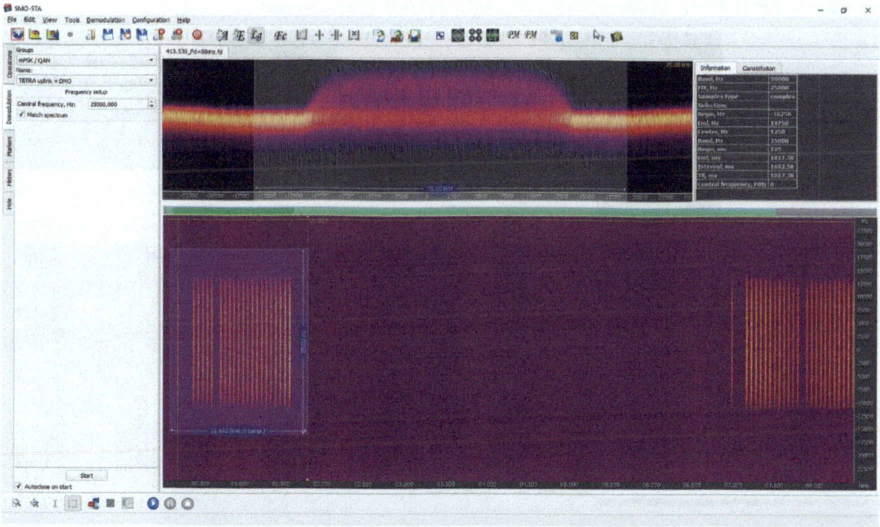

(b)

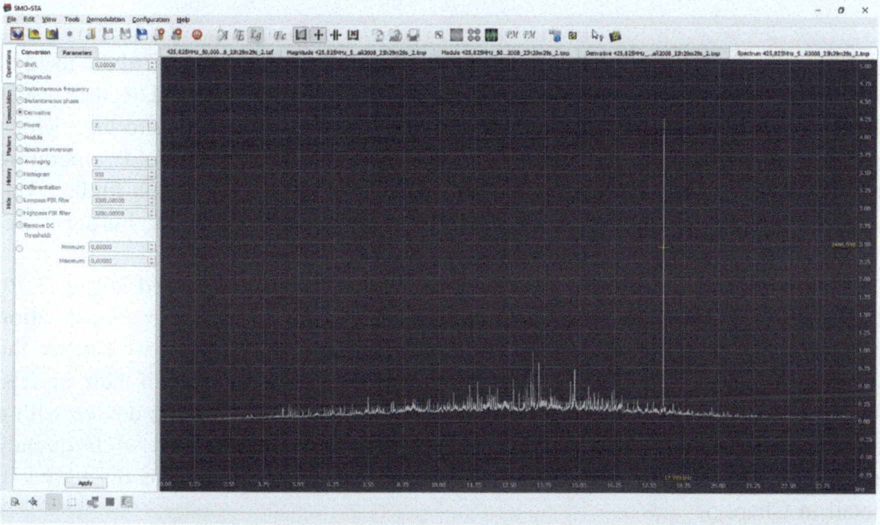

Fig. 2.24 Rejection of frequency and time signal fragments (**a**) and its involution (**b**) to determine modulation type and parameters

The SMO-STA package provides signal demodulation with the help of amplitude, phase, frequency (including multi-position) demodulators as well as series of specific.

All applications provide tasks solution both at local and network control of equipment.

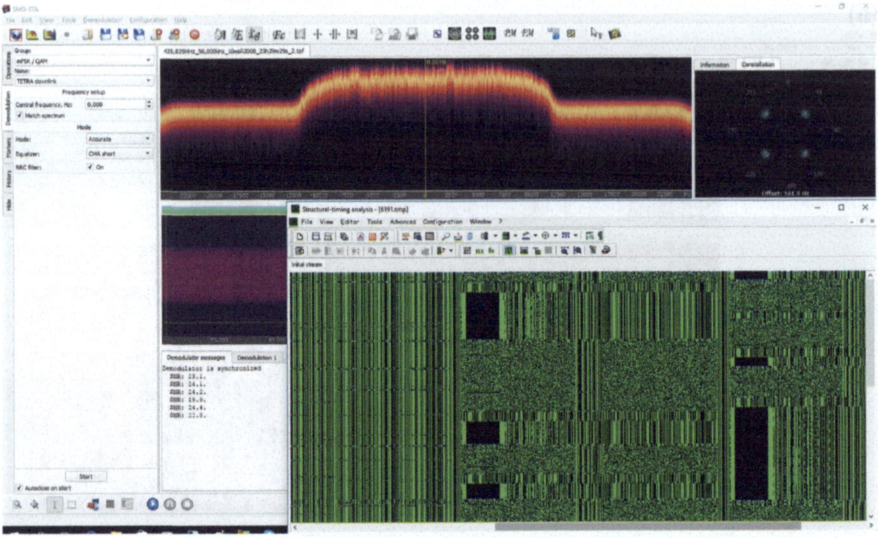

Fig. 2.25 Demodulation of the $\pi/4$-DQPSK signal and window of the bit-flow analysis

The software packages SMO-ANDROMEDA and SMO-RP [15, 16] are used to control the manpack radio direction-finder ARC-RP3 M. SMO-ANDROMEDA is the software package of manual direction-finding and radio monitoring, which operates on the tablet computers and smartphones under “Android” operating system. It provides panoramic spectral analysis, detection, direction-finding (including “address” one), and measurement of radio signal parameters. Furthermore, it allows equipment control via wireless Bluetooth interface.

The probabilistic spectrum of Wi-Fi and GSM signals is presented in Fig. 2.26.

Figure 2.27 shows an example of the software interface for address localization of devices of the DECT communication system. On the left picture, we can see the list of detected identifiers of the DECT standard devices and a level of their signals. In the center, the graph of signal level variation is presented for the device with a chosen identifier. On the right, we can see a displaying variant of frequency channels occupancy degree. Issues of address direction-finding are discussed in detail in Chap. 5.

The software package SMO-RP for manual direction-finding provides fulfillment of the search, the signal spectral analysis and the direction-finding, the revelation of short, single, and periodic signals, the signals of DECT standard at non-standard frequencies.

If necessary, equipment for unauthorized RES detection in the room can be accompanied by specialized software packages, for instance, by the SMO-BS package for identifiers analysis and determination of service parameters of the base stations [17].

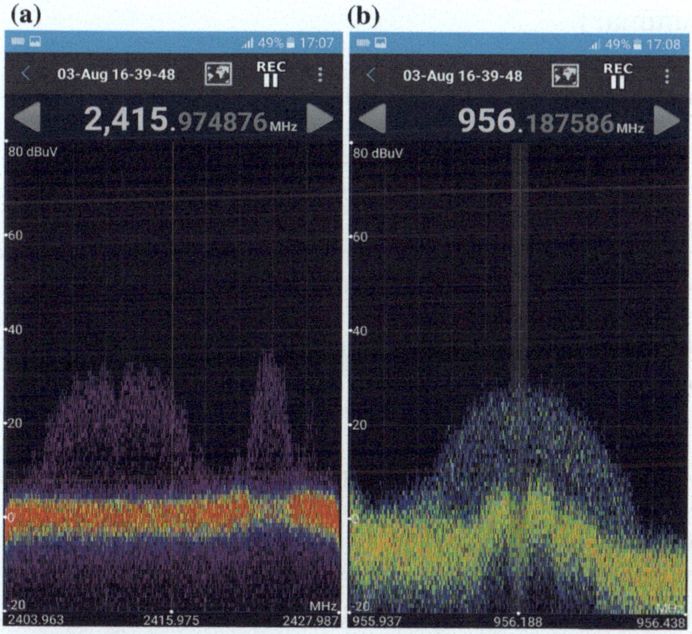

Fig. 2.26 Probabilistic spectrums for signals of **a** Wi-Fi and **b** GSM communication systems

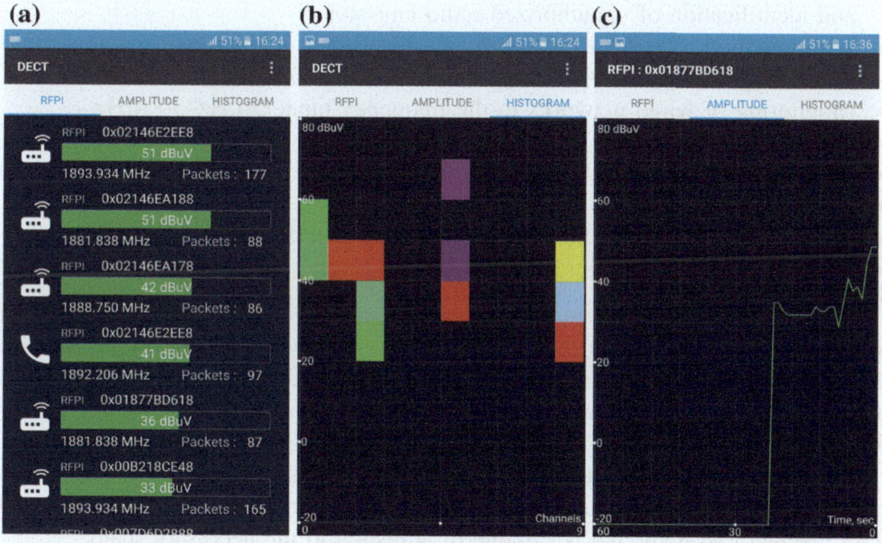

Fig. 2.27 Address direction-finding of the unauthorized RES of the DECT communication system according to the RFPI number

2.14 Summary

Integration of equipment of radio monitoring into unified automated system enables to solve different radio monitoring tasks of unauthorized RES revelation in the automatic mode. Centralized control of radio monitoring equipment allows minimization of a number of workstations, unification of account database by the radio emission sources.

This chapter describes application of typical variants of automated radio monitoring system AREAL used to reveal unauthorized radio emissions and to localize their sources in the restricted territory, in the buildings and structures, in the transport vehicles, and other objects; it also testifies that structure, functions, and system contents depending on the task.

The digital receivers of the ARGAMAK-family are the basis of radio monitoring equipment used in the AREAL ARMS. They provide high merits of analogous reception path and dynamic range, high performance of digital processing, possibility of recording, and technical analysis of radio signals in the frequency range up to 40 MHz. All equipment is multi-functioning and provides solution for the following tasks:

- Fast detection of "new" radio emissions with arbitrary modulation types including ultra-wideband or similar signals and frequency hopping signals in the range 9 kHz–40 GHz;
- Continuous radio signal recording in the 40 MHz bandwidth, technical analysis, and identification of unauthorized radio emissions;
- Determination of radio emission source location at internal and external radio monitoring of objects in the range 0.3 MHz–18,000 MHz;
- Monitoring of wired networks in the frequency range 50 Hz–30 MHz;
- Automatic direction-finding in the frequency range 1.5 MHz–8000 MHz;
- Automated direction-finding in the frequency range 8 GHz–40 GHz;
- Multi-channel recording of radio signals;
- Technical analysis of radio signals in the deferred processing mode;
- Possibility of remote control.

Utilization of described equipment ensures effective solution of unauthorized RES revelation problems in the territory and objects under monitoring as well as their sources localization in premises, in the surrounding district and at outside object inspection under complicate electromagnetic conditions, which is typical for the big cities and industrial centers.

To control ARMS operation, the SMO-ARMADA software is used, which has client/server structure, organizational peculiarities, and function of unauthorized RES revelation system, and is capable of adjusting to the necessary scale.

The software enables to control radio monitoring equipment, which are situated in the controlled objects or outside of them, to solve radio monitoring tasks and tasks of search for discrete units of information leakage together with immediate notification of employees about revelation of sources of potential risk or equipment fault using e-mail or SMS, in the round-the-clock mode including without operator.

Dividing of all possible situations of the AREAL ARMS usage to the described AREAL-1…AREAL-5 variants are rather synthetical, but enables to select the structure of the system and included equipment correctly.

The AREAL-1 automated system variant is used when operating along lengthy boundary of the monitored territory and in the adjacent territory. Stationary and mobile radio direction-finders in maintainable and unmanned implementations, and control centers are included in the system.

The AREAL-2 systems are used at single or periodical radio monitoring in the restricted area. The system includes mobile, transportable, and portable radio monitoring equipment. To fulfill enhanced study of radio electronic environment and detailed technical analysis of radio signals, the AREAL-3 system is used, into which the complexes of recording and radio signals technical analysis are included.

The AREAL-4 system supports permanent radio monitoring in the closed territory with fixed boundaries. Equipment of this system consists of fixed, mobile, transportable, and manpack units. In this case, the control can be executed by mobile equipment of radio direction-finding, video monitoring, and SMO-SECTOR software package.

To reveal unauthorized radio emission sources inside the premises and transport vehicles, the AREAL-5 automated system can be used. Standard distributed systems of remote control ARC-D13R, ARC-D15R, portable complexes ARC-D11, and manpack direction-finders ARC-RP3 M, which enable detection and localization of unauthorized RES operating in one radio frequency band and using GSM, Wi-Fi, and DECT technologies, have application in the AREAL-5 system.

References

1. Rembovsky A, Ashikhmin A, Kozmin V, Smolskiy S (2009) Radio monitoring. In: Problems, methods and equipment. Lecture notes in electrical engineering. Springer, p 507
2. Rembovsky AM (2016) Automated system for unauthorized radio emission revelation AREAL—tasks, a structure, components (in Russian). Spetstehnika i svyaz (1):26–37
3. Alekseev DA, Bogdanov AY, Rembovsky AM (2016) Detection of unauthorized radio emission in the controlled objects (in Russian). Spetstehnika i svyaz (4):2–13
4. Handbook Spectrum monitoring (2011). ITU-R
5. The Catalogue 2017 of IRCOS JSC. http://www.ircos.ru/zip/cat2017en.pdf. Accessed 28 Nov 2017
6. Digital radio receivers. http://www.ircos.ru/en/rsv_main.html. Accessed 28 Nov 2017
7. SMO-ARMADA software package of automated spectrum monitoring systems. http://www.ircos.ru/en/sw_armada.html. Accessed 28 Nov 2017
8. SMO-SEKTOR software package for detection of radio emission sources indoors. http://www.ircos.ru/en/sw_sector.html. Accessed 28 Nov 2017
9. Automated localisation of electromagnetic emission sources at sites. http://www.ircos.ru/zip/art_vkl1_eng.pdf. Accessed 28 Nov 2017
10. Clark L, Algaier WE (2007) Surveillance detection, the art of prevention. Cradle Press LLC, 197 p

11. SMO-PA/PAI/PPK panoramic analysis, measuring and direction finding software package. http://www.ircos.ru/en/sw_pa.html. Accessed 28 Nov 2017
12. Localization by a mobile monitoring station of radio signal transmitters in city. http://www.ircos.ru/zip/art_vkl2_eng.pdf. Accessed 28 Nov 2017
13. SMO-DX software package for indoor radio monitoring and radio signal parameter evaluation. http://www.ircos.ru/en/sw_dx.html. Accessed 28 Nov 2017
14. SMO-STA software package radio signal technical analysis. http://www.ircos.ru/en/sw_sta.html. Accessed 30 Nov 2017
15. SMO-ANDROMEDA software package for manual direction finding and radio monitoring for tablets and smartphones. http://www.ircos.ru/en/sw_andromeda.html. Accessed 28 Nov 2017
16. SMO-RP software package for manual direction finding. http://www.ircos.ru/en/sw_rp.html. Accessed 28 Nov 2017
17. SMO-BS software packages for wireless data communication system analysis. http://www.ircos.ru/en/sw_bs.html. Accessed 28 Nov 2017

Chapter 3
SMO-ARMADA Software System

3.1 Introduction

Software is one of the three basic components of the automated radio monitoring systems. Integration of radio monitoring manifold units used for the combined solution of the assigned tasks, support of the users' interaction with the system hardware, control of the system conditions, generation of accounting and reporting information, interaction with external information systems and other functionality, which is necessary to improve efficiency of radio monitoring equipment operation, are the software main purposes [1, 2].

As the radio monitoring equipment, the software developed by IRCOS Company has several generations. The first generation was used to control several equipment and operated under MS-DOS operating system. At present, the first-generation software is not practically used.

The second-generation software provided operation under MS Windows operation system and allowed controlling of a few networks of several radio monitoring equipment simultaneously. There are examples of the second-generation software: SMO-PA I, SMO-PPK, SMO-KN STALKER, SMO-DX, etc. [3–5]. The software of the second generation is maintained and modified up to the present day and is successfully used for radio monitoring task solution.

The third-generation software provides operation with equipment of the "+" family with the bandwidth of simultaneous analysis up to 24 MHz, and it supports remote control of the equipment according to the unified open protocol. Implementation of labor-intensive computing algorithms was transferred to the lower software layers (equipment drivers) and to the local control servers.

The SMO-ARMADA software system can be concerned as the fourth-generation software [6]. It is the cross-platform software that is developed using Web technologies and is the software kernel of the produced automated systems. At present, the software of the second, third, and fourth generations is used jointly complementing each other.

A. M. Rembovsky et al., *Radio Monitoring*, Signals and Communication Technology, https://doi.org/10.1007/978-3-319-74277-9_3

The main requirements to the software support of the automated radio monitoring system (ARMS) are defined by its purpose. Software support provides radio monitoring task solution at the local, regional, or national levels assuming the control of the set of manifold equipment, which are located at large distance.

To reduce expenses on radio monitoring equipment operation, the ARMS should support automation of typical task solution except cases, when operator has to deal with supernumerary situations or tasks, which are difficult to be formalized. In addition, the software support of the ARMS should provide the equipment functioning both in the automatic mode in accordance with the assigned schedule and under operator direct control.

Software implementation of the risk-oriented approach, which supposes notification of employees, that operates with the ARMS, about situations that happened during task execution in the automatic mode and required operator's interference, enables to enlarge the equipment amount, that is maintained by the only one operator, and to reduce automated workstations amount.

At typical use, the ARMS represents the distributed systems, in which elements can be located at large distance, and the ARMS structure mainly repeats the structure of organization management, which performs the radio monitoring; i.e., the ARMS has the hierarchical structure. Amount of hierarchy levels depends on the ARMS usage scale.

Different tasks can be assigned for system elements at different hierarchy levels. Elements of the lower level (radio monitoring equipment), as a rule, perform the tasks of measurement, search for interferences, spectrum occupancy data acquisition, and report generation. Control points should provide radio transmitters database support, assignment of tasks for subordinate radio monitoring (RM) stations, generalization of the data acquired, and, in case of presence of control points of the higher level, generation of report documents. Since sizes and requirements on data safety essentially differ at different hierarchy levels, there are differences in requirements to the hardware. Computer servers should be used at the control center and control points, but usage of computer servers may be inadmissible at the nodes with radio monitoring equipment due to restrictions on the location conditions, power consumption, and cost. Significant difference in tasks to be solved and in available computing resources leads to modularity of the software for ARMS, allowing extraction of functional components and flexible configuring of the module structure used at various system hierarchy levels.

The availability of application programming interface (API) description allows developing and integrating of the user's own modules, which widen the general functionality and solve the narrow-specified tasks, to the system structure independently.

Integration of radio monitoring equipment, which is spaced on large distance, into the unified system raises the question of data exchange organization between them. Used communication channels may have significantly differed capacities: from gigabit connections in the Ethernet local network up to data transmission with the help of HF radio communication with baud rate of several kilo bauds. Nodes may have several active network connections for reservation or make connections to the system only episodically. All this leads to requirements not only on the traffic minimization, but also on the automatic adaptation of the transmitted data volume depending on the communication channel capacity.

A possibility to use nodes, with which there are episodic connections only, in the system structure leads to requirement of the independent operation, which assumes a possibility of automatic task solution in accordance with the assigned schedule. It means that the software should provide storage and timely execution of the assigned tasks, saving and preliminary processing of the acquired results, and the results transfer to the higher node of the system at appearance of communication.

ARMSs, as a rule, are multi-user ones. To avoid conflicts of the data access for users, which act at various hierarchy levels, the software for ARMS should anticipate priority system and access rights, which are adjustable for different users or user's groups.

For effective ARMS usage, the interaction possibility with other specialized information systems is required, for instance with control system for spectrum usage, with systems of electromagnetic compatibility calculation.

As a rule, ARMSs are developed by manufacturers of the radio monitoring equipment. However, users may be provided by the equipment of different manufacturers, and lack of opportunity of integration into the structure of the technical system of other manufacturers may essentially restrict the ARMS implementation area and reduce effectiveness of its usage. Thus, software support of ARMS should have universal interface in its structure for various equipment to be connected to, with purpose to provide share usage.

Variety of hardware and software units used in the system structure makes the task of automatic control of the system efficiency as a whole as well as its separate elements an essential one. Therefore, the software support of ARMS should provide monitoring of system element conditions and of communication channels between them and urgently inform the system users about problems that happen.

At last, software support of ARMS should provide possibility of automatic update, because manual software support update in large system with great number of nodes and potential opportunity of the permanent absence of the communication channel may require significant time, during which, as minimum, the system part will be disabled.

To sum up the above mentioned, the development of software support for ARMS is considered not only with the task solutions, which directly relate to functions of radio monitoring, but also with tasks, which relate to system functioning as a single whole. In this chapter, we shall consider ways of fulfillment of mentioned requirements for the SMO-ARMADA software system and describe the architecture and main engineering solutions [1].

3.2 Architecture of the SMO-ARMADA

The SMO-ARMADA represents the cross-platform scaling fault-tolerant client/server system with "thin" client, which is developed in accordance with International Telecommunication Union (ITU) recommendations [7–9].

The system software is divided into the server and the client parts. The server software support, main parts of which are the application server and the data server, is installed on the nodes, to which the radio monitoring equipment is connected to, as well as on the servers at the control center and the control points. It fulfills the following functions:

- Assignment of the tasks to the radio monitoring equipment and results handling;
- Execution of the tasks according to the schedule;
- Execution of the tasks with respect to their priority;
- Authorization of users and restriction of their rights;
- Storage of the tasks and results;
- Diagnostics of the radio monitoring equipment and infrastructure;
- Getting tasks from the external information systems;
- Formulation of the missions for the personnel and account of the requests for search for interferences;
- Sending task execution results to the external information systems;
- Control of the equipment in the real time;
- Representation of the current status of the radio monitoring equipment condition;
- Representation of the equipment location on the digital area map;
- Urgent notification of the operator about different situations arising;
- Automatic synchronization of the account database between all nodes.

An example of the client–server architecture of the SMO-ARMADA is presented in Fig. 3.1.

Operation in the system is carried out with the help of automatic workstation (AWS) operators, where the client part of the ARMS is presented by the Web interface in the browser and by desktop applications to operate with radio monitoring equipment in the real time (Fig. 3.2). Using of the server side for common computing operations reduces requirements to the characteristics of the controlling computers at the operator workstations and essentially reduces expenses on deployment and maintaining of the system efficiency.

At present, Web-based GUI for operative radio monitoring functions is under development in order to make it possible to stop using installation of the additional software on AWS and to obtain completely cross-platform ARMS client.

The system is built on the J2EE (Java 2 Enterprise Edition) platform representing de facto the industrial standard of the complicate distributed control system constructions. The user access to the system is provided from any modern browser with HTML5 and WebSocket support, and the efficiency is tested in the Google Chrome browser.

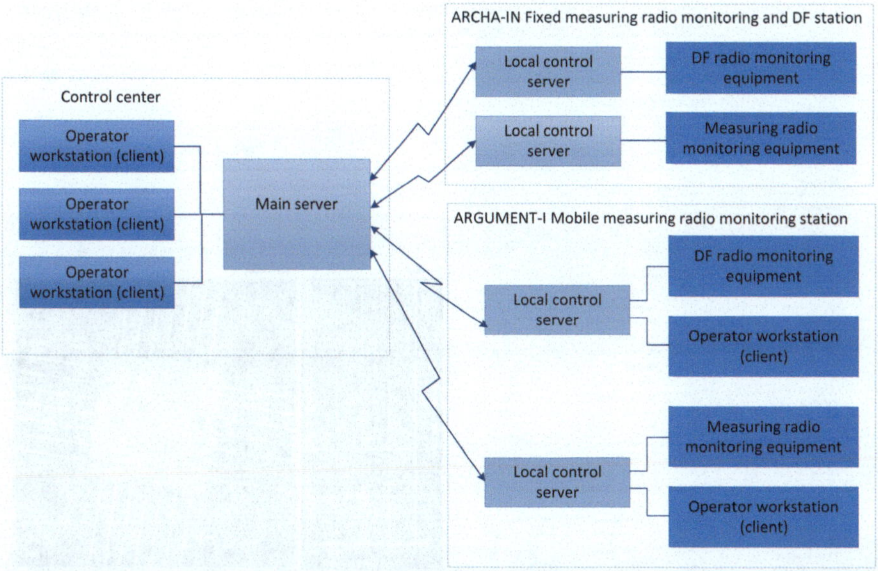

Fig. 3.1 Client–server architecture of the ARMADA ARMS

In the browser window, the full-value multi-window user interface is realized with usual control elements, context menus, and "hot" key support (Fig. 3.3). The GWT (Google Web Toolkit) technology and the Sencha GXT component library are used as basis of the user Web interface. The graphic information displaying is realized with the help of the GFlot library.

The GWT technology gives the programmer a possibility to interact with browser and be abstracted from JavaScript, HTML, and DOM (Document Object Model) and supports the basing elements of the user interface, the serialization (transformation) mechanism of objects, and RPC (Remote Procedure Call) for remote request of the server methods from the client and for acquiring of the execution results, as well as the cross-compiler from the Java development language to the JavaScript and component for the code debugging.

Besides the GWT-RPC-basing mechanism for the client/server interaction, the WebSocket mechanism is sometimes used in SMO-ARMADA. In contrast to the classical RPC, which assumes the execution of some operations by the server according to the client request, the WebSocket allows sending of the messages by the server to the client on its own initiative. This mechanism is used in SMO-ARMADA to display spectrum in the real time as well as sometimes to inform the client using table data about current data obsolescence and the necessity of data refreshment from the server.

The mentioned methods, technologies, and libraries provide user interface realization, which satisfies the modern requirements and also simplifies the interface revision during development of SMO-ARMADA.

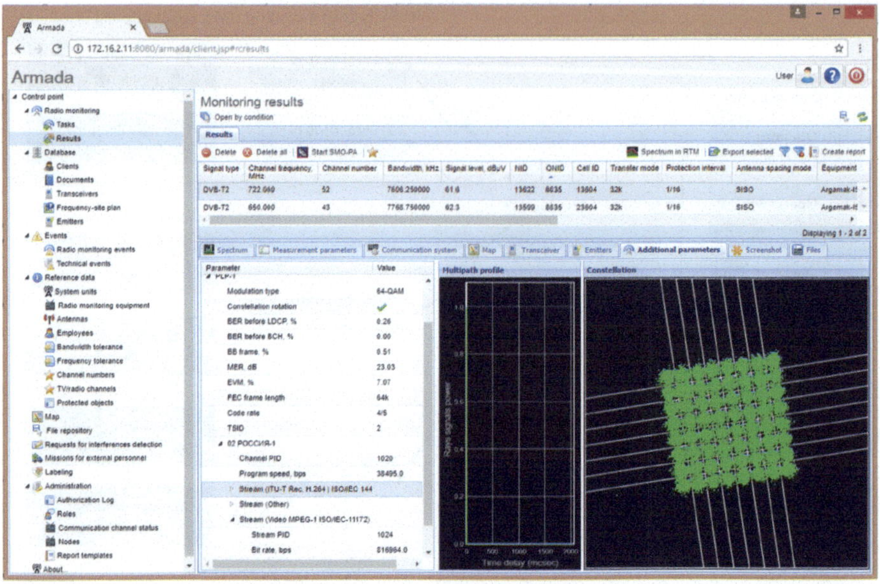

(a) SMO-ARMADA in web-browser

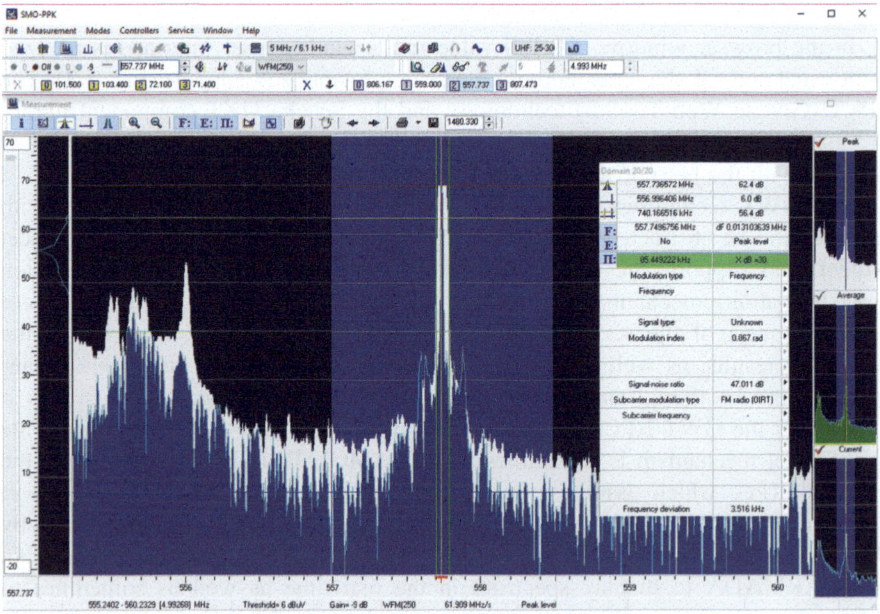

(b) SMO-PAI application in the "Spectrum" mode

Fig. 3.2 Web interface in the browser and by desktop

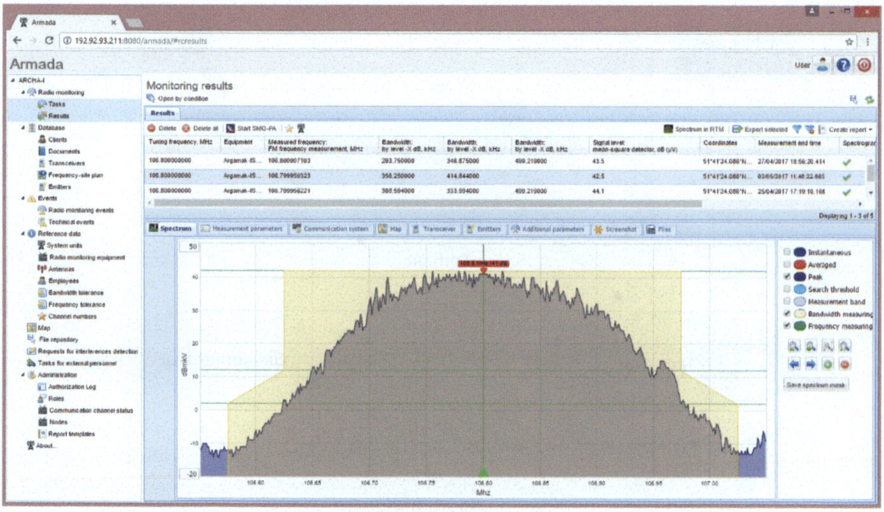

Fig. 3.3 SMO-ARMADA interface

The approach assuming maximal usage of typical solutions and components was used at development of server part of SMO-ARMADA, which provides the fulfillment of the following requirements:

- Cross-platform structure—the possibility of system functioning under Windows and Linux OSs;
- Scalability—the possibility of system deployment both on the separate computer and at the national level scale;
- Fault tolerance—the maintenance of the system efficiency under conditions of the partial equipment fault and/or essential degradation (up to loss) of communication channels, automatic control of the equipment, and communication channel statuses;
- Openness—the possibility of radio monitoring data exchange with other systems by means of exchanging formats and/or protocols, as well as the possibility of usage of radio monitoring equipment of different manufactures;
- Simplicity of the software updating—the central system server is updated firstly, and then, subordinate servers are updated automatically;
- Separate usage of different agencies data—each operator of the system should be associated with the definite agency (or several agencies). It excludes a possibility of non-authorized access to the data from the other agency sections.

In Fig. 3.4, we can see interaction of the server part components of SMO-ARMADA with the client parts, which are realized in the browser. The following components can be attributed to the server ones:

- Java Web server (the container of Web application) Tomcat;
- Data base management system (DBMS) PostgreSQL;

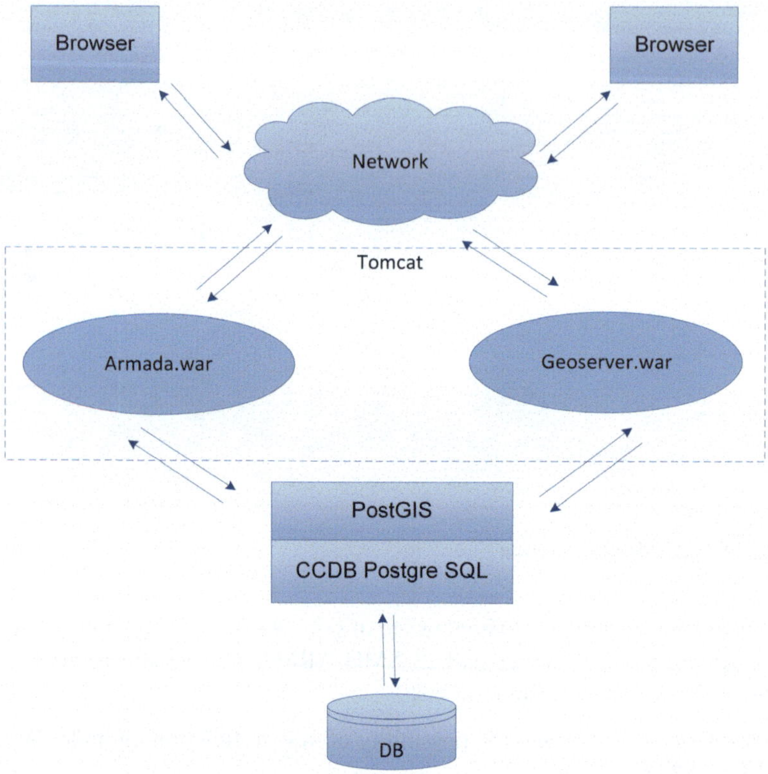

Fig. 3.4 Server and client parts of the SMO-ARMADA

- Armada.war and geoserver.war Web applications;
- Auxiliary components (launcher.jar loader and others).

In order to simplify the system deployment under Windows OS, the Java, Tomcat, and PostgreSQL are not installed as separate services, but they are integrated into the unified distributive with the graphic user interface, which installs the only Armada Service in the operation system. To launch the Java application as the Windows service, the auxiliary executable file is used, which launches the JavaVM virtual machine with the required parameters. At JavaVM start, the specific developed loader launches the integrated PostgreSQL and then the integrated Tomcat, which, in turn, launches the supplied Web applications.

In Linux OS, from compatibility consideration, the PostgreSQL is used, which is installed preliminarily, and other components of SMO-ARMADA are supplied in the form of the installation DEB-packet.

The life cycle of Web application is controlled by the multi-functional Spring Java framework, which represents the "light" and essentially more flexible alternative of the standard EJB (Enterprise Java Beans) technology. To increase system performance, the interaction with DBMS is also provided by means of the "light-weight" MyBatis library, not using traditional ORM frameworks (Object-Relational Mapping) of the Hibernate type. Interaction with the client code is mainly performed by means of GWT–RPC services, as well as (sometimes) by means of servlets.

3.3 Components of the Software System

For further consideration of the SMO-ARMADA architecture, it is reasonable to extract main system components and to examine the architecture of each of them separately. The following sub-systems are included into the SMO-ARMADA structure:

- Radio monitoring sub-system purposed to solve tasks of control of radio monitoring equipment in the operative (under operator's control) and planned (automatic according to the tasks and schedules) modes;
- Accounting and reference data sub-system purposed for storage and visualization of information about registered radio electronic equipment, licenses, as well as reference data necessary for the ARMS functioning;
- Information about events sub-system purposed for the operators' prompt notification about arising situation, which requires involving of the operator;
- Formulation of missions for personnel and account of requests for search for interferences sub-system;
- Cartography sub-system purposed for representation of registered and reference information on the electronic map as well as visualization of the acquired radio monitoring results;
- Data transmission sub-system fulfilling the tasks of automatic data synchronization between the system servers;
- Generation of report sub-system purposed for generation of reports, based on the samples from the data base, or documents of the approved forms;
- Data storage sub-system providing storage, reservation and recovery of the system data;
- Data exchange sub-system providing interfaces for interaction with external systems;
- Administration sub-system used to control efficiency of the system, to set account users' data and to control access to the system data;
- Update sub-system intended for updating of some system components or the software as a whole;
- Self-diagnostics sub-system checking the system efficiency.

Let us consider the mentioned sub-systems in detail.

3.4 Radio Monitoring Sub-system

In the operative mode, tasks are solved with the help of the specialized software packets, which represent necessary possibilities for solution of one or several specific tasks. To provide further usage of the data acquired (e.g., for their further deferred processing), the operation results are stored in the SMO-ARMADA database.

The SMO-PAI software packet for the panoramic analysis and measurement, the SMO-BS software packet for digital signals analysis, the SMO-STA software packet for signal technical analysis, and some others can be attributed to the main software programs, which provide task solution in the operative mode [4, 10, 11].

In the planned mode, the radio monitoring task execution is provided in accordance with a schedule defined by the operator. The base of automatic (planned) mode is the radio monitoring task, which determines the set of measurements to be made and its parameters, the radio monitoring equipment, frequencies or ranges, acquired results, as well as a schedule and the execution priority. The task can be formed by the operator using Web interface or can be loaded using the data exchange sub-system.

The control of equipment of radio monitoring in the planned and operative modes is performed according to the open unified protocol (UP) of exchange. For this, the "driver of radio monitoring server" software packet is used. This driver provides command reception, their processing, and sending of the acquired results to the sender. Exchange with the driver is performed according to the TCP/IP protocol.

Task for the radio monitoring equipment can be assigned from the any system unit taking into account the hardware opportunities (the supported set of measurements, the antenna inputs, and some other parameters). The task, which was successfully checked, is stored in the database of the node, which formulated the task, and is sent to the node-executor and to the control point (Fig. 3.5, the stage 1, the stage numbers are marked by ciphers).

After that, the task at the node-executor passes to the planner, which defines further actions (stage 2) on the base of the schedule and the new task priority, its own task priority queue, and the current task being executed by the equipment: to wait, to execute the task, to interrupt the task currently being executed.

Interaction of the planner with the driver of radio monitoring server is performed according to UP, which provides commands and radio monitoring data serialization. To transform the task from Java objects to the UP command sequence as well as for reverse transformation of the results, the special module translator is anticipated in the software (stage 3).

During the task execution (stage 4) results are acquired (stage 5), which are transferred to the node-task originator using the transport sub-system.

The node-task originator sends results to the control point (stage 6) and visualizes them in the browser. Other operators, which are connected to the control database via the browser, may also review the task and the acquired results (stage 7).

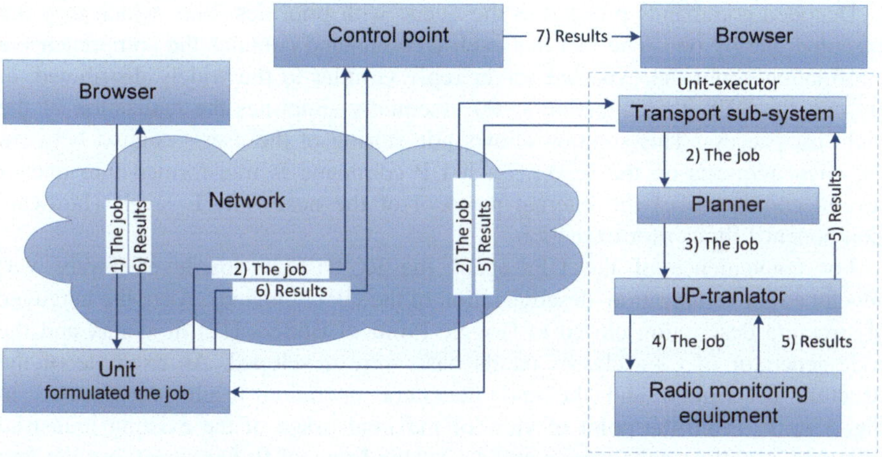

Fig. 3.5 Formulation and execution of the tasks in the planned mode

3.5 Driver of the Radio Monitoring Server

The structural diagram of the RC2HWCL radio monitoring server driver is presented in Fig. 3.6. The object handler, which receives and decodes the input commands of UP, is created inside the driver for each incoming connection.

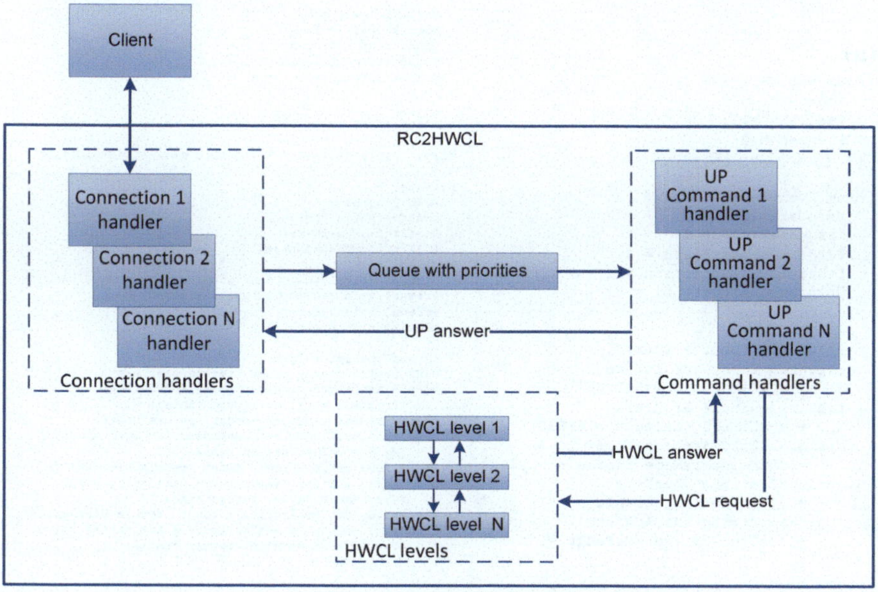

Fig. 3.6 Structural diagram of the driver of the radio monitoring server

Decoded commands proceed to the queue with priorities, from which they are executed one by one. The fact that each UPcommand contains the comprehensive information about the expected server reply (similar to the widely distributed, at present, the RES T architecture style) essentially simplifies the realization of the multi-user access. This solution allows non-storing of the user's context between the client requests on the server. Each UP command is transformed into one or several commands of the internal protocol of the next HWCL level (Hardware Component Library) for execution.

For convenience of the UPusage, the RC2HWCL low-level library was developed for serialization–deserialization of the UP commands. Also, the language of structure description closed to Google Protocol Buffers [12] in syntax and the code-generator of C++ classes on its base were developed. An example of the structure description and the code-generator operation result is presented in Fig. 3.7a, b. From the point of view of maximal usage of the existing industrial solutions, it will be rather reasonable to use the Protocol Buffers itself, but the first public available version of this Google project appeared only in one and a half years after finishing development of the first version of the unified protocol.

(a)

```
enum HardwareMode {
  Idle = 0;
  RealTime = 1;
  Planned = 2;
}

enum EventType {
  HardwareReady = 0;
  HardwareFail  = 1;
  USBHardwareNotFound = 2;
  EthernetCommunicationFail = 3;
  UnknownConfiguration = 4;
}

message RepTechnicalEvent {
  repeated EventType EvType
      = upvTechnicalEventType;
  required uint16 Progress
      = upvDonePC [default = 65535];
  optional HardwareMode Mode
      = upvHardwareMode [default = 0];
  optional string UserName
      = upvUserName [default = ""];
  optional string UserAddress
      = upvUserAddress [default = ""];
}
```

(b)

```
enum nHardwareMode
{
  Idle = 0,
  RealTime = 1,
  Planned = 2
};

typedef std::vector<nHardwareMode> nHardwareModeList;

enum nEventType
{
  HardwareReady = 0,
  HardwareFail = 1,
  USBHardwareNotFound = 2,
  EthernetCommunicationFail = 3,
  UnknownConfiguration = 4
};

typedef std::vector<nEventType> nEventTypeList;

class nRepTechnicalEvent : public nBaseStreamedObject
{
private:
  WORD Progress;
  nHardwareMode Mode;
  AnsiString UserName;
  AnsiString UserAddress;
public:
  virtual void load(upItem* item);
  virtual void save(upOutParam* item);
  virtual void clear();
  nRepTechnicalEvent();

  nEventTypeList EvType;
  WORD getProgress() const {return Progress;}
  nRepTechnicalEvent& setProgress(WORD value)
      {Progress = value; return *this;}
  nHardwareMode getMode() const {return Mode;}
  nRepTechnicalEvent& setMode(nHardwareMode value)
      {Mode = value; return *this;}
  AnsiString getUserName() const {return UserName;}
  nRepTechnicalEvent& setUserName(AnsiString value)
      {UserName = value; return *this;}
  AnsiString getUserAddress() const {return UserAddress;}
  nRepTechnicalEvent& setUserAddress(AnsiString value)
      {UserAddress = value; return *this;}
};
```

Fig. 3.7 An example of structure description and the code-generator operation result. **a** Description of "technical event" structure using "structure description language". **b** Header of the automatically generated "technical event" C++ class

Usage of the code-generator provided essential increasing of the code readability and decreasing of number of development errors, as well as simplifying of the communications between developer teams of the SMO-ARMADA and of the radio monitoring server driver due to the UP formalized description.

3.6 HWCL Protocol

At all advantages, from the point of view of higher-level software, UP turns out to be unsuitable for implementation of direct control of the hardware. The hardware represents the system with large number of possible states and unified protocol implementation (without additional abstraction level), which represents the typical stateless protocol and turns out to be either not-optimal in performance, or unjustifiably complicated and difficult in debugging and testing. The way out from this situation was found out by means of usage of the intermediate protocol of medium type, which, in its logic, is close to the control commands for the hardware.

At a point in time, when development of UPbegan , there were vast elaborations on the software used to control hardware in the operative mode, which used the hardware-independent HWCL protocol developed in the early 2000s. This protocol was created as a replacement of the protocol family for control of the hardware with the LPT interface; however, after some developments, it turned out to be fully suitable for usage as a protocol of the medium level. The high percent of the inherited code reuse and also already developed control modules both for manufactured (in thuds years) hardware of the fourth generation and for equipment of the second and third generations with the LPT interface produced at the end of 1990s are the additional advantages of the HWCL protocol usage.

The HWCL library is intended for arrangement of hardware-independent exchange between user interfaces and hardware complexes. Usage of this library allows creation of the software basing on multilayer structure principle (the additional layers realizing intermediate computing and controlling functions can be used between the user interface and the equipment driver). The software using the HWCL consists of layers with the same interface. These layers can be linked in the linear or tree-type structure. The data exchange between layers can be provided both in the framework of the same address space and with usage of the network connections.

Each layer may be stored in the separate software module (DLL library), which allows changing of the software configuration without recompilation of its component parts. There are embedded debugging possibilities in the library, so, for instance, the layer formed as DLL, may be loaded into the address space of the separate process (transparently for the remaining system). It is convenient for diagnostics of such problems as leakage or fault of memory. The possibility of automatic testing of the separate layers with the help of auxiliary software, which sends the testing data, presented as appropriate HWCL commands, to the layer input and checks command results obtained from the layer, is an additional adventure. This technology, in combination with the server of continuous integration, is widely used for verifying measuring algorithms, direction-finding algorithms, signal analyzers of communication systems, and so on.

The library supports the programmer with information about configuration possibilities of the software–hardware complex ("complex description") owing to which the development of the unified software, which can operate in systems with various configuration, is possible. The complex description is a tree-type list, and it includes information about hardware available operation ranges and bandwidths of analysis, list of antenna inputs, the presence of the embedded navigation system. The descriptions of available algorithms for signal analysis, beginning from the level measurement algorithms and finishing by DVB, LTE, etc., analyzers, are also sent via the complex description to the higher-level software.

Figure 3.8 shows the simplified typical scheme of layer connections at usage of radio monitoring driver. The layers used here may be relatively divided into the following groups:

- Realization of signal analysis algorithms and gathering of statistics: The module of occupancy estimation, the module of statistics gathering, and so on can be attributed to them;
- Realization of the basic complex functions: the module of file recording, the search module;
- Modules of network operation and integration with software support in the operative mode;
- Support of various external devices, such as rotator with USB control, external GPS-receiver using NMEA protocol [13] or proprietary GARMIN GPS 18 protocol [14];
- Modules for hardware control: the hardware controller, which directly controls the device, and optional additional modules, such as accurate the temperature calibration module (used for hardware with requirements on amplitude–frequency response irregularity of ± 1 dB), the module-provider of information for the self-diagnostics service (with support of the SNMP protocol) etc.;
- Modules for planned mode support (they realize expansion of the HWCL command system for convenience of RC2HWCL development);
- Module for operative mode support, for example, module for sound reproduction. In case, when operating mode application is connected simultaneously to several RC2HWCLs, the modules that process data simultaneously from several software–hardware complexes automatically become available, for instance the module of localization, which uses the time difference of arrival method.

The presence of the intermediate layer support in HWCL enables quite flexible control of various types of hardware, which differ significantly in their hardware possibilities. In case, when some functionality is not supported at the hardware level, the additional intermediate layers can imitate this functionality by the software. For example, if the equipment has no embedded implementation of the root-mean-square detector, the module for signal analysis can request time sample

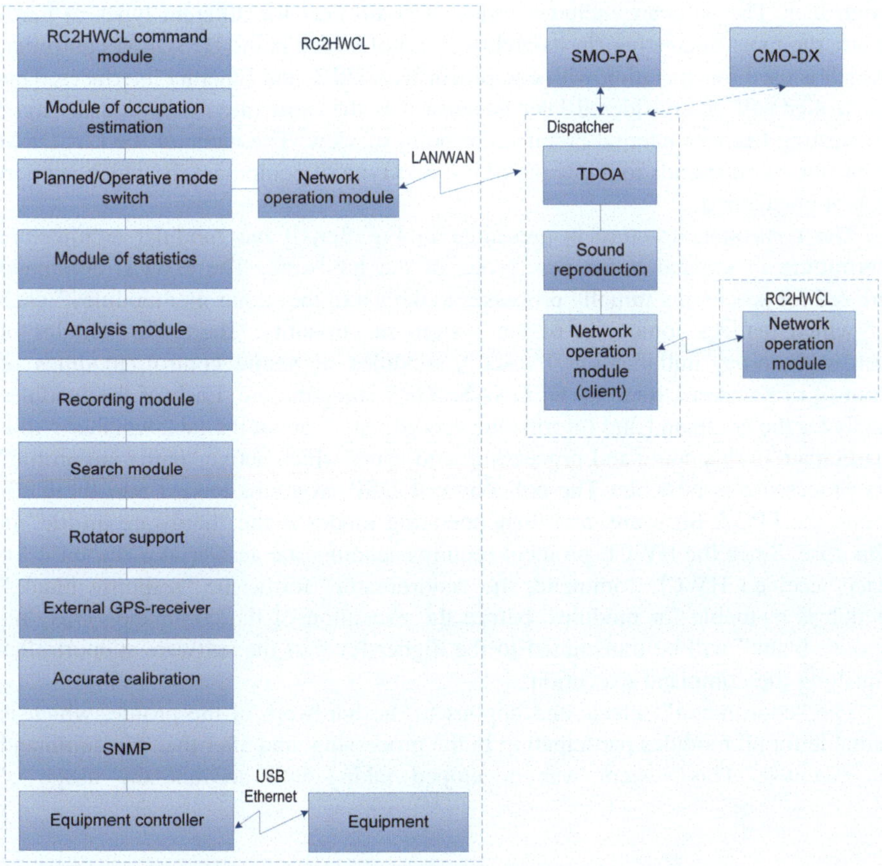

Fig. 3.8 Structure of the radio monitoring server driver

and provide level estimation calculated by the software. At that, there will be no difference for the higher-level software how exactly this estimation was made. The module of search for signals may operate in the similar manner: Some hardware types can provide search independently, but for those, in which this option is not realized, the signal search is performed by the software. It is also possible to meet the situation when the hardware may perform the search according to the fixed or floating threshold, but it does not support the search according to exceeding over the panorama (for instance, due to restricted size of the embedded memory). In this case, the search module will use hardware search for the first two algorithms, while the search over the panorama will be realized independently in the software.

The hardware controller, which transforms the HWCL commands directly into the hardware control commands, is the important part of the HWCL. Controller usage allows other layers of the HWCL operating independently from the used type of the hardware, being oriented only on the complex description presented by the controller. The software includes series of controllers for different types of hardware, the most interesting (in its architecture) of which is the D7SDRM controller, which is used for operation with equipment with USB and Ethernet interfaces. This controller will be considered later because it is the most often used and the most interesting from its internal architecture point of view. The name of the D7SDRM controller corresponds to the name of the first type of equipment, in which support was implemented.

The controller structure is presented in Fig. 3.9. It has modular architecture permitting to expand it for new types of the hardware. The HWCL command arrived for processing initially processes to the list of the "controller modules," each of which realizes some part of the system functionality: There are modules of hardware initial initialization ("reset"), modules of sound control, modules of control of receivers, rotators, GLONASS/GPS compass, etc. Each of the modules analyzes the command and informs the "coordinator" about the fact, whether will it participate in this command processing, and using which active "hardware profile" its processing is possible. The collection of DSP programs loaded into the hardware, the FPGA firmware, and their operating modes is the "hardware profile" in this case. Since the HWCL protocol requires sending one answer as a response for each received HWCL command, the "coordinator" forms the "response blank," which is available for modules, before the execution of the command. This "response blank" will be transmitted to the higher level of the software support after finishing the command execution.

The "coordinator" selects and applies to the hardware of the profile, which is suitable for all modules participating in the processing, and after that, the command is executed. This system was developed taking into account the usage of

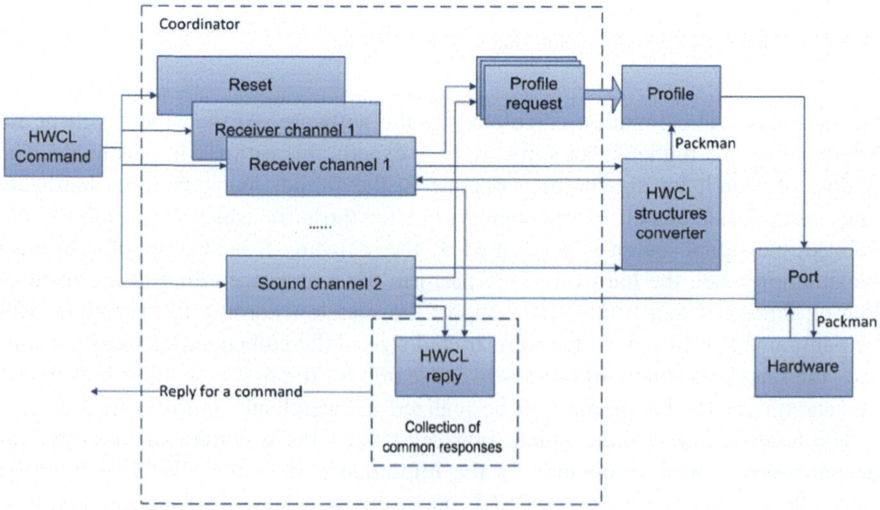

Fig. 3.9 Structure of the D7SDRM hardware controller

low-consuming digital processors, in which computing power is rather restricted, for the portable complexes in the product line. It forces to redistribute the resources flexibly depending on the requested data received from the client interface. The typical example of usage is the acquiring of the synchronous spectrum from two channels, which can be done either using two DSPs simultaneously (at that, there are no free remaining DSP in the hardware for other tasks), or using the only one DSP (the performance is lower, but the second DSP can be used, for instance for sound demodulation). In this case, the module responsible for spectrum acquiring will inform the "coordinator" about possibility of implementation in two "profiles": the fast one using two DSPs and slow one using the only one DSP, and the "coordinator" makes a decision about implementation method depending on the current operation of other modules.

After fitting the "profiles," modules transform the received HWCL commands into the low-level control commands with the help of the "converter." The "converter" consists of several tens of class-processors, each of which is responsible for transformation of some structure from the HWCL command. This solution allows forming of the new "converters" easily in case of appearance of new types of the hardware, which partially use PCBs, developed earlier. For instance, at appearance of the "new receiver with new command system" combination, it is enough to implement and to register the classes only, which are responsible for receiver tuning to the frequencies, the control of its attenuators, etc., in the "converter." For all other functions, the already existing software modules will be used.

The hardware control commands received as a result of "converter" operation are preliminarily processed in classes that encapsulate the support of the hardware current "profile"; then, they are proceeded to the object-"port," which is responsible for sending of these commands directly into the hardware and for response receiving from it. Physically, the "port" may transmit data using various communication channels depending on the hardware type. For equipment produced at present, the USB or UDP packets of the Ethernet network are used, but potentially, this may be, for instance, PCIe or any other communication channel. The packets sent to the equipment are formed in accordance with the "Packman" protocol, which, in essence, represents the set of agreements on the data block header, which is sent to the hardware.

Responses from the hardware pass on interpretation into the modules, which interpret them depending on their contents either as data or as confirmation of fulfillment of the commands sent to the hardware. Each module fills up its "own" fields by information obtained in the "response blank." When all modules complete the blank filling and inform the "coordinator" about their readiness, the response to the HWCL command will be sent, and the hardware controller will become ready to receive the next command.

3.7 Operation with Accounting and Reference Data

The sub-system of accounting and reference data is intended for entering, storage, editing, and deleting of information in catalogues, which are used at radio monitoring task solution. The operator operation with the sub-system is performed using the graphic interface, in which data are mainly presented in the table form. New data entering and editing are carried out with the help of the dialog window shown in Fig. 3.10.

The sub-system of accounting and reference data provides the control of the following data groups:

- Accounting data (information about frequency assignment, owners of radio electronic equipment, licenses);
- Reference information (information about equipment and radio monitoring stations, calibration data of antennas, norms on bandwidth and frequency, channels of digital communication systems, etc.).

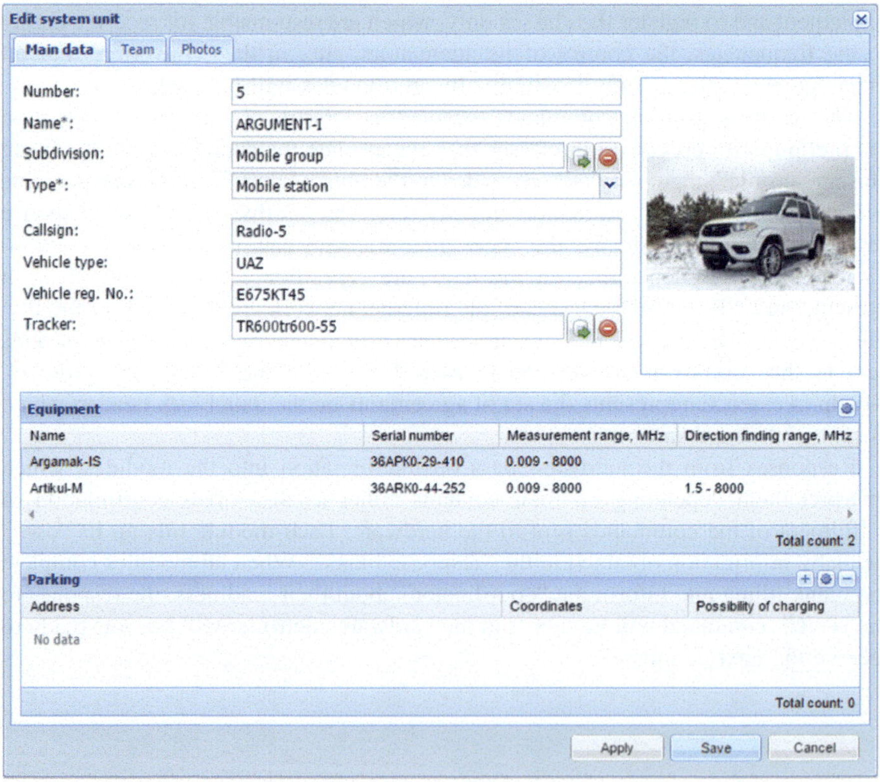

Fig. 3.10 Reference information about mobile radio monitoring complex

The set of accounting and reference data for ARMS may change depending on the system purposes and application conditions.

3.8 Accounting of Requests for Search and Formulation of Missions

This sub-system is intended to account requests for search for interferences and to control execution procedures of the search for interferences.

The request form for search for interferences is presented in Figs. 3.11, 3.12, 3.13 and contains the following information:

- Registration number, which is formed automatically during saving of the request in the database;
- Priority;
- Information about applicant;

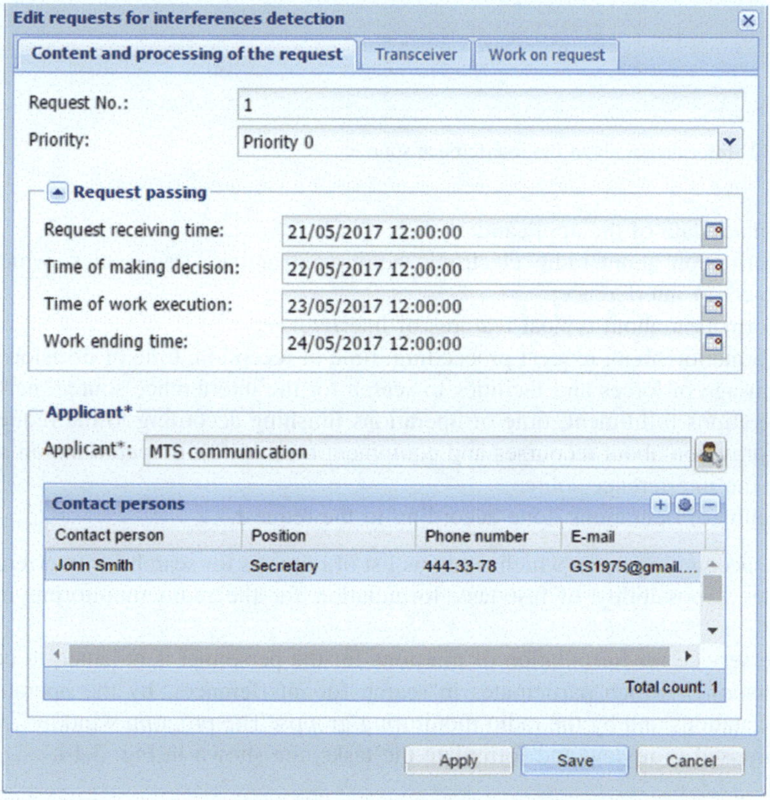

Fig. 3.11 Information about applicant and request time

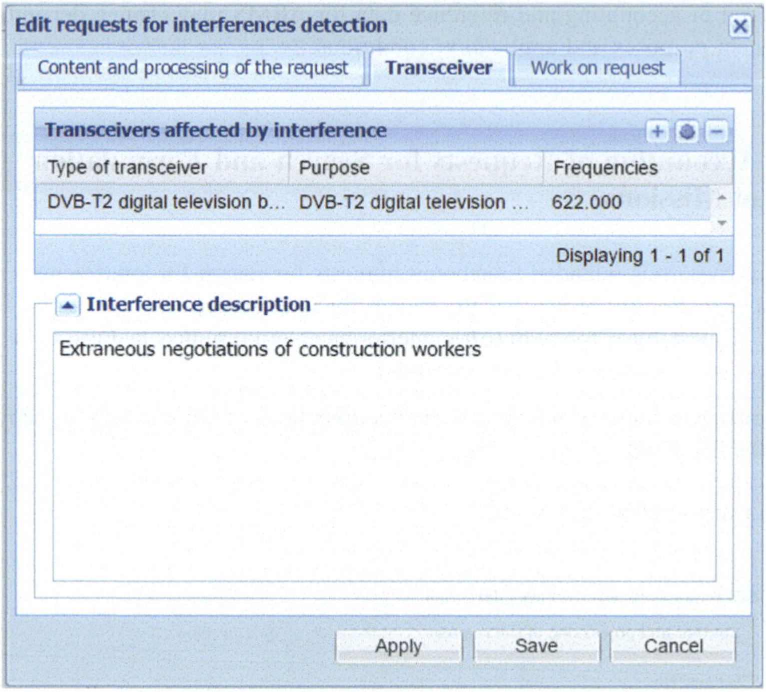

Fig. 3.12 Information about the interference source

- Contact data of the applicant;
- Information about radio electronic equipment and its frequencies, which are subject to interference;
- Information about typical features of interference;
- Information about request proceeding: time of reception, time of decision made on usage of forces and facilities to search for the interference source, period of operations fulfillment, time of operations finishing according to the request;
- Information about recourses and equipment used to fulfill operations on search for the interference source;
- Results of operations done according to the request.

The accounting form, which contains list of requests for search for interferences, provides a possibility of fast task formulation for the radio monitoring mobile groups.

The service for formulation of missions for the personnel is indented to control the personnel, which participates in search for interferences, by the operators of mobile stations and by the radio monitoring groups. The program windows, which are purposed to review and formulate the tasks, are shown in Fig. 3.14.

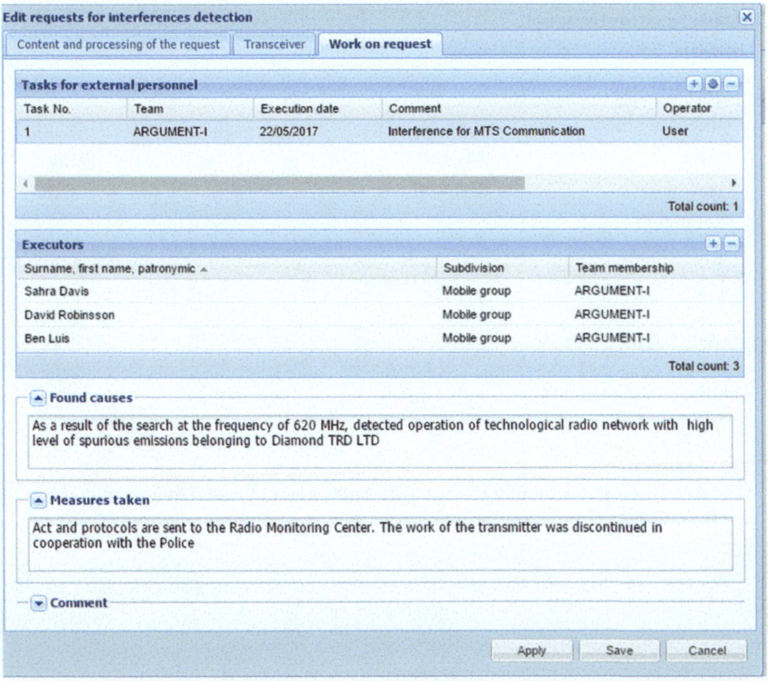

Fig. 3.13 Results of operations according to the request

The task contains the following fields:

- № of task—filled automatically;
- Information about mobile group, to which this task is assigned;
- Date of task fulfillment;
- List of the assigned missions, their contents, and fulfillment periods;
- Time of task reception;
- Results of the assigned mission's fulfillment;
- Information about employees included in the mobile group and its radio monitoring equipment;
- Results of the task fulfillment.

The service provides execution of the following functions:

- Visual check of fulfillment period of the assigned tasks: Tasks containing the expired missions are marked by red color.
- Automatic notification of the mobile group employees about appearance of new task or deleting of earlier assigned missions.
- Automatic notification of the employee, who assigned the mission, about its receiving and fulfillment.
- Prompt review of the information about request for search for interferences from the search (for interferences) tasks.

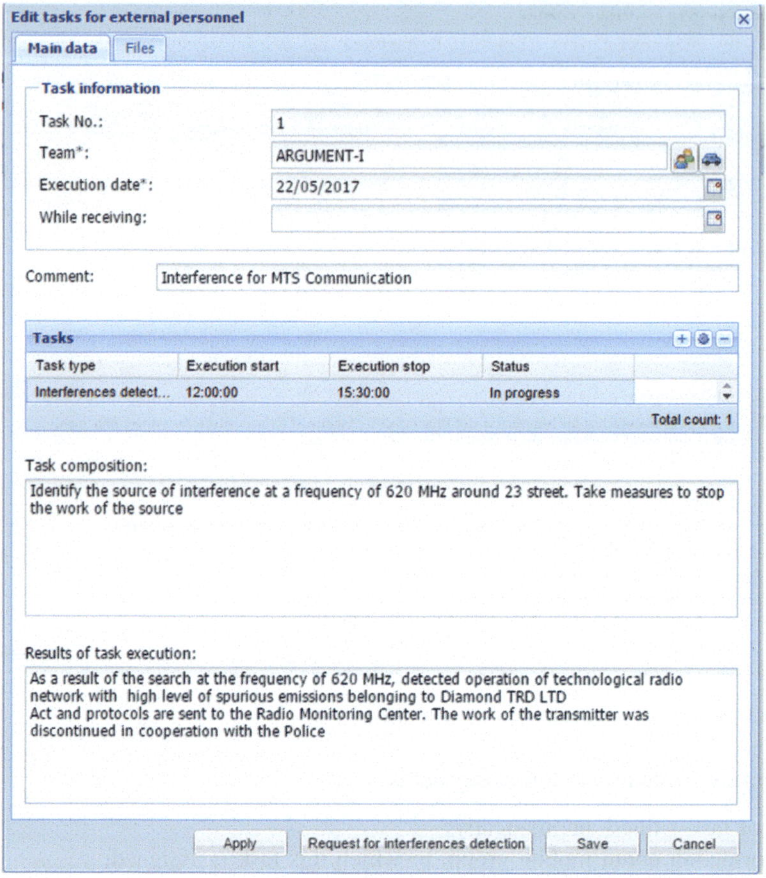

Fig. 3.14 Window of the personnel mission

3.9 Cartographic Support

The cartography is the important part of any ARMS. The cartography sub-system of the SMO-ARMADA is developed in Java using GeoServer geospatial data server [15]. It provides software interface for access to vector and raster maps and to data for its displaying by the external clients (Fig. 3.15).

The OSM (Open Street Map) maps [16] are applied as the cartographic basis. For visualization of the data stored in SMO-ARMADA, the standard set of the cartographic layers is supplemented by the problem-oriented layers: radio electronic equipment, radio monitoring stations, mobile stations and their tracks, etc.

To support Web cartography, the Leaflet library is used in the browser window, which provides map displaying according to the data received from the cartographic server and the interactive operation of the operator with the map.

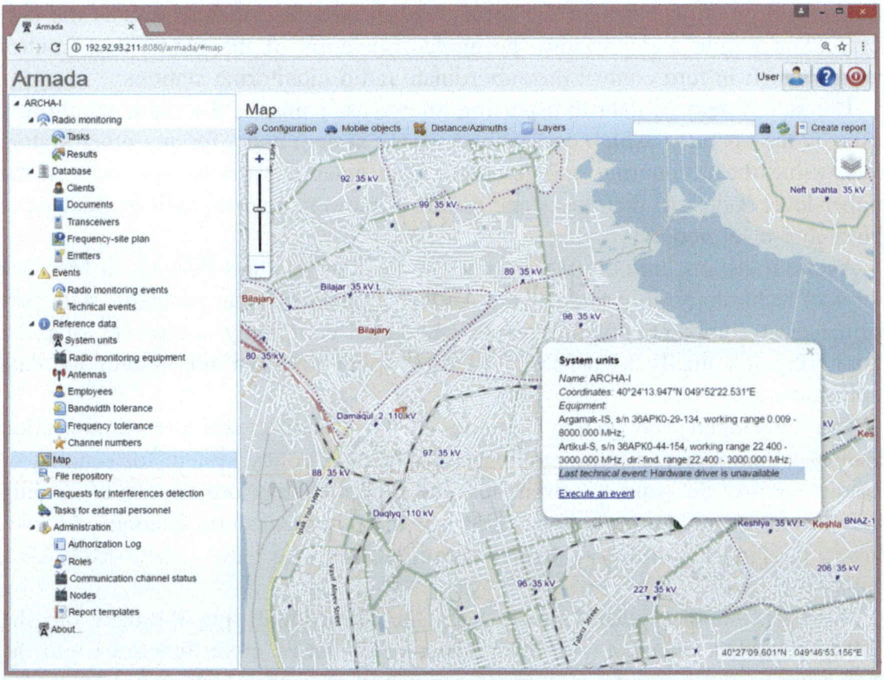

Fig. 3.15 Information about control objects on the map

To represent current location of the mobile stations on the electronic map, the interaction with GLONASS/GPS trackers of various manufacturers is implemented. Two alternative approaches are realized to obtain data from the trackers:

- Acquiring of the data about coordinates via the GSM/3G-channel in the NMEA-similar format directly from the tracker;
- Request of the data about coordinates via the interface of the system for mobile object monitoring.

The cartography sub-system provides possibility of addition of user layers and objects to the map.

3.10 Organization of Data Transporting

The sub-system of data transporting is intended to synchronize (replicate) the changed data between all databases of the distributed system. This sub-system provides the required scalability of the SMO-ARMADA in the "transparent" mode for the other sub-systems. The SMO-ARMADA scalability is realized at the expense of the nodes' integration as the hierarchical structure with replication of the selected information between the servers. Each control center maintains the corresponding region and provides the agreed operation of the subordinate control points, which in turn control the subordinate radio monitoring stations.

The sub-system of data transporting allows assigning of the radio monitoring tasks by the operator, which can connect to any system node, for any other system node using the accounting and reference data entered by other operators at the higher-level nodes. At that, tasks and results of radio monitoring will be doubled at the higher-level nodes.

As the transport, initially we assumed to use the standard JMS (Java Message Service) technology, in particular the ActiveMQ platform, but problems with performance at comparatively low-power embedded computing complexes of radio monitoring led finally to necessity of developing the own sub-system for data replication.

The developed sub-system provides data resending when communication channel is temporary absent, anticipates the flexible policy system for sending of various types of data and system nodes, and supports the priority queue of sending and accounting of the current condition of the communication channel. At first, these possibilities fully satisfied requirements for the SMO-ARMADA sub-systems.

However, during development of the functional possibilities of the SMO-ARMADA, namely with the transfer of operative mode functions into the Web browser, the necessity in alternative transport aroused, which would ensure the data transition with small delays without usage of databases both between the client and the server and between the servers. The appropriate transport was developed from the first and includes the following:

- The client serialization on the base of GWT serialization;
- Client/server interaction via WebSocket;
- Server serialization via standard ObjectOutputStream of the Java language;
- Interaction between the servers directly via Socket according to the own protocol.

At present, both variants of the date transport are used in the SMO-ARMADA in accordance with its purposes (fast delivery of small data packets without delivery guarantee or guaranteed delivery of big data packets with permissible delay).

3.11 Eventual Mechanism

The sub-system of information about events is intended for the ARMS personnel notification about arisen situations, which require the operator participation. Notifications are adjusted individually for each user and include information about the type of events to be informed about, the equipment, and the notification channel.

Three main types of events are used in the system:

- Radio monitoring events, for instance violation of the frequency-spatial plan or emission parameter deviation from requirements of licenses;
- Technical events, for instance radio monitoring hardware fault or changing of the communication channel condition;
- Service events, for instance launching the task for execution, interruption, or finishing of the task execution.

Events content, which is generated by the system, is illustrated in Fig. 3.16.

The registered event is stored in the database of the node, where this event was formed; after that, it is transmitted to all system nodes by the sub-system of data transport.

Notification of the operator can be performed by displaying information about event on the monitor screen, by sending a letter via e-mail or with the help of SMS message. Besides, the list of events arisen during the system functioning is stored in the database, and it is presented in the table form in the appropriate section of the user interface. Data samples about events, if necessary, can be exported as a report.

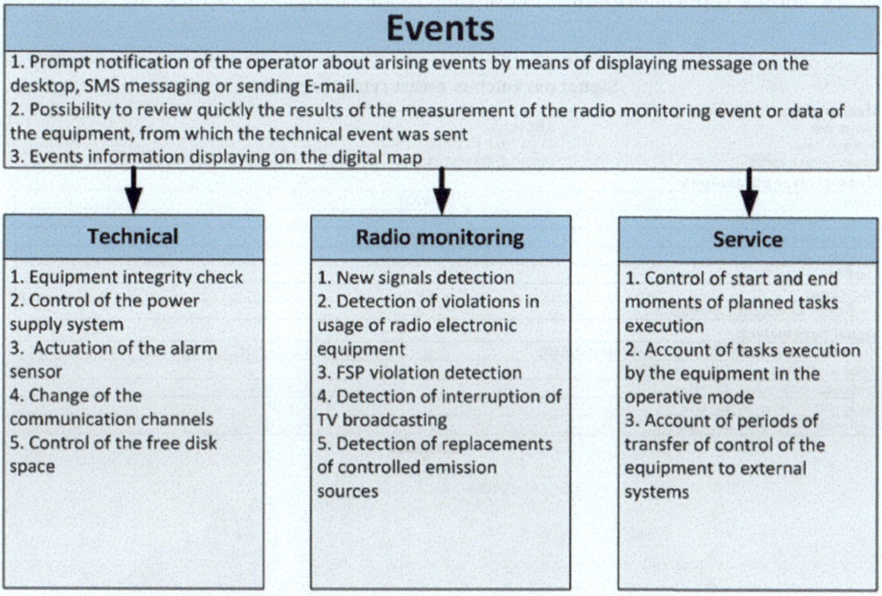

Fig. 3.16 Types of events generated in the system

3.12 Report Generation Sub-system

The sub-system of report formation is implemented on the base of the BIRT (Business Intelligence and Reporting Tools) library, which provides report generation in the popular HTML, DOCX, XMLX, PDF, and other formats according to given templates. The set of preformed templates is supplied with the SMO-ARMADA as well as the visual template editor. All reports are conditionally divided into two groups:

- Formalized documents: The sub-system forms the document blank of the approved form and adds data into the necessary fields. Measurement documents, recommendations on radio electronic equipment usage, etc., are the examples of such a type.
- User data samples: The data set, adjusted by the operator using sorting, filtering, and selecting of the fields to be represented, is added to the report. The document containing data in the same form as in the application interface is the result of the report formation.

Figure 3.17 shows the report on signal parameter measurement. The BIRT library allows report formation according to the templates, which are formed in

Signal parameters measurement report

Measurement conditions:

System unit:	ARCHA-I
Date and Time:	25.04.2017 17:19:10
Measurement location:	51°41'24"N 39°11'31"E

Measurement parameters:

Owner	Lion FM
Tuning frequency, MHz	106.800000000
Measurement bandwidth, kHz	500.000000000
Attenuator, dB	20
Amplifier	No
Polarization	Vertical
Resolution: Accumulated spectrum, Hz	781.25

Signal parameters:

Measured frequency: FM frequency measurement, MHz	106.799956221
Signal level: mean-square detector, dB (µV)	44.1
Bandwidth: by level -30 dB, kHz	308.594000
Bandwidth: by level -40 dB, kHz	333.594000
Bandwidth: by level -50 dB, kHz	499.219000

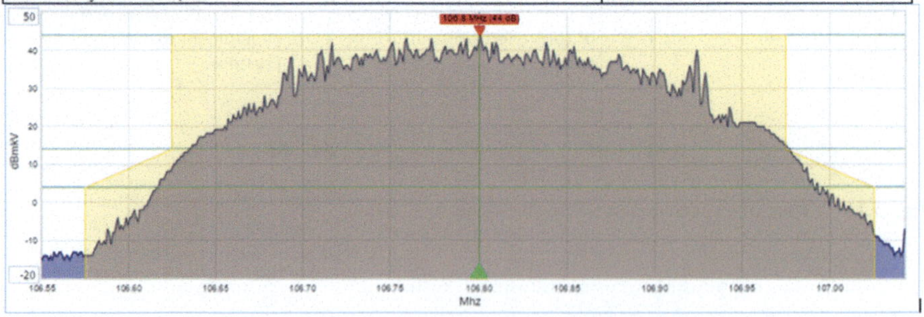

Measured by (user) **User**

Fig. 3.17 Report example of signal parameter measurement

advance. The template contains information about the form of the report (results, information, tables, and graphs to be included), order of report sections, lines and columns to be included into the table, format of graphs to be used, for instance, full names of authorized persons and so on.

3.13 Data Storage

The sub-system of data storage is implemented by using PostgreSQL DBMS. To optimize the storage and extraction of cartography data, the PostGIS extension is used. The initial database structure and also the contents of the reference catalogues are formed by the installer at system deploying.

In order to update support of already deployed database without information loss, the database versioning is anticipated. At that, each new version of the SMO-ARMADA includes the set of SQL-scripts for transformation of the "database of version N into the database of version N + M." These scripts are automatically executed by the installer at new version deployment.

Other sub-systems use the sub-system of data storage by means of SQL-requests, which are implemented in the code of appropriate sub-systems. In special cases, the stored procedures are used to eliminate the SQL-code doubling.

Owing to the fact that SMO-ARMADA is the distributed system, the unique UUIDs (Universally Unique Identifiers) are used as the identifiers of the majority of the objects stored in the database. Synchronization of data stored in the database is carried out by using the sub-system of data transport.

Results acquired from the hardware are exposed to the multi-level processing. One of the stages of such a processing, which is optionally performed in SMO-ARMADA, provides fulfillment of the following procedures at interaction with the sub-system of accounting and reference data:

- To check correspondence of the detected stations to the licenses, i.e., to control emissions correspondence to norms on frequency and bandwidth tolerances;
- To compare detection results acquired during search in the frequency range with data stored in the database for the purpose of checking correspondence with the registered stations;
- To compare results of the signal analysis of the digital signal base stations on correspondence to the frequency-spatial plan (FSP);
- To form radio monitoring events and to inform users about them in accordance with the notification adjustments.

The task and the radio monitoring results can be exported (via the sub-system of data exchange) into the exchange format file or can be used in the sub-system of report formation.

3.14 Data Exchange with External Systems

The sub-system of data exchange is intended to provide interaction of the SMO-ARMADA with external information systems. The sub-system of data exchange realizes the following functions:

- Export and import of tasks and radio monitoring results using files of the exchange format on the base of XML (eXtensible Markup Language) that are used at interaction with other information systems;
- Import of accounting data from the files of the exchange format MDB v.15 of FAIAS (Federal automated information-analytic system of radio frequency spectrum usage and in mass media). Files can contain information about licenses on the usage of radio electronic equipment and about its owners;
- Web services basing on WSDL (Web Services Description Language) and SOAP (Simple Object Access Protocol) technologies in HTTP that are used at interaction with operative mode applications;
- Auxiliary functionality representing the set of Web servlets, which are used for interaction with trackers, signaling systems, etc.

3.15 Sub-system of Administration

The sub-system of administration is intended to control the operator's access to the system functions and to control the system structure, including for addition modification and deletion of system nodes.

Access to the system is possible for the registered users only. Each operator has his own login and password. Inputs and outputs of the users from the system are registered in the authorization log.

Access rights to the work with the system are defined by the roles, which are assigned to the users. On the base of rights available for the operator, some sections and/or control elements are hidden or blocked in the system interface. The check of rights is fulfilled independently both on the client side and on the server side, which excludes a possibility of unauthorized elevation of rights made by correction of JavaScript-code on the client side.

The control of system nodes is also performed by the sub-system of administration. The operator can add new nodes, change parameters of existing nodes, and delete the not-used nodes using the system software. Changes of the system composition are automatically taken into account by the sub-system of data transport, and the data, available at the node directly after its connection into the system, become an actual one due to the possibility of remote updating of the node database.

The possibility of dividing data of different agencies, which are stored in the node database and in the control center of the system, is anticipated. Each system operator must be associated with the definite agency (or several agencies). This excludes a possibility on unauthorized access to the data from the sections of other agencies.

Besides the isolated sections in databases, there are sections available for all agencies including the accounting and reference data (e.g., FSP), as well as data about configuration and workload of the radio monitoring equipment, conditions of system nodes, etc. This enables any system operator, when having appropriate rights, to execute radio monitoring tasks at any system node.

The radio monitoring tasks are executed taking into account the established priorities and schedules. Tasks and radio monitoring results are stored in the database section, which is associated with the operator agency that created the task, and they are unavailable for employees of other agencies.

Figure 3.18 shows the access organization of each of four agencies to their own section of the database.

The radio monitoring tasks formed by the operator using the automatic synchronization of the database (which is anticipated in the system) are sent for execution to the nodes defined by the operator. The task execution results are returned to the node-installer. The node-installer sends the task and results to the control center.

Figure 3.19 shows an example of data exchange in the system at radio monitoring task assignment. The task assigned via the control center by the operator "1" is sent to the fixed and mobile stations mentioned in the task. Results are sent from nodes to the control center into the database section of the agency "1", in which operator assigned the task.

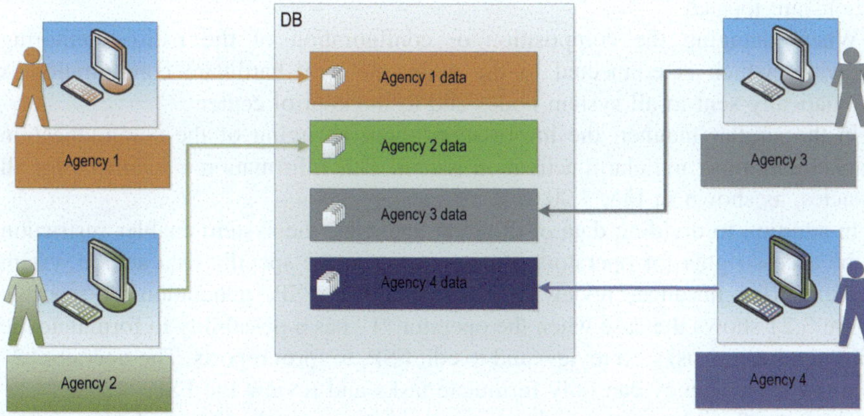

Fig. 3.18 Data division in the system

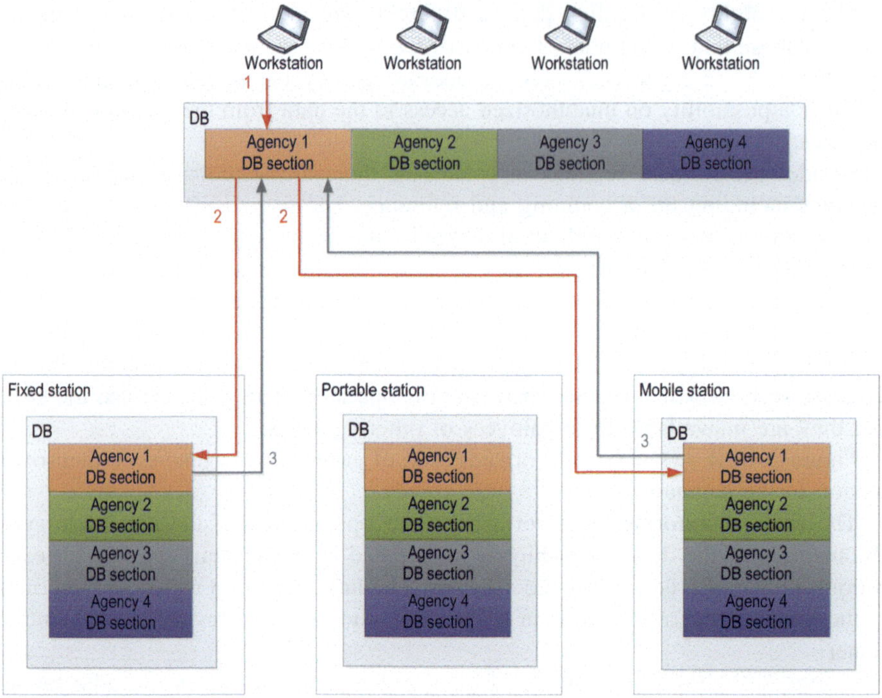

Fig. 3.19 Example of data exchange at task setting on the fixed and mobile stations for the agency "1"

Radio monitoring tasks can be assigned by the operators not only from the ARMS control center, but directly from the nodes themselves, where the workstations are located.

When changing the composition or configuration of the radio monitoring hardware, which is connected to the node, the new hardware configuration is automatically sent to all system nodes and to the control center.

In the similar manner, the information about changing of the communication channel condition and alarm activation is sent. This information is available for all agencies, as shown in Fig. 3.20.

In addition to dividing data of different agencies, the system enables restriction of the access rights for operators of one agency to the specific data and/or system functions (for instance, review and/or editing of FSP, generation of reports). Figure 3.21 shows the case when the operator "1" has a possibility to formulate the radio monitoring tasks, to review and to edit FSP, to form reports. The operator "2" from the same agency can only formulate tasks and review the FSP.

Thus, the flexible control and access management mechanisms, which are used in the system, provide the effective simultaneous control by several agencies. At that, the possibility of separate storage of confidential data is implemented.

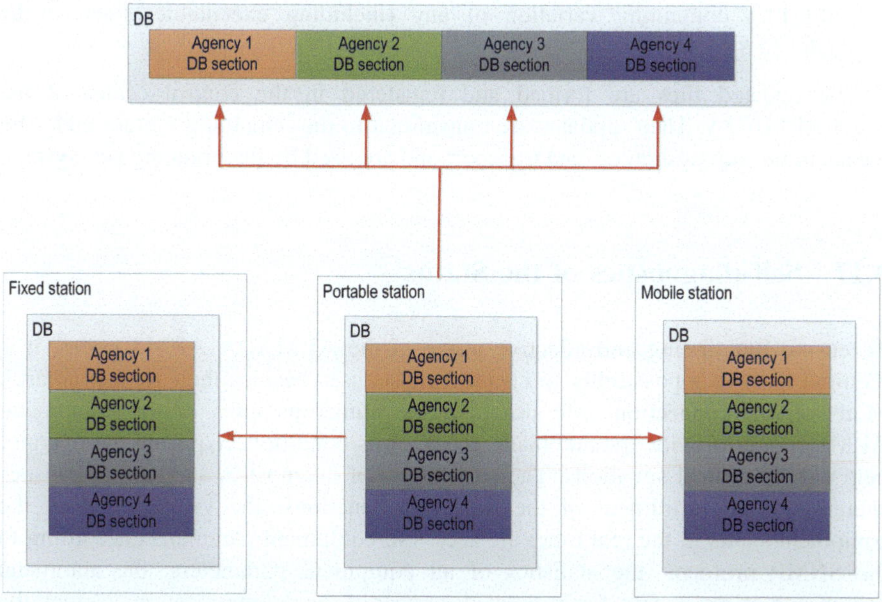

Fig. 3.20 An example of data updating at changing of the configuration of radio monitoring equipment on the portable station

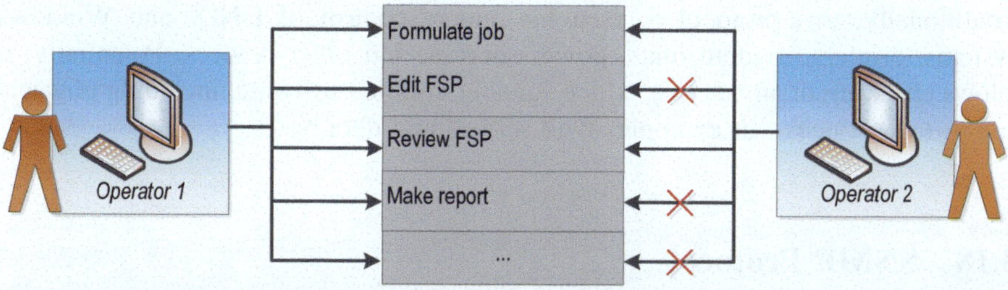

Fig. 3.21 Restriction of the users' access to the system functions

3.16 Updating of the Software Sub-system

The sub-system of updating is intended for centralized automated updating of the SMO-ARMADA servers. Several types of updating are anticipated:

- SQL-files containing requests to be executed in the updating database;
- DUMP-files containing full dump of the database, which replaces the updating database;
- CMD-files containing scripts to be executed in the updating system;

- DIFF-files containing variation of any (including executable) files of the deployed system.

The updated files are formed and registered in the control center of the SMO-ARMADA. Then, updates are transmitted to the remaining system nodes by means of the sub-system of data transport and are used by the updating sub-system.

3.17 Self-diagnostics of the Status

To ensure functioning and effective usage of the ARMADA ARMS nodes, it is required to have a possibility to check their statuses besides their remote control. Moreover, in connection with usage of the numerous units of the equipment, ARMS should provide the automatic check of its condition, requiring the operator help only in critical situations. The sub-system of diagnostics and self-diagnostics should provide fulfillment of the following functions: the visualization of the equipment status in the real time; the access to equipment parameters according to the SNMP protocol; the statistics of all equipment parameters; the automatic response to arisen events, for instance disappear of the communication channel, the hardware fault; the acquiring and storage of diagnostic information for remote diagnosis of the fault reason [2].

The SNMP (Simple Network Management Protocol) protocol was chosen as the basis of the self-diagnostics sub-system operation in the SMO-ARMADA. Traditionally, this protocol is used for remote control of UNIX and Windows systems, printers, modem units, power sources, and other devices. Potentially, it allows checking of any device, which suitable installed software, including physical and virtual devices, for example, Web servers and data bases.

3.18 SNMP Protocol

The Simple Network Management Protocol was at first used in 1988 to control large number of network devices. From that moment, this protocol gained popularity, and at present, it is a standard one [17].

Besides the remote control, the SNMP protocol gives a possibility to monitor different devices and allows information acquiring from any network devices, whether it is the router or simply the computer. The contents of acquired information may be various, for instance equipment operation time, different counters of central processing unit performance, temperature sensors.

The network, which uses the SNMP for control, contains three main components:

- SNMP manager—the software installed on the administrator computer of the ARMS;
- SNMP agent—the software launched on the network node, after which the monitoring is carried out;
- SNMP MIB (Management information base)—the database in the form of text files. This system component provides structuredness of data, which is exchanged between the agents and the managers.

Factually, the SNMP manager is an interface between the operator and the node with the loaded SNMP agent. The SNMP agent is an interface between the SNMP manager and the equipment at the network node. If to carry out the analogy of the SNMP with the client/server architecture, for example, with the Web server, then the Web server operates as the service at some port, and the user is addressed to the Web server as the client using the browser. In the case of the SNMP, roles of the client and the server are somewhere degraded. For instance, the SNMP agent is the service of the device being monitored, and it listens for the requests at the configured port, so it is factually working as a server. The SNMP manager is a kind of the client, which addresses to the server of the SNMP agent. In SNMP, the interruption (trap) is realized in the form of the notification from the agent to the manager. At sending of this notification, the agent and the manager change their roles. In this case, the manager is a server, which uses the definite port, and the agent is a client. In last versions of the SNMP, the trap may be called a notification.

Interaction of the agent and the manager on the level of the SNMP protocol is organized by means of PDU (Protocol Data Unit) object packets, which are encapsulated into the transport protocol. Although, the SNMP supports the various transport types, usually UDP (User Datagram Protocol) is used. At that, each PDU message contains the definite command (to read the variable, to set the variable value, or the response/trap of the agent). Agents acquire information about devices and write it to the MIB database, thus making it available.

For system operation, the MIB-files are used, which contain description of the device parameters, which are available for polling according to the SNMP. At that, to check the equipment, the external software may be used, for example, the Zabbix software [18], which is popular among the network administrators. Figure 3.22 shows the Zabbix window, which is connected to the self-diagnostics system. The graph of the power supply voltage changing is presented in this figure during three operation hours.

3.18.1 Architecture of the Self-diagnostics Sub-system

As SMO-ARMADA as a whole, the self-diagnostics sub-system has the client–server architecture. Its organization is presented in Fig. 3.23.

The external sub-system clients are divided into two types: SNMP clients and Web clients. Any standard software, which supports the SNMP, for instance

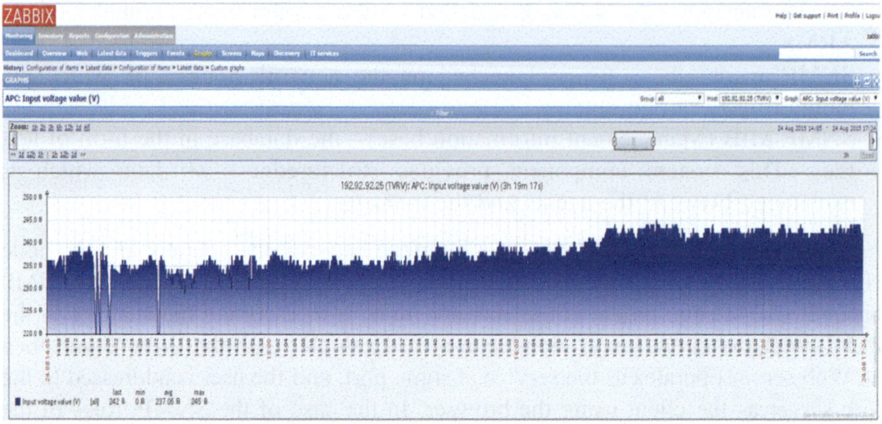

Fig. 3.22 Window of the Zabbix software

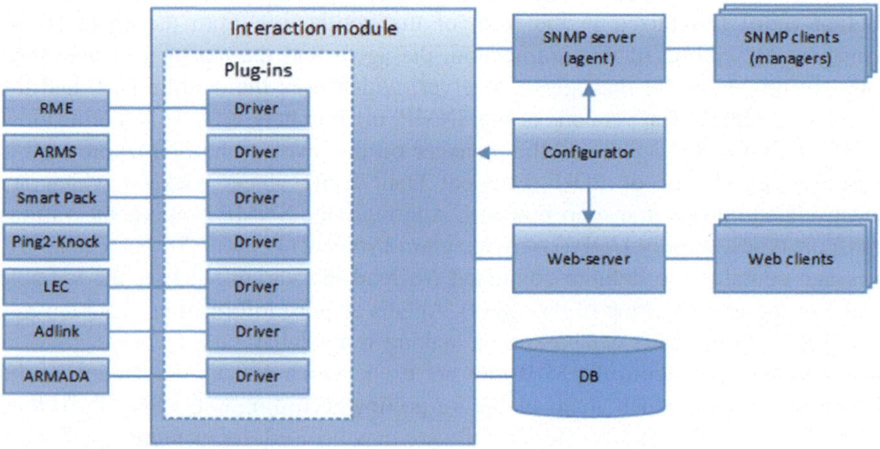

Fig. 3.23 Self-diagnostics sub-system

Zabbix, Nagios, may play role of the SNMP client. Any Web browser may be the Web client. The self-diagnostics sub-system can function independently from the SMO-ARMADA. At joint operation, the sub-system interface is integrated into the SMO-ARMADA interface.

Integration of all devices included in the ARMS structure with the self-diagnostics sub-system is realized with the help of the drivers, which are implemented in the form of the plug-ins. Such architecture enables support of the new components both of the own manufacture and of external manufacturers with minimal labor costs. To date, the following components are supported:

- Radio monitoring equipment;
- Uninterrupted power supplies;
- Devices of network monitoring;
- Industrial computers;
- Data communication channels.

Drivers provide acquisition of the detailed information from each module included in the radio monitoring equipment. The current and voltage values in the checking points, the temperature of components and environment, exploitation time are included into the list of the gathered data. Owing to information collected, the support service may determine the unit operating in the critical mode, may find the disabled module, and explain the fault reason. For instance, for uninterrupted power sources, the check of the main parameters, such as temperature, battery charge, external power source, load current, is performed. Besides, the direct access to the SNMP interface is possible, if it is anticipated by the manufacturer. For network monitoring devices, such as Ping2-Knock, the information is gathered and transmitted. The monitoring of computers fulfilling functions of local or central servers, for example, LEC, Adlink and others, is carried out. Besides, drivers provide support of different sensors, which are connected via GPIO (general-purpose input-output) bus. The driver presents information about the communication channel status and sends notifications about events happened.

The interaction module, which provides initialization of the drivers, the information storage in the database, and also the request processing from the SNMP agent or the Web server, is the central unit of the self-diagnostics sub-system. The database of the sub-system stores list of the connected devices, the numeration and structure of available parameters and the configuration of each device, as well as the data acquired from the devices. The SNMP agent represents the service, which provides support of the SNMP, and the Web server is responsible for the support of the user interface. The configurator is intended for adjustment of the self-diagnostics system. It provides addition, deleting of plug-ins and their adjustment.

3.18.2 Adjustment of the Self-diagnostics Sub-system

The "Configurator of SNMP plug-ins" software is used to adjust the self-diagnostics sub-system. The configurator window is presented in Fig. 3.24. The configurator allows addition and deletion of plug-ins, thus widening or restricting the list of the equipment, which are controlled by the system. The following functions are implemented in the configurator: possibility of visual adjustment of the device configuration plug-ins and representation of the data acquired from the devices in graphic and text interpretations.

Let us consider the operation with the configurator on the example of the uninterrupted power source (UPS). After adding the UPS plug-in to the system, the window appears with main adjustments, which is shown in Fig. 3.25.

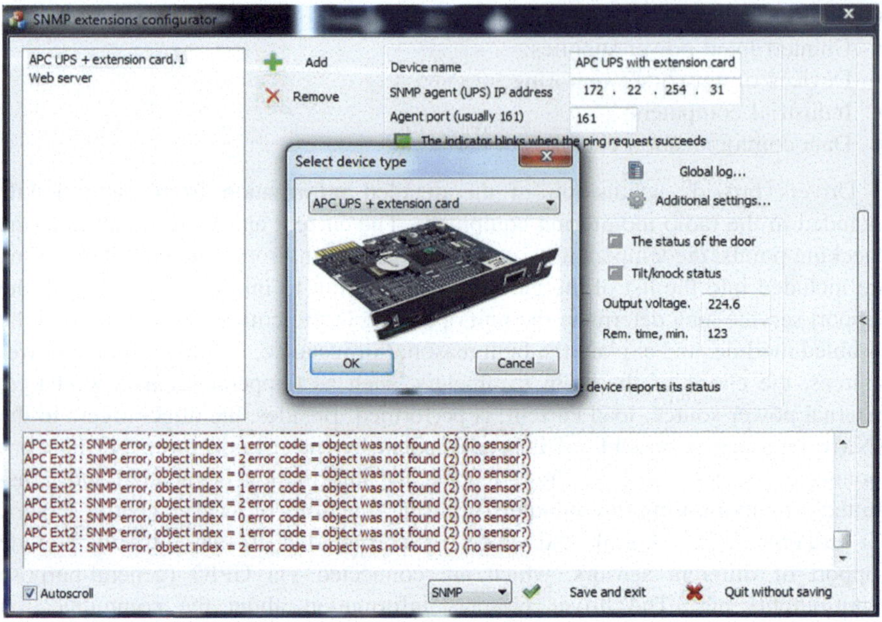

Fig. 3.24 Window of the configurator software

After setting the UPS address and port, the values of the current temperature, the battery charge, etc., will appear in the appropriate fields. The following parameters are available in the UPS additional adjustment menu:

- Interval of the sensors polling;
- Selection of the parameters supporting the trap notification according to the SNMP;
- Threshold of events arising;
- Possibility of correct personal computer shutdown at achievement of the threshold value at operation from the battery;
- Location (name of the monitoring station or the mobile station) in the ARMADA ARMS;
- Request for additional parameters values from the ARMS (the version of firmware, etc.).

3.18.3 Check of the Hardware Parameters

The Web interface gives a possibility to check parameters in the form, which is convenient for the operator. In the left upper part of the window, the menu is displayed, which lets to select one of the three review modes. Each of these modes represents the separate sections.

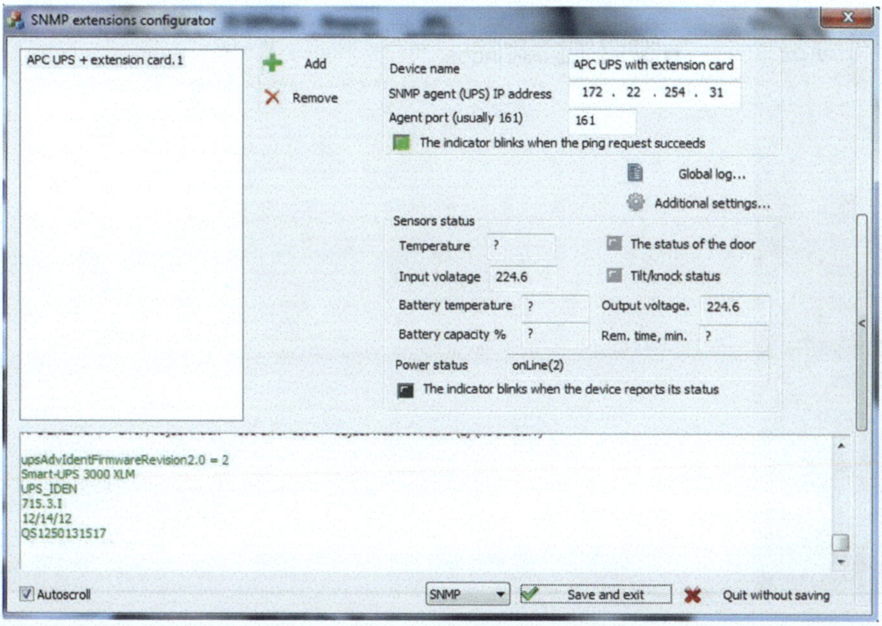

Fig. 3.25 Adjustments of the UPS

The section "History" shown in Fig. 3.26 presents the access to the history of the parameter variation in the table form. The table consists of the following columns: "date and time," "description," "response from the device." The treelike list is displayed to the left of the table. The list consists of the hardware units. The hardware, in turn, may consist of the modules. The fault of any module is represented visually in the treelike list. There is a possibility of sorting and filtering. The data export from the table into the "Comma Separated Values" format is realized. Exported files can be reviewed and edited both in text and in table forms (Fig. 3.27).

The section "Current state" shown in Fig. 3.28 contains a list of parameters, which are typical for the hardware. In this section, we can see the table of the parameters at the current time moment.

The section "Charts" is intended for visual analysis of the history of numerical parameter variations of the hardware. This section content represents the set of sections (one section for the hardware). When section is opened, the interactive graph of parameter variations during the last several days will appear. In this graph, with the help of the mouse, the navigation is possible along the abscissa axis. One may choose also the initial and final data manually. Automatic and manual adjustments are realized along the ordinate axis. Figure 3.28 shows the window of the "Charts" section, in which the input and output voltages of UPS for radio monitoring hardware are presented.

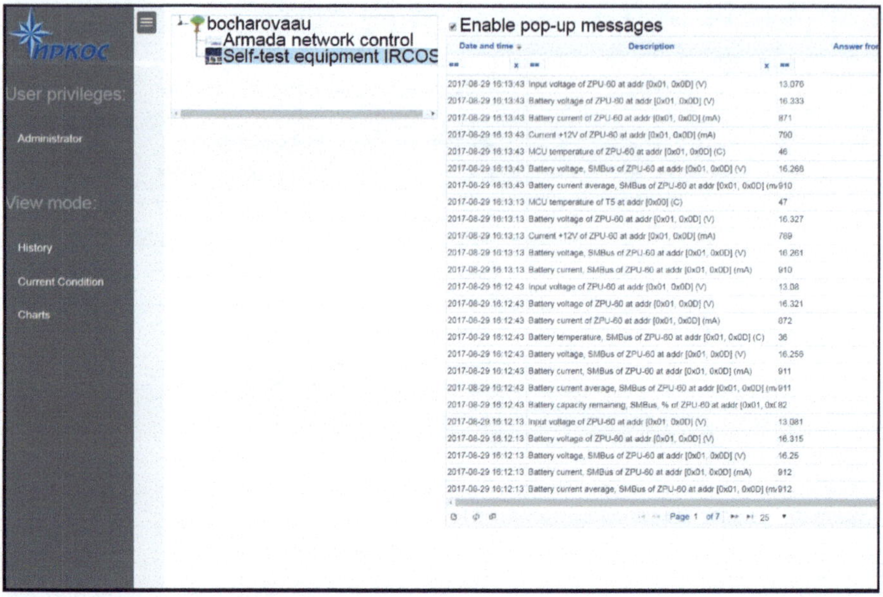

Fig. 3.26 "History" section

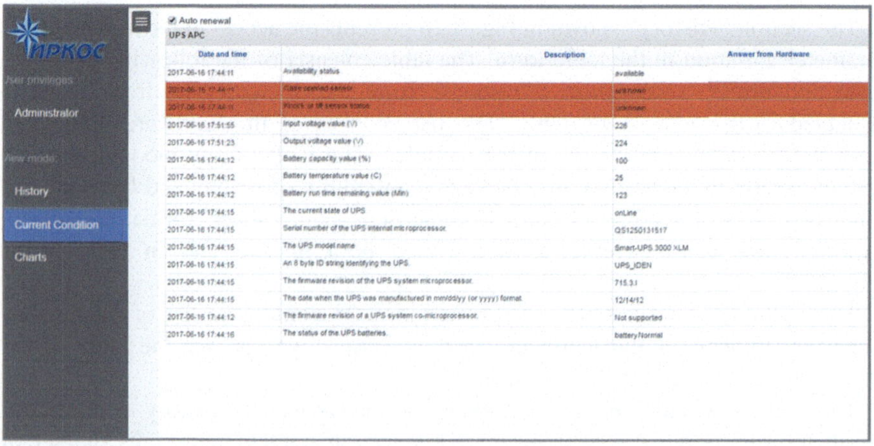

Fig. 3.27 "Current condition" section

Disappear of communication with the hardware being monitored is also registered by the sub-system, which forms notifications to the operator and is represented in the graphs. Figure 3.28 shows the graph window with the time interval, in which hardware was unavailable.

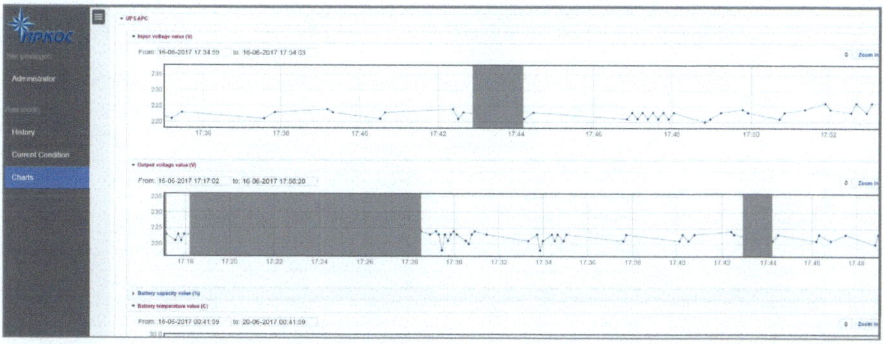

Fig. 3.28 "Charts" section

Besides the statistics gathering, the self-diagnostics sub-system provides the permanent comparison of the hardware parameters with the threshold values, if they are defined. In case of exceeding the defined value, the notification to the operator is formed. An example of the diagram, on which the case of exceeding of permissible temperature threshold for the UPS battery is set, is shown in Fig. 3.29.

Notifications to the operator can be sent by three methods: SMS messages to the preset phone numbers, interpretation of the events in the SMO-APMADA window as pop-up controls and information in the database table, and messaging via e-mail. Figure 3.30 shows the SMO-ARMADA window with the section of event interpretation registered by the self-diagnostics sub-system.

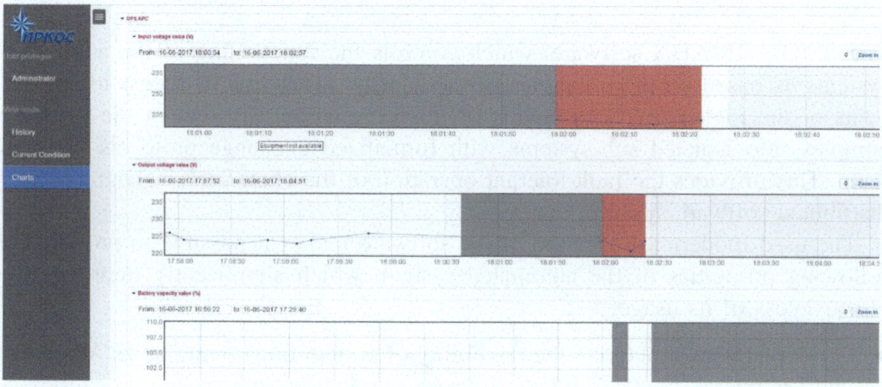

Fig. 3.29 Values exceeding limits of the set threshold

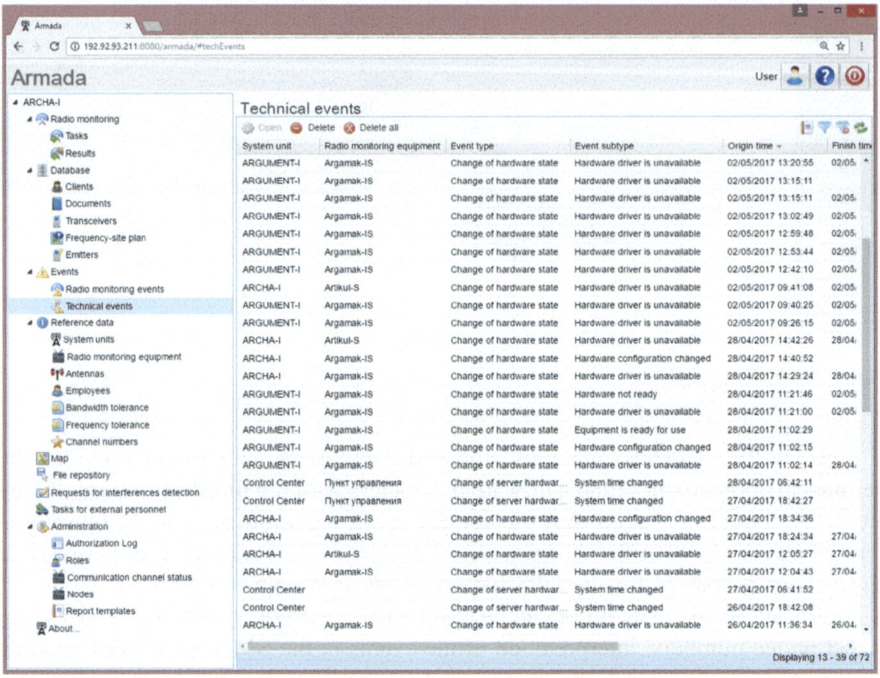

Fig. 3.30 Displaying of the technical events in the SMO-ARMADA window

3.19 Summary

The SMO-ARMADA software, which controls the functioning of the automated systems, is based on the principles of modularity, restriction of access to components of the higher levels to details of implementation, division of the software complex into isolated sub-systems with formalized exchange protocols between them. This provides the fault-tolerant operation of the client/server geographically distributed software system.

The used modern architecture of the software makes it possible to provide the following properties of the automated system, which significantly improves the effectiveness of its usage:

- Cross-platform structure—the possibility of system functioning under Windows and Linux;
- Scalability—the possibility of system deployment both on the separate computer and at the national level scale;
- Fault tolerance—maintenance of the system efficiency under conditions of the partial equipment fault and/or essential degradation (up to loss) of communication channels, automatic control of the equipment, and the communication channel statuses;

- Openness—the possibility of radio monitoring data exchange with other systems by means of exchanging formats and/or protocols, as well as the possibility of usage of radio monitoring equipment of different manufactures;
- Simplicity of the software updating—it is executed on the central system server, and further, the update is carried out automatically on the subordinate servers;
- Separate usage of different agencies' data. Each system operator should be associated with the definite agency (or several agencies), which excludes a possibility of non-authorized access to the data from the other agency sections.

The controlling software is obeyed to the modularity principle. We can single out the following main components of the SMO-ARMADA:

- Radio monitoring;
- Accounting and reference data;
- Information about events;
- Formulation of missions for personnel and registration of requests for search for interferences;
- Cartography;
- Data transmission;
- Generation of reports;
- Data storage;
- Data exchange;
- Administration;
- Update;
- Self-diagnostics.

As the main directions of development of the SMO-ARMADA software, we are planning in the nearest future to modernize the radio monitoring operative mode based on Web technologies, providing the possibility of radio monitoring equipment usage in the real time (in the operative mode) simultaneously by several users, as well as to support the equipment of new generation by our both own development and the external manufacturers.

References

1. Avdushin AS, Alekseev DA, Ashikhmin AV, Zhukov AA, Kozmin VA, Pyatunin AN, Sysoev DS (2016) Software architecture of the spectrum monitoring system (in Russian). Spetstehnika i Svyaz (4):115–128
2. Bocharov DN, Kozmin VA, Korochin SV, Provotorov AS (2016) Self-test system for equipment of automated radio monitoring system ARMADA (in Russian). Spetstehnika i Svyaz (4):90–97
3. SMO-DX software package for indoor radio monitoring and radio signal parameter evaluation. http://www.ircos.ru/en/sw_dx.html. Accessed 28 Nov 2017
4. SMO-PA/PAI/PPK panoramic analysis, measuring and direction finding software package. http://www.ircos.ru/en/sw_pa.html. Accessed 28 Nov 2017

5. SMO-KN navigation and cartography software package. http://www.ircos.ru/en/sw_kn.html. Accessed 30 Nov 2017
6. SMO-ARMADA software package of automated spectrum monitoring systems. http://www.ircos.ru/en/sw_armada.html. Accessed 28 Nov 2017
7. Recommendation ITU-R SM.1537. Automation and integration of spectrum monitoring systems with automated spectrum management
8. Recommendation ITU-R SM.1139. International monitoring system
9. Recommendation ITU-R SM.1370. Design guidelines for developing advanced automated spectrum management systems
10. SMO-BS software packages for wireless data communication system analysis. http://www.ircos.ru/en/sw_bs.html. Accessed 28 Nov 2017
11. SMO-STA software package radio signal technical analysis. http://www.ircos.ru/en/sw_sta.html. Accessed 30 Nov 20
12. Developer Guide. Protocol buffers. Google developers. https://developers.google.com/protocol-buffers/docs/overview. Accessed 28 Nov 2017
13. Garmin Proprietary (2006) NMEA 0183 sentence technical specifications. Garmin International, Inc.
14. GPS 18 (2005) Technical Specifications. Garmin International, Inc.
15. GeoServer documentation. http://docs.geoserver.org. Accessed 28 Nov 2017
16. OpenStreetMap foundation. http://wiki.osmfoundation.org/wiki/Main_Page. Accessed 28 Nov 2017
17. Mauro DR, Schmidt KJ (2005) Essential SNMP, 2nd edn. O'Reilly Media, p 442
18. Rihards O (2016) Zabbix Network Monitoring. 2nd edn. Packt Publishing, p 754

Chapter 4
Engineering-Technical Infrastructure

4.1 Introduction

The engineering-technical infrastructure consists of engineering constructions necessary for the ARMS functioning, in particular, control centers, control points, service rooms for units arrangement, prepared special platforms for antenna-mast devices mounting, systems, and lines for data transmission, server equipment, etc. This chapter is devoted to a description of the components of the ARMS engineering-technical infrastructure including an arrangement of control centers and control points, radio monitoring points, communication, and data transmission [1].

4.2 Control Center and Control Points

The automated radio monitoring system is obeyed to the hierarchical principle and may have (except radio monitoring points) control centers and control points, which are responsible for the ARMS elements interaction, as the nodes in its structure. The own specifics of engineering infrastructure correspond to each level of this hierarchy.

The control center represents the complexity of the hardware–software units providing organizational and information interaction and functioning of the ARMS elements. It gives a possibility to analyze the current system state, to plan operation scenarios, to make management decisions, and to assign tasks for radio monitoring and data processing. The control center interacts with control points (which are the same control centers in essence, but they have a lower lever in the hierarchy and always subordinate to the higher-level node) and, if necessary, directly with separate radio monitoring equipment.

The control center can be fixed or have a mobile implementation. For fixed control center, the separate building (or its part), where hardware equipment can be

A. M. Rembovsky et al., *Radio Monitoring*, Signals and Communication Technology, https://doi.org/10.1007/978-3-319-74277-9_4

Fig. 4.1 Mobile control center

arranged and operating personnel can be situated, is required. The mobile control center can be implemented on the base of a mechanical vehicle, in which possible external view is presented in Fig. 4.1. The mobile control center requires the availability of wireless channels of communication with nodes.

The control center of the automated system necessarily has the sub-systems of control, communication and data transmission, and electric power supply in its structure. The possible structural scheme of the control center is presented in Fig. 4.2.

The sub-system of control includes the operators' automated workstations (AWS) and control servers. The AWS represents the control devices and displaying devices (notebooks, fixed and tablet computers), which have access to the ARMS via the wired or wireless communication channels. The possible view of the AWS is shown in Fig. 4.3. The number of ARMS workstations depends on the ARMS scale, the number of control organizations and agencies, which use this system, as well as the number of the employee. The central element of the SMO-ARMADA [2] client/server software, the main database, in which measurement results from all system nodes are stored, as well as additional services (server of the state

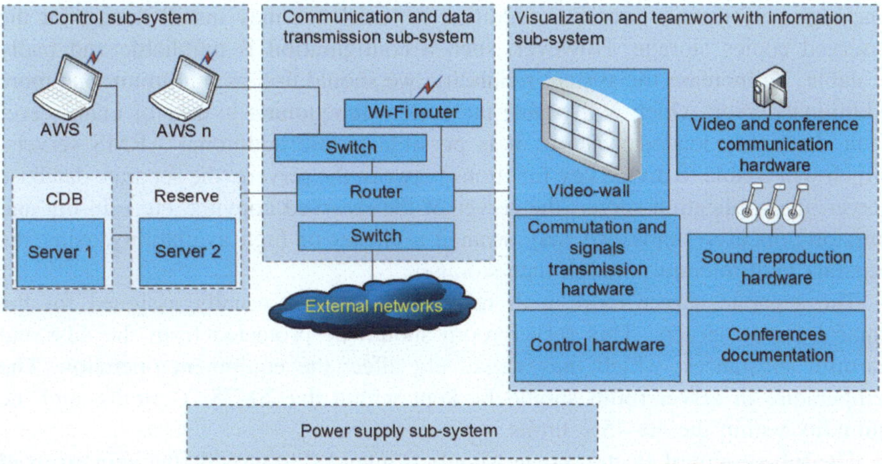

Fig. 4.2 Structural scheme of the fixed control center

Fig. 4.3 Operator's automated workstation

monitoring, server of the exact time, etc.) are located on the ARMS control center servers (see Chap. 1). The software of the ARMS servers provides coordinated operation of the subordinated nodes of the system as well as allows controlling each of them.

The number of the ARMS servers, their configuration, and destination may be different depending on the system scale, data volumes and requirements on reliability and fault tolerance. In the simplest case, all main functions are performed by one high-performance server, in addition to which we may install device for the reserved copies storage. However, such a configuration is unreliable and badly scalable. To increase the system reliability, we should use, as minimum, one more additional server, which will enable the system functioning in case of main server fault. In highly loaded systems, it is possible to use additional ARMS servers, which will enable to divide key functions between the servers (the separate database server, the application server, the server of the reserved copying, etc.). In big and mission-critical systems, we may arrange a cluster of high availability using the separate fault-tolerant network data storage.

The separate server room with restricted access is usually assigned for the control center servers. This server room should be protected from the dust and harmful substances, which may negatively affect the equipment operation. The temperature in server room should be kept within the 20–25 °C limits, and the humidity within the 40–45% limits.

The sub-system of electric power supply is intended to provide the uninterrupted operation of the ARMS servers and AWSs in the control center; therefore, the equipment of the reserved power supply is its necessary component. Depending on the maximal power, it may be the diesel-fuel electric stations, uninterrupted power sources on accumulators, including renewable ones from the solar stations. The separate rooms or buildings should be anticipated for the equipment of the reserved power sources. The diesel generators are vibration sources; therefore, they negatively affect the operation of the active server equipment, the contacts, and the connections (the oscillation amplitude should not exceed 0.1 mm in the frequency range below 25 Hz). Therefore, we must provide their maximal remoteness, as well as generator location on the special platforms, at the control center designing. When accumulators are used, the air conditioning and air ventilation must be provided in the server room as well as temperature within the 15–25 °C limits. These conditions extend the life cycle of rechargeable batteries and provide their maximal output capacity.

The sub-system of communication and data transmission plays a crucial role. It fulfills the following functions:

- High-speed traffic routing (commutation);
- Reservation at the level of the equipment and channels;
- Load balancing on parallel channels;
- Fast switching between main and reserve channels;
- Effective usage of the connection bandwidth.

If it is necessary to protect the transmitted data additionally, the central node may appear as VPN-server for the nodes of the next level. The high-performance routers with the ciphering hardware support are the best for usage as VPN-servers. The IPSec, OpenVPN, and PPTP protocols are the examples of the protected VPN protocols.

To increase the reliability of the data transmission system, the key elements of the network and communication lines must be reserved. Figure 4.4 shows the possible variant of the fault-tolerant network connection organization of the central node communication equipment.

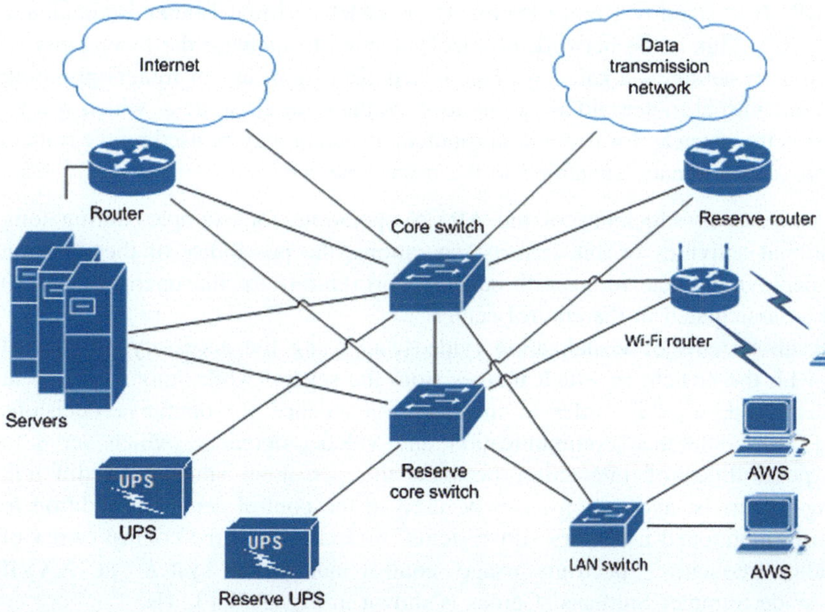

Fig. 4.4 An example of the connection scheme of the data transmission node network equipment with the hardware reservation

From the presented scheme, we see that the main network equipment has a "hot" reservation and additional links, which enable the communication central node to continue its operation in case of some system element fault, whether it is the network equipment or one of the communication lines. So, for instance, core switches and routers are duplicated, and the AWSs connections may be performed in two ways: via the wired connection and Wi-Fi.

From the software point of view, the following protocols and technologies are used at the development of the similar schemes:

- IEEE 802.3ad (Link Aggregation) allows joining of several physical connections into the one or several logical ones in the Ethernet. It is used in switchers, routers, and servers for reservation of physical connections and increasing of the connection capacity.
- STP (Spanning Tree Protocol) is the protocol allowing avoidance of the loops in the Ethernet networks, in which there are one or more network bridges excessively linked. It may be used to duplicate connections as an alternative to 802.3ad.
- IEEE 802.1Q is the standard describing the traffic tagging procedure for transmission of the information about belonging to the virtual network (VLAN). It is used for transmission of the data of several networks via one logical connection.

- HSRP (Hot Standby Router Protocol) or VRRP (Virtual Router Redundancy Protocol). This is the network protocol intended to increase the availability of the routers acting as a gateway. This is achieved by means of routers grouping into one virtual router and assigning to them the same IP-address, which will be used as the gateway for network computers. Protocol may be used by the routers to provide automatic switching to the reserve device.

To increase the efficiency of the ARMS operation, for example, during long international activities or anti-terrorist operations, the possibility of the situation center deployment (the room with several workstations for the operative group) should be anticipated in the control center.

The sub-systems of visualization (video-wall using the necessary number of panels with the switch, in which images from the several workstations may be at once displayed, used in order to increase convenience for operative decisions making), of conference communication and videoconferences, which serve to widen possibilities of interaction between the personnel situated in different buildings, districts, and regions, can be used in the control center in addition to previously mentioned necessary sub-systems. An example of the control center of the radio frequency spectrum usage control automated system at XXVII World-Wide Summer Students' Games is shown in Fig. 1.31 [3, 4].

4.3 Radio Monitoring Station

Radio monitoring stations represent the engineering constructions for deployment of the fixed equipment of radio monitoring, which include radio receiving equipment (antenna systems and receivers) to be located outdoors, and equipment for processing and control (controller, communication equipment, and system for electric supply) to be placed both inside the technical rooms and outdoors in the temperature-controlled waterproof secure cabinet (Fig. 4.5a). If the radio monitoring station is maintainable, the additional rooms for the personnel and the AWSs should be anticipated (Fig. 4.5b). The radio monitoring equipment, which is not intended for outdoor installation, can be located in special service rooms where necessary temperature conditions for the normal functioning are kept.

In big towns, we have to use already existing buildings for equipment arrangement (Fig. 4.6a). Beyond the complicated urban conditions, the masts with antenna systems can be mounted directly on the ground (Fig. 4.6b) or on the cellular tower (Fig. 4.6c). In this case, the additional constructions can be raised for arranging other equipment (barriers, boxes, and buildings with rooms for personnel in case of maintainable radio monitoring stations).

The uninhabited buildings are preferable to be used as the radio monitoring stations, because their usage will essentially reduce different agreements with its owners. The search for a suitable place in the urban conditions is a complex task, since we must take into consideration the following factors:

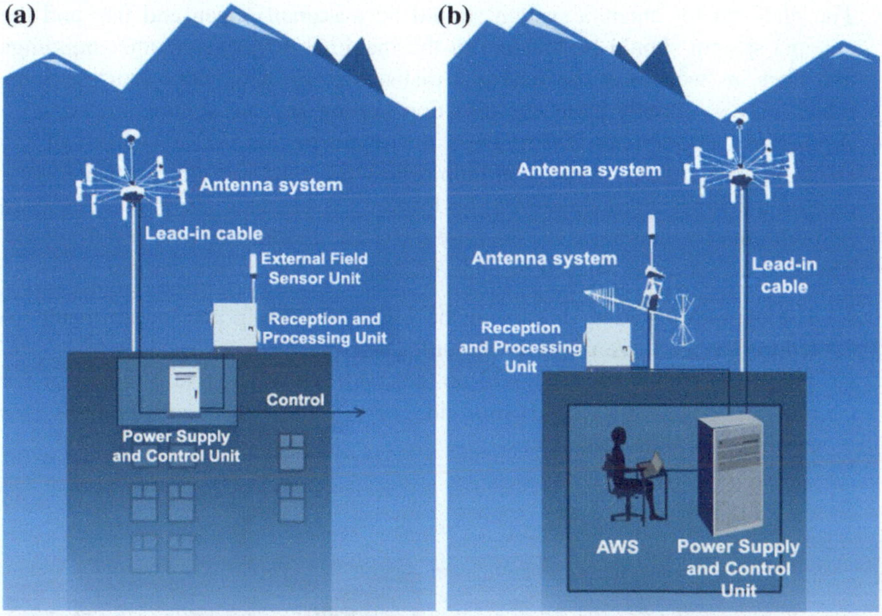

Fig. 4.5 Scheme of the radio monitoring station: **a** unmanned, **b** maintainable

(a) (b) (c)

Fig. 4.6 Variants of antenna system location in the radio monitoring stations: **a** on the building roof, **b** on the ground, **c** on the cellular operator tower

- The platform for antenna system should be maximally open and flat, and the antenna system should be located on the maximal height over other buildings and trees, as well as at the maximal distance from other metal constructions, which may essentially affect the direction-finding and measurement accuracy.
- At distance closer than 200 m from the platform, the active sources of the electromagnetic field (radio communication transmitters, electric generators, electric machines, etc.) should not be located [5]. Besides, the platform should be preliminarily examined for the absence of the electromagnetic interferences.
- The roof construction should provide installing of the masts, namely:
  - The permissible load should not be less than 400 kg per square meter (to reduce the load, we may use the unloading platforms—Fig. 4.7).
  - The mast bracing wires should be attached to the concrete blinding of the roof, the concrete (bricked) roof side, or to rigidly fixed constructions (see examples in Fig. 4.8);

Fig. 4.7 Variants of the unloading frames

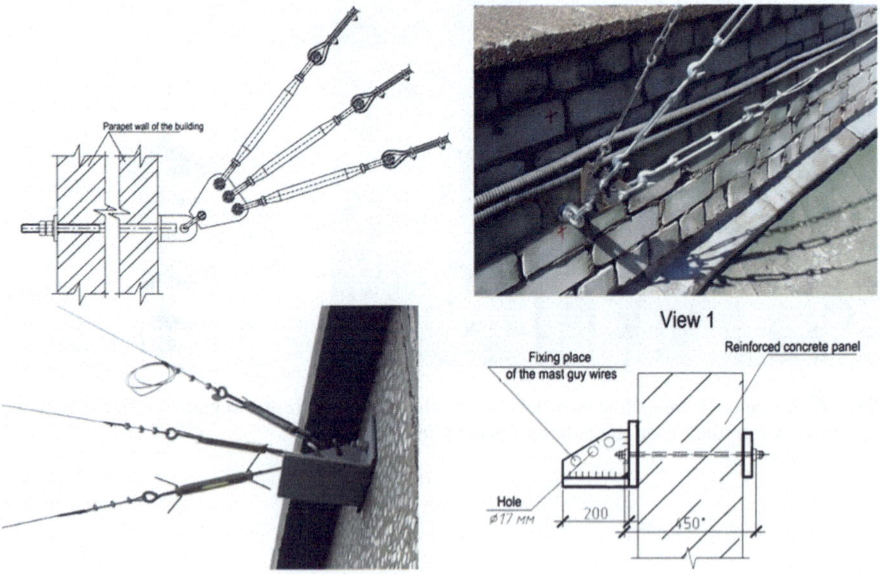

Fig. 4.8 Variants of mast bracing wires attachment

- The angle within the 30–45° limits between the bracing wires of the upper tier and the mast vertical axis should be kept;
- The appropriate azimuth angle (for example, 120° if the mast is kept by three bracing wires) should be kept.

The mobile, portative, and manpack equipment are deployed at the temporal posts; therefore, no engineering constructions are required for them. For their operation in the system, it is enough to have communication channels with other nodes.

In any radio monitoring station, the stations of connection to the electric network and to the LAN for the radio monitoring equipment and the grounding of all equipment should be anticipated. Additionally, it is possible to connect the mast body and all bracing wires to the protective ground. To protect operator against the electric trauma caused by the lightning hit, the antenna systems are connected to the circuit of lightning protection ground. To reduce a probability of equipment damage caused by lightning, the lightning rod can be installed on the additional mast near the antenna system. However, it should be taken into account that installation of the lightning rods in close quarters to the antenna-feeder systems leads to degradation of the radio monitoring equipment performance.

The radio monitoring stations should provide protection from access and actions of the unauthorized persons. For example, fixed complexes and ARCHA stations have the embedded sensors for alarm signaling (hit, tilt, openness/closeness) in their structure, at which activation the technical event is formed in the database of the ARMADA ARMS, and the operator urgently receives the notification via e-mail and/or SMS.

4.4 Data Transmission System

The data transmission system represents a group of nodes, which are united by the data transmission network, capable to be controlled remotely and to function autonomously in accordance with the stated requirements. All interaction within the limits of the data transmission system is based on the TCP/IP protocols stack and can be realized with the help of the standard network components.

Depending on the dimensions and complexity, the ARMS data transmission system may represent both the simple LAN within the frameworks of the separate radio monitoring complex and the complicate multi-level network with a number of nodes of different types, as well as with complicated structure of communication channels and configuration of the server equipment.

The data transmission system may contain the following types of nodes:

- The central node (the control center);
- The control point;
- The stationary communication node (the fixed radio monitoring station);

- The mobile node (the mobile radio monitoring station);
- The remote automated workstation.

In general case, the connection scheme has a tree-like structure shown in Fig. 4.9. All nodes of the lower level are linked to the higher-level node grouping in a star way. The main advantage of this topology is that such a network does not require additional communication lines and is easily scaled and controlled and provides troubleshooting.

To increase the connection reliability or in the cases, when connection between the nodes of the same level can be realized easier or more economically profitable than with the higher-level node, we may use the combined topology and integrate nodes of the same level into a ring, as it is shown in Fig. 4.10.

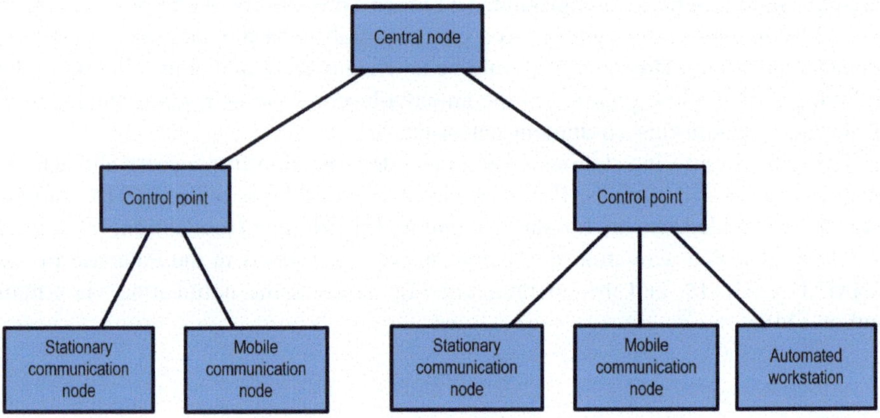

Fig. 4.9 Integration of nodes into the network with the tree-like structure

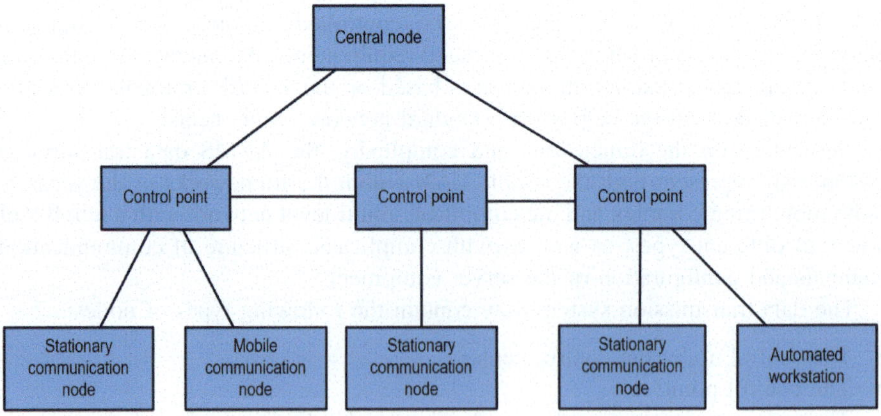

Fig. 4.10 Integration of control points into the ring

To provide high reliability, scalability (possibility to expand or to adjust the network with minimal expenses), and high performance at the development of the data transmission network, we use the multi-level architecture based on the following principles:

- The hierarchy: the network is divided into several levels, and each level fulfills the definite functions.
- The modularity: levels are constructed on the base of separate modules, and each module represents the functionally completed unit, which fulfills functions of the appropriate level.

To provide independence of the data transmission system nodes, the equipment of each node is located in the separate routed sub-network. The network router, which performs the network equipment high-speed switching and provides the packet delivery to the system remote elements, is located at each of the nodes. Besides the main router of the communication node, the radio monitoring equipment may be equipped with additional equipment for the data transmission. Wired channels, as a rule, are connected directly to the router or, in case of connection to the optical communication links, using media converters. To arrange wireless data transmission, the complexes are equipped with communication equipment including the wireless router or VSAT terminal when using the satellite communication. Each radio monitoring station can be connected to the common ARMS network with usage of one or several channels for data transmission. When several communication channels are used at one complex of the radio monitoring unit, one of the communication channels is the main and others are reserved. At that, the main router of the communication node provides automatic switching to reserve channels in accordance with the given priority.

In complicate multi-level schemes, the protocols of dynamic routing (RIP, OSPF, EIGRP, etc.) can be used for network route exchange between the nodes.

4.5 Data Transmission Channels

The communication channel, which provides traffic transmission between the nodes, is the main component that links the nodes of the data transmission network. Usage of widely spread protocols for data transmission of the TCP/IP stack enables to develop networks basing on the standard network components and, if necessary, to use already existing network infrastructure.

Both wired and wireless communication lines can be used for data transmission between the system nodes.

The most common wired channels can be relatively divided into the following groups:

- Dedicated communication lines—the optical or copper cables, which connect user network nodes (these may be both own lines and leased communication lines);
- Dedicated data channels—the data channels, which are provided by the communication operator as well as the own data transmission network (frame relay (PVC), ATM (PVC), E1/E3/STM-1, Ethernet VLAN);
- Services on connection basing on the "group" access, for instance IP VPN, Virtual Private LAN Service (VPLS), and the Internet network.

Wireless technologies are used in the cases, when wired technology usage is difficult or impossible due to technical reasons (for example, for connection of the mobile nodes), or with the aim to arrange reserve channel for data transmission. At that, the following solutions can be used:

- The cellular GSM communication (GPRS, EDGE, 3G, and LTE);
- The HF/UHF communication channel;
- The Wi-Fi radio channel;
- The satellite radio communication.

Utilization of cellular public networks for the operation of the data transmission automated systems is highly attractive since it does not require serious initial financial and temporal expenses for communication infrastructure deployment. So, the data transmission via the cellular networks (using high-speed 3G/4G modems) can be used as the main method for mobile radio monitoring stations. For stationary nodes located in areas with urban cable channels, the GSM/3G-channel can be used as the reserve one. In areas with poor reception, we can use specially directed and non-directional antennas and signal amplifiers. If the communication node is located in the coverage area, the data transmission channels arranged via the cellular networks of third and fourth generations enable to provide high-speed data exchange with remote radio monitoring stations and to solve all necessary tasks. In particular, the LTE standard owing to MIMO technology provides high-performance indices both for fixed and mobile users, which move with the speed up to 120 km/h. To increase reliability of the data transmission channel, we can apply the simultaneous connection to various cellular networks and support the automatic transition to the SIM-card of the reserve operator with automatic return to the SIM-card of the main operator in case of cellular networks operation faults.

Figure 4.11 shows an example of organization of communication with mobile radio monitoring complexes on the base of compact LTE routers of iRZ Company (RL11w). These devices provide a stable connection to modern mobile networks and support main standards of wireless high-speed data transmission via the cellular networks (LTE/HSPA+/UMTS/EDGE/GPRS). They have a wide range of operation temperatures and small dimensions and are suitable for location in the automobile.

The main advantages of communication channel arrangement via the cellular network are:

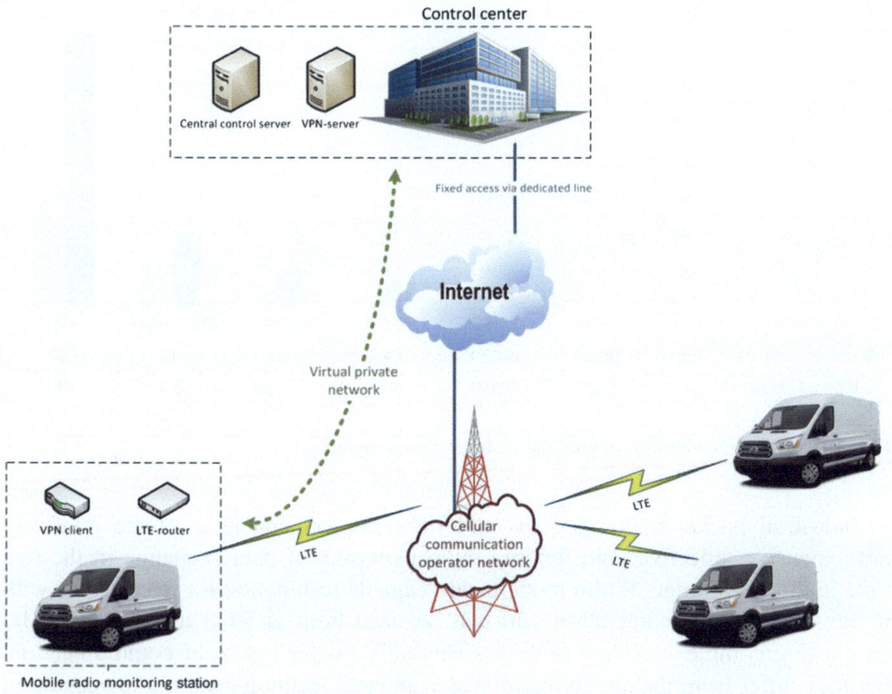

Fig. 4.11 An example of communication organization on the base of the LTE routers

- Low cost of deployment;
- Low weight and small dimensions of data transmission devices;
- Simplicity of installation and adjustment;
- High capacity (up to 75 Mbit/s in LTE networks).

Disadvantages of this variant of the wireless access are:

- Territorial restriction by the cellular networks coverage area;
- Additional exploitation expenses on mobile operator services;
- Dependence of the channel capacity to be achieved maximally upon the standards and technologies, which are used by the cellular communication operator (see Fig. 4.12);
- Data transmission is performed not directly between the system nodes, but via the operator networks and, in general case, via the Internet, which requires adjustment of the protected VPN network with the location of one or several VPN-servers at the control stations and the control center.
- The absence of guarantee of communication continuity due to the fact that operation of the radio network of the cellular communication depends on the unpredictably changed current load (a number of simultaneously active mobile users) to a great extent.

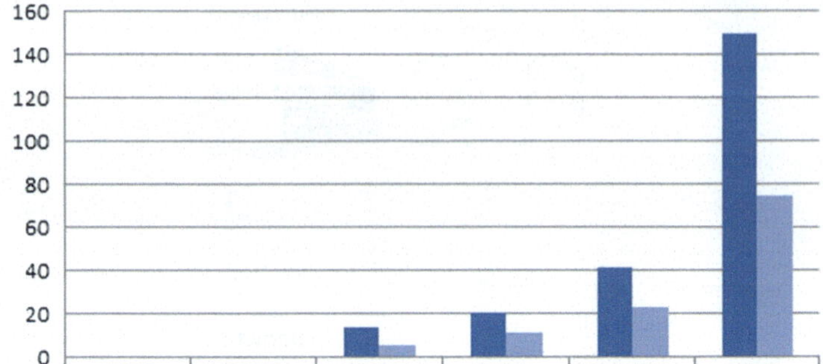

	GPRS	EDGE	HSDPA	HSPA+	DC HSPA+	LTE
Download rates (Mbps)	0.02	0.059	14.4	21	42	150
Upload rates (Mbps)	0.02	0.059	5.76	11.5	23	75

Fig. 4.12 Data transmission rate of the cellular communication

Industrial packet radio modems and radio routers enable to create relatively not-expensive, effective, and flexible radio networks of data exchange in the real time in the UHF range. Radio modems are capable to function for many years with minimal technical maintenance and can be used both at fixed and mobile radio monitoring complexes. They provide essentially bigger radius of communication, but they differ from the above-mentioned wideband technologies in a lower rate of data transmission. The highest operation rate and reliability are achieved at the direct radio visibility between the objects (at distance up to 30 km in the open country and up to 10 km under urban conditions with average housing density). At utilization of fixed radio networks of exchange in the structure, radio modems function under conditions of minimal variations of signal reception conditions. It helps to manage without special methods, which can be used to increase noise immunity, and to use devices, which function simpler.

An example of radio network created with usage of Viper-100/400 radio modems and capable to function at a rate up to 128,000 bit/s is presented in Fig. 4.13. At providing of electromagnetic availability between the positions of all communication nodes, this scheme turns out to be very reliable, since Viper-100/400 radio modems enable to adjust the automatic route change mode of information transmission in case of any system element fault.

Radio modems, which anticipate the automatic transfer of mobile objects from the operation zone of the one base station to the operation zone of another base station, the guaranteed delivering of transmitted messages by their own embedded units, and the automatic distribution of radio network load and many other functions related to the network operation, are used in the structure of mobile radio networks. Technical problems, related to increasing of data exchange rate, have got a solution in the modern radio modems, which use the technology of the "Parallel Decoding/Smart Combining" for radio signals. They use two receivers with two spaced antennas to manage with signal fading. This principle is implemented,

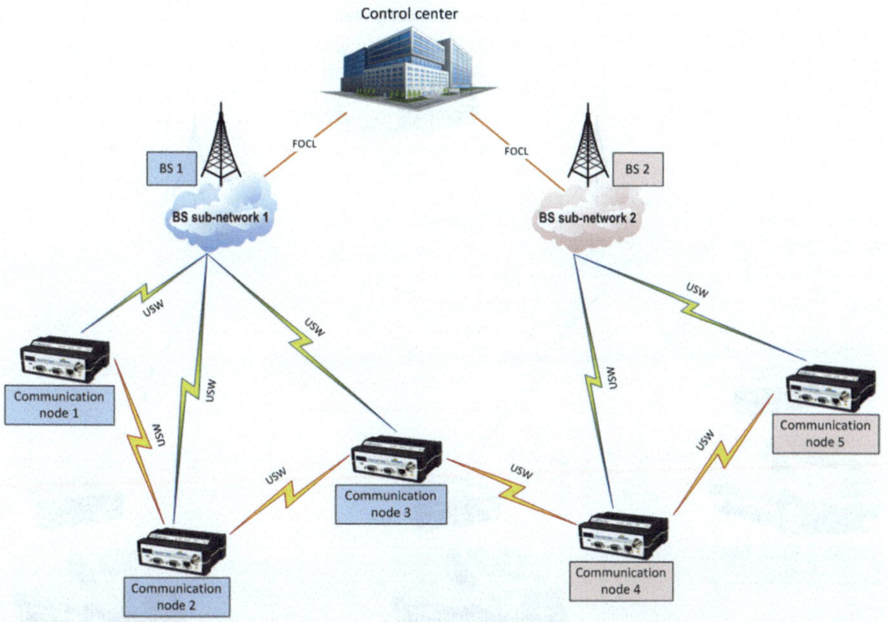

Fig. 4.13 Scheme of stationary radio network for data exchange with increased reliability and survivability, based on the Viper-100/400 radio modems

in particular, in ParagonG3/GeminiG3 radio modems, which provide data exchange with maximal rate of 57.6 kbit/s (the UHF range), 64 kbit/s (800 MHz), and 128 kbit/s (700 MHz, the channel with radio frequency grid step of 50 kHz). Operation in the radio network is arranged according to the UDP or TCP/IP protocols with automatic data compression. Usage of OOB (Out-of-band signaling) allows significant increase of the number of employees, which operate in one radio channel, at the expense of effective assignment of the frequency resource and automatic transmission of navigation information at each communication session. The protocol of data exchange anticipates a possibility of initiative transmission both from the base station side and from the mobile node. In case of loss of communication with the base station, the onboard radio modem automatically searches for another base station and commutates with it. An example of the network created on the base of ParagonG3/GeminiG3 radio modems is presented in Fig. 4.14.

The main advantage of narrow-band radio networks is their independence from the third party communication infrastructure and a possibility to develop it starting from the actual requirements (the radio network belongs to the user himself, its operation parameters and operation zone may be changed by him independently). However, low capacity and peculiarities of signal propagation in UHF ranges do not enable to use such a kind of communication as the main channel of data transmission in big high-loaded ARMSs. The UHF communication will rather be

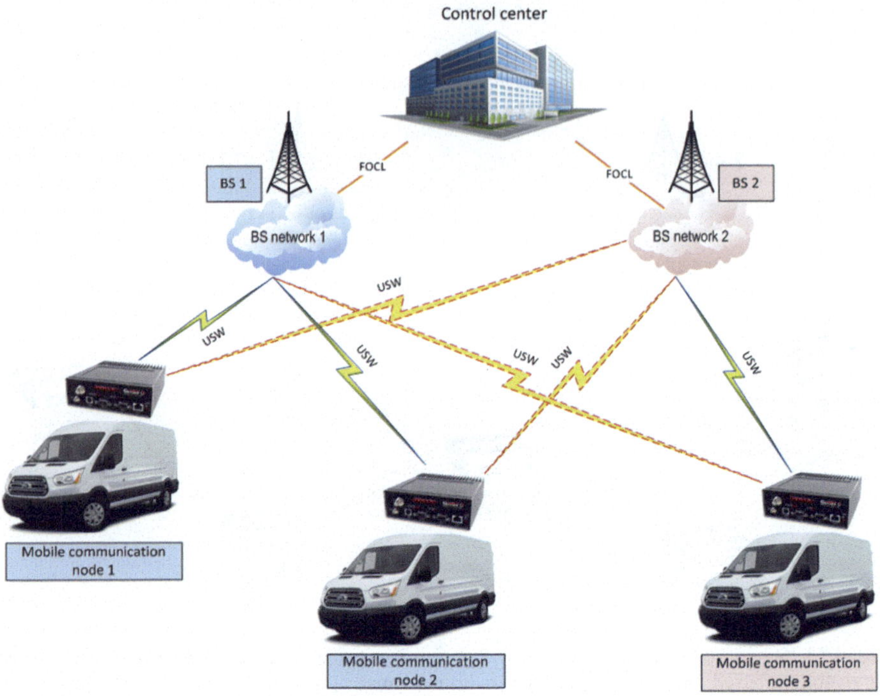

Fig. 4.14 Scheme of the mobile radio network with increased reliability and survivability, based on the ParagonG3/GeminiG3 radio modems

convenient to organize emergency channel in such systems or to be used as main channel for small systems or systems with a limited set of functions.

The wireless Wi-Fi technology is used for organization of local computer networks of small operation radius. Such widely spread protocols of wireless data transmission as IEEE 802.11b, IEEE 802.11g, IEEE 802.11a, and IEEE 802.11n are suitable for organization of the wireless computer network inside the rooms or in the streets with operation radius up to 300 m, but problems arise at these protocols usage for data transmission at large distances. The CSMA/CA (Carrier Sense Multiple Access with Collision Avoidance) media access method, which is used in IEEE 802.11a/b/g/n standards, assumes monitoring of the carrier frequency by means of broadcasting before starting of data transmission. At that, the mechanism of delivery confirmation of each transmitted packet is used to avoid conflicts. The more transmitters listen to each other and the less distance is between them, the more accurately they can synchronize their actions, not allowing collisions at transmission. But when transmitters are spaced by large distances, the complicated situation arise, known as the problem of the hidden node, when clients communicate with the base station only, but cannot communicate with each other, which seriously complicate the reservation.

For wireless data transmission at large distances, the Nstream and NV2 polling protocols developed by Mikrotik Company are suitable. Polling is the alternative method of the media access, which solves the hidden node problem. At its usage, the client devices start data transmission only after receiving the special permission packet (marker) from the base station. The adaptive algorithm minimizes a number of service information transmitted in the network, adjusting the marker transmission rate for each mobile user depending on its activity and the network total load. As a result, the Nstream protocol provides:

- Low level of overhead in the packet title, which enables to increase data transmission rate;
- The absence of limitations in distance and rate of data transmission;
- Dynamic protocol adjustment depending on the type of transmitted data and used resources and, as a consequence, increasing of factual data transmission rate at the expense of improved traffic compression.

The NV2 protocol is a logical continuation of the Nstream protocol. It is based on the modern digital technology of TDMA (time division multiple access) signal transmission, which uses parallel access with time distribution, when unique time slots within the limits of one channel are assigned to each client. This technology enables to access one radio frequency channel simultaneously by a large number of clients. The base station transmits right away the schedule, according to which the operation time for all clients is assigned, thus remaining more time for factual data transmission. In essence, this is the group polling. As a result, the NV2 protocol compared to Nstream protocol:

- Allows schedule creation for clients on the base of assumed distances (the delay of signal propagation) and the QoS (Quality of Service) function;
- Monitors the operation of adjacent wireless networks and tries to reduce collision by adjusting the schedule for clients;
- Provides the own embedded system of network safety based on the hardware ciphering, which uses the AES-CCM protocol of block ciphering with the 128-bit key.

All this provides effective radio channel usage and significantly increases capacity, especially in complex conditions. Although, it does not make the data transmission a protected one from the unauthorized access. To ensure the wireless network safety, it is necessary to use special methods of authentication and ciphering.

We should note also that owing to better penetrating capability in the household or office computer Wi-Fi networks, the 2.4 GHz frequency range is often used, while to create the wireless networks for large distances, the 5 GHz frequency range is more suitable. Waves with 5 GHz frequency are worse at overcoming such obstacles as tree foliage. However, in this frequency range:

- There is a big quantity of relatively non-overlapping channels.
- The data transmission rate is higher due to the increasing of informational density of packets.
- The antenna patterns are narrower, which increases the signal propagation distance and the system noise immunity.

As an example, we consider organization of communication in field conditions between three radio monitoring stations with the help of four wireless routers of the external implementation Mikrotik RB911G-5HPnD-QRT (QRT5) with the transmitter power of 1 W and integrated antennas 23dBi MIMO 2 × 2 in 5 GHz range with antenna pattern 10° × 10°, operating in the 802.11 b/g/n MIMO 2 × 2 mode, having the gigabit network Ethernet 10/100/1000M ports and supporting the NV2 protocol. The scheme of radio monitoring stations location is shown in Fig. 4.15. The photograph of the relief at the radio wave propagation path is presented in Figs. 4.16 and 4.17 show the results of Fresnel zones analysis for the A-B segment at antenna height of 10 m.

To connect B-A and A-B points, the scheme of the "point-to-point" type is used in order to arrange the communication at distances up to 30–40 km, using narrow-directed antennas if necessary. Testing of the rate characteristics showed that average packet delay was 7 ms at the B-A segment, near 3 ms at the A-C segment; the rate for transmission (tx) and reception (rx) was of 190 Mb/s order. At that, the time for deployment of B and C points and for communication settling with A point was 10 min at average.

When planning to use the wireless technologies to ensure the operative communication between mobile and fixed radio monitoring stations, we should take into consideration the following definite circumstances complicating their usage:

- The direct or partial visibility is necessary for stable connection.
- The unstable operation is possible under strong interferences conditions.

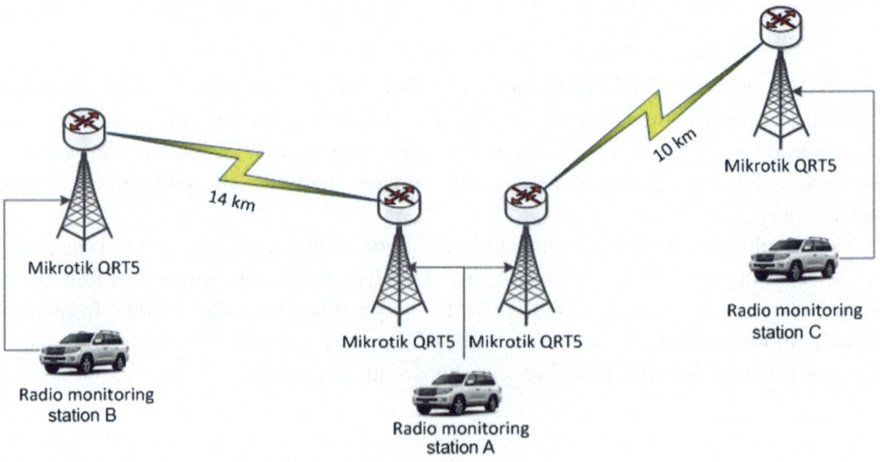

Fig. 4.15 Scheme of location of radio monitoring stations

Fig. 4.16 C-A segment of the wireless communication path

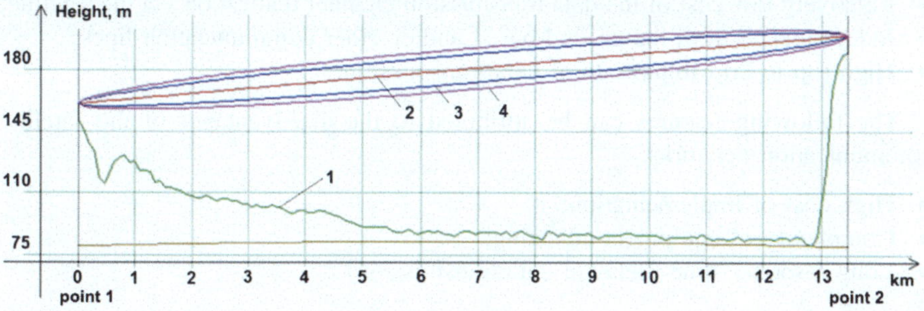

Fig. 4.17 Comparison of the Fresnel zones for ranges of 2.4 and 5 GHz. 1—the relief of Earth surface; 2—the direct visibility between two antennas; 3—the Fresnel zone for the 5 GHz range; 4—the Fresnel zone for the 2.4 GHz range

- The difficulties for direction positioning between network nodes arisen under the condition of communication at large distances and when using narrow-directed antennas.
- It is difficult to achieve the communication stability under conditions of the strong fog or moist snow.

The satellite communication channels can be used for high-speed data transmission between geographically distributed objects in regions, where there are no other suitable communication systems. An example of communication channel realization via the satellite networks is shown in Fig. 4.18.

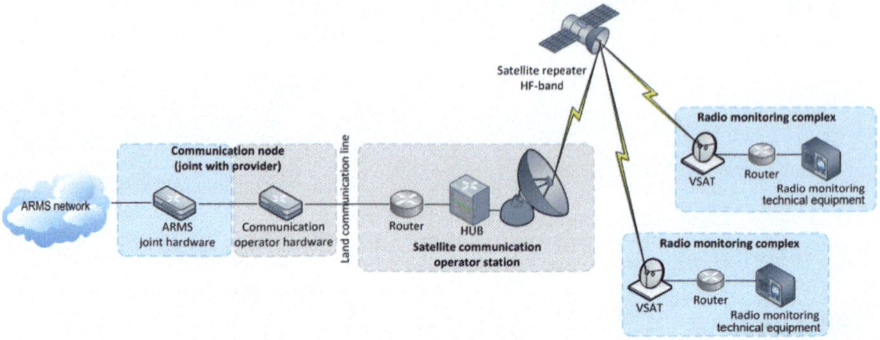

Fig. 4.18 An example of communication realization via the satellite network

The advantages of the satellite communication networks are:

- Quickness of the data transmission (via the satellite) channel deployment;
- Wideness of geographical coverage of territories, in which data transmission via the satellite is possible;
- Relatively low cost of the data transmission channel realization via the satellite;
- Independence from the ground-based and/or other communication lines;
- High (up to 20 Mbit/s) rate of data transmission.

The following features can be attributed to the disadvantages of the satellite communication networks:

- High cost of implementation;
- Complexity of equipment adjusting;
- Long response time (delay at data transmission).

4.6 Summary

Effectiveness of the automated radio monitoring system operation depends not only on perfection of existing radio monitoring equipment, but on the features of functioning of numerous auxiliary sub-systems, which comprise the engineering infrastructure.

The automated radio monitoring system is constructed basing on the hierarchical principle. The control center has stationary or mobile implementation. It interacts with the control points and, if necessary, directly with the separate radio monitoring equipment. The control center of the automated system has necessarily the sub-systems of control, communications and data transmission, and the electric power supply, etc., in its structure. The network equipment of the control center should have "hot" reservation and additional links, which help central communication node to continue operating in case of fault of some system element, whether it is the network equipment or communication lines equipment.

Radio monitoring stations represent the engineering constructions for deployment of fixed radio monitoring equipment, in which structure we can single out the radio receiving equipment, mounted outdoors, and the equipment of processing and control, which can be mounted both inside the technical rooms and outdoors in the temperature-controlled waterproof secure cabinets. It is preferable to use the uninhabited buildings as the radio monitoring stations because it will essentially reduce different agreements with its owners. The search for a suitable place in urban conditions is often a difficult task.

The system of data transmission represents the assembly of the nodes, which are united by the data transmission network and are capable to be remotely controlled and to function independently in accordance with the assigned requirements. All interaction is based on the TCP/IP protocols within the frameworks of the data transmission system. To provide high reliability, scalability, and high performance, the multi-level architecture is used at data transmission networking.

The communication channel, which provides the traffic transmission between the nodes that may be wired and wireless, is the key component connecting the nodes of the data transmission network. Wireless technologies are used where application of wired technologies is difficult or impossible due to technical reasons, for instance, at connection of mobile stations or with purpose of reserve channel organization for information transmission.

References

1. Barkalov SV, Kozmin VA, Sysoev DS, Tokarev AB (2016) Engineering infrastructure of the automated radio monitoring systems (in Russian). Spetstehnika i Svyaz (4):69–80
2. SMO-ARMADA software package of automated spectrum monitoring systems. http://www.ircos.ru/en/sw_armada.html. Accessed 28 Nov 2017
3. Alekseev DA, Ashikhmin AV, Kobelev SG, Kozmin VA, Rembovsky AM, Sysoev DS, Tsarev LS (2014) Features of automated spectrum management system at 27th Summer Universiade in Kazan (in Russian). Electrosvyaz (4):34–41
4. Report ITU-R SM.2257-2 (2014) Spectrum management and monitoring during major events. SM Series. Spectrum management. p 64
5. Handbook on Spectrum Monitoring (2011) ITU-R, Geneva, p 659

Chapter 5
Digital Radio Receivers and Direction-Finders

5.1 Introduction

At present, the receivers, in which the received analog radio signal is converted into the digital form after preliminary processing, are most widely distributed to execute radio monitoring. The further processing is performed in the digital signal processing (DSP) path: in digital filter of the main selectivity, digital demodulators, digital units of recognition, signal parameters measurement, etc. Digital radio receiving devices are based on principles of the software defined radio (SDR). The structural diagram of DRR is presented in Fig. 5.1.

The analog path receives signal from antenna and filters it against interference. Depending on the structure variant, the input signal spectrum can be transferred to the intermediate frequency (IF), at which the analog-to-digital converter (ADC) is executed, or the direct signal conversion at the baseband frequency is carried out. The auxiliary units: automatic gain control (AGC) system, attenuators, limiters, etc., which have an influence on the amplitude characteristic of the amplifying path, but not introducing distortions into the information being received, can be included in the path structure.

The frequency synthesizer converts external or internal reference oscillator frequency and forms frequency grids from it, which are necessary for receiver operation, including frequencies necessary for operation of digital devices for signal processing.

In the DSP path, the main signal processing on the base of SDR is provided. Digital radio signals are processed by the embedded software in digital signal processors (DSP) and field programmable gate arrays (FPGA). Depending on the goals, the digital filtering, the fast Fourier transform, the changing of the sampling frequency, the auto-tuning, the demodulation and decoding, the various types of after-detection processing, etc., are performed.

With the help of the control device, the necessary operation mode of the receiver can be set: powering on/off, search for signal and selection of it, adaptation to

A. M. Rembovsky et al., *Radio Monitoring*, Signals and Communication Technology, https://doi.org/10.1007/978-3-319-74277-9_5

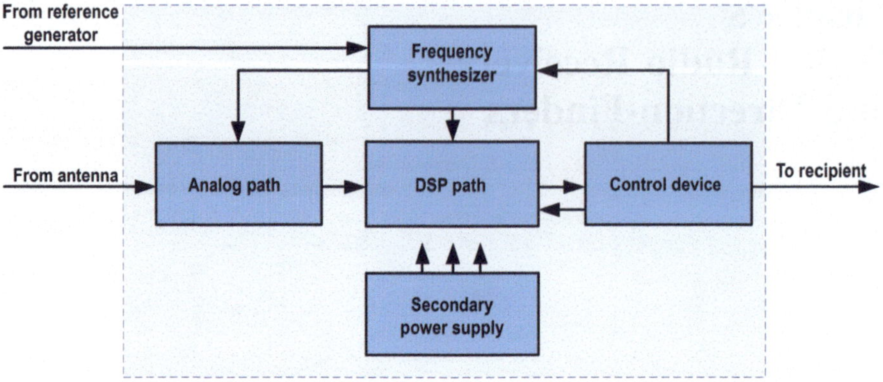

Fig. 5.1 Structural diagram of the digital radio receiver

changing operation conditions, etc. It allows operator to control the receiver remotely or directly.

The secondary power source is intended for conversion of primary power source energy, for example, of the stationary net of 220 V or onboard net into the form suitable for usage directly in the receiver.

Spectrum analyzers, panoramic radio receivers, measuring receivers, and signal analyzers find application for radio monitoring tasks solution among digital radio receivers. Characteristics of digital radio receivers, which are the products of the IRCOS Company, are described in detail in [1, 2].

The spectrum analyzer is the universal measuring device intended for investigation of signals spectral distribution and measurement of their parameters. The structure of modern spectrum analyzer usually agrees with the structure of super-heterodyne receiver with DSP. Depending on the input signal sensor type, the spectrum analyzer can be used in various areas of science and engineering. In particular, at antenna application, the spectrum analyzer becomes the panoramic receiver. Such an analyzer can perform tasks of the panoramic measuring receiver at additional connection of preselection circuits between the antenna and the spectrum analyzer. In the base structure, the spectrum analyzer, as a rule, has no units for preselection of signals; therefore, its utilization for radio monitoring of signals in ether is not always effective.

The panoramic digital receiver is the DRR with wide pass-band from hundreds of kilohertz to several tens or even hundreds of megahertz with a possibility of representation of signal spectral distribution and high performance of spectral analysis (from hundreds of megahertz to tens of gigahertz per second). The panoramic receiver, as a rule, contains different signal demodulators and decoders.

The panoramic digital measuring receiver is the panoramic receiver with high precision measurement of level, frequency, bandwidth, and other radio signal parameters, which is certified as the measuring unit. The digital measuring receiver is the "heart" of the modern radio monitoring complex. The set of peak, quasi-peak, and RMS detectors intended for measurement of signal levels is usually included in the structure of the measuring receiver. The measuring receiver should provide frequency resolution from several hertz to tens kilohertz at the signal spectrum analysis and operate under computer control.

The first type of the measuring receiver is the meter of voltages and powers of signals and interferences; it will operate as a meter of field strength if parameters of the receiving antenna are known. It represents the special super-heterodyne receiver with measuring device at output, which has high sensitivity and selectivity, and is the multi-range receiver with pass-band adjustment by intermediate frequency (IF). The linear reception path enables to use different types of measuring detectors, which gives a possibility to measure average rectified, RMS, peak values of harmonic and noise signals.

The second type of the measuring receiver is the spectrum analyzer. Besides measurements of voltages and powers, it provides measurement of various parameters of modulation and coding. Signal analyzers have, as a rule, high-quality characteristics of wide-band analog radio frequency path and sufficient computing power of DSP path, which allows implementation of necessary set of signal analysis functions in time, frequency, and code domains. As a rule, working characteristics of such devices in various frequency ranges are optimized taking into account the signal type of communication systems under investigation.

The panoramic digital radio receiving device is the main component of the modern radio direction-finder structure. At present, the automatic radio direction-finders, which operate according to the principle of correlation interferometers [2], have the wide distribution. The structural scheme of such radio direction-finder is shown in Fig. 5.2. The radio direction-finder consists of the following main parts: antenna system, analog radio receiving path, DSP path, and controller.

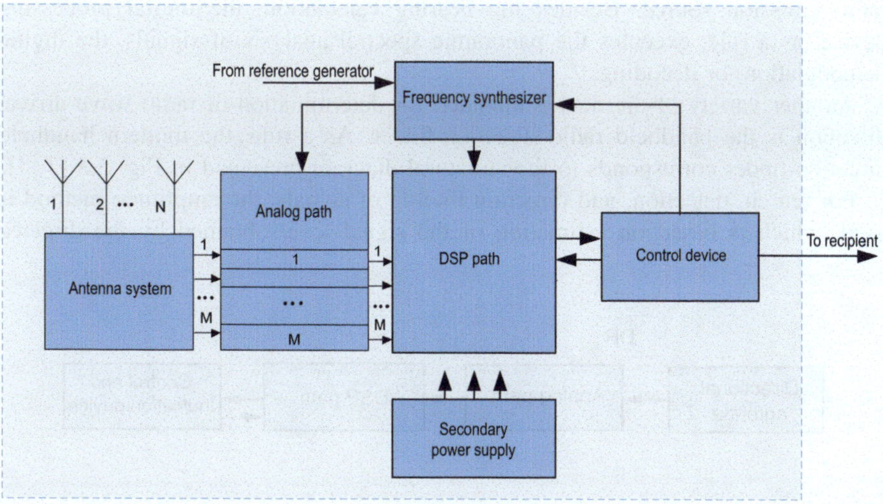

Fig. 5.2 Structural diagram of the automatic radio direction-finder

Receiving elements of the antenna system are located according to the definite order, for example, along the circumference forming the so-called antenna array. The coil antennas, the conical and bi-conical vibrators, the spike antennas, the disk-cone antennas, the directed antennas, etc., may be used as antenna elements.

The operation principle of the correlation interferometer consists of the sequential comparison of measured levels and phase differences between the elements of the antenna array for the received radio signal with the sets of levels and phase differences, which are preliminary calculated theoretically for the given antenna array and for the possible directions of signal arrival. Comparison is carried out by means of calculation of the correlation coefficient. The "theoretical" direction, for which the correlation coefficient is maximal, is considered as the bearing direction.

To achieve the direction-finding accuracy of the correlation interferometer not worse than 1° and to provide wide working range, for instance, from 20 to 1300 MHz, it is necessary to have nine antenna elements located along the circumference with about one-meter radius.

The radio receiving unit is intended for selection, amplification, and conversion of the input signal frequency. In so-called monopulse direction-finders, the number of receiving sections M equals to the number of antenna elements: $M = N$. In this case, the maximal rate of direction-finding is ensured. However, the engineering implementation of such a solution is complicated, and the cost of the direction-finder turns out to be rather high. Therefore, the radio receiving path having from one to three channels ($M = 1, 2, 3$) is often used. If the number of antenna elements is more than the receiving channels amount, then we can use high-frequency switches, which connect antenna elements to the receiver paths in turn.

At the output of the analog path, signals are converted into the digital form and further are exposed to digital processing with the aim of calculation of azimuth onto radio emission source. Besides the bearing calculation, the digital processing device, as a rule, executes the panoramic spectral analysis of signals, the digital demodulation, or decoding.

Another variety of the device intended for determination of radio wave arrival direction is the handheld radio direction-finder. As a rule, the modern handheld direction-finder corresponds to the structural diagram presented in Fig. 5.3 [3, 4].

For search, detection, and direction-finding of signals, the amplitude method is used, which is based on estimation of the signal level obtained by the directed

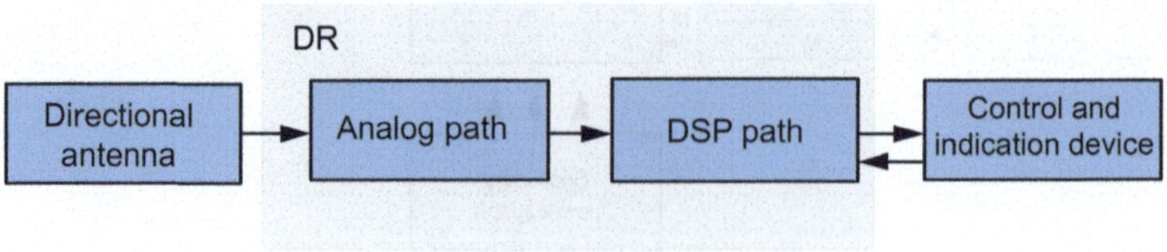

Fig. 5.3 Structural diagram of the handheld direction-finder

antenna. For required source signal selection, the central frequency and the receiving bandwidth are specified.

Two types of digital radio receivers: Super-heterodyne receivers and direct conversion receivers will be described in the present chapter. The effectiveness of digital radio receivers integrated into the antenna system will be discussed. Besides, the main DSP algorithms will be presented, the constructive features and characteristics of digital radio receiving devices of the ARGAMAK family and radio direction-finders of the ARTICUL family, which are delivered by IRCOS Company at present [1], will be discussed.

5.2 Constructive Variants of Digital Radio Receiving Devices

In the modern professional radio equipment, the heterodyne receivers of two types are mostly distributed: super-heterodyne receivers working with the received signal frequency transfer to the intermediate frequency, and direct conversion receivers, which transfer the receiving signal to the baseband ("zero") frequency. Functional schemes of these types of receivers are presented in Figs. 5.4 and 5.5, relatively [5].

Receivers of the super-heterodyne type have more complicated structure of the analog path including preselector, which provides main and image receiving channels selectivity, as well as one or several units for frequency conversion to IF

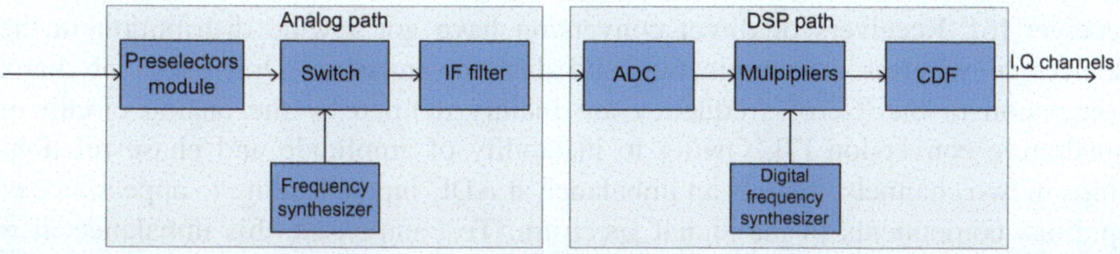

Fig. 5.4 Functional diagram of the super-heterodyne DRR

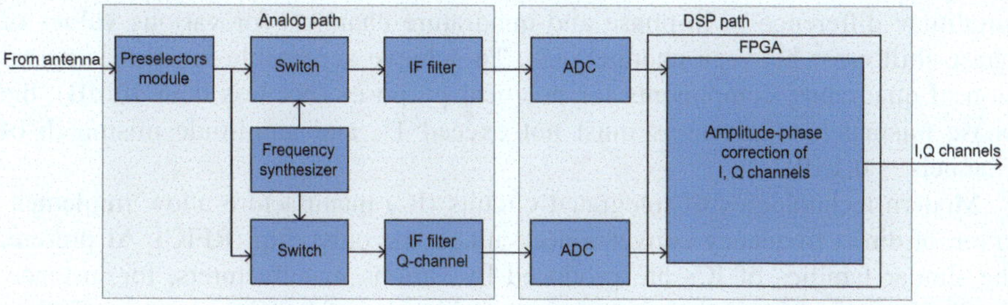

Fig. 5.5 Functional scheme of the direct conversion DRR

with high-selectivity IF filter of the main selectivity. As a rule, receivers of super-heterodyne type have two or three stages of frequency conversion.

The wide dynamic range free from spurious spectral components, the good stability to out-of-band interferences owing to high squareness of IF filters, the absence of DC component offset and quadratic distortions since the only one ADC is used, and the conversion of the real signal into the complex form is carried out in FPGA with the help of the digital step-down converter, are the advantages of the super-heterodyne structure. Among drawbacks, we can mention the following: due to multiple frequency conversion, receivers of this type have a significant power consumption and have a large number of spurious reception channels, which requires high demands to the filtering system; due to high squareness of filters and complicated switch scheme the noise factor increases; at multiple frequency conversion, the phase noise of local oscillators is summed; the relatively high IF passes to the ADC input of the wide-band super-heterodyne DRR for radio monitoring, which requires high clock rate of ADC. This imposes limitations on the dynamic range and defines high requirements to performance of the ADC path.

Receivers of direct conversion have less complicated structure of the analog part. In general case, it includes the only frequency conversion unit with local oscillator, which has quadrature outputs with LPFs. Such a structure allows obtaining less power consumption level at the same bandwidth of simultaneous analysis and provides possibility to achieve less mass-dimension indices, comparatively low level of noise and intermodulation components. To improve characteristics on the intermodulation distortions of second and third orders and sensitivity, the unit of preselection with wide-band or adjustable filters and low-noise amplifiers is included, as a rule, into the structure of the direct conversion receiver [6]. Receivers of direct conversion have got a wide distribution in the devices of wireless communication and data transmission. Drawbacks of direct conversion to the "zero" frequency are mainly defined by the analog circuit of quadrature conversion [7]. Owing to instability of amplitude and phase relationships in two channels, there is an imbalance at ADC inputs leading to appearance of spurious components in the signal spectrum. To compensate this imbalance, it is necessary to perform correction of in-phase and quadrature reception channels. To achieve the best results under varying conditions of exploitation, the correction must be repeated periodically.

Figure 5.6 shows functions of spurious components levels suppression versus amplitude difference in in-phase and quadrature channels for various values of phase-shift error between channels [8]. To achieve acceptable level of suppression of quadrature components for practical purposes (not less than 40 dB), the phase mismatch of channels must not exceed 1°, and amplitude mismatch of channels − 0.1 dB.

Modern technologies of integrated circuits (IC) manufacture allow implementation of direct frequency conversion circuit on the only chip (RFIC). At present, the similar families of ICs are produced by various manufacturers, for instance, Analog Devices, Lime Microsystems, etc. Calibration of RFIC during manufacture provides typical value of the signal/interference ratio (SINR) of 40 dB, which is

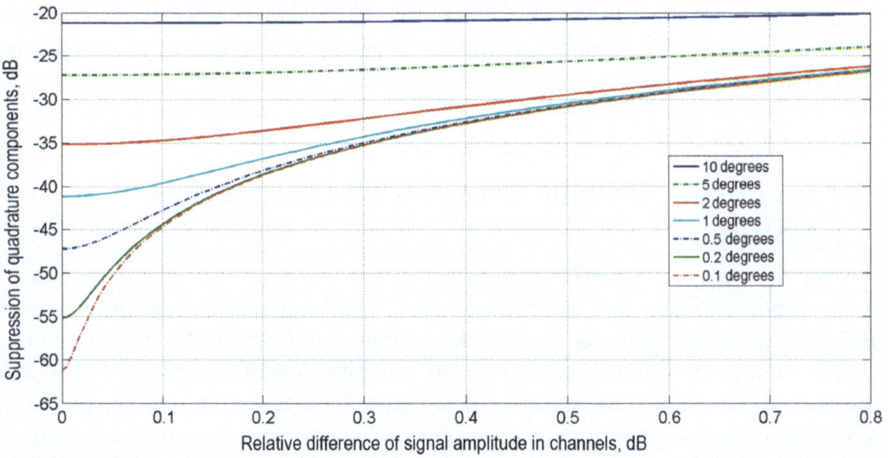

Fig. 5.6 Dependence of suppression level of spurious components upon amplitude–phase relationships in the receiver channels

enough for reception of most digital communication systems signals. Additional software calibration allows improvement of SINR up to 60 dB.

Utilization of compact and economical modules is effective at execution of radio monitoring by mobile devices, including pilotless aircraft, and in portable radio monitoring complexes, which can be used for solution of problems requiring urgent estimation of radio channel occupancy, determination of signal presence at the non-licensed frequencies, search and identification of radio interference sources.

At the modern stage of radio monitoring equipment development, the receivers of super-heterodyne type have got a wide distribution in tasks related to measurements of signal parameters. Among such devices, we can mark out: ESMD by Rohde&Schwarz, ComCat Tuner by Medav, the series M93XX by Agilent, the series PXIe-56XX by National Instruments, ARGAMAK-IS, and ARGAMAK-M N [1, 4]. At the same time, at solution of radio monitoring tasks, which do not require accurate measurement of signal parameters, it is admissible to apply circuits with direct frequency conversion, which allows essential reducing of the mass-dimension indices of DRRs.

The super-heterodyne architecture is preferable for measuring receivers with high dynamic range and high accuracy of measurements. The architecture with frequency transfer to "zero" frequency allows implementation of compact economic devices with wide functional possibilities, which are able to solve a lot of radio monitoring tasks. In particular, the super-heterodyne receiver ARGAMAK-MN provides spurious channels selectivity not less than 70 dB and dynamic range on intermodulation of third and second orders not less than 75 dB, and power consumption is 15 W only. The digital receiver ARC-CPS3 with conversion to the "zero" frequency has twice less power consumption, however, essentially worse characteristics of spurious channels selectivity (40 dB) and in the dynamic range on intermodulation of third and second orders (60 dB), which can negatively manifest itself on the operation quality of this receiver under conditions of complicate electromagnetic situation.

5.3 Integration of Radio Receivers into the Antenna System

At development of territory-distributed radio monitoring system, the provision of maximally possible operation zone of the system at utilization of minimal number of equipment is the significant requirement since the cost of equipment defines the total cost of the whole system in the most cases. Therefore, the users try to install equipment of radio monitoring on the objects, which provide maximal height of antenna systems, and also to use equipment with high sensitivity and wide dynamic range [9, 10].

In the present section, an applicability of digital radio receiving devices integrated into the structure of antenna system will be shown.

5.3.1 Feeder Influence on Monitoring Equipment Characteristics

To extent the operation zone of monitoring equipment, the receiving antenna system is mounted on the available heights, for example, on the roofs of multi-floor buildings. At that, to reduce influence of local obstacles, the receiving antenna system is usually mounted on the mast. Figures 5.7 and 5.8 show two possible variants of monitoring equipment installation on the high-rise building. In the first variant, the receiver is in the special container installed directly on the building roof, at that, the length of the feeder from antenna system to the receiver input is about 15–20 m. In the second variant, the receiver is located in the service room; the length of the feeder significantly increases and may achieve 80–100 m.

Usually, the coaxial cables are used as feeders in equipment of radio monitoring. As any other conductor with losses, the coaxial feeder is the noise source. At the room temperature, the noise factor of the feeder is equal approximately to attenuation in it [4, 11]. In Table 5.1, parameters of several distributed types of coaxial cables are given, and Fig. 5.9 shows functions of its loss L_a per 1 m unit of length versus frequency.

The loss per unit length increases with frequency growth, and the lengthy cable has a significant noise factor, especially at frequencies more than several hundred megahertz. For example, the linear attenuation of the cable RK 50-3-14 in the frequency range from 1000 to 3000 MHz grows from 0.5 to 1 dB/m; therefore, the noise factor of the 30 m cable takes values from 15 to 30 dB. With growth of the

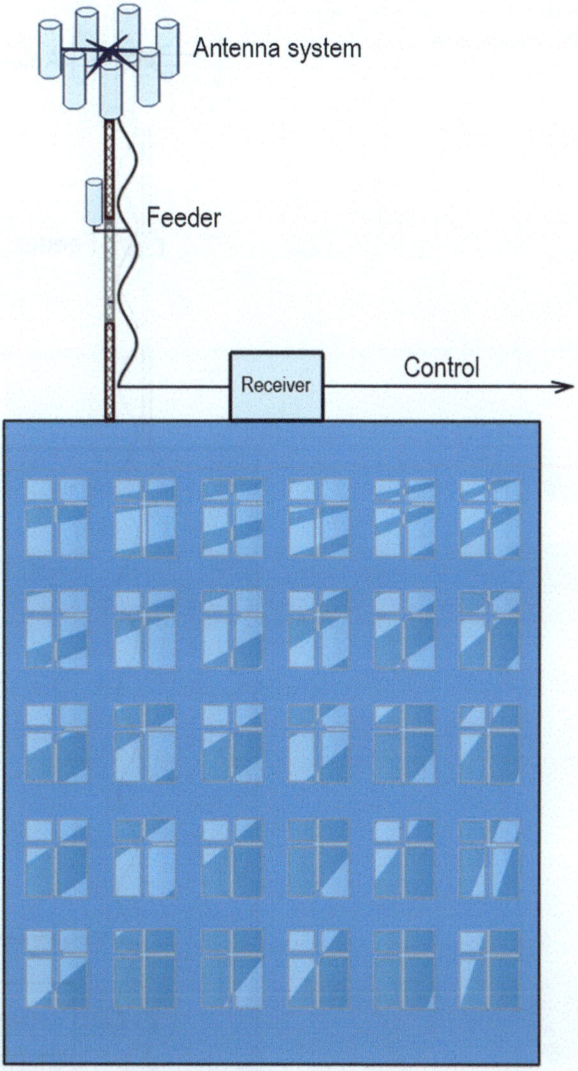

Fig. 5.7 Receiver location on the mast basis

noise factor, the monitoring equipment sensitivity proportionally decreases and can become inadmissibly low.

The radio monitoring equipment (especially, portable and transportable) should provide a possibility of repeatable deployment and uninstalling of antenna systems; therefore, the minimal radius of cable bending characterizing its flexibility is the significant parameter. As a rule, the smaller the cable loss per unit length is, the smaller its flexibility is. So, the native cable RK 50-3-35 and foreign SPUMA 400 have less loss per unit length, but larger radius of bending compared to other types included in the table.

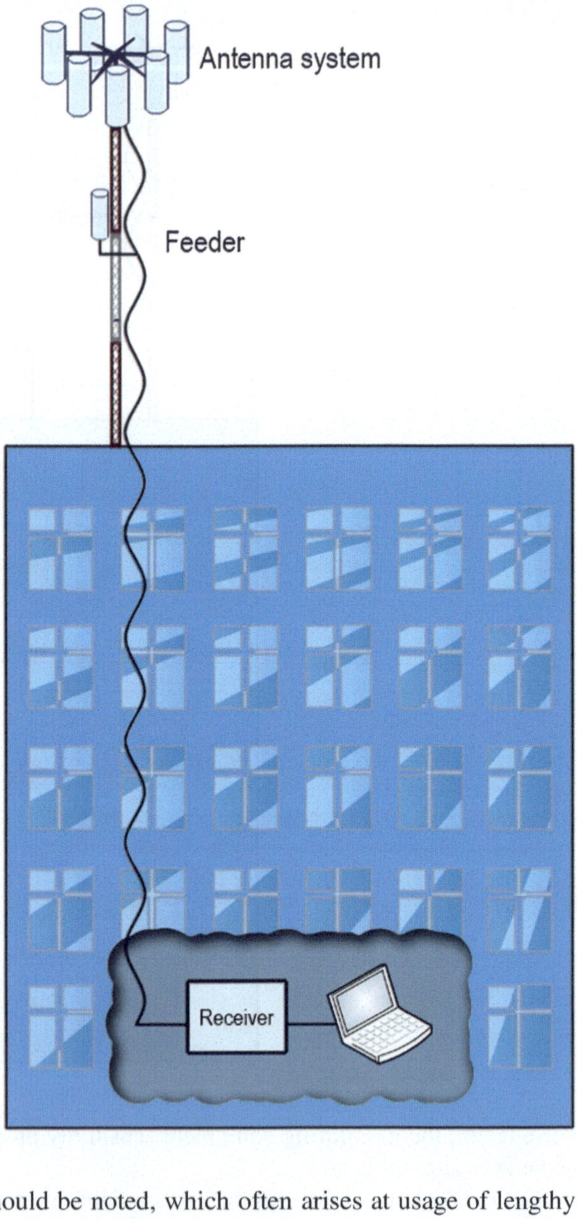

Fig. 5.8 Receiver location in the service room

One more phenomenon should be noted, which often arises at usage of lengthy feeder. This is the antenna effect, at which the conducted interferences are induced by external electromagnetic fields in the feeder. Interference to reception caused by the antenna effect arises not only from the influence of powerful or near-located radio stations, but from illuminating, power and computer nets, various household and industrial electric devices located near the monitoring equipment installation place and its antenna feeder. For direction-finding units, the antenna effect degrades

Table 5.1 Constructive parameters of flexible coaxial cables

Name	RG 58 C/U	RK 50-3-14	RK 50-3-35	SUCOFLEX 104 P	SUCOFLEX 106 P	SPUMA 400
External diameter, mm	5	5	5	5.5	7.9	10.3
Minimal bend radius (single), mm	25	25	45	16	24	25
Minimal bend radius (multiple), mm	50	50	90	25	40	100

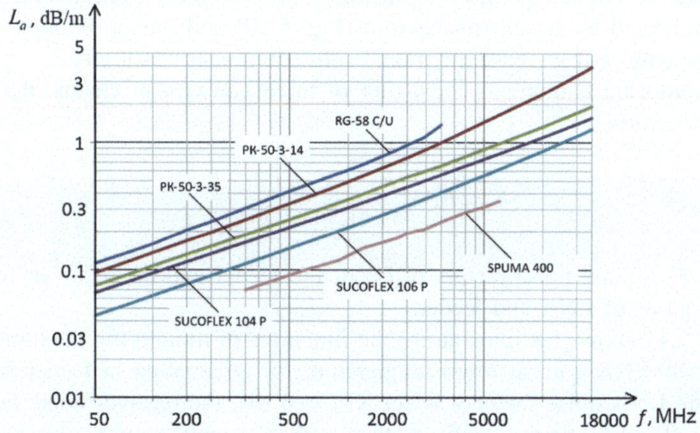

Fig. 5.9 Functions of loss per unit length versus frequency

an accuracy and sensitivity of direction-finding since the phase and amplitude relationships in reception channels are disturbed due to induced signals. For measuring units, the antenna effect increases the field strength measurement errors, distorts the antenna patterns, and degrades the sensitivity.

5.3.2 *Preamplifiers Application*

The direct method of monitoring equipment sensitivity growth consists of application of high-quality feeder with low losses and high shielding properties. However, at large length of the feeder, the signal attenuation at frequencies above several gigahertz will be significant in this case and hence, the noise factor will be high. In addition, coaxial cables with low losses have, as a rule, larger diameter and lesser flexibility, which complicates deployment and exploitation of monitoring equipment (see Table 5.1).

The known expression for the noise factor of chains connected in series has a form:

$$F = F_1 + \frac{F_2 - 1}{g_1} + \frac{F_3 - 1}{g_1 g_2} + \cdots + \frac{F_M - 1}{\prod_{m=1}^{M-1} g_m}, \tag{5.1}$$

where F_i, g_i are the noise factor and gain of ith chain, $i = 1, 2, \ldots M$, M is chains quantity. The noise factor of real radio engineering devices is always more than 1 due to internal noises. In engineering practice, one usually uses noise factor expressed in decibels. It is related with noise factor as NF = 10lg(F).

It follows from formula (5.1) that gain growth of the first chain reduces noise of all consequent circuits. Hence, application of low-noise preliminary amplifier (LNA) connected to the antenna output (Fig. 5.10) and having enough gain and low-noise factor will increase the monitoring equipment sensitivity.

For monitoring equipment consisting of three consequent chains, the formula (5.1) takes a form

$$F = F_1 + \frac{F_2 - 1}{g_1} + \frac{F_3 - 1}{g_1 g_2}, \tag{5.2}$$

where F_1, F_2, F_3 are noise factors of LNA, the feeder and the receiver, relatively; g_1, g_2 are gains of LNA and feeder.

Figure 5.11 shows (as illustration) the functions of monitoring equipment noise factor versus LNA gain at room temperature for attenuation in feeder from 5 to 40 dB. The LNA noise factor was set $\mathrm{NF}_1 = 3$ dB, the receiver noise factor was $\mathrm{NF}_3 = 10$ dB. The figure confirms that with G_1 gain growth, the monitoring equipment noise factor asymptotically achieves the LNA noise factor, at that, monitoring equipment sensitivity increases and its dynamic range extends from the side of low signal values. But the additional gain introduced by LNA will decrease the dynamic range from the side of large signal values.

Let us estimate an influence of nonlinear properties of chains included in the monitoring equipment structure. We shall use the interception points on third-order intermodulation in the logarithmic scale by chain output and input OIP_3 and IP_3 measured in mW, and in the linear scale oip_3 and ip_3 measured mW [4, 12]. It is known that interception point by chain output OIP_3 is related to the interception point by input IP_3 via the chain gain $\mathrm{OIP}_3 = \mathrm{IP}_3 + G$. The value of the interception point characterizes the linearity degree of the chain. The value of OIP_3 is usually

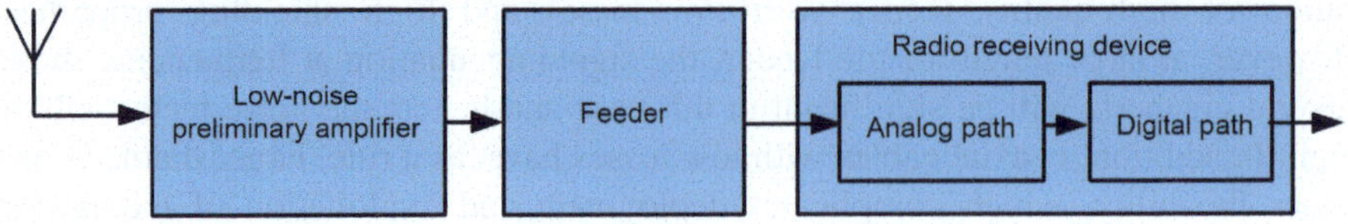

Fig. 5.10 An example of LNA application

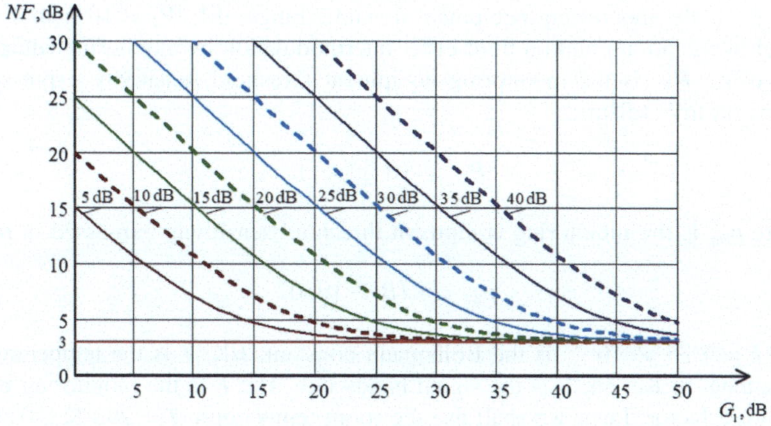

Fig. 5.11 Functions of monitoring equipment noise factors versus LNA gain

from 20 to 45 dBm for wide-band LNAs presented in the international market at the noise factor from 0.5 to 3 dB.

The interception point on third-order intermodulation by output for three chains in series is defined by formula [12]:

$$\mathrm{oip}_3 = \frac{1}{\frac{1}{\mathrm{oip}_{33}} + \frac{1}{\mathrm{oip}_{32} g_3} + \frac{1}{\mathrm{oip}_{31} g_2 g_3}}, \tag{5.3}$$

where $\mathrm{oip}_{33}, \mathrm{oip}_{32}, \mathrm{oip}_{31}$ are levels of interception points of third order by output for the receiver, the feeder, and the LNA, relatively. The feeder is the linear passive chain, and its oip_{32} level is much bigger than oip_{33} and oip_{32}. Therefore, we can neglect the medium term in the denominator of (5.3). As a result, we have

$$\mathrm{oip}_3 = \frac{1}{\frac{1}{\mathrm{oip}_{33}} + \frac{1}{\mathrm{oip}_{31} g_2 g_3}}. \tag{5.4}$$

Proceeding from interception points by output to interception points by input, we obtain

$$\mathrm{ip}_3 = \frac{1}{g_1 g_2 \left(\frac{1}{\mathrm{ip}_{33}} + \frac{1}{\mathrm{ip}_{31} g_1 g_2}\right)}, \tag{5.5}$$

and then we determine the monitoring equipment dynamic range free from products of third-order intermodulation [4, 12]:

$$D_3 = \frac{2}{3}(\mathrm{IP}_3 - P_{\mathrm{thr}}), \tag{5.6}$$

where D_3 is the monitoring equipment dynamic range, dB; $\mathrm{IP}_3 = 10\lg(\mathrm{ip}_3)$ is the level of interception point on third-order intermodulation by monitoring equipment input, dBm; P_{thr} is the monitoring equipment threshold sensitivity expressed in decibels on mW (dBm):

$$P_{\mathrm{thr}} = 10\ \lg(p_{\mathrm{thr}}). \tag{5.7}$$

Here p_{thr} is the monitoring equipment threshold sensitivity expressed in mW:

$$p_{\mathrm{thr}} = kTBF \cdot 1000, \tag{5.8}$$

where $k = 1.38 \times 10^{-23}$ is the Boltzmann constant, J/K; T is the temperature of environment in Kelvin; B is the signal bandwidth, Hz; F is the monitoring equipment noise factor. Later we shall use the room temperature T = 293 K (20 °C).

The dynamic range in the linear scale is

$$d_3 = (\mathrm{ip}_3/p_{\mathrm{thr}})^{\frac{2}{3}}. \tag{5.9}$$

Substituting formulas (5.5), (5.8), and (5.2) into (5.9), we obtain

$$d_3 = \left(\frac{g_1 g_2 \mathrm{ip}_{31} \mathrm{ip}_{33}}{kTB(g_1 g_2 F_1 + g_2(F_2 - 1) + F_3 - 1)(g_1 g_2 \mathrm{ip}_{31} + \mathrm{ip}_{33})}\right)^{\frac{2}{3}}. \tag{5.10}$$

Figure 5.12 shows functions of the monitoring equipment dynamic range according to (5.10) versus the gain G_1 at variation of signal attenuation in the feeder from 5 to 40 dB. Values of interception points on third-order intermodulation for LNA and the receiver were taken $\mathrm{IP}_{31} = -5$ dBm, $\mathrm{IP}_{33} = 10$ dBm. The input signal bandwidth is B = 10 kHz. Noise factors for the receiver and LNA had the same values as used at Fig. 5.11, i.e., $\mathrm{NF}_1 = 3$ dB, $\mathrm{NF}_3 = 10$ dB.

As it follows from the figure, the monitoring equipment dynamic range firstly increases (caused by growth of sensitivity due to reduction of the noise factor), and after achieving the maximum, decreases (due to further gain growth) at LNA gain growth.

Let us find gain g_1^*, at which the dynamic range's maximum is achieved for the given attenuation of the feeder. Since the function $x^{2/3}$ is monotonous, the abscissa of the maximum can be determined from the following equation

$$\frac{\partial}{\partial g_1}\left(\frac{g_1 g_2 \mathrm{ip}_{31} \mathrm{ip}_{33}}{kTB(g_1 g_2 F_1 + g_2(F_2 - 1) + F_3 - 1)(g_1 g_2 \mathrm{ip}_{31} + \mathrm{ip}_{33})}\right) = 0. \tag{5.11}$$

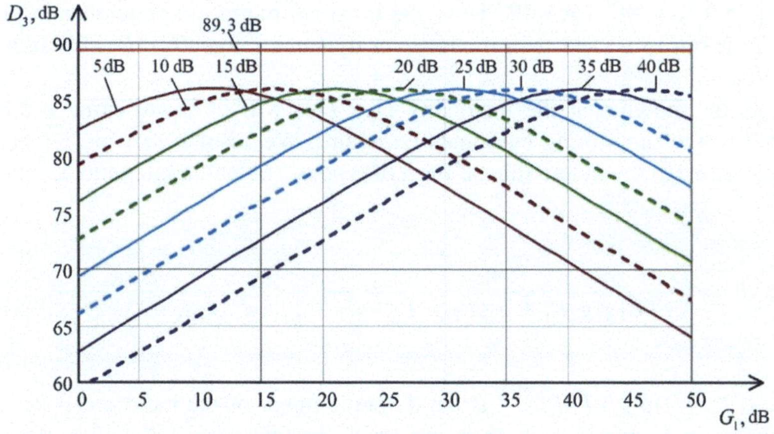

Fig. 5.12 Dependence of the dynamic range on LNA gain

From this, it follows that

$$g_1^* = \frac{1}{g_2}\sqrt{\frac{\mathrm{ip}_{33}(F_3 + g_2(F_2 - 1) - 1)}{F_1 \mathrm{ip}_{31}}}. \tag{5.12}$$

Taking into account that the noise factor of the cable is approximately equaled to its attenuation $F_2 \approx 1/g_2$, the formula (5.12) is simplified

$$g_1^* = \frac{1}{g_2}\sqrt{\frac{\mathrm{ip}_{33}(F_3 + g_2)}{F_1 \mathrm{ip}_{31}}}. \tag{5.13}$$

The last expression shows: the larger attenuation $1/g_2$ and lesser the LNA linearity ip_{31} are, the larger LNA gain must be for achieving of the dynamic range maximum.

Let us determine the relationship between the monitoring equipment dynamic range and the receiver dynamic range included in its structure. We preliminary calculate the receiver dynamic range at above-accepted values of parameters:

$$\begin{aligned} D_{33} &= \frac{2}{3}(\mathrm{IP}_{33} - 10\lg(p_{\mathrm{thr3}})) \\ &= \frac{2}{3}(10 + 174 - 10\lg(10{,}000) - 10) = 89.3\,\mathrm{dB}, \end{aligned} \tag{5.14}$$

where $p_{\mathrm{thr3}} = kTBF_3 \times 1000$ is the threshold sensitivity of the receiver, mW.

Fulfilling the similar calculations for all monitoring equipment paths at $G_2 = -25$ dB, $\mathrm{IP}_{31} = -5$ dBm, and $G_1^* = 36$ dB, we obtain the monitoring equipment dynamic range $D_3 = 82$ dB. At that, the monitoring equipment noise factor (see Fig. 5.12) is NF = 4.5 dB. Thus, the maximal monitoring equipment dynamic range turns out to be less than the receiver dynamic range $D_3 < D_{33}$ by more than 7 dB.

It can be shown that the inequality $D_3 < D_{33}$ will be always true if LNA is included in the monitoring equipment structure. We assume that $\mathrm{ip}_{31} \to \infty$, then opening in (5.10) the indeterminate form of type ∞/∞ and taking into account that $1/g_2 \approx F_2$, we obtain

$$\lim_{\mathrm{ip}_{31}\to\infty} d_3 = \left(\frac{\mathrm{ip}_{33}}{\mathrm{kTB}(g_1 g_2 F_1 + g_2(F_2 - 1) + F_3 - 1)}\right)^{\frac{2}{3}} = \frac{d_{3\,\mathrm{Rec}}}{\left(\frac{g_2(g_1 F_1 - 1)}{F_3} + 1\right)^{\frac{2}{3}}}, \quad (5.15)$$

where $d_{3\,Rec} = (\mathrm{ip}_{33}/kTBF_3)^{2/3}$ is the dynamic range of the radio receiver.

Since $g_1 > 1$ and $F_1 > 1$, then $g_1 F_1 > 1$. In this case $g_1 F_1 - 1 > 0$ and the denominator $\left(\frac{g_2(g_1 F_1 - 1)}{F_3} + 1\right)^{\frac{2}{3}} > 1$, hence, even in hypothetical case when LNA has infinite linearity, the monitoring equipment dynamic range as a whole will be less than the dynamic range of the receiver. From expression (5.15), we see that monitoring equipment dynamic range is equal to the dynamic range of the receiver $D_3 = D_{33}$ only at absence of LNA in the monitoring equipment structure, i.e., at $g_1 = F_1 = 1$.

Thus, application of LNA decreases the noise factor and increases monitoring equipment sensitivity, extends the dynamic range at the side of small signal values and, at the same time, narrows it at the side of large values. In order to eliminate the reduction of the dynamic range, the LNA gain should not be too large. Besides, this is desirable to match the LNA gain with frequency-depended losses in the feeder. LNAs presented in the international market have usually constant or decreasing amplification (with frequency growth), however, to keep the monitoring equipment dynamic range, it is necessary to have the reverse dependence, i.e., the LNA gain should increase with the frequency growth. For instance, for the feeder of RK 50-3-14 type with the length of 30 m in the frequency range from 1000 to 3000 MHz, the noise factor will change from 15 to 30 dB. From Fig. 5.12, it follows that to support the maximal dynamic range, it is necessary to have the LNA gain increased from 26 to 42 dB. The provision of such dependence in wide frequency range keeping the small noise factor and high linearity complicates the LNA construction.

To reduce unfavorable influence of the lengthy feeder, one can try to choose its length a minimally necessary one, starting from the factual distance from antenna system to the receiver. But at that, to keep maximal possible dynamic range, it is necessary to use the LNA with individually chosen gain, which complicates the serial production.

At LNA application, the amplified signal is transmitted via the feeder, which decreases the unfavorable influence of the antenna effect, but cannot eliminate it completely. Special measures applied to minimize the antenna effect, for example, differential cables with the triple shield, ferrite filters in the form of the latch, essentially complicates the monitoring equipment construction.

5.3.3 Integration of Radio Receiver into the Antenna System

The modern level of electronic component development allows production of digital panoramic radio receivers with small dimensions promoting their direct integration into the antenna system construction. At receiver integration into the circular antenna array of the direction-finder, the feeder length from antenna elements to the receiver input will not exceed the units of meters, at that, attenuation will be within the limits of tenth parts of decibel to several decibels depending on the type and working frequency. For instance, the cable of RK 50-3-14 type with the length of 2.5 m at frequency of 3000 MHz will have losses of 2.5 dB order. The small feeder length allows utilization of cables with high-quality parameters without essential monitoring equipment cost growth, for instance, instead of RK 50-3-14 one can use SUCOFLEX 106 P, which attenuation is less by half.

If losses in the feeder are small and it is not necessary to have better monitoring equipment sensitivity than the receiver sensitivity, then LNA is not required. At that, the maximal possible dynamic range of monitoring equipment is provided. If we need to increase the monitoring equipment sensitivity, requirements to LNA become less strict.

Figures 5.13 and 5.14 show functions of noise factor and monitoring equipment dynamic range versus LNA gain. Attenuation in the feeder was varied from 0.5 to 4 dB, at LNA $\mathrm{IP}_{31} = 15$. From figures, it follows that at attenuation in the cable of 2.5 dB, the maximal dynamic range of monitoring equipment is $D_3 = 88.1$ dB, the LNA gain is $G_1^* = 3.6$ dB. To ensure the monitoring equipment noise factor of 4.5 dB, the LNA gain should be increased up to 13 dB, at that, the monitoring equipment dynamic range will be 85.9 dB. Thus, application of the integrated receiver, compared to the above-mentioned example, extended the maximal dynamic range of monitoring equipment by more than 6 dB. Under conditions of the same sensitivity, when noise factor NF = 4.5 dB, the benefit in the dynamic range also exceeds 3 dB.

In the previous example, at the feeder length of 25 m and for achieving the maximal monitoring equipment dynamic range for $\mathrm{IP}_{31} = -5$ dBm, the LNA gain is $G_1^* = 36$ dB and the interception point on third-order intermodulation by output is $\mathrm{OIP}_{31} = \mathrm{IP}_{31} + G_1 = -5 + 36 = 31$ dBm. At application of the integrated receiver, to achieve the same parameters, it is enough to take the LNA with $\mathrm{IP}_{31} = 15$ dBm, $G_1 = 13$ dB, or (the same) with $\mathrm{OIP}_{31} = 28$ dBm. Such a LNA will be simpler in implementation.

There are two variants of digital receiver integration. In the first variant, the only analog path of receiver signal processing is embedded into the antenna system, and the signal is transmitted via the feeder into the DSP path at the low IF. In the second

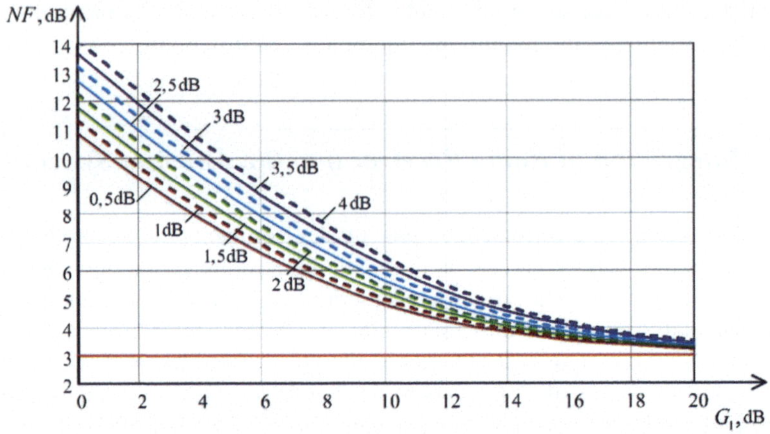

Fig. 5.13 Function of the monitoring equipment noise factor versus LNA gain at application of the integrated receiver

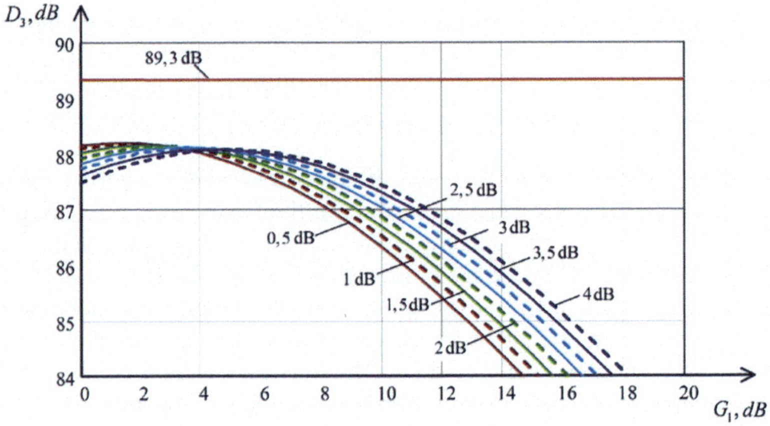

Fig. 5.14 Dependence of the dynamic range on LNA gain with $IP_{31} = 15$ dBm at application of the integrated receiver

variant, the receiver is embedded as a whole into the antenna system including the DSP path. In this case, the signal is transmitted via the feeder in the digital form.

The analog path of signal processing, which is embedded into the antenna system, allows utilization of cheap coaxial cable as the feeder of necessary length since the signal is transmitted at the comparatively low IF, when the linear cable losses are small. The IF signal is limited in bandwidth by several tens or hundreds megahertz, it is simpler technically to be amplified and to be matched in the transmission path, than the signal of high frequency in the whole monitoring equipment working range; therefore, essential decrease of antenna effect influence

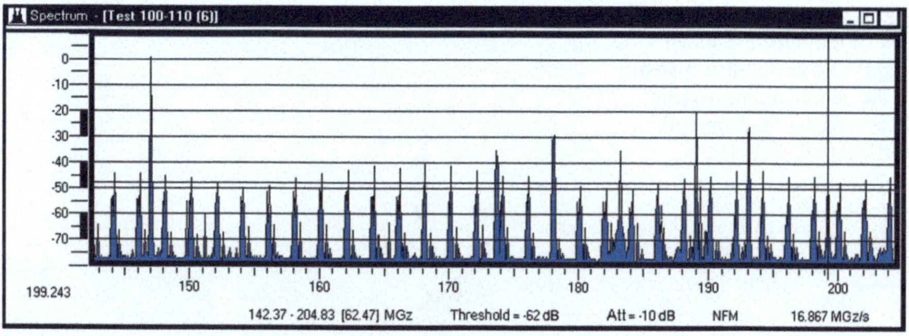

Fig. 5.15 Manifestation of the antenna effect from the powerful HF radio station

is observed. The calibration of measuring units is simplified as well, because it is enough to calibrate feeder at one frequency (equaled to IF) only.

Unfortunately, the transmission of the analog signal to IF significantly reduces the antenna effect influence, but cannot eliminate it completely. Figure 5.15 shows an example of demonstration of the powerful closely located HF radio station signal, which emission frequency falls in the IF bandwidth. The interference has a line spectrum with period, equal to the receiver tuning step.

The second variant of the receiver integration is the cardinal method of antenna effect elimination in the feeder, when both analog and digital paths of signal processing are embedded into the antenna system, and signal and control commands transmission via the feeder is carried out in the digital form only. It enables to use Ethernet or fiber-optical cables of large length, and the necessity of its calibration for measuring units is completely excluded.

5.3.4 Examples of Equipment with Integrated Receivers

The antenna system of the radio direction-finder ARTIKUL-S with the circular antenna array AS-PP4 and the antenna system of the digital panoramic measuring receiver ARGAMAK-IS [1] are shown in Fig. 5.16. The first variant of the receiver integration—the analog path only is embedded into the antenna system. The ADC unit is located in the service room or in the special container. For feeder, the cheap coaxial cable RG-58 is used in products; its length can achieve 80–100 m.

Figure 5.17 shows another monitoring equipment example—the multi-functional radio monitoring station ARCHA-INM [1], which combines functions of the measuring receiver and the correlation-interferometric radio direction-finder. The second variant of receiver integration is used in this station. The digital measuring receiver ARGAMAK-IS is embedded into the basis of the bearing antenna array (Figs. 5.18 and 5.19), the data and control commands are transmitted via the feeder according to Ethernet technology.

Fig. 5.16 Antenna array AS-PP4 of the direction-finder (right) and the antenna system of the panoramic measuring system of the field strength ARGAMAK-IS (left)

Technical characteristics of radio monitoring equipment considered in examples, in spite of small dimensions of the used integrated receivers, satisfy and even exceed values mentioned in specifications of ITU [10]. In particular, for the ARCHA-INM station, the typical values of sensitivity and direction-finding accuracy, linearity of the radio receiving path in the range 20–3000 MHz are 3 μV/m, 1^0, $IP_{33} = 10$ dBm, relatively.

Thus, application of the integrated receivers has a series of technical advantages compared to traditional arrangement of the receiver outside of the antenna system, for which lengthy high-frequency feeder is required. We can attribute the following to technical advantages:

- Extension of the dynamic range of equipment, achievement of its value to the width of the used receiver dynamic range;
- Reduction or complete elimination of antenna effect influence, which, in turn, increases sensitivity, reduces errors of weak signal parameters measurement and direction-finding;
- Calibration simplification of measuring unit's paths;
- Utilization of the cheap flexible coaxial cables of large length, Ethernet and fiber-optic lines as the feeder, which simplifies deployment and uninstalling of the antenna system, especially at location on high height.

Fig. 5.17 ARCHA-INM station

Fig. 5.18 Antenna basis of the ARCHA-INM station

Fig. 5.19 Digital receiver ARGAMAK-IS integrated into the basis of the antenna system ARCHA-INM (open access to the receiver)

The modern level of radio electronics development allows creation of integrated digital radio receivers with technical characteristics completely satisfying and even exceeding the characteristics recommended by ITU in connection with radio monitoring equipment.

5.4 Selection of Sampling Frequency

Problems of sampling frequency selection for the digital radio receiver are described fully enough in many books on DSP, for instance [8, 13–16]. In the present section, we shall consider the order of sampling frequency selection for the general case when overlapping of spectral components of low-frequency or band-pass signals, which do not belong to the useful informative signal [17], is admitted.

Time sampling of the continuous process $x(t)$ consists in determination of its values $x(k) = x(\mathrm{kT})$ in equidistant points $t_k = \mathrm{kT}$, where $k = 0, 1, 2, \ldots$. Here $T = \frac{1}{F_s}$ is the sampling period, F_s is the sampling frequency. It is known that discrete signal spectrum has a form

$$S_s(\omega) = \frac{1}{T} \sum_{n=-\infty}^{+\infty} S_x\left(\omega - \frac{2\pi n}{T}\right). \tag{5.16}$$

Proceeding from the circular frequency to the linear, we obtain another form

$$S_s(f) = F_д \sum_{n=-\infty}^{+\infty} S_x(f - F_s n) \tag{5.17}$$

Thus, the discrete signal spectrum represents the result of summation of infinite number of copies of the initial signal spectrum with accuracy within the scale multiplier F_s. These copies are formed by means of transfer of initial signal spectrum up and down along the frequency axis with the step $\pm nF_s$, where n is the integer number. In other words, at signal $x(t)$ sampling, its spectrum, consisting of the lobe A_0^+ in the positive frequency $f > 0$ region and the lobe A_0^- in the negative frequency $f < 0$ region with a period equaled to the reverse sampling frequency, is repeated up and down along the frequency axis. Evidently that if $F_s > 2f_h$, the sampling frequency meets the Nyquist–Kotelnikov theorem, lobes $S(f)$ will be located around frequencies nF_s, where $n = 0, \pm 1, \pm 2, \ldots$, in such a manner that frequency zones $nF_s \pm f_B$ do not overlap, and the lobes itself do not overlap each other. The discrete signal spectrum is presented in Fig. 5.20.

The Nyquist–Kotelnikov theorem has generalization to the case of narrowband signals, according to which at observance of the definite conditions the sampling frequency may be chosen below the $2f_h$ frequency. The main initial requirement at sampling frequency choice according to this method consists in the fact, that the narrowband spectrum is concentrated between the low boundary frequency f_l and the high boundary frequency f_h. The spectrum of the analog bandpass signal is shown in Fig. 5.21. At its sampling by the frequency $F_s \geq 2f_h$, the discrete signal spectrum will have a form shown in Fig. 5.21b. Let us begin to decrease the sampling frequency. Obviously, that at $2f_l < F_s < 2f_h$, we shall observe the superposition of spectrum lobes (or aliasing) of the discrete signal. It becomes impossible to recover the initial analog signal spectrum by the pass-band filter. Nevertheless, the further frequency decrease $F_s < 2f_l$ will provide again non-aliasing of spectrum lobes of the discrete signal to each other and recovering of the analog signal. Thus, below the zone $2f_l < F_s < 2f_h$, which is forbidden for sampling frequency choice, there is a zone, permitted for such a choice.

Since the energy spectrum of Fig. 5.21 discrete signal is periodical and even function, in the general case the location of spectrum lobes of the discrete signal near the lobe A_0^+ of the initial signal will correspond to Fig. 5.22. The condition defining boundaries of permitted zones for frequency choice is enough to be formulated for any spectrum lobe. We perform it for initial lobe A_0^+:

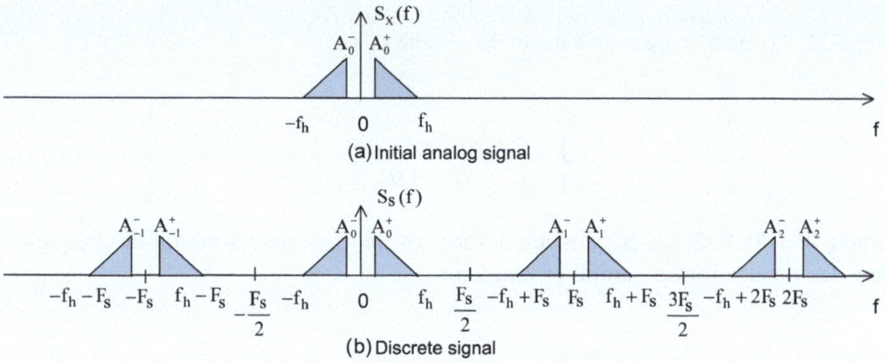

Fig. 5.20 Sampling of the low-frequency signal

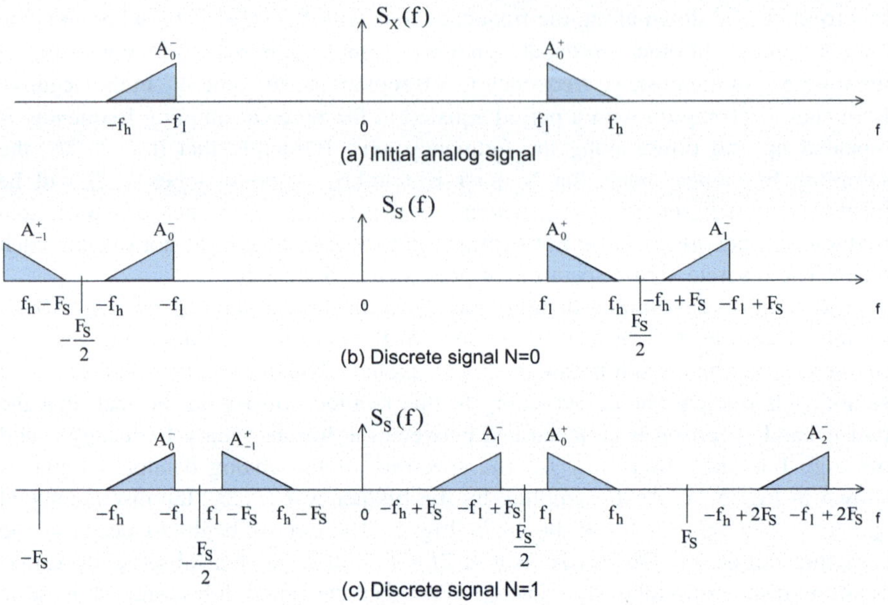

Fig. 5.21 Sampling of the bandpass signal

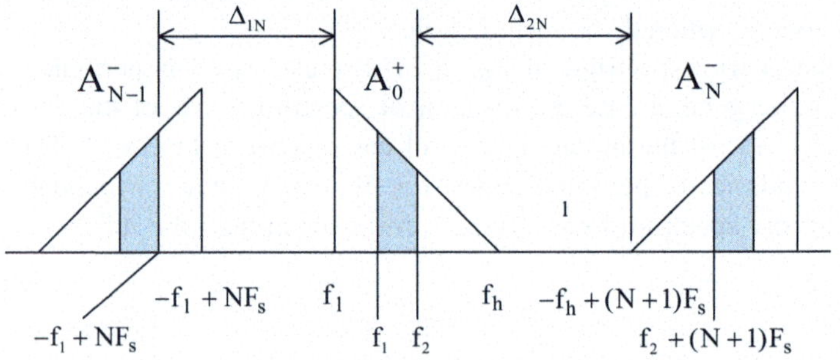

Fig. 5.22 Location of spectrum lobes of the discrete signal

$$\begin{cases} -f_l + NF_s < f_l; \\ -f_h + (N+1)F_s \geq f_h, \end{cases} \tag{5.18}$$

where $N = 0, 1, 2, \ldots, N_{\max}$ is the number of the permitted zone on the frequency axis for a selection of sampling frequency.

From inequality system (5.18), we obtain an expression for sampling frequency selection:

$$\frac{2f_h}{N+1} < F_s < \frac{2f_l}{N}. \tag{5.19}$$

From the following evident inequality

$$\frac{2f_h}{N_{\max}+1} \leq \frac{2f_l}{N_{\max}} \tag{5.20}$$

We can determine the number $N_{\max}$, which corresponds to minimally possible value of the sampling frequency $F_{s\,\min}$.

Let us transform the last expression as

$$2f_h N_{\max} \leq 2f_l N_{\max} + 2f_l. \tag{5.21}$$

At last, we have:

$$N_{\max} \leq \frac{f_l}{f_h - f_l} \tag{5.22}$$

or

$$N_{\max} = \operatorname{int}\left(\frac{f_l}{f_h - f_l}\right), \tag{5.23}$$

where int(a) is an operation of taking integer part from the number a.

The expression (5.19) corresponds to the Nyquist–Kotelnikov theorem generalized to the case of narrowband signals and provides accurate recovery of the analog signal $x(t)$ according to the discrete signal $x(\mathrm{kT})$ [15]. The minimal sampling frequency is defined from the condition

$$F_{s\,\min} \geq \frac{2f_h}{N_{\max}+1}. \tag{5.24}$$

The last inequality will enhance if the following expression will be substituted in it:

$$F_{s\,\min} \geq \frac{2f_h}{\frac{f_l}{f_h - f_l} + 1}. \tag{5.25}$$

From here we obtain the known expression

$$F_{s\,\min} \geq 2(f_h - f_l) = 2B. \tag{5.26}$$

In other words, the minimal possible sampling frequency cannot be less than the doubled bandwidth of the signal spectrum $B = f_h - f_l$.

Selection of the sampling frequency in the permitted zones provides some free gaps between the spectrum lobes of the discrete signal $x(\mathrm{kT})$. From Fig. 5.22 we see, that the gap, which value is higher and lower in frequency from the lobe A_0^+, is defined as

$$\begin{cases} \Delta_{1N} = 2f_h - \mathrm{NF}_s; \\ \Delta_{2N} = (N+1)F_s - 2f_h. \end{cases} \tag{5.27}$$

The above-mentioned expressions correspond to the case without aliasing of the discrete signal spectrum lobes. At the same time, if the linearity of the receiver analog path is rather high and intermodulation distortions are small, then it seems to be possible to admit overlapping (aliasing) of non-informative and suppressed (at further digital filtering) spectrum fragments at the ADC output with the purpose of sampling frequency minimization. The initial requirement to the sampling frequency selection, as before, consists in the fact that the signal spectrum at input of the digital path is concentrated between the lower boundary frequency f_l and the upper boundary frequency f_h, and the spectrum of the informative signal is located between the lower frequency f_1 and upper frequency f_2 inside the spectrum having the bandwidth $B_{\inf} = f_2 - f_1$. In the case of sampling without aliasing of the whole signal spectrum at the output of the IF path, as earlier, the spectrum of the discrete signal will have the form shown in Fig. 5.22. The condition (5.18) defining boundaries or permitted zones must be now formulated in connection with the "useful" informative frequency zone located within any spectral lobe, for instance, for the lobe A_0^+

$$\begin{cases} -f_l + \mathrm{NF}_s < f_1; \\ -f_h + (N+1)F_s \geq f_2, \end{cases} \tag{5.28}$$

where $N = 0, 1, 2, \ldots, N_{\max}$ is a number of the permitted zone on the frequency axis to select the sampling frequency.

Performing transformations, which are similar to the above-fulfilled ones for the case without spectrum lobes aliasing, we obtain the expression for permitted value of the sampling frequency

$$\frac{f_2 + f_h}{N+1} < F_s < \frac{f_l + f_1}{N}, \tag{5.29}$$

and the maximal zone number permitted for this choice

$$N_{\max} = \text{int}\left(\frac{f_l + f_1}{B + B_{\text{gap}}}\right), \tag{5.30}$$

and the value of free gaps higher and lower in frequency from the informative signal spectrum:

$$\begin{cases} \Delta_{1N} = f_l + f_1 - \text{N}F_s; \\ \Delta_{2N} = (N+1)F_s - f_h - f_2, \end{cases} \tag{5.31}$$

as well as the total free gap

$$\Delta_N = \Delta_{1N} + \Delta_{2N} = F_s - B - B_{\text{inf}}. \tag{5.32}$$

Figure 5.23 shows zones on the frequency axis forbidden and permitted for sampling frequency selection, and also the value of frequency gaps versus sampling frequencies taken in Nth and N + 1th permitted zones.

If the spectrum aliasing of the suppressed signals is inadmissible due to some reasons, then $f_1 = f_l$, $f_2 = f_h$, and expressions (5.29) and (5.31) take a form of formulas (5.19) and (5.27).

Digital signal processing should include digital filtering of the useful informative signal. If the digital filter, which is matched to the spectrum of useful signal by the bandwidth, is used for selection, then its amplitude–frequency response is practically symmetric with regard to a middle of its pass-band. To provide the same suppression of non-informative spectrum fragments located up and down in frequency from the spectrum fragment of the useful signal, it is expedient to have the frequency gaps up and down in frequency, which are equal between each other

$$\Delta_{1N} = \Delta_{2N}, \tag{5.33}$$

since this will decrease the possibility of the spectrum lobes aliasing. Substituting in formula (5.33) the values of frequency gaps defined by (5.31), we obtain

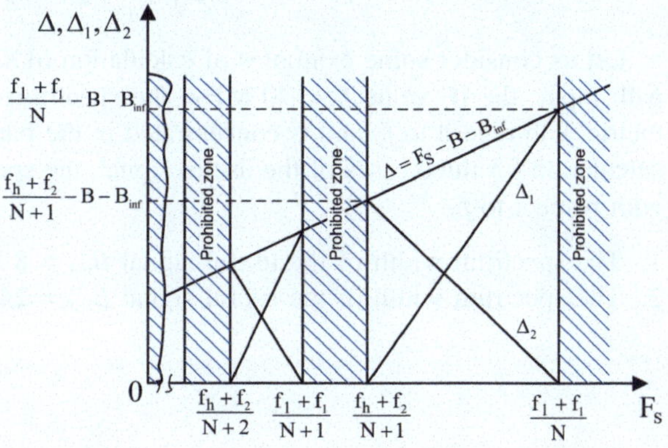

Fig. 5.23 Selection of the sampling frequency of the informative signal

$$F_s = \frac{f_l + f_h + f_1 + f_2}{2N + 1}. \tag{5.34}$$

If the spectrum interval (f_1, f_2) of the useful informational signal has the central frequency $f_0 = \frac{f_1 + f_2}{2}$, which coincides with the central frequency of the full signal spectrum at ADC input, which is within the interval (f_l, f_h), then the last formula can be transformed to the form:

$$F_s = \frac{\frac{2(f_l + f_h)}{2} + \frac{2(f_1 + f_2)}{2}}{2N + 1} = \frac{2f_0 + 2f_0}{2N + 1} = \frac{4f_0}{2N + 1}. \tag{5.35}$$

The N parameter, as earlier, defines a number of the permitted zone for a selection of the sampling frequency. Its maximal value N_{max} is defined by formula (5.30).

As it will be shown later, selection of sampling frequency in accordance with (5.35), besides providing the same gaps between spectrum lobes of the sampled signal, essentially simplifies the formation of quadrature components.

Evidently that for symmetric spectrums, the gaps higher and lower in frequency are expedient to be equal in value, i.e.,

$$\Delta_{1N} = \Delta_{2N}, \tag{5.36}$$

since this will reduce the aliasing possibility for the spectrum. From two last expressions, we obtain the condition for F_s value selection ensuring the equal frequency gaps:

$$2f_l - \mathrm{N}F_s = (N + 1)F_s - 2f_l. \tag{5.37}$$

Consecutively simplifying the last expression, we obtain

$$(2N + 1)F_s = 2(f_l + f_h); \tag{5.38}$$

$$F_s = \frac{2(f_l + f_h)}{2N + 1} = \frac{4f_0}{2N + 1}. \tag{5.39}$$

Let us consider some examples of calculation of sampling frequency. At digital path input, the IF value f_0 = 70 MHz; the spectrum of the analog signal is symmetric with regard to f_0 and is concentrated in the band B = 48 MHz. We perform calculation for three cases of the useful signal, the spectrum of which is symmetric with respect to f_0:

1. The spectrum width of the useful signal B_{inf} = 8 MHz.
2. The spectrum width of the useful signal B_{inf} = 24 MHz.

Table 5.2 Calculation of possible values of the sampling frequency

Number of permitted zone	$F_{s\max} = \frac{f_l + f_1}{N}$, MHz	$F_{s\min} = \frac{f_2 + f_h}{N+1}$, MHz	$F_s = \frac{4f_0}{2N+1}$, MHz
$B_{\inf} = 8$ MHz			
0	∞	168	280
1	112	84	93.333
2	56	56	56
$B_{\inf} = 24$ MHz			
0	∞	176	280
1	104	88	93.333
$B_{\inf} = B = 48$ MHz			
0	∞	188	280

3. The spectrum width of the useful signal equals to the spectrum width of the whole signal at the digital path input $B_{\inf} = B = 48$ MHz.

Calculation results are presented in Table 5.2.

At sampling with aliasing of the spectrum fragments to be suppressed, the value of the sampling frequency may be taken essentially smaller, since the width of the permitted zone is significantly wider than at sampling without such an aliasing.

5.5 Formation of the Quadrature Components

If the digital radio receiver has a super-heterodyne architecture, then spectrum transfer is usually performed into the baseband area (to the "zero" frequency) in the DSP path. Conveniently, the similar transfer is carried out by means of multiplying of discrete time samples $x(\mathrm{kT})$ by the complex reference modulating sequence

$$\alpha(\mathrm{kT}) = A \exp[j\,(k\omega T + \eta)], \tag{5.40}$$

where $k = 0, 1, 2\ldots$ is a number of the time sample; $T = 1/F_s$ is the period of the time sampling; A is an amplitude; η is the initial phase.

In order to reduce calculations amount related to generation of the reference sequence, it is expedient to select the sampling frequency F_{sampl}, the amplitude A, and the initial phase η in such a manner, that values of the $\alpha(\mathrm{kT})$ sequence elements would be exactly repeated at each period and would be suitable for executing of multiplication operations, for example, correspond to the set of numbers 0, ±1/2, ±1, ±2, and so on.

Application of "time" representation for the reference sequence does not give the general answer how to select its samples. Besides, it is desirable to give the clear

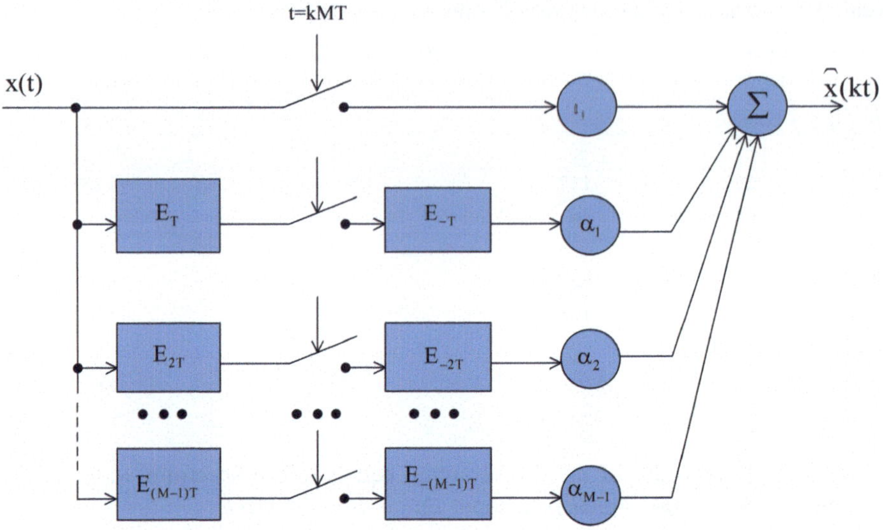

Fig. 5.24 Formation of the discrete signal

representation of spectrum transformation of the discrete signal at periodic modulation of its time values. Now we obtain relationships for the spectrum of the sampled signal $\hat{x}(\mathrm{kT})$ with the period MT modulated by the sequence $\alpha(\mathrm{kT})$, i.e., $\hat{x}(\mathrm{kT}) = x(\mathrm{kT})\alpha(\mathrm{kT})$ [18]. For this, we form the $\hat{x}(\mathrm{kT})$ signal from the continuous signal $x(t)$ with the help of the scheme presented in Fig. 5.24. This scheme consists of the parallel branches containing the ideal switches, which simultaneously closes and right now opens with the MT period. The continuous signal $x(t)$ is applied to each branch, and output discrete signals are summed with each other.

Then, for the discrete signal $x(\mathrm{kT})$ obtaining, which has the sampling period T differing from the period of MT switching on, each channel should contain the advancing elements of the continuous signal at input of its switch by time $0, T, 2T, \ldots, (M-1)T$, and at switch output should have elements of delay of discrete signals by time $0, -T, -2T, \ldots, -(M-1)T$. If the signal $x(\mathrm{kMT})$ in each branch is multiplied by the α_m coefficient, the discrete signal $\hat{x}(\mathrm{kT})$ can be written as

$$x(\mathrm{kT}) = \sum_{m=0}^{M-1} \alpha_m E_{-\mathrm{mT}}\{D\{E_{\mathrm{mT}}\{x(t)\}\}\}, \tag{5.41}$$

where E_{mT} is the operator of advancing by time mT (signal shift to the left); $E_{-\mathrm{mT}}$ is the delay operator by time mT (signal shift to the right); D is the operator of signal sampling with the MT period.

If the continuous signal $x(t)$ has the Fourier spectrum $X(\omega) = F\{x(t)\}$, the signal $E_{\mathrm{mT}}\{x(t)\}$ will have the spectrum $X(\omega)e^{j\omega \mathrm{mT}}$. According to expression

(5.16), the spectrum of the sampled (with the period MT) signal $E_{\mathrm{mT}}\{x(t)\}$ will have a form

$$F\{D\{E_{\mathrm{mT}}\{x(t)\}\}\} = \frac{1}{\mathrm{MT}}\sum_{n=-\infty}^{\infty} X\left(\omega - \frac{2\pi n}{\mathrm{MT}}\right)e^{j\left(\omega-\frac{2\pi n}{\mathrm{MT}}\right)\mathrm{mT}}. \quad (5.42)$$

The spectrum of delayed discrete signal $E_{-\mathrm{mT}}\{D\{E_{\mathrm{mT}}\{x(t)\}\}\}$ in each branch will be equal

$$F\{E_{-\mathrm{mT}}\{D\{E_{\mathrm{mT}}\{x(t)\}\}\}\} = \frac{1}{\mathrm{MT}}\sum_{n=-\infty}^{\infty} X\left(\omega - \frac{2\pi n}{\mathrm{MT}}\right) e^{j\left(\omega - \frac{2\pi n}{\mathrm{MT}}\right)\mathrm{mT}}e^{-\mathrm{jmT}}, \quad (5.43)$$

or, which is the same

$$F\{E_{-\mathrm{mT}}\{D\{E_{\mathrm{mT}}\{x(t)\}\}\}\} = \frac{1}{\mathrm{MT}}\sum_{n=-\infty}^{\infty} X\left(\omega - \frac{2\pi n}{\mathrm{MT}}\right)e^{-j\frac{2\pi nm}{M}}. \quad (5.44)$$

The spectrum of the sum discrete signal $\widehat{x}(\mathrm{kT})$ will have a form

$$\widehat{X}_D(\omega) = \frac{1}{\mathrm{MT}}\sum_{m=0}^{M-1}\alpha_m\sum_{n=-\infty}^{\infty} X\left(\omega - \frac{2\pi n}{\mathrm{MT}}\right)e^{-j\frac{2\pi nm}{M}}. \quad (5.45)$$

Let us change the summation consequence presenting the last expression as:

$$X_D(\omega) = \frac{1}{T}\sum_{n=-\infty}^{\infty}\frac{1}{M}\sum_{m=0}^{M-1}\alpha_m e^{-j\frac{2\pi nm}{M}} X\left(\omega - \frac{2\pi n}{\mathrm{MT}}\right) \quad (5.46)$$

or

$$\widehat{X}_D(\omega) = \frac{1}{T}\sum_{n=-\infty}^{\infty}\sigma_n X\left(\omega - \frac{2\pi n}{\mathrm{MT}}\right), \quad (5.47)$$

where the σ_n coefficient with accuracy to the constant multiplier is defined as the discrete Fourier transform of the modulating sequence $\alpha_0, \alpha_1, \ldots, \alpha_{M-1}$:

$$\sigma_n = \frac{1}{M}\sum_{m=0}^{M-1}\alpha_m e^{-j\frac{2\pi nm}{M}}, \quad (5.48)$$

or in the matrix form

$$\vec{\sigma} = \frac{1}{M}\boldsymbol{F}\vec{\alpha}, \quad (5.49)$$

where F is the square matrix of rotational multiples of the direct FFT with dimension M × M, which elements are

$$f_{\rm iq} = e^{-j\frac{2\pi iq}{M}}, \quad 0 \le i \le M-1, \; 0 \le q \le M-1; \tag{5.50}$$

$\vec{\alpha} = \begin{bmatrix} \alpha_0 \\ \alpha_1 \\ \cdots \\ \alpha_{M-1} \end{bmatrix}$ is the modulating vector with dimension $M \times 1$.

If modulation is absent, all $\alpha_m = 1$ and coefficients σ_n equal to zero for all n except $n = 0, \pm M, \pm 2M, \ldots$, when they take values equaled to 1. Therefore, at $\alpha_m = 1$ the expression (5.47) after variable change $n = $ mi transfers to the usual expression for the spectrum of the discrete x(kT) signal:

$$X_D(\omega) = \frac{1}{T}\sum_{i=-\infty}^{\infty} X\left(\omega - \frac{2\pi i}{T}\right). \tag{5.51}$$

Let us divide the sum (5.47) to M partial sums with numbers $k = 0, \pm M, \pm 2M, \ldots$ and introduce new variable $n = m + \mathrm{iM}$, and as a result we have

$$\widehat{X_D}(\omega) = \frac{1}{T}\sum_{m=0}^{M-1}\sum_{i=-\infty}^{\infty} \sigma_{m+iM} X\left(\omega - \frac{2\pi\,(k+iM)}{\mathrm{MT}}\right). \tag{5.52}$$

Because coefficients $\sigma_{m+\mathrm{iM}}$ are periodic with the period M, i.e., $\sigma_{m+\mathrm{iM}} = \sigma_m$, they can be taken outside of the sign of summing by i. Then, changing the variable m by n and transferring to the cyclic frequency $f = 2\pi\omega$, we obtain

$$\widehat{X_D}(f) = \frac{1}{T}\sum_{n=0}^{M-1} \sigma_n \sum_{i=-\infty}^{\infty} X\left(f - \frac{F_D(n+iM)}{\mathrm{M}}\right). \tag{5.53}$$

The expression (5.53) allows formulation of the next rule of spectrum transformation of non-periodic discrete signal x(kT) (with the T sampling interval) at its multiplication by the periodic sequence (with the MT period): Each spectrum lobe of the discrete signal is divided into M parts (sub-lobes) falling in the interval equaled to the sampling frequency F_s, equidistantly with the step F_s/M, i.e., near frequencies $\mathrm{rF}_s + \mathrm{mF}_s/M$, where r and m are integer numbers, at that $m < M$. Each formed in such a manner sub-lobe repeats the initial spectrum lobe of non-modulated signal in form, but it is taken with the complex weighting coefficient σ_n.

Changing a size and elements of the modulating vector $\vec{\alpha}$, we may change the vector $\vec{\sigma}$, and hence, the power and phases of spectral lobes concentrated near points $\mathrm{rF}_s + \mathrm{mF}_s/M$.

The circular shift of $\vec{\alpha}$ vector elements by n positions leads to the fact that modules of weighting coefficients remain constant, and arguments are changed by the value $\frac{2\pi}{M}n$, which makes possible to form the digital sequences with the given phase shift, quadrature, and multi-phase signals.

Starting from the above-formulated rule of spectrum transformation at periodic modulation to provide the central frequency f_0 transfer into the baseband ("zero") frequency, the sampling frequency and converted central frequency should be related by the relationship

$$\mathrm{r}F_s + \frac{m}{M}F_s = f_0, \tag{5.54}$$

where r and m are integer numbers, at that, $m < M$.

To form quadratures, the cyclic shift of the modulating vector should provide the argument variation of weighting frequency coefficients σ_n by the $\pi/2$ angle. At shift to one position, the argument will change by the value $2\pi/M$, hence, the minimal possible length of the modulation vector $M = 4$. From expression (5.54) it follows that to provide transformation of the central frequency into zero point, the sampling frequency and the central frequency of the useful signal should meet the condition:

$$\mathrm{r}F_s + \frac{F_s}{4} = f_0, \tag{5.55}$$

or the condition

$$\mathrm{r}F_s + \frac{3F_s}{4} = f_0. \tag{5.56}$$

Figure 5.25a shows the spectrum of the initial analog signal, and Fig. 5.25b, c show the corresponding signals spectrums sampled with the sampling frequency according to (5.55) and (5.56). In the first case, the signal spectrum located around the point $\frac{F_s}{4}$ is not inverted by the frequency, while in the second case it has the inverted character.

If to impose the condition, that the signal spectrum should not have the inversed character after quadrature formation, then evidently, the procedure of periodic modulation will be different depending on evenness or oddness of the permitted zone number.

In accordance with the above-formulated rule of spectrum transformation at periodic modulation, from Fig. 5.25b, which shows the spectrum of the signal sampled with the sampling frequency F_s, located in the even permitted zone, it follows that for a result of modulation, in order to locate the spectral lobes around points $0, \pm F_s, \pm 2F_s, \ldots$, as it is shown in Fig. 5.25d, the weighting vector is necessary in the frequency domain

$$\vec{\sigma} = [0 \quad 0 \quad 0 \quad 1]^T. \tag{5.57}$$

The modulating vector in the time domain corresponds to it

Fig. 5.25 Formation of the quadrature signal

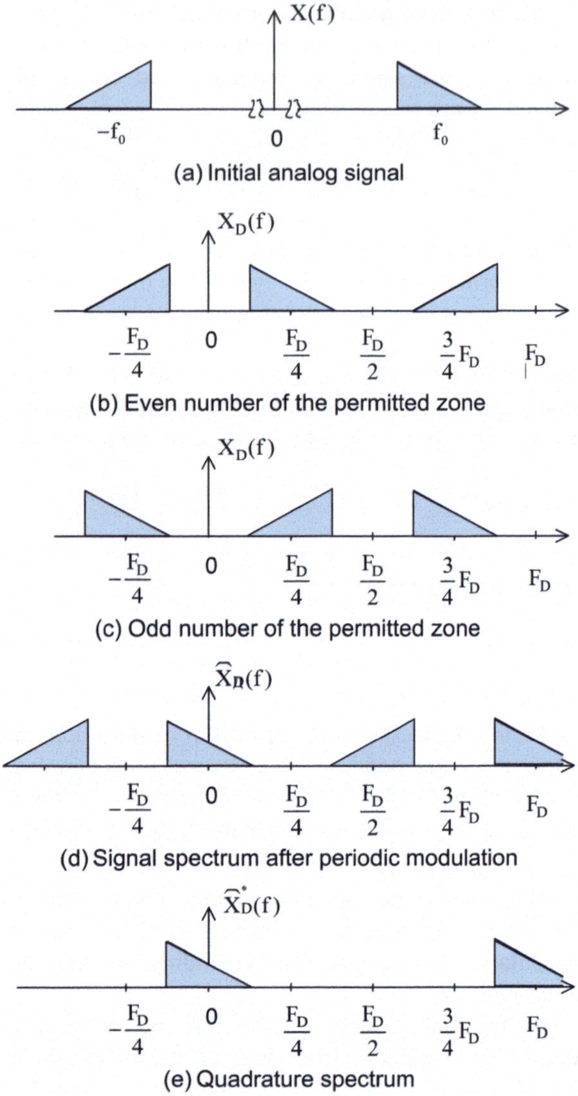

$$\vec{\alpha} = \mathbf{MF}^{-1}\vec{\sigma} = [1 \quad -j \quad -1 \quad j]^{\mathrm{T}}. \tag{5.58}$$

Similarly, for the signal sampled with the frequency F_s in the chosen zone with the odd number (see Fig. 5.25c), the weighting vector is

$$\vec{\sigma} = [0 \quad 1 \quad 0 \quad 0]^{\mathrm{T}} \tag{5.59}$$

and the modulating vector in the time domain corresponds to it

$$\vec{\alpha} = [1 \quad j \quad -1 \quad -j]^{\mathrm{T}}. \tag{5.60}$$

The spectrum of the modulated signal $\widehat{x}(\mathrm{kT})$ will have a form shown in Fig. 5.25d. The real and imaginary parts of this signal are not the quadrature components, because in vicinities of points $n\frac{F_s}{2}$ (n is the integer number) there are spectrum lobes. To form quadratures, these spectral components must be suppressed, for instance, by means of low-frequency filtering.

Usually, it is recommended to use the periodic complex sequence, which is the sample from the discrete complex exponent at frequency $F_s/4$, to form the quadrature components. In connection with the periodic modulation, this sequence at the period is defined by the modulating vector (5.58). In this connection, we should note that modulating sequences, being samples from the complex exponent, are not the only possible ones to form quadratures. For example, vector

$$\vec{\alpha} = [2 \quad -1-j \quad 0 \quad -1+j]^{\mathrm{T}} \tag{5.61}$$

provides also the formation of quadrature components since the following weighting vector corresponds to it

$$\vec{\sigma} = [0 \quad 0 \quad 1 \quad 1]^{\mathrm{T}}. \tag{5.62}$$

But the modulation vector, being the sample from the complex exponent, does not give additional lobes in the spectrum of the modulated signal because in this case, the one element of the weighting vector in the frequency domain is equal to 1 and the others are zeros, which simplifies further digital processing.

The formulated rule enables determination of dimension and elements of the modulating vector, which provides formation of quadrature components for DSP algorithms of radio monitoring equipment, which are built in accordance with the super-heterodyne architecture.

5.6 Complex Digital Filtering

In those cases when it is necessary to provide reception of different types of signals by the only one receiver, one can use DSP with quadratic sample pairs.

Formation of quadratic components is performed in analog or digital paths of the radio receiver. Formation in the receiver analog path is based on application of quadratic local oscillators as it was examined in Sect. 5.1, in which multiplication of the input analog signal takes place by the sine and cosine components of the local oscillator signal with further low-frequency filtering. Instead of quadrature local oscillators, the analog Hilbert converters or phase-shifters by 90° can be used.

One of possible receiver structures with analog formation of quadrature components is presented in Fig. 5.26. The output signal of the main receiving path is

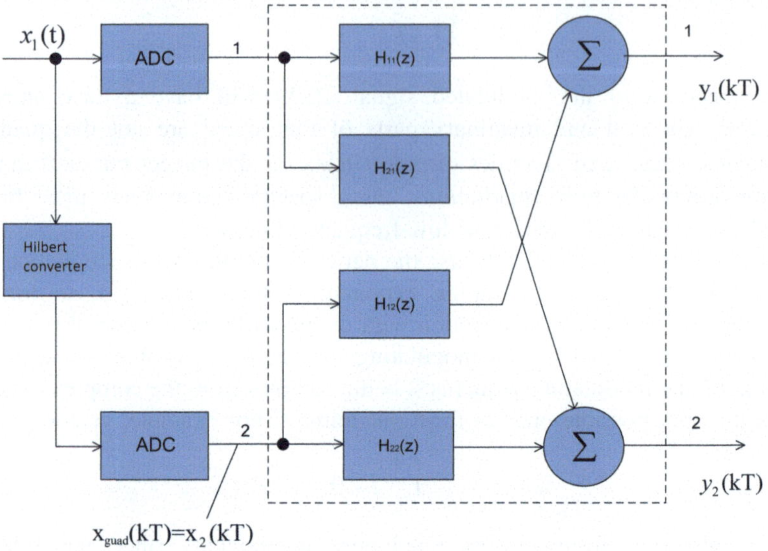

Fig. 5.26 Dual-channel system for DSP with analog quadrature formation

divided into the in-phase and quadrature components with the help of the Hilbert converter (HC). Signal extraction of the working transmission channel is carried out by the dual-channel digital filter having (in general case) the direct and cross-couplings.

The transfer matrix of the dual-channel filter

$$\boldsymbol{H}(z) = \begin{bmatrix} H_{11}(z) & H_{12}(z) \\ H_{21}(z) & H_{22}(z) \end{bmatrix} \tag{5.63}$$

connects z-representations of output sequences $\mathrm{Y}_1(\mathrm{z})$ and $\mathrm{Y}_2(\mathrm{z})$ with z-representation of input sequences $\mathrm{X}_1(\mathrm{z})$ and $\mathrm{X}_2(\mathrm{z})$:

$$\begin{bmatrix} Y_1(z) \\ Y_2(z) \end{bmatrix} = \boldsymbol{H}(z) \begin{bmatrix} X_1(z) \\ X_2(z) \end{bmatrix}. \tag{5.64}$$

At formation of quadrature components in the analog receiver path, it is necessary to use two ADCs, as shown in Fig. 5.26, but at that the value less by half of the sampling frequency is required comparing to the single-channel, since the signal having the single-side spectrum is exposed to sampling. At quadrature component formation in the DSP path, the digital quadrature local heterodynes can be used, which reproduce the complex exponential sequence, or methods of periodic modulation considered in Sect. 5.4.

The complex filtering of the sequence $x(\mathrm{kT}) = x_1(\mathrm{kT}) + jx_2(\mathrm{kT})$ in the complex digital filter (CDF) is the possible algorithm of quadrature components processing.

The CDF is the unit with two inputs and two outputs, which z-representations of output sequences $Y_1(z)$ and $Y_2(z)$ are connected with z-representation of input sequences by the following expression

$$Y_1(z) + jY_1(z) = H_{CF}(z)[x_1(\mathrm{kT}) + jx_2(\mathrm{kT})], \tag{5.65}$$

where $H_{CF}(z) = H_1(z) + jH_2(z)$ is the transfer function of the complex filter; $H_1(z)$ and $H_2(z)$ are real and imaginary parts of the CDF transfer function, which are functions with real coefficients.

Expression (5.65) has the following view in the matrix form:

$$\begin{bmatrix} Y_1(z) \\ Y_2(z) \end{bmatrix} = \boldsymbol{H}_{CF}(z) \begin{bmatrix} X_1(z) \\ X_2(z) \end{bmatrix}, \tag{5.66}$$

where $\boldsymbol{H}_{CF}(z)$ is the transfer matrix of CDF

$$\boldsymbol{H}_{CF}(z) = \begin{bmatrix} H_{11}(z) & H_{12}(z) \\ H_{21}(z) & H_{22}(z) \end{bmatrix} = \begin{bmatrix} H_1(z) & -H_2(z) \\ H_2(z) & H_1(z) \end{bmatrix}. \tag{5.67}$$

Thus, the complex filter is the dual-dimension filter with antisymmetric cross-coupling [19].

The CDF frequency response $H_{CF}(e^{j\bar{\omega}})$ connects the output complex sequence $\dot{y}(\mathrm{kT})$ with input exponential complex sequence $\dot{x}(\mathrm{kT}) = e^{j\bar{\omega}k}$, where $\bar{\omega} = \omega T$. In contrast to frequency responses of real digital filters, the module and argument of this characteristic may have functions of general form, i.e., may not to be even or odd with respect to points $\mathrm{nF}_s = n/T$. Therefore, the CDF can be matched with the spectrum of the useful complex signal, which is also non-symmetric relatively to these points providing its extraction on the background of spurious signals of adjacent channels.

At Hilbertian and conjugated Hilbertian input signals, the CDF output signal will be Hilbertian or conjugated Hilbertian as well. Therefore, both outputs of such a filter are informatively equivalent. The real and imaginary components of the output signal have usual double-sided spectrum, and therefore can be exposed to usual digital-to-analog conversion. Figure 5.27 illustrates operations of useful signal extraction by the complex digital filter on an example of a signal with single-sideband modulation (SSB). In this case, the SSB signal of the direct transmission channel having the amplitude spectrum $|X_d(f)|$ is considered as the useful signal. The signal of the inverse channel with the amplitude spectrum $|X_{\mathrm{inv}}(f)|$ is spurious. Figure 5.27 shows the spectrum of the analog signal $x(t)$ with the central frequency f_0 equaled to suppressed carrier of the single-sideband signal. With the help of the analog Hilbert converter, the quadratic component $x_{\mathrm{quad}}(t)$ is formed as shown in Fig. 5.26. The in-phase and quadrature components form the analytic signal $x_{an}(t) = x(t) + x_{\mathrm{quad}}(t)$ having the single-sideband spectrum, as shown in Fig. 5.27b. After analog-to-digital conversion with the frequency F_s multiple to the central frequency f_0, the discrete signal is formed which amplitude

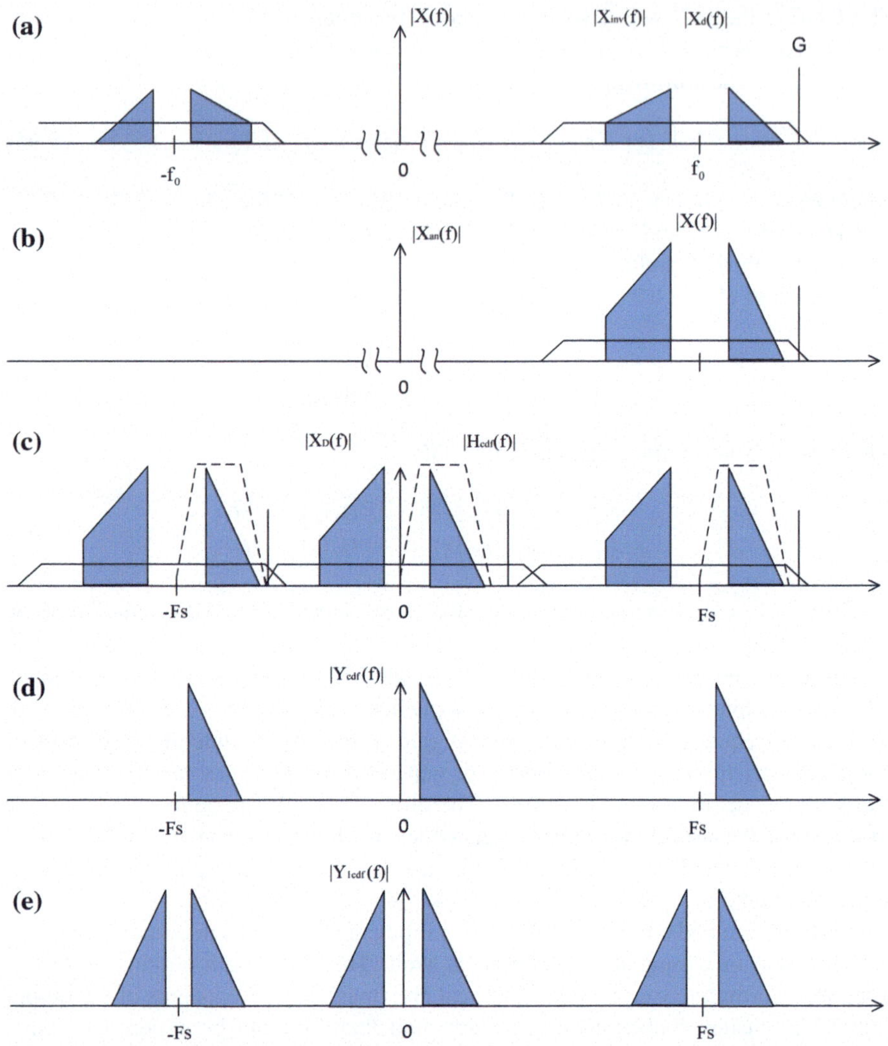

Fig. 5.27 Extraction of the useful signal by the complex filter **a** analog signal spectrum, **b** analog analytical signal spectrum, **c** discrete signal and CDF amplitude–frequency characteristic, **d** spectrum of extracted signal, **e** double-sided spectrum of imaginary or real component

spectrum $|X_D(f)|$ is shown in Fig. 5.27c. With the help of CDF having the amplitude–frequency characteristic $|X_{\text{inv}}(f)|$ matched in the bandwidth with the spectrum of the useful signal $|X_d(f)|$, the filtering is executing. As a result, the useful signal is extracted, which spectrum is shown in Fig. 5.27d. At that, interference of the adjacent channels, which are out of pass-band of CDF, is suppressed. The real or imaginary component has the usual double-sided spectrum, as shown in

Fig. 5.27e, and can be exposed to usual digital-to-analog conversion, for instance, with the purpose of audio reproduction.

Switching of the radio receiver to the transmission channel below "zero" frequency can be performed by sample inversion of the in-phase and quadrature components because at that spectrums of $x_d(\mathrm{kT})$ and $x_{inv}(\mathrm{kT})$ signals will interchange.

Another variety of the dual-channel scheme of the useful informational signal extraction is based on the spectrum shift of the informational signal in order to its middle transfer to the "zero" frequency. Spectrum transfer is executed by multiplication on samples of the complex exponent. After such a shift, the spectrum of the informational signal will be symmetric with regard to the "zero" frequency, and, hence, the CDF tuned to this signal will represent two similar not-coupled low-pass digital filters, which will correspond to the CDF transfer matrix without cross-coupling

$$\mathbf{H}_{\mathrm{CF}}(z) = \begin{bmatrix} H_{11}(z) & 0 \\ 0 & H_{22}(z) \end{bmatrix} = \begin{bmatrix} H_{\mathrm{LP}}(z) & 0 \\ 0 & H_{\mathrm{LP}}(z) \end{bmatrix}, \tag{5.68}$$

where $H_{11}(z) = H_{22}(z) = H_{\mathrm{LP}}(z)$ is the transfer function of the low-pass digital filter.

After filtering, the spectrum of extracted signal can be transferred to the necessary frequency area or be exposed to other processing according to the operation algorithm of the radio monitoring equipment.

For calculation of the CDF transfer function, one can use the following method [19]. At first, the transfer function $W(z)$ of usual real single-channel low-frequency prototype of the digital filter is calculated, which frequency response exactly repeats the required frequency characteristic of the complex filter, but is symmetric with regard to the "zero" frequency. After this, the shift of prototype frequency response is executed by means of variable z change in $W(z)$ to $z \times \mathrm{e}^{-j2\pi f_c/F_s}$, where f_{sh} is the required shift of the frequency response defined by the useful signal spectrum.

At such a calculation method, the CDF transfer function can be presented as

$$H_{\mathrm{CF}}(z) = \frac{\sum_{i=0}^{N} a_i z^{-i}}{\sum_{i=0}^{L} b_i z^{-i}}, \tag{5.69}$$

where

$$a_i = \alpha_i \times e^{j2\pi f_c/F_s}; \; b_i = \beta_i \times e^{j2\pi f_c/F_s}. \tag{5.70}$$

In the last formula, α_i and β_i are coefficients of the transfer function of CDF prototype. For the non-recursive CDF, coefficients in the denominator equal to zero besides the coefficient $b_0 = 1$.

Having written coefficients as $a_i = a_{\text{ri}} + ja_{\text{imi}}$, $b_i = b_{\text{ri}} + jb_{\text{imi}}$ and having represented a numerator and a denominator of $H_{\text{CF}}(z)$ in the form of complex sums of real polynomials of z, we obtain

$$H_{\text{CF}\Phi}(z)\frac{A_r(z) + jA_{\text{im}}(z)}{B_r(z) + jB_{\text{im}}(z)}, \tag{5.71}$$

where

$$A_r(z) = \sum_{i=0}^{N} a_{ri}z^{-i};\ A_{\text{im}}(z) = \sum_{i=0}^{N} a_{\text{imi}}z^{-i}; \tag{5.72}$$

$$B_r(z) = \sum_{i=0}^{N} b_{\text{ri}}z^{-i};\ B_{\text{im}}(z) = \sum_{i=0}^{N} b_{\text{imi}}z^{-i}. \tag{5.73}$$

Using (5.66), we have the equation system

$$\begin{cases} B_r(z)Y_1(z) - B_{\text{im}}(z)Y_2(z) = A_r(z)X_1(z) - A_{\text{im}}(z)X_2(z); \\ B_{\text{im}}(z)Y_1(z) - B_r(z)Y_2(z) = A_{\text{im}}(z)X_1(z) - A_r(z)X_2(z), \end{cases} \tag{5.74}$$

which defines the CDF possible structure shown in Fig. 5.28.

According to structural diagram in Fig. 5.28, using real polynomials of z in the numerator and the denominator of the CDF transfer function, we obtain expressions for elements of CDF transfer matrix $\mathbf{H}_{\text{CF}}(z)$

$$\begin{aligned} H_{11}(z) = H_{22}(z) &= \frac{A_r(z)B_r(z) + A_{\text{im}}(z)B_{\text{im}}(z)}{B_r^2(z) + B_{\text{im}}^2(z)}; \\ H_{21}(z) = -H_{12}(z) &= \frac{A_{\text{im}}(z)B_r(z) - A_r(z)B_{\text{im}}(z)}{B_r^2(z) + B_{\text{im}}^2(z)}. \end{aligned} \tag{5.75}$$

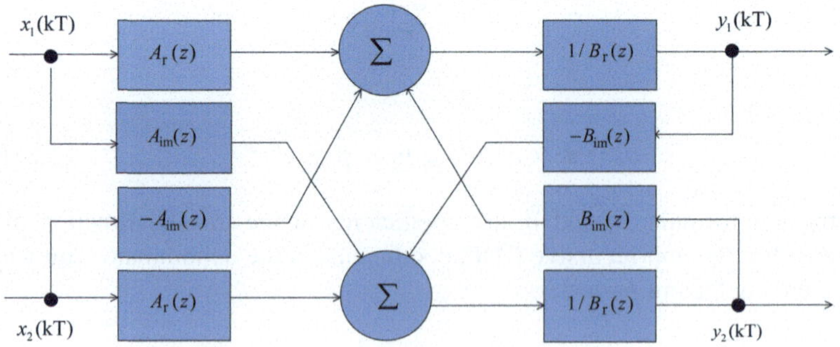

Fig. 5.28 Implemented CDF structure

For the non-recursive digital filter, the polynomials of the denominator $B_r(z) = 1$, $B_{\text{im}}(z) = 0$ and two last formulas take forms

$$\begin{aligned} H_{11}(z) &= H_{22}(z) = A_r(z); \\ H_{21}(z) &= -H_{12}(z) = A_{\text{im}}(z). \end{aligned} \tag{5.76}$$

In order to have concept of achieved selectivity and complexity of obtained CDF, we consider two examples. Figure 5.29 shows the amplitude–frequency A(f) and the phase-frequency $\Theta(f)$ characteristics of recursive CDF of sixth order containing three biquadratic sections calculated according to typical requirements to devices for SSB signal filtering: amplitude–frequency characteristic (AFC) irregularity in the pass-band is not more than 1 dB, attenuation in the filter attenuation band not less than 60 dB at frequency offset from boundary frequencies of the pass-band of 600 Hz; the spectrum of reproduced frequencies of the informational signal with the bandwidth 300–3400 Hz has higher frequencies relative to the frequency of the converted suppressed carrier. This filter provides not only filtering, but demodulation of the useful SSB signal.

Figure 5.30 shows the amplitude–frequency *A(f)* and the phase-frequency $\Theta(f)$ characteristics of the non-recursive CDF synthesized with the help of the Parks–McClellan algorithm [14] according to the same initial requirements to selectivity, as in the previous example.

In contrast to the recursive CDF of sixth order, the non-recursive CDF has the 70th order, but its phase-frequency characteristic is linear. If together with filtering the frequency downsampling is not made, then non-recursive filter implementation will require high computing expenses compared to the recursive filter. This drawback can be essentially reduced or even eliminated at all by using

Fig. 5.29 Frequency characteristics of the complex recursive digital filter of the sixth order

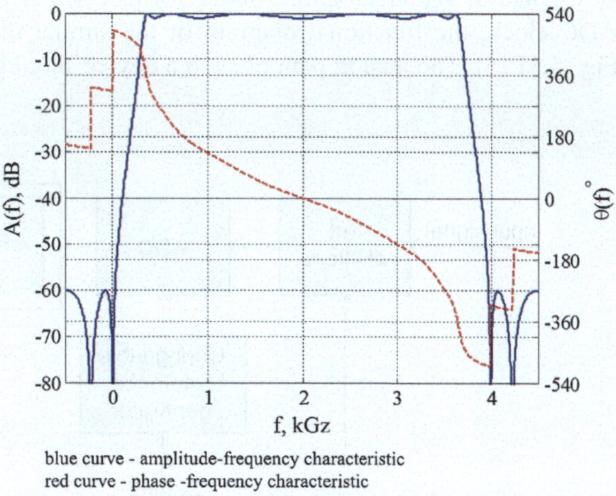

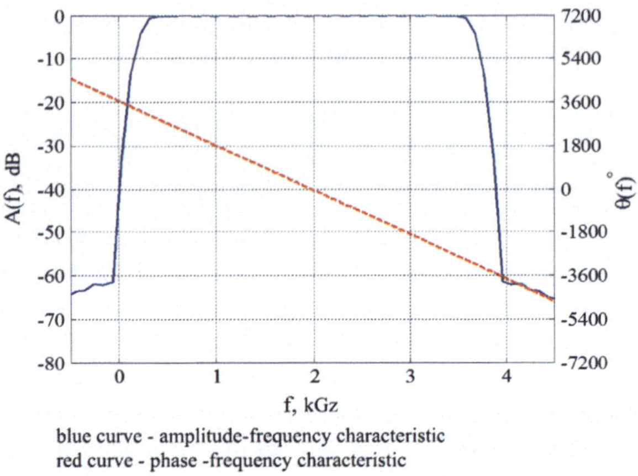

Fig. 5.30 Frequency characteristics of the complex non-recursive digital filter of the 70th order

non-recursive filtering algorithms on the base of fast convolutions, for example, with the help of FFT [13].

5.7 Resampling of Digital Signal

The receiver for radio monitoring should provide signal processing and demodulation of signals of digital communication standards, which spectrum bandwidth does not exceed bandwidth of the digital receivers. For signal processing with different symbol rates, it is necessary to be able to change frequency, with which the signal sampling is executed. Devices for sampling frequency changing are often referred to as oversampling or resampling devices.

Change of signal sampling frequency can be performed due to changing of the ADC clock; the functional diagram of the similar digital receiver is presented in Fig. 5.31 [5]. The analog path of such a device should contain a set of filters, which

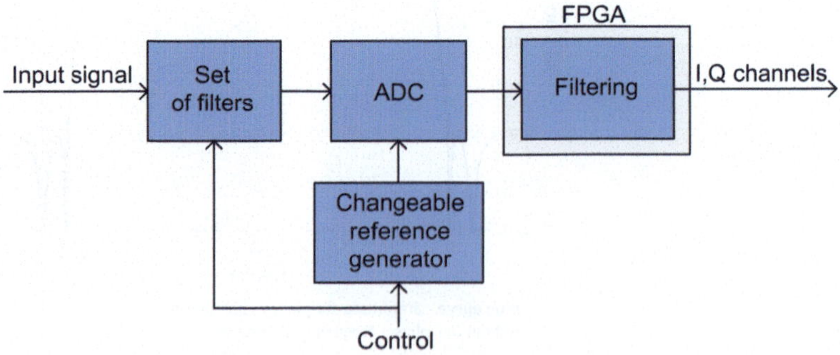

Fig. 5.31 Functional diagram of the digital receiver with ADC changeable clock

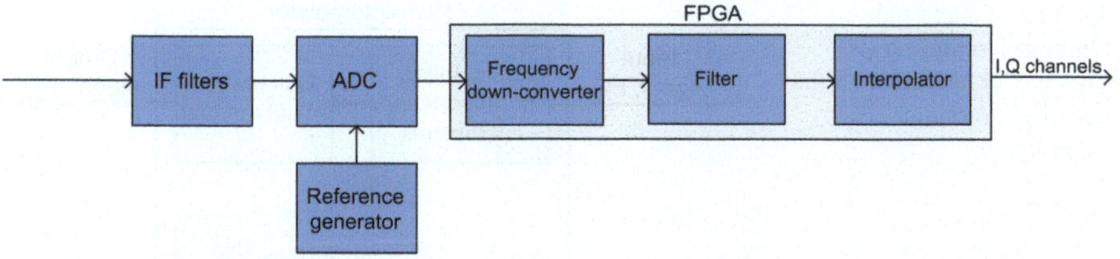

Fig. 5.32 Functional diagram for sampling frequency changing by the filter method

limit the signal bandwidth in accordance with the preset sampling frequency. The necessity of utilization of large set of analog filters is the significant disadvantage of such an engineering solution.

The alternative approach supposes usage of ADC fixed clock and sampling frequency changing by digital methods in FPGA. The appropriate circuit of the digital receiver is presented in Fig. 5.32. The analog path contains one IF filter with bandwidth defined in advance, because the ADC clock is fixed. It provides receiver high spurious response selectivity and stability of its characteristics.

Changing of sampling frequency in FPGA can be performed by different methods. The filter approach supposes utilization of filter-decimators and interpolators with integer-valued upsampling and downsampling factors [20].

As a result of filtering, the signal bandwidth at the digital filter output of the main selectivity becomes much smaller than the signal bandwidth at its input, and downsampling after filtering will not cause aliasing of digital signal spectrum lobes. At integer-valued downsampling, unnecessary samples are not calculated, which saves the computing resources. For non-recursive filter, the computing economy will be proportional to the decimation coefficient.

This solution enables provision of the given distortion level by selection of filters characteristics, but has limitations related to restricted maximal clock of FPGA. Besides, application of a large number of intermediate stages of upsampling and downsampling may become inexpedient due to excessive utilization of FPGA computing resources.

Another approach of sampling frequency changing is application of interpolating (approximating) algorithms, which allow obtaining of new samples with required sampling frequency on the base of signal samples at the initial sampling frequency. Figure 5.32 shows the functional diagram of sampling frequency changing in FPGA using interpolating algorithms.

The digital signal after ADC passes to FPGA at the digital frequency converter, where complex signal formation is executed; then the preliminary digital filtering is executed with the purpose of signal bandwidth limitation. The complex signal limited in the bandwidth passes to the interpolator providing the sampling frequency changing with the help of polynomial interpolation.

The functional diagram of the unit for sampling frequency changing in FPGA constructed on the base of the interpolating algorithm is presented in Fig. 5.33. The input signal has the sampling frequency F_{s1}, and the output signal has a required

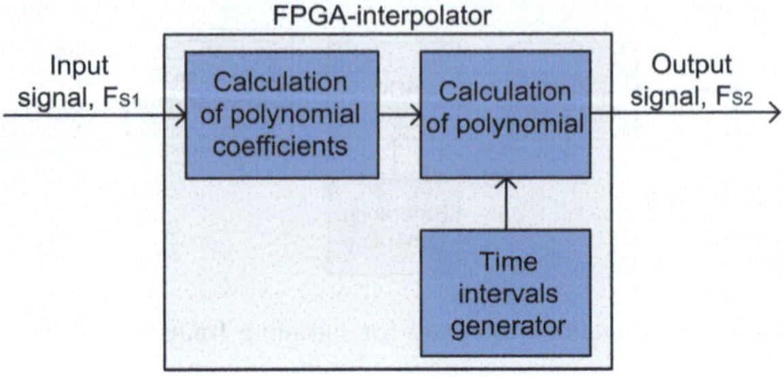

Fig. 5.33 Functional diagram of sampling frequency changing with the help of polynomial interpolation

frequency F_{s2}. To execute interpolation, it is necessary to calculate the polynomial value of n order. Polynomial coefficients are determined in agreement with the input signal samples according to approach described in [21, 22]. The time interval generator performs the formation of difference intervals characterizing position of signal samples at new frequency F_{s2} with respect to samples at the initial frequency F_{s1}. The dynamic range provided by the resampler depends on the interpolating polynomial order and on the relationship of input and output sampling frequencies. As an example, the AFC of resamplers with interpolating Lagrange polynomials of third, fifth, and seventh orders is presented in Fig. 5.34; at that, the frequency axis is normalized with respect to the input sampling frequency. Figure 5.35 allows comparison of calculated and obtained at modeling dependences of the resampler dynamic range versus the normalized bandwidth of the input complex signal; results are presented for different orders of the interpolation polynomial [21].

It should be noted that with the help of resampler built on the base of interpolating polynomial, it is possible to perform not only sampling frequency changing, but clock synchronization of the digital signal.

Thanks to possibilities of optimization of computing expenses and output signal quality, the system of resampling on the base of the interpolating polynomial is

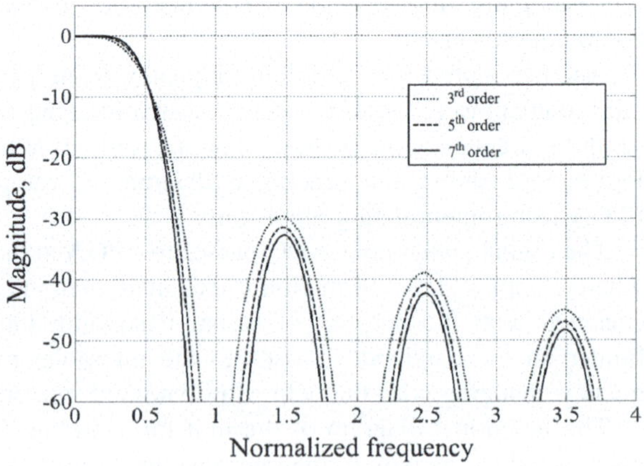

Fig. 5.34 Amplitude–frequency characteristic of interpolators used in the manufactured equipment

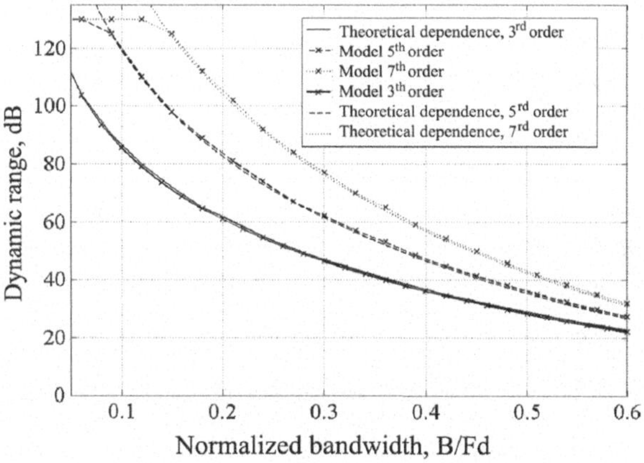

Fig. 5.35 Dynamic range of interpolators applied in the manufactured equipment

universal. In below-presented algorithms, the similar resamplers are used for solution of specific tasks typical for radio monitoring, including detection of pulse signals with small duration, direction-finding of the signal sources, multi-channel signals filtering in the wide band.

5.8 Multi-channel Narrowband Filtering

At monitoring of the modern digital communication systems signals such as DMR, APCO, TETRA, the necessity in simultaneous reception of a quantity of narrow-band signals distributed in the wide band up to 20 MHz and higher arises. At solution of the similar tasks, the receiver is tuned to the average frequency of the range, and the digital multi-channel filter with downsampling enables tuning to the preliminarily revealed channels and perform their selected analysis or recording at total number of such channels up to 128 [5].

The algorithm of the narrowband filtering is based on usage of preliminary overlapping vector summation of signal samples and calculation of FFT in the real time. Owing to such an approach, in the wide-band reception channel there is a possibility to form and to record at the sampling frequency of 50 kHz up to 128 channels arbitrary in frequency but restricted by the bandwidth of 25 MHz with the channel pass-band width of 25 kHz and with the grid step of 25 kHz (the AFC of the separate channel filter is presented in Fig. 5.36), or at sampling frequency 38.4 kHz (the resampler is used) up to 32 channels with the bandwidth of 12.5 kHz and with the grid step 12.5 kHz (for standards DMR and APCO 25).

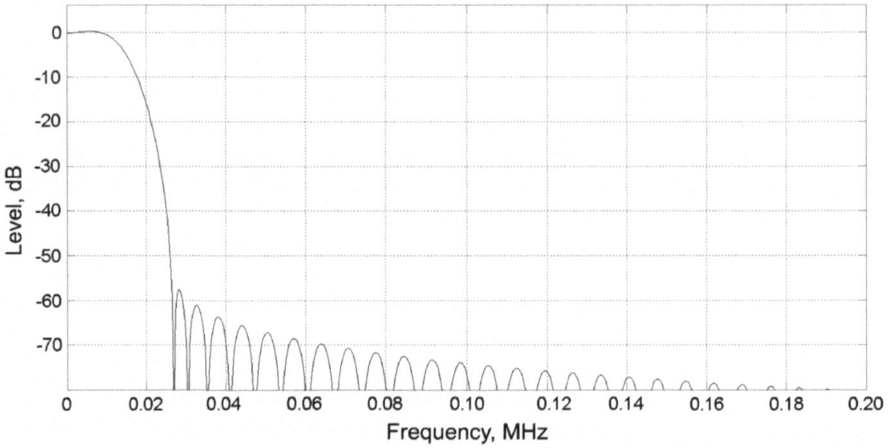

Fig. 5.36 Amplitude–frequency characteristic of the channel filter with the pass-band width of 25 kHz

5.9 Direction-Finding of Signals with Small Duration

For direction-finding of small duration signals, it is necessary to perform their detection and processing in the real time. The synchronization of antenna pairs switching, selection of signal sources according to definite signs, and individual calculation of bearing information are provided at realization of the phase direction-finding method on the base of dual-channel correlation interferometer. The similar possibilities of signal detection and processing in the real time are necessary for the amplitude direction-finder too.

For short signal direction-finding in FPGA path of the digital receiver of the ARGAMAK family, the signals universal detectors are implemented in the given bandwidth by the energy threshold [5]. Two-level meters operate in parallel in two channels with the given frequency bands: 24, 10, 5, 2 MHz and 250, 120, 50, 25, 12.5 kHz. At active conditions of detectors, data are begun to be transmitted to the

Table 5.3 Detection delays

Receiver bandwidth, MHz	Delay, ms	Receiver bandwidth, kHz	Delay, ms
24	0.01	250	0.14
10	0.05	120	0.142
5	0.12	50	0.145
2	0.28	25	0.9
		12.5	1.2

computer of bearing information only at signal exceeding the threshold defined by the user. Detection delays in different reception bands are shown in Table 5.3.

Direction-finders using the above-described approach on the base of digital receivers of the ARGAMAK family allows obtaining up to 1800 bearings per second at detection of signals with duration from 530 μs with spectrum resolution of 12.5 kHz in the bandwidth of 24 MHz, or to form up to 6000 bearings per second at detection of signals with duration from 140 μs and spectrum resolution of 800 kHz in the bandwidth of 24 MHz.

5.10 Address Direction-Finding of Digital Signals

Direction-finding of mobile or fixed devices of digital communication is complicated by the following factors [3, 22]:

- Complicate packet structure of transmission protocols;
- Presence of signal in ether during limited time, for instance, for GSM packets the duration of the packet is 577 μs, for DECT the minimal duration is 96 μs;
- Possibility of several transmitters operation at the same frequency in the mode with code and time division;
- Possibility of transmitter frequency change, frequency hopping, which is typical for GSM and DECT sources;
- Several types of modulation within the limits of one data transmission, for example, transmitters of Wi-Fi IEEE 802.11b standard admit information transmission of the physical level with the help of differential binary phase-shift keying, and the MAC level with the help of more complicated modulation types;
- Under conditions of the modern town, a great amount of packets, emitted by a quantity of different sources, are broadcasted in the working frequency range of GSM, DECT or Wi-Fi communication systems in the same bandwidth.

Taking this information into account, the task solution of the direction-finding of digital communication devices requires necessary identification of sources, which signals had been detected. Time sign, identification signs, and signal parameters should be associated with each detected packet. The software analysis of information obtained according to totality of signs should extract the packet groups belonging to different sources. Calculation of parameters for direction-finding of the data for each detected source should be performed in address manner, i.e., individually for each source according to the packets totality, which are identified with it. At that, detection and packets decoding must be performed in the real time.

Thus, for detection, dividing, and direction-finding of short packet signals from different transmitters, we need to use features of communication standard and determination of available identification signs of the signal transmitter in the real time. Such a mode is referred to as address direction-finding [3, 5].

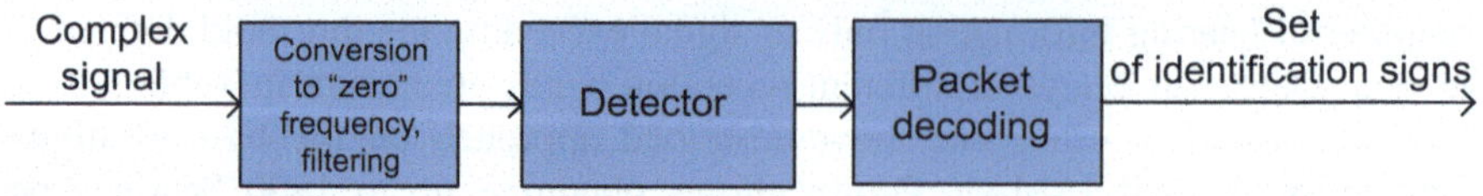

Fig. 5.37 Structural scheme of the digital processor

Detection and identification of the received packets is carried out in the real time; therefore, the main computing load accounts for the digital processor, which structural diagram is presented in Fig. 5.37. The digital processor should use the SDR principle permitting to launch the software realizing the required algorithms of signal processing. The digitized signal passes to the processor input, then its transfer to "zero" frequency and filtering are provided. Data after filtering passes to the detector, in which the packet extraction and time synchronization are performed. The detector can be implemented both by parametric scheme taking into account some information known in advance about the signal (for instance, the preamble view), and by simpler energy scheme, where the signal level exceeding the threshold is the detection criterion. The threshold value can be specified by the constant or defined adaptively depending on the situation in ether (the adaptive threshold). After fulfilling necessary adjustments (amplitude, frequency, clock, or time), the information decoding and determination of identification signs are executed. Thus, the message containing a set of information about decoded packet on the base of decoded information [3] is the output signal of the digital processor.

The MAC address of the transmitter may be taken as identification sign for the Wi-Fi standard devices, for devices of DECT standard—RFPI stations and transmission direction (uplink and downlink channels), for mobile GSM devices—the number of time slot and signal level.

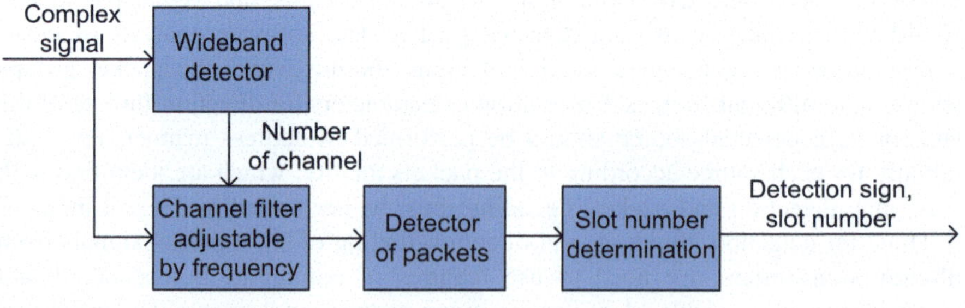

Fig. 5.38 Structural scheme of GSM packets processing

5.10.1 Direction-Finding of GSM Mobile Stations

The structural diagram of digital processing for packets detection of GSM mobile stations and for determination of identification signs is presented in Fig. 5.38.

Detection of GSM mobile stations packets is performed with the help of the wide-band detector. Its main part is the unit performing computation of fast Fourier transform (FFT). Decision of packet detection in the frequency channel is made in the case if the sum of FFT samples (bins) corresponding to the frequency channel exceeds the threshold. The threshold value is calculated with account of previous FFT realizations. According to FFT calculation results, the power level of detected packet is determined. The number of frequency channel, in which at the current moment the information transmission is provided, is transmitted to the frequency synthesizer. The received signal conversion to "zero" frequency and filtering with the help of the channel filter are carried out. After filtering, the time synchronization is executed according to the packet leading edge. Then, signal demodulation and determination of the slot number according to matched filter response to the training sequence are performed. If the matched filter response level exceeds the threshold, the received packet will be considered as detected, the message about packet detection with the number of the frequency channel, the estimation of the signal level and the number of the time slot will appear at the digital processor output. According to this set of identification signs, the information accumulation is performed and then direction-finding of the signal source by the amplitude method becomes possible using the directed antenna.

5.10.2 Direction-Finding of Wi-Fi Devices

The structural diagram of the digital processor for detection of IEEE 802.11b devices packets and determination of identification signs is presented in Fig. 5.39.

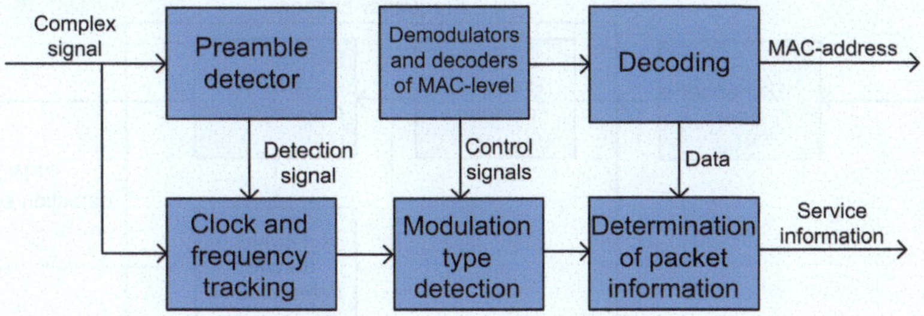

Fig. 5.39 Structural diagram of IEEE 802.11b packet processing

In this processor, the parametric detector on the base of known preamble is used. The frequency and clock tracking are fulfilled according to the detector activation. The preamble detector forms the signal of time synchronization for all further units of the processor.

At algorithm realization in the digital receiver ARGAMAK-M N [1], the one of the tasks of clock synchronization system consists in variation of the sampling frequency of the initial signal from 25 MHz (1/4 from the ADC clock rate of 100 MHz) to the Wi-Fi frequency—11 MHz [3]. Resampling is performed using the interpolating polynomial of fifth order. This system fulfills the clock change and adjustment of the leading edge to accuracy up to one-fourth of the cycle (the equivalent sampling frequency is 44 MHz).

The modulation type of MAC level is determined from the heading of the physical level; selection of appropriate DBPSK, DQPSK, or CCK demodulator is fulfilled. As a result of signal decoding at output of the processor, the MAC address of the packet transmitter appears. Reliability of address value is checked with the help of the control sum (CRC) calculation, and then the packet type is determined.

Thus, in the case of packet detection at processor output, the following information appears at detector output: the value of MAC address of the transmitter, which emitted the data packet; the level of received packet and additional service information permitting to obtain a presentation about type; and transmission direction of the received packet (from the access point or to the access point). Additional service information is used at topology construction of the wireless Wi-Fi network.

5.10.3 Direction-Finding of DECT Devices

Figure 5.40 shows the structural diagram of the digital processor presented for packet detection of DECT devices and determination of identification signs.

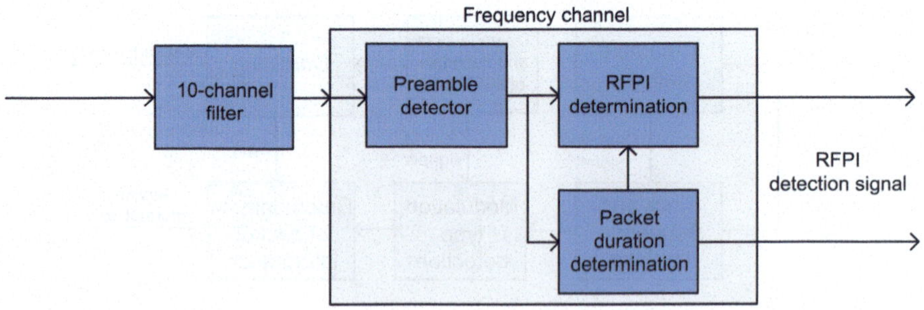

Fig. 5.40 Structural scheme of DECT packet detection

Ten independent frequency channels are anticipated by the standard of DECT technology; therefore, the parallel multi-channel processing is implemented in FPGA. In the frequency band of 24 MHz, the channel frequency division is performed. In each frequency channel, the signal transfer to "zero" frequency and filtering are carried out. After filtering, each channel is processed separately. Signal detection is fulfilled with the help of the parametric preamble detector. The clock tracking and time synchronization are performed according to the detector response. Determination of RFPI station and the signal transmission direction (uplink and downlink) are performed; the packet duration is defined. At output of the digital processor, messages containing RFPI stations, signal level, and transmission direction appear.

The above-examined algorithms of address amplitude direction-finding are implemented in the digital receiver ARGAMAK-MN. The presented approach to packet signal processing allows direction-finding of other standards signals of digital communications operating in the time division mode.

5.11 Radio Receivers of the ARGAMAK Family

Digital radio receivers are the base of radio monitoring equipment; all twenty-five years of the history of the IRCOS Company work they were in the permanent development. At present, Company produces the fifth-generation devices [1, 2].

The imported communication radio receiving devices with DSP units of the own development and manufacture were the digital receivers of the first generation. Modernization of imported receivers was reduced to minimum: The buffer amplifiers to pass signals to DSP unit were set at the analog outputs of IF, and the switch of AGC circuit was mounted with the aim to provide operation in the panoramic observation mode with the constant transfer function of the path.

The imported analog receivers with DSP unit of the own manufacture were the digital receivers of the second generation. However, now the analog part was exposed to more substantial improvements. Besides mounting of additional buffer amplifiers at the analog outputs of IF and the switch AGC circuits, the units of frequency synthesizers of the own development and manufacture with the aim of reduction of the receiver tuning time to the given frequency and increasing of the panoramic observation rate were applied.

The experience of operation of the first and second equipment generations showed the significant drawbacks and restrictions caused by utilization of communication receivers of import manufacture. High irregularity of the transfer gain, high noise level, insufficient dynamic range, presence of large number of self-cut frequencies, retuning performance not more than 50 MHz/s made the development of the receiver of the development a necessary one.

The first such receiver was produced in 1999. It was the digital receiver of the third generation, which had got the name ARC-CT1. This receiver had the analog and digital parts of the own development; it provided acceptable for those times engineering characteristics: the working range 20–2020 MHz, the pass-band 2 MHz, the dynamic range on intermodulation of third order not less than 70 dB, retuning time of the frequency synthesizer not more than 10 ms, which made possible the fulfillment of digital panoramic spectral analysis with the rate up to 140–150 MHz/s at the pass-band of 2 MHz.

Permanently increased requirements to the radio monitoring complexes from consumers in combination with successfully managed production of the single-channel and dual-channel digital receivers ARC-CT1 and ARC-CT2 of the third generation formed the basis for development of the digital receivers of the fourth generation.

At designing of the fourth-generation equipment, the special attention was paid to development of functionally and constructively completed modules, on the base of which it could be possible to create different variants of radio monitoring systems taking into consideration requirements of the specific consumer.

In 2003, the digital radio receiving device of the fourth-generation ARC-CT3 appeared at the international market. The working frequency range of the new receiver was 9 kHz–3 GHz, the retuning time of the frequency synthesizer was not more than 5 ms, the receiver pass-band increased up to 5 MHz, which gave a possibility to calculate the spectral panorama with rate more than 700 MHz/s.

Taking into consideration the accumulated experience in development and manufacture of digital receivers, the portable direction-finders, and the equipment for handheld direction-finders, as well as mounting of the equipment on different ground-based and airplane carriers, in 2003 the development of the portative digital receiver of the fifth generation was started. The new digital receiving device named ARGAMAK appeared in the middle of 2004.

The digital radio receiving device ARGAMAK compared to receivers of third and fourth generations has several times reduced weight and dimensions. At that, its parameters in accuracy, sensitivity, and dynamic range were not worse, and some even better, than ARC-CT3 parameters. For instance, the retuning time of the frequency synthesizer was not more than 2 ms. One more useful feature was possibility of pass-band selection from a series of 2, 5, or 10 MHz values. The receiver had stable characteristics, a possibility of external reference oscillator connection. On the base of this receiver, the series of digital panoramic measuring receivers ARGAMAK-I, ARGAMAK-IM, ARC-KNV4, ARC-D1TP, ARGAMAK-2K and other were developed.

Further development of digital receivers of the ARGAMAK family passed in two directions. The task of the first direction was creation of radio receiving device effective for autonomous utilization in the manpack variant. Determining requirements in this case are small mass, dimensions, and energy consumption at keeping the main working characteristics. The second direction was defined by requirements in receiver operation with signals of modern communication and data transmission systems. The main problems of this direction consisted in extension of the receiver pass-band up to 24 MHz and increasing of the DSP unit performance.

As a result of research in the first direction, in 2009 the dual-channel digital receiver ARGAMAK-2K and the single-channel digital receiver ARGAMAK-M with the pass-band up to 10 MHz appeared, which have the embedded rechargeable battery and, at that, small dimensions and weight. During research in the second direction, in 2011 the manufacture of digital receivers ARGAMAK+ with pass-band of reception path up to 24 MHz started.

Receivers ARGAMAK-M and ARGAMAK+ have the performance permitting digital signal processing in the real time, i.e., without omission of input radio signal fragments in time.

Development of digital radio receivers before 2015 proceeded in the way of functionality, reliability, and stability growth of the produced radio receivers. In 2012, on the base of the digital receiver ARGAMAK+, the serial production of the digital measuring radio receiver ARGAMAK-IS became commercialized, in 2015 the modernized ARGAMAK-M N with the pass-band up to 22 MHz and with the panoramic spectral analysis up to 6.5 GHz/s at resolution of spectral samples 6.25 kHz became produced serially.

In 2016, the serial manufacture of the miniature digital radio receiving module ARC-CPS3 and the digital radio receiving device on its base with power consumption near 6 W was started. At that, the pass-band is 40 MHz, the rate of panoramic analysis more than 20 GHz/s. In the same year, the portable multi-functional radio receivers ARGAMAK-RS (radio sensors) were developed and manufactured.

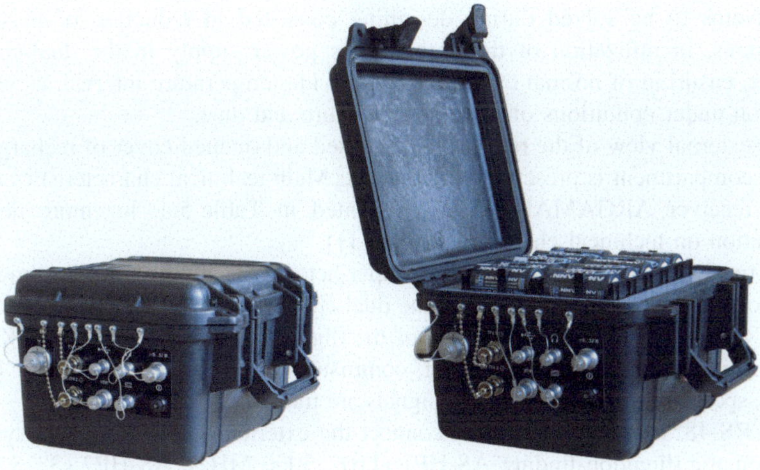

(a) Closed cover (b) Opened cover

Fig. 5.41 Digital receiver ARGAMAK-2K

Table 5.4 Technical characteristics of the digital receiver ARGAMAK-2K

Parameter	Value
Working frequencies range	0.009–3000 MHz
Bandwidth of simultaneous analysis	5 MHz
Noise factor	
In the range 25–465 MHz	12 dB
In the range 465–3000 MHz	12–14 dB
Spurious response selectivity, not less than	70 dB
Dynamic range on intermodulation of third and second orders	75 dB
Rate of panoramic spectral analysis, not less than	3500 MHz/s
Phase noise of the local oscillator at 10 kHz offset	
In the range 25–1000 MHz	−95 dBc/Hz
In the range 1000–3000 MHz	−85 dBc/Hz
Frequency instability of the internal reference oscillator	$\pm 5 \times 10^{-7}$
Power consumption, not more than	20 W
Overall dimensions, not more than	330 × 250 × 170 mm
Mass with rechargeable batteries, not more than	5.5 kg
Interval of working temperatures	–20 °C … +55 °C

5.12 Digital Radio Receiver ARGAMAK-2K

The goal of the digital receiver ARGAMAK-2K was creation of portable digital radio receiver with two coherent reception channels, autonomous power supply, which is able to operate in manpack automatic correlation-interferometric radio direction-finders, in complexes of search for mutual interferences, in fast-deployed systems for search for non-licensed radio emissions.

Problems to be solved during designing consisted in reduction of mass and dimensions, in utilization of the autonomous power supply in the dual-channel receiver, ensuring of normal operation in the wide temperature interval, as well as operation under conditions of increased moisture and dust.

The external view of the receiver with closed and opened cover of rechargeable battery compartment is presented in Fig. 5.41. Main technical characteristics of the digital receiver ARGAMAK-2K are presented in Table 5.4, for more detailed information on technical characteristics see [1].

The receiver ARGAMAK-2K has the super-heterodyne architecture. Two modules of analog conversion ARC-PC5 and one dual-channel module of digital processing ARC-CO10 are the main components of the digital receiver ARGAMAK-2K. The USB 2.0 interface is used for control; commands and digital data streams: Radio signals, spectrums, and demodulated signals are transmitted via it.

The RS-485 interface is used to connect the external devices; antenna systems for automatic direction-finding AS-HP1 (110…3000 MHz), AS-HP2 (3…8 GHz) and AS-HP0 (1.5…30 MHz), the signal converter ARC-KNV4 and other devices can be connected via it.

The receiver ARGAMAK-2K contains the embedded power source with ten NiMH-rechargeable batteries, one set of which provides not less than four hours of autonomous operation. Any DC source with the voltage of 10–32 V can be used as power supply and for charging of rechargeable batteries.

The receiver has the durable protected body in the form of the case, which has the protection degree not worse than IP64 (dust-proof, waterproof against water splashes falling form any direction). All external electric connections are provided through the waterproof connectors located on the side surface of the case. The receiver weight with a set of rechargeable batteries does not exceed 5.5 kg. Overall dimensions are not more than 330 × 250 × 170 mm.

Each channel of the receiver has the working frequency range from 9 kHz to 3000 MHz. The channel operation is possible both in the coherent mode, when two channels are synchronized from the same heterodynes and independently of each other. The coherent mode allows utilization of ARGAMAK-2K in the automatic correlation-interferometric direction-finders, in systems of search for mutual interferences, in complexes of search for non-licensed radio emission. At the same time, independent channel operation provides high scanning rate, at that, reception channels are tuned with offset by the bandwidth of the simultaneous analysis. This permits to almost double increasing of the panoramic spectral analysis rate.

At panoramic spectral analysis in the frequency band exceeding the bandwidth of simultaneous processing, the unit of analog radio signals conversion retunes by the frequency. At average retuning time of the synthesizer of 2 ms and at spectrum resolution of 6.25 kHz, the panoramic analysis rate is not less than 3500 MHz/s. Besides solution of tasks of signal detection and direction-finding, the receiver allows analysis of GSM, UMTS, IS-95, cdma2000, EV-DO, TETRA and DECT signals.

Table 5.5 Main technical characteristics of ARGAMAK-MN receiver

Parameter	Value
Working frequencies range	0.009–3000 MHz
Maximal bandwidth of simultaneous analysis	22 MHz
Noise factor, not more than	12 dB
Spurious response selectivity, not less than	70 dB
Dynamic range on intermodulation of third and second orders, not less than	75 dB
Rate of panoramic spectral analysis in the working frequencies range (at spectrum resolution of 6.25 kHz), not less than	6500 MHz/s
Power consumption, not more than	15 W
Overall dimension, not more than	110 × 60 × 245 mm
Weight with rechargeable batteries, not more than	1.5 kg
Interval of working temperatures	–40 °C ... +45 °C

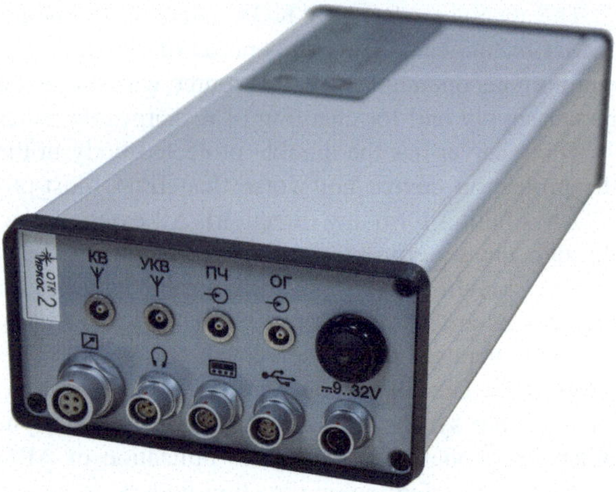

Fig. 5.42 Digital receiver ARGAMAK-MN

5.13 The Digital Radio Receiver ARGAMAK-MN

The digital radio receiver ARGAMAK-M N has the super-heterodyne architecture and has the analog signal converter, the module of digital processing, and charging power source with embedded rechargeable batteries in its structure. Constructively, the digital receiver ARGAMAK-MN is made as mono-block in the compact aluminum case with dimensions 110 × 60 × 245 mm and the weight 1.5 kg. The receiver is intended for panoramic spectral analysis, recording of radio signal to the embedded or external data storage, technical analysis of radio signals, and parameter analysis of analog and digital data transmission, communication, TV and broadcasting systems, for operation in the handheld direction-finding structures. The main technical characteristics of this receiver are presented in Table 5.5; the detailed information is in [1].

The embedded autonomous power source provides continuous operation with duration not less than four hours. With the help of additional set of rechargeable batteries, the time of continuous operation may be increased. The external view of this receiver is shown in Fig. 5.42; its structural diagram is presented in Fig. 5.43.

The embedded self-contained power source provides operation of the digital receiver and charging of embedded rechargeable batteries from the DC source with the voltage 10 … 32 V. The power source from AC net of 90 … 230 V and a set of rechargeable batteries for autonomous operation are included into the digital receiver delivery.

The analog path of the receiver, for the frequency range 20 … 3000 MHz, has the super-heterodyne architecture with double frequency conversion and matchable filters in the unit of preliminary selection. The IF signal of 70 MHz is the output signal of the analog converter. For the range 0.009 … 30 MHz, the radio signal just after preselectors passes to the ADC input. Further processing is performed in the digital form.

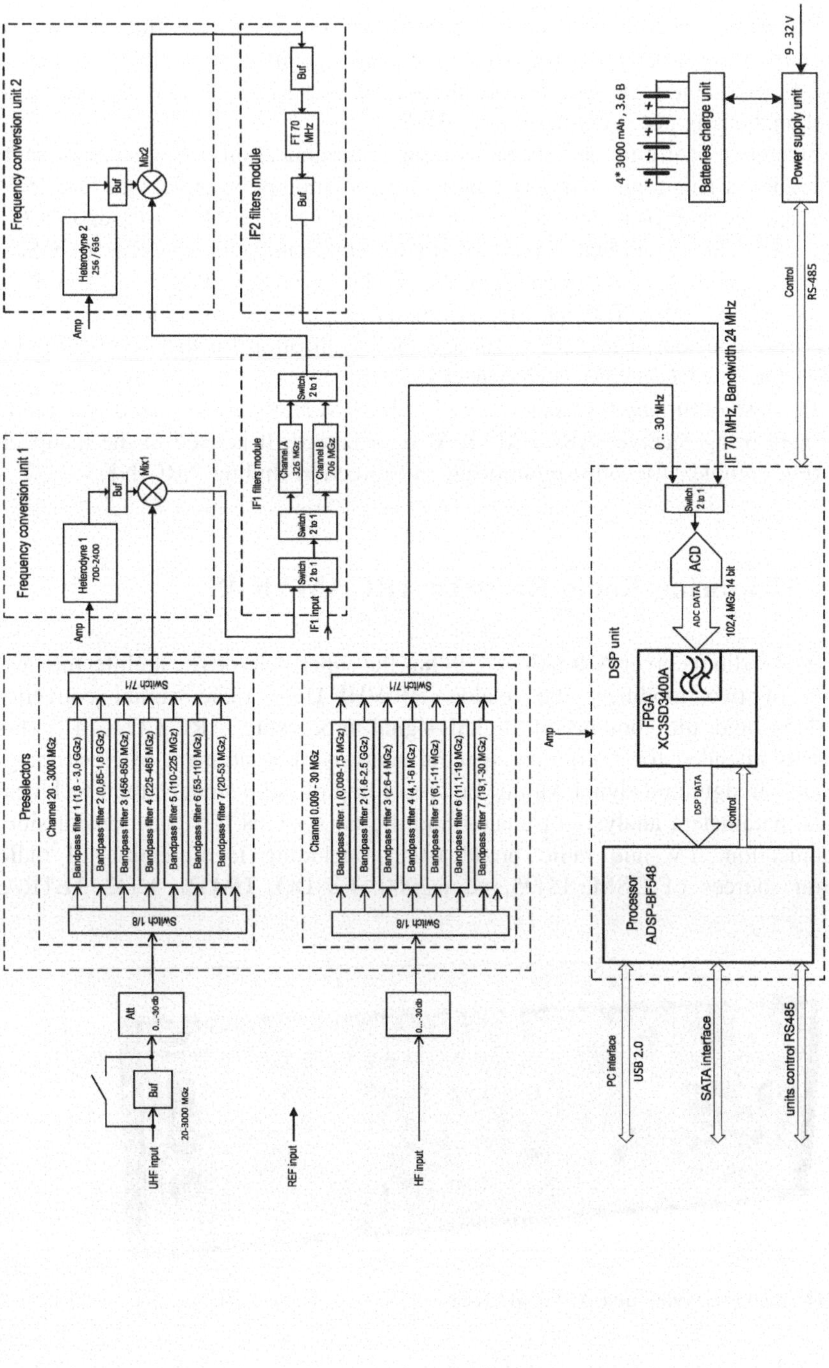

Fig. 5.43 Structural diagram of ARGAMAK-MN

At connection of the external converter, the frequency range can be extended upward to 8 or 18 GHz. Tuning of preselection filters, frequency synthesizers, parameters of other units of the converter are performed from the DSP module via the control bus RS-485.

At panoramic spectral analysis of the frequency band, which width exceeds the bandwidth of simultaneous processing, the analog converter of radio signals is retuned by frequency. At spectrum resolution of 6.25 kHz, the rate of panoramic spectral analysis is not less than 6500 GHz/c.

At reception of digital signals of wireless communication, the system of sampling frequency changing on the base of the interpolating polynomial (see Sect. 5.6) is used. The receiver provides address direction-finding of GSM, Wi-Fi, and DECT signals, decoding of service signals of digital communication systems of GSM, IS-95, cdma2000, EV-DO, UMTS, LTE, TETRA, DECT, DMR, APCO P25, Wi-Fi, WiMAX, DVB-T/T2/H, etc. standards. Algorithms of address direction-finding were considered in Sect. 5.10, and their realization for the handheld radio direction-finder ARC-RP3M is shown in [3].

On the base of the digital receiver ARGAMAK-M N, the modernized panoramic digital measuring receiver ARGAMAK-M is produced. It is used in the manpack measuring complex for radio monitoring and direction-finding ARC-NK5.

5.14 Measuring Radio Receiver ARGAMAK-IS

The digital radio receiver ARGAMAK-IS has the super-heterodyne architecture and consists of two modules: the analog HF-VHF-UHF radio signal converter ARC-PS5+ and the module of digital signal processing ARC-CO+ [4]. This receiver is intended for panoramic spectral analysis, measurement of electromagnetic field strength and signal parameters, recording of radio signals, their technical analysis, parameters analysis of analog and digital systems for data transmission, communication, TV and radio broadcasting, including identification of radio emission sources of GSM, IS-95, cdma2000, EV-DO, UMTS, LTE, TETRA,

Fig. 5.44 Radio receiving unit ARGAMAK-IS

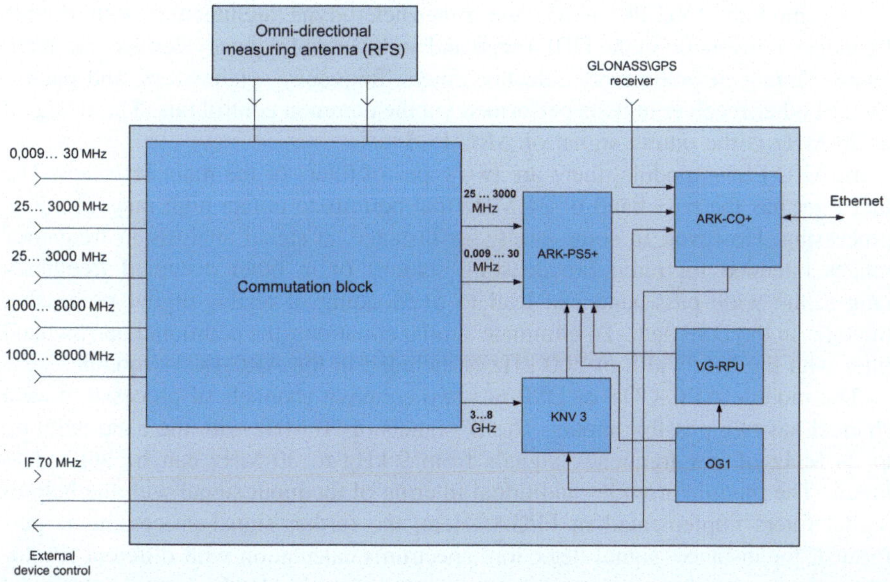

Fig. 5.45 Structural diagram of the radio receiving unit ARGAMAK-IS

DECT, DMR, APCO P25, Wi-Fi, WiMAX, DVB-T/T2/H, etc. systems. Measurements are performed in accordance with attested approaches [1]. This receiver is used in the structure of measuring mobile and fixed radio monitoring stations. The external view of the radio receiving unit ARGAMAK-IS is presented in Fig. 5.44, and its structural diagram is shown in Fig. 5.45.

There are the following modules in the structure of the radio receiving unit:

- Module of signal reception and conversion ARC-PC5+;
- Module of digital processing ARC-CO+;
- Converter of radio signals ARC-KNV3;
- High-stability rubidium reference generator ARC-OG1;
- Frequency synthesizer;
- Switch unit.

The analog HF-VHF-UHF radio signal converter ARC-PS5+ and the module of digital signal processing ARC-CO+ are made in the form of separate multilayer printed boards with dimensions 160 × 100 mm.

Application of the switchable preamplifier in ARC-PS5+ allows extension of the dynamic range of the received signals: If this preamplifier is switched on, the noise factor is significantly reduced that allows reception and processing of weak signals; if this preamplifier is switched off, the device has high linearity, which allows reception and processing of powerful signals. This module provides spurious response selectivity not less than 80 dB, which reduces to minimum the aliasing effects at further analog-to-digital conversion.

The module ARC-PC5+ has the super-heterodyne architecture with double frequency conversion in the UHF range and with single frequency conversion in HF range. Tuning of preliminary selection filters, frequency synthesizers, and parameters of other receiver units is performed via the common control bus. The IF signal of 70 MHz is the output signal of ARC-PC5+.

In ARC-PS5+ module, there are two types of filters of the main selection. The first filter has the pass-band of 24 MHz that permits to obtain high rates of signal processing. However, in some cases, for instance, at signal analysis in frequency ranges intended for radio broadcasting stations or in other occupied frequency ranges, the wide pass-band can lead to overloading of analog-digital processing modules in the DSP path. To eliminate similar situations, the additional narrowband filter with the bandwidth of 500 kHz is included in the ARC-PC5+ module.

The module ARC-CO+ of DSP has two coherent channels of processing; each channel has two possible inputs. The IF signals of 70 MHz with the bandwidth up to 24 MHz of low-frequency signals from 9 kHz to 30 MHz can be applied to inputs. The module provides additional filtering of the input signal with the help of digital filters implemented in FPGA. Here, the further signal processing is performed, for instance, signal detection, spectrums calculation with different resolution in frequency, spectrums averaging, search for maximal values, accumulation of probabilistic spectrum, etc. Besides, control of external devices via sequential interface RS-485 is provided, for instance, by the converter ARC-KNV4, support-rotated devices ARC-UP1 M, ARC-UP2, ARC-UP3D [1].

The module of the radio signal converter ARC-KNV3 is intended for radio signal reception in the range 3000–8000 MHz. It has the super-heterodyne architecture with the single frequency conversion. It has the pass-band up to 24 MHz and provides suppression of spurious channels not less than 70 dB. IF signal of 70 MHz is the output signal of the ARC-KNV3 module.

Besides mentioned modules, the module of the high-stable reference generator ARC-OG1 and the module of the external local heterodyne VG-PRU are included in the structure of the measuring receiver ARGAMAK-IS.

The module of high-stable reference oscillator ARC-OG1 performs function of additional reference oscillator with relative frequency error not more than $\pm 1 \times 10^{-9}$. The nominal value of the output frequency is 10 MHz.

The module of the external local oscillator VG-RPU forms the signal of the local oscillator for the signals converters ARC-PC5+ with decreased value of the phase noise not more than −100 dBc/Hz (at offset from the central frequency of 10 kHz).

There is a possibility to connect the external reference oscillator and also the timing navigational receiver for synchronization from GNSS GLONASS/GPS, in this case, the error of frequency measurement is less than 5×10^{-11}.

The precision switch unit contains preamplifier and high-quality attenuators; this makes possible to measure signals with power up to 2 W with accuracy not worse than ±1 dB, and realization of several modes of amplification in the receiver provides its dynamic range up to 150 dB. Up to five additional measuring antennas mounted on the rotation and polarization changing device can be connected

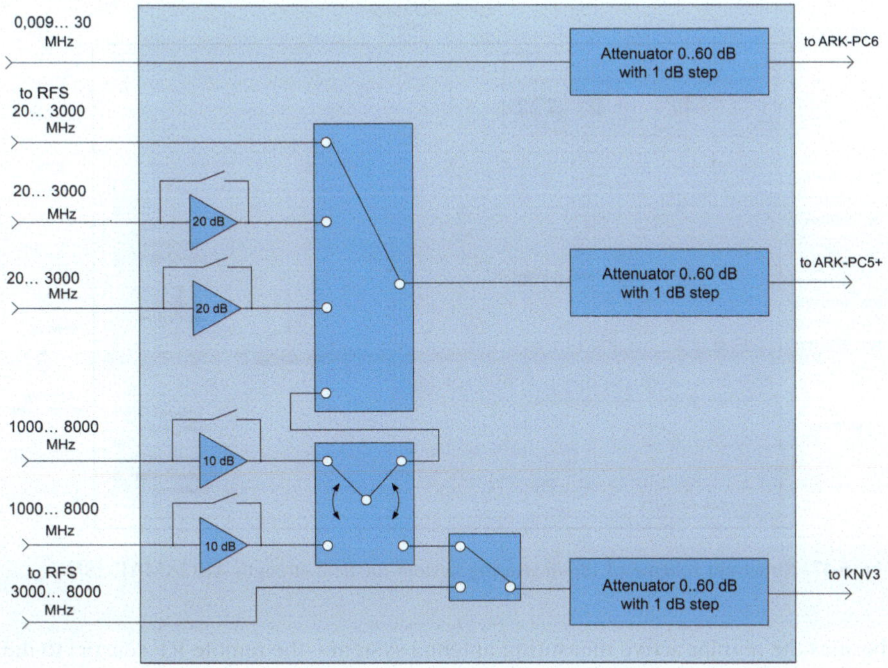

Fig. 5.46 Structural diagram of the switch unit

Table 5.6 Technical characteristics of the digital receiver ARGAMAK-IS

Parameter	Value
Working frequencies range	0.009–8000 MHz
Maximal bandwidth of simultaneous analysis	22 MHz
Noise factor	
• in the range 0.09–20 MHz	15 dB
• in the range 20–8000 MHz	12 dB
Spurious response selectivity, not less than	80 dB
Dynamic range on intermodulation on third and second orders	75 dB
Rate of panoramic spectral analysis, not less than	10,000 MHz/s
Phase noise of the local oscillator at 10 kHz offset	−100 dBc/Hz
Frequency instability of the internal reference oscillator	$\pm 1 \times 10^{-9}$
Power consumption, not more than	30 W
Overall dimensions, not more than	490 × 280 × 150 mm
Interval of working temperatures	+5 °C … +45 °C

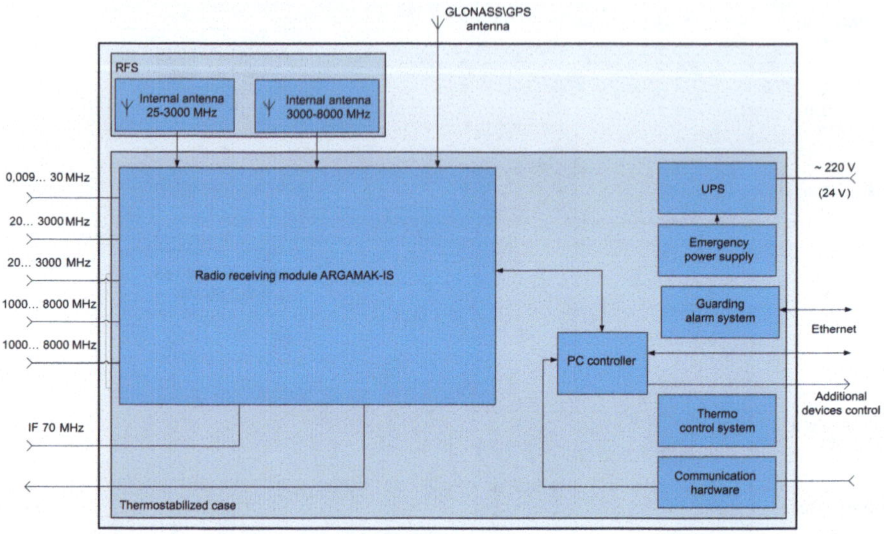

Fig. 5.47 Structural scheme of the measuring system for field strength ARGAMAK-ISN

(besides the regular active measuring antenna system—the remote RF sensor) to the switch unit. The structural diagram of the switch unit is shown in Fig. 5.46.

The possibility of radio signal switching from antenna inputs 1000…8000 MHz to the input of the ARC-PC5+ module, which has the working frequency range 20…3000 MHz, is the useful feature of the switch unit. Such a connection allows utilization of the same directional antenna, for instance, P6-23, for signal reception from 850 to 8000 MHz.

The control and data transmission are provided via the Ethernet cable, since signals are transmitted in the digital form, and antenna effect and any attenuation are absent.

The main technical characteristics of radio receiving unit ARGAMAK-IS are presented in Table 5.6 [1].

Technical characteristics of the radio receiving unit ARGAMAK-IS correspond to recommendations of ITU [10].

As it was already mentioned above, the radio receiver ARGAMAK-IS is used in the structure of mobile and fixed radio monitoring stations; it is the main component of the panoramic measuring system for the field strength ARGAMAK-ISN [1]. For operation in the structure of fixed stations, it has unmanned implementation. At that, the unit ARGAMAK-IS, the local control server (industrial computer), communication equipment are located in thermostabilized and waterproof case. Such an implementation does not require service rooms for work; the measuring system can be mounted directly in the open air, for example, on the roofs of high buildings, since the range of working temperatures is within the limits of −40 to +50 °C. The implementation variant is possible with low boundary of working temperature down to −55 °C. The structural diagram of the panoramic measuring system for the

Fig. 5.48 External view of ARGAMAK-ISN

Fig. 5.49 Variant of antenna system of ARGAMAK-ISN

field strength is shown in Fig. 5.47. The external view of measuring system and the variant of its antenna system are presented in Figs. 5.48 and 5.49.

The panoramic measuring receiver ARGAMAK-IS equipped with additional sync channel for signal reception, is the base of measuring stations of radio monitoring ARCHA-INM and ARCHA-IT. Measuring stations are intended not only for parameters measurement, but also for direction-finding for radio signals. External view of the ARCHA-INM station is shown in Fig. 5.17. The measuring receiver, which case has the specific implementation, is mounted directly into the basis of the antenna array.

5.15 The Radio Receiver ARGAMAK-RS

The miniature digital radio receiver ARGAMAK-RS is intended for spectral and technical analysis of radio stations in the range 9 kHz…8000 MHz, for measurement of field strength and modulation parameters, identification of radio signal sources of digital technologies. At synchronization from GNSS, the receiver in the structure of radio monitoring network provides determination of radio emission source location.

The digital receiver ARGAMAK-RS has the super-heterodyne architecture, which practically repeats the structure of the panoramic measuring receiver ARGAMAK-IS. It has the same radio frequency modules in its structure:

- Module of signal reception and conversion ARC-PC5+;
- Module of digital processing ARC-CO+;
- Converter of radio signals ARC-KHV3;
- Frequency synthesizer.

Fig. 5.50 Implementation variant of the miniature measuring radio receiver ARGAMAK-RS

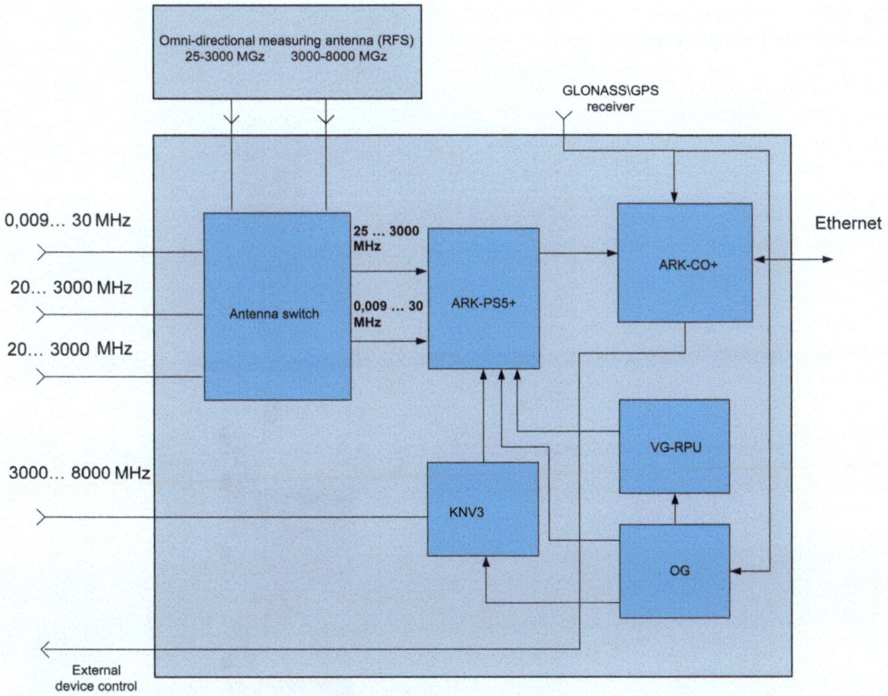

Fig. 5.51 Structural diagram of ARGAMAK-RS

Table 5.7 Technical characteristics of the digital receiver ARGAMAK-RS

Parameter	Value
Working frequency range	0.009–8000 MHz
Maximal bandwidth of simultaneous analysis	22 MHz
Noise factor	
• in the range 20–3000 MHz	12 dB
• in the range 3000–8000 MHz	16 dB
Spurious response selectivity in the range 20–3000 MHz, not less than	80 dB
Dynamic range on intermodulation of third and second orders	75 dB
Rate of panoramic spectral analysis, not less than	10,000 MHz/s
Phase noise of the heterodyne in the range 20–3000 MHz at 10 kHz offset	−100 dBc/Hz
Frequency instability of the internal reference oscillator	$\pm 5 \times 10^{-8}$
Power consumption, not more than	35 W
Overall dimensions, no more than	490 × 280 × 150 mm
Weight with rechargeable batteries, not more than	7 kg
Interval of working temperatures	−40 °C … +55 °C

Fig. 5.52 Fixed measuring complex ARGAMAK-RSS

The main differences between ARGAMAK-RS and ARGAMAK-IS consist in absence of the precision switch unit and utilization of more compact quartz reference oscillator instead of the rubidium frequency standard oscillator. On the base of the improved module arrangement, it became possible to reduce the device case overall dimensions to 300 × 300 × 150 mm. At that, the thermostabilized and waterproof case satisfies requirements to the protection degree IP67 and shock resistance degree IK10. The external view of implementation variant of ARGAMAK-RS is shown in Fig. 5.50, and its structural diagram is presented in Fig. 5.51.

The technical characteristics of the digital receiver ARGAMAK-RS are presented in Table 5.7 [1].

The receiver ARGAMAK-RS is used in the structure of miniature measuring complexes of radio monitoring ARGAMAK-RSS and ARGAMAK-RSM. The external view of the fixed measuring complex ARGAMAK-RSS is shown in Fig. 5.52.

Measuring complex ARGAMAK-RSS includes digital receiver ARGAMAK-RS and the following units:

- Remote RF sensor unit (omnidirectional measuring active antenna);
- Navigational timing equipment GLONASS/GPS;
- Power supply (located downward).

The power supply, as the radio receiver ARGAMAK-RS, can be used in the open air, has a case satisfying requirements IP67 and IK10. In the power supply, there are emergency rechargeable battery, controller, communication equipment providing remote control of the receiver via wire, and wireless 2G/3G/4G channels. The receiver and the power supply have sunscreens.

5.16 Radio Receiving Module ARC-CPS3

The radio receiving module ARC-CPS3 [1, 5] has the direct frequency conversion architecture; the integrated circuit RFIC AD9364 is used for signal conversion to "zero" frequency and for analog-to-digital conversion. This module provides:

- Panoramic spectral analysis of radio signals, accumulation of spectrum panorama in the given frequency range;
- Continuous recording of I, Q channels to the embedded data storage;
- Technical analysis, determination of modulation type and measurement of radio signal parameters;

Table 5.8 Technical characteristics of ARC-CPS3

Parameter	Value
Working frequencies range	70–6000 MHz
Maximal bandwidth of simultaneous analysis	40 MHz
Noise factor, not more than	15 dB
Spurious response selectivity, not less than	40 dB
Dynamic range on intermodulation of third order, not less than	60 dB
Rate of panoramic spectral analysis in the working frequencies range (at spectrum resolution of 12.5 kHz), not less than	20 GHz/s
Phase noise of the heterodyne at 10 kHz offset, not more than	−100 dBc/Hz
Frequency instability of the internal reference oscillator	$\pm 5 \times 10^{-7}$
Level of out-of-band emissions of the transceiver, not more than	−40 dBc
Power consumption, not more than	7 W
Overall dimensions, not more than	170 × 100 × 15 mm
Weight, not more than	0.5 kg
Range of working temperatures	
Without embedded data storage	–40 °C ... +55 °C
With embedded data storage	−25 °C ... +55 °C

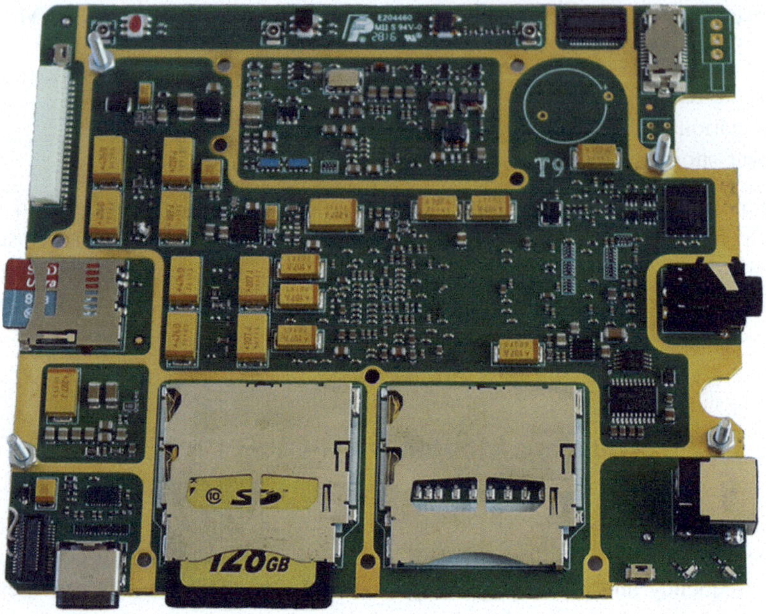

Fig. 5.53 Radio receiving module ARC-CPS3

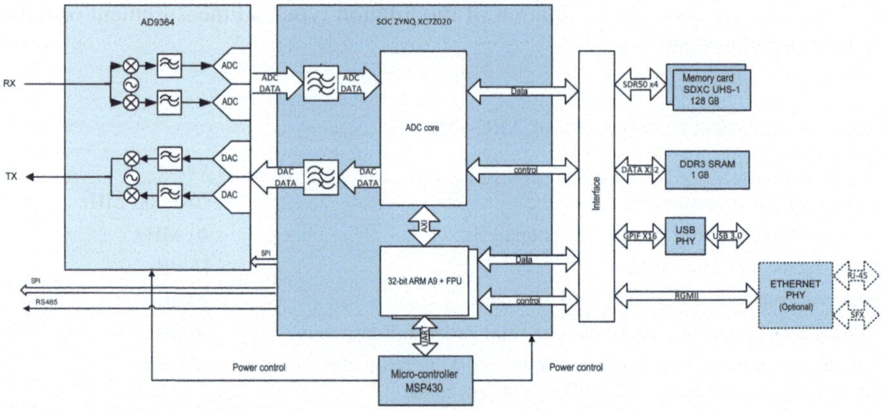

Fig. 5.54 Structural diagram of the radio receiving module ARC-CPS3

- Identification of radio signal sources;
- Formation of radio signals with the given characteristics.

The main technical characteristics of the ARC-CPS3 module are presented in Table 5.8. The external view is shown in Fig. 5.53; the structural diagram is presented in Fig. 5.54.

The radio frequency part of the digital receiver ARC-CPS3 is based on the one-chip digital transceiver AD9364. The system-on-chip SoC ZYNQ XC7Z020 provides functions of transceiver control and DSP. Power control of all board units and loading control of embedded software are implemented on the microcontroller MSP430FR5869.

The transceiver AD9364 has radio receiving and radio transmitting paths with quadrature conversion to "zero" IF. Embedded ADC and DAC have digital capacity of 12 bits and support the sampling frequency up to 112 MHz. The bandwidth of simultaneous processing may achieve 56 MHz. The transceiver is built on the SDR principle and allows operative control of all radio path parameters including tuning frequencies of the receiver and the transmitter, sampling frequencies, parameters of filters, parameters of manual, and automatic gain adjustment.

The system-on-chip SoC ZYNQ XC7Z020 contains dual-kernel processing system on the computing kernels ARM Cortex-A9 with co-processors for calculations with floating point FPU NEON and programmable logic matrix FPGA of Artix-7 family in its structure. The processor system operates under control of OS Linux. In the logic matrix, the support of digital interfaces to the external devices and algorithms of DSP on the base of fast parallel computations with minimal time and energy expenses are realized. The DSP projects include digital down frequency converters, digital filters, synchronization devices, demodulators, decoders, and other digital modules and units necessary for signal processing of different communication systems.

A set of functions performed by the ARC-CPS3 module is defined by the embedded software. The ROM on the board allows storage of several tens variants of the embedded software for operation in the panoramic and spectral analysis modes, measurement of radio signal parameters, monitoring of various communication systems, such as GSM, UMTS, LTE, Wi-Fi, for solution of special radio monitoring and radio suppression tasks.

The microcontroller MSP430FR5869 provides loading of the necessary software complex for solution of the specific task. The loading control in the operative mode is performed by the external control and representation device (computer, tablet, smartphone) via interfaces USB 3.0 or Ethernet. The Ethernet interface is implemented in two variants: 1000BASE-T on the copper twisted-pair cable or 1000BASE-Sx with optical interface in the format of exchangeable modules of SFP (Small Formfactor Pluggable—the industrial standard of module compact transceivers). The real-time clock embedded into the controller allows activating of the task assigned in the memory in the autonomous mode at the preset time. Results of radio monitoring task fulfillment are stored in the exchangeable memory card of the format SDXC (Secure Digital eXtended Capacity). Parameters of memory cards allow recording of the continuous radio signal in the bandwidth up to 10 MHz.

Digital transmission path makes it possible to form radio signals with the given characteristics for calibration and testing of radio monitoring complexes, interactive monitoring of various communication systems, testing of signal propagation conditions in locality and in premises. The embedded radio signal transceiver provides a possibility of radio signals reproduction, which were earlier written to the exchangeable memory cards or loaded from the control device.

5.17 Automatic Radio Direction-Finders of the ARTIKUL Family

Effectiveness of direction-finding and radio emission sources (RES) localization are defined by technical characteristics of the hardware–software components of the radio monitoring complex. At that, parameters of the hardware part, in particular, radio receivers and antennas switches, directly influence on quality of further digital signal processing. Development of digital receivers ARGAMAK, ARGAMAK-M N, and ARGAMAK-IS made possible launching of automatic and manual radio direction-finders with high technical characteristics in weight and dimensions, with

Fig. 5.55 Antenna system AS-PP4 of the fixed radio direction-finder ARTIKUL-S (working range 1.5–8000 MHz)

Fig. 5.56 Antenna system ARTIKUL-M1 in non-exchangeable implementation (working range 1.5–8000 MHz)

Fig. 5.57 Antenna system ARTIKUL-M1 in exchangeable implementation (working range 1.5–8000 MHz)

extended bandwidth of IF path, higher performance and working bandwidth, to the international market [1, 4].

Stationary (deployed) and mobile automatic radio direction-finders—correlation-interferometers are included in the ARTIKUL radio direction-finding family. In Figs. 5.55, 5.56, and 5.57, the external view of antenna systems of fixed and mobile radio direction-finders ARTIKUL-C and ARTIKUL-M1 are displayed. Table 5.9 shows their technical characteristics.

The antenna system, which contains not only lettering antenna arrays and switch, but also dual-channel digital receiver, is included in the structure of fixed direction-finder ARTIKUL-C on the base of the modules ARGAMAK+. The dual-channel digital receiver is embedded into the antenna systems of the mobile direction-finder ARTIKUL-M1 in changeable and non-changeable implementations.

Direction-finder control and signal transmission are provided in the digital form via the Ethernet interface (twisted-pair cable or fiber-optical cable); it is used for commands and digital data streams transmission: time samples of radio signals,

Table 5.9 Main technical characteristics of ARTIKUL-C and ARTIKUL-M1

Parameter	Value
Working frequencies range	1.5–8000 MHz
Maximal bandwidth of simultaneous analysis	24 MHz
Dynamic range on intermodulation of third and second orders, not less than	75 dB
Rate of multi-channel direction-finder in the range 1.5–20 MHz	Up to 10 MHz/s
Rate of multi-channel direction-finder in the range 20–8000 MHz	Up to 2500 MHz/s
Direction-finding sensitivity of ARTIKUL-C in the range, not more that	
1.5–20 MHz	15 μV/m (typ. 10 μV/m)
20–3000 MHz	6 μV/m (typ. 2 μV/m)
3000–8000 MHz	12 μV/m (typ. 5 μV/m)
Direction-finding sensitivity of ARTIKUL-M1 in the range, not more than	
1.5–20 MHz	15 μV/m (typ. 6 μV/m)
20–3000 MHz	8 μV/m (typ. 3 μV/m)
3000–8000 MHz	12 μV/m (typ. 5 μV/m)
Instrumental accuracy (RMS error) of direction-finding of ARTIKUL-C in the range, not more than	
1.5–20 MHz	5° (typ. 2.5°)
20–3000 MHz	2° (typ. 1°)
3000–8000 MHz	4° (typ. 2°)
Instrumental accuracy (RMS error) of direction-finding of ARTIKUL-M1 in the range, not more than	
1.5–20 MHz	5° (typ. 2.5°)
20–3000 MHz	2.5° (typ. 2°)
3000–8000 MHz	3° (typ. 2°)
Minimal duration of the signal under direction-finding	
Single	1 ms
Repeated	10 ms
Range of working temperature	−40 … +55 °C
Overall dimension and weight of ARTIKUL-C	
Dimensions (diameter × height)	2110 × 2820 mm
Weight, not more than	40 kg
Overall dimension and weight of ARTIKUL-M1	
Dimensions (length × width × height), not more	1550 × 1100 × 550 mm
Weight, not more than	50 kg

spectrums, demodulated signals, etc. Such a technical solution completely eliminated the antenna effect in the feeder; increased sensitivity and accuracy of direction-finding at keeping of the dynamic range (see Sect. 5.3).

The generalized structural diagram of antenna systems of direction-finders is presented in Fig. 5.58. The ether signals are received by the antenna array of the appropriate range. Elements of antenna arrays are active, which provides high

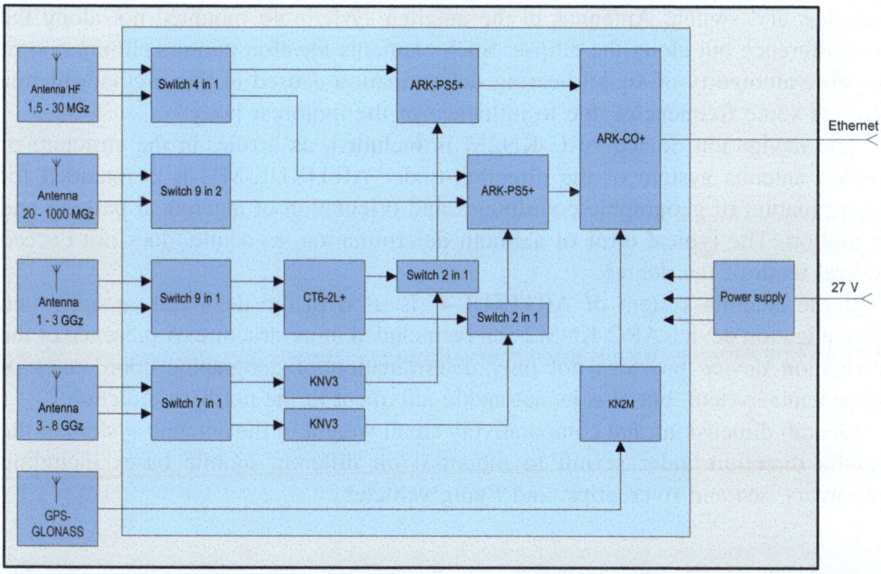

Fig. 5.58 Structural diagram of the antenna system ARTIKUL-C

sensitivity at small antenna size. Signals from the antenna system output pass to the appropriate range switch. All antenna switches are equipped with switchable preamplifiers with low-noise factor that also increases sensitivity of direction-finders. From the output of antenna switches, signals pass to the radio signal converters of the appropriate frequency range.

Modules of radio signal converters ARC-PS5+ and KNV3, the dual-channel signal converter CT6-2L+, are included in the structure of direction-finders.

The module of radio signal converter CT6-2L+ has the super-heterodyne architecture with single frequency conversion. The signal of IF at 706 and 280 MHz is the output signal of CT6-2L+. The signal at IF of 706 MHz is the output signal of KNV3.

Signals of IF at 70 MHz from the output of converters ARC-PC5+ pass to dual-channel module of digital processing ARC-CO+. Selection of IF signals or LF-HF from two coaxial inputs to the appropriate circuit of preliminary analog filtering and amplification is provided in each channel. Then, with the help of ADC, the signal is converted to the digital form and passes to FPGA, where digital frequency selection, frequency conversion, parallel digital processing are performed.

It is necessary to distinguish separately the antenna receiving unit of lettering 2, which includes the antenna system, the antenna switch 9 in 1, and the radio signal converter CT6-2L+. This unit is made in the form of the unified constructive block. The antenna switch and the converter are located in the direct closeness to the antenna system, which enables minimization of connecting cable lengths between antennas and switch. Antennas in the antenna system are mounted not along the circumference but along the ellipse: Such elements location permits elimination of possible ambiguity of signal bearing determination caused in the circular antenna array at some frequencies due to influence of the motorcar body.

The navigation device ARC-KN2M is included, as a rule, in the structure of mobile antenna system of the direction-finder ARTIKUL-M1. It is intended for determination of geographic coordinates and orientation of antenna at parking and in motion. The typical error of azimuth determination, as a rule, does not exceed several tenth of the degree.

If the antenna system of ARTIKUL-C is used in the deployed version, then the navigation device ARC-KNM2 can be included in its structure. A presence of the navigation device provides not only determination of geographic coordinates of the antenna system, but also its automatic affixment to the northern direction.

Overall dimensions and comparatively small weight of the antenna system of the mobile direction-finder permit to mount it on different mobile bases including motorcars, sea and river ships, and flying vehicles.

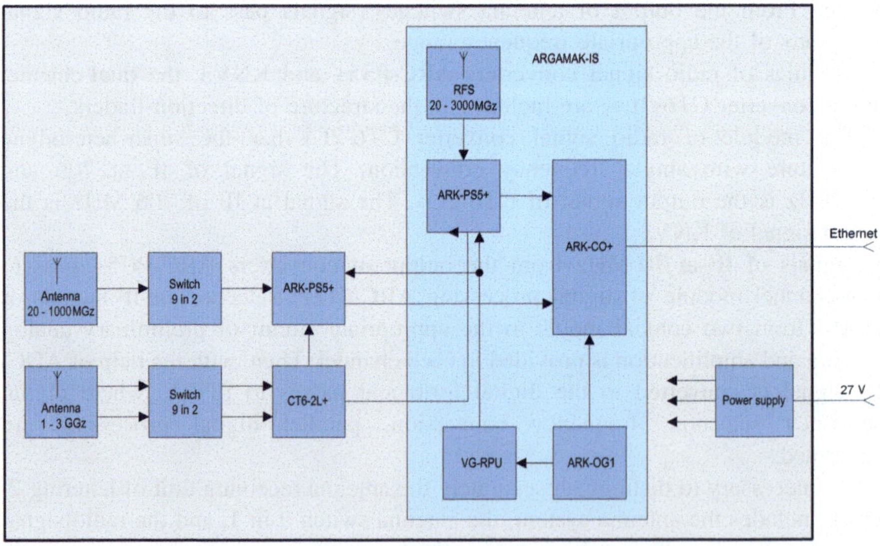

Fig. 5.59 Structural diagram of the antenna system of the ARCHA-INM station

5.18 Multi-functional Radio Monitoring Station ARCHA-INM

Among units of direction-finding, we can separately distinguish the multi-functional radio monitoring station ARCHA-INM. This station feature is utilization of measuring receiver ARGAMAK-IS with additional sync reception channel embedded directly into the direction-finding antenna system. The structural diagram of this station is shown in Fig. 5.59; its external view is presented in Fig. 5.17.

Such a solution allows integration of the measuring receiver and the correlation interferometer into one device. Functional possibilities of this station provide solution of several tasks: execution of signal parameters measurements, direction-finding by correlation-interferometric method, and also solving of the problem of sources localization by the time difference of arrival method being in the structure of a system with several stations.

The ARCHA-INM station does not require the separate technical room. All equipment including the power reserve source can be mounted in the open air.

5.19 Transportable Measuring Station ARCHA-IT

The radio monitoring station ARCHA-IT, as the radio monitoring station ARCHA-INM, is built on the base of the measuring ARGAMAK-IS with additional sync reception channel, located in the case for mounting on the mast. But in contrast to the ARCHA-INM station, in ARCHA-IT station, the changeable antenna systems RF sensor antenna, AS-HP-KV, AS-HP5, AS-HP2 are used. Such a solution, on the one hand, permits also to integrate properties of the measuring receiver and the correlation interferometer in one device. On the other hand, this station has high mobility, and can be used both at parking and in motion, if it is situated at the transportable base. External view of this station with the direction-finding antenna system AS-HP5 was shown in Fig. 1.16 in Chap. 1.

5.20 Automatic Radio Direction-Finder ARTIKUL-H1

The peculiarity of the automatic correlation interferometer ARTIKUL-H1 is its compactness and the small weight that allows its usage in regions difficult to traverse in the manpack variant. The digital receiver ARGAMAK-2K and a set of changeable direction-finding antennas are included in the structure of the direction-finder. External views of this station were shown in Figs. 1.19, 1.20, 1.21, and 1.22 in Chap. 1.

It is close to mobile and fixed direction-finders by its functional possibilities. Its working range is from 25 to 3000 MHz (in extended version from 1.5 to 8000 MHz).

Table 5.10 Some technical characteristics of the direction-finder ARTIKUL-H1

Antenna	Variants of usage	Range of work. freq., MHz	Direction-finding sensitivity, not more	Accuracy (RMS error) of dir.-finder, not more	Weight, kg	Dimensions (diam. × height), mm
AS-HP-KV	Mobile, fixed	1.5–30	15 μV/m (typ. 9 μV/m)	5° (typ. 2.5°)	6.2	465 × 170
AS-HP0	Manpack, mobile, fixed	20–300	15 μV/m (typ. 7 μV/m)	5° (typ. 2.5°)	3.9	465 × 150
AS-HP1	Manpack, mobile, fixed	110–3000	8 μV/m (typ. 4 μV/m)	4° (typ. 3°)	2.7	465 × 135
AS-HP5	Mobile, fixed	20–110 110–3000	10 μV/m (typ. 6 μV/m) 8 μV/m (typ. 4 μV/m)	5° (typ. 2.5°) 2.5° (typ. 2°)	8.5	580 × 250
AS-HP2	Manpack, mobile, fixed	3000–8000	15 μV/m (typ. 5 μV/m)	5° (typ. 2.5°)	2.2	345 × 155

The direction-finding error does not exceed 5 degrees in the whole range of working frequencies. The weight of the base set in the manpack variant is less than 15 kg. The continuous operation time is not less than 4 h, and it may be increased at the expense of the additional set of rechargeable batteries. The direction-finder is equipped with navigation units, which provide automatic antenna system affixment to the northern direction and determination of the direction-finder coordinates. Characteristics of the radio direction-finder depend on the used antenna system; main technical characteristics are presented in Table 5.10.

The direction-finder ARTIKUL-H1 can be used in manpack, mobile, and fixed variants, as it was described in Chap. 1.

In manpack variant, the equipment of the direction-finder ARTIKUL-H1 is located in the special bag for carrying. In its structure, the antenna system, the dual-channel panoramic receiver ARTIKUL-2K, the set of rechargeable batteries are included. The control of the direction-finder is provided with the help of the portable notebook.

The special plan-table attached to the operator's chest is anticipated for portable notebook attachment. Construction of this plan-table anticipates adjustment of keyboard location and slope angles of the screen with account of individual features of the operator. Antenna transfer from the transport position into the operation position and back is performed by the special mechanism on the bag, which allows operator pulling down the antenna system into the transport position independently, to overcome obstacles, and then again to pull up the antenna system in the operation position at passage of complicated segments, bushes, low-growing bushes. At that, it is not required to switch off the equipment.

In the fixed variant, the antenna system is fastened to the mast with height up to four meters. This extends the detection zone and the zone of radio emission source direction-finding.

In the mobile variant, for antenna system mounting of the direction-finder on the motorcar roof, the special magnet holder is used, which provides reliable attachment of the antenna system at speed up to 70 km/h, another variant—the antenna is located in the plastic luggage box.

5.21 Transportable Automatic Direction-Finder ARTIKUL-MT

The transportable automatic correlation direction-finder ARTIKUL-MT is a hybrid of the fixed direction-finder of the ARTIKUL family and the radio direction-finder ARTIKUL-H1 intended for operation at parking and in motion. It has advantages of both types of direction-finders: compactness, small weight, and also improved characteristics of modules of the ARGAMAK family.

In the base variant, the direction-finder is delivered with the antenna system AS-HP5 with the frequency range from 20 to 3000 MHz. In the extended variant, the range of working frequencies achieves values from 1.5 to 8000 MHz. Connection of the high-stable reference oscillator with relative frequency error not more than $\pm 1 \times 10^{-9}$ is also possible in the extended variant. The external view of the direction-finder with the antenna system AS-HP5 is presented in Fig. 1.18 in Chap. 1.

Control of the direction-finder and signal transmission is carried out in the digital form via the Ethernet interface (the twisted-pair cable or fiber-optic cable), commands, and digital data streams: Time samples of radio signals, spectrums, demodulated signals, etc., are transmitted via it.

5.22 Handheld Radio Direction-Finder ARC-RP3M

The panoramic digital receiver ARGAMAK-M N serves as the hardware base for manual direction-finder ARC-RP3M and the manual measuring complex for radio monitoring and direction-finding ARC-NK5I.

The direction-finder ARC-RP3M is intended for manual direction-finding and determination of RES location including sources of wireless communication system signals and data transmission signals, interference sources, for panoramic spectral and technical analysis. It permits to perform identification and address direction-finding of base stations and mobile GSM phones, base and subscriber DECT stations, wireless Wi-Fi devices. The external view of the direction-finder ARC-RP3M were shown in Fig. 1.25 in Chap. 1 and Fig. 2.15 in Chap. 2.

In the composition of the direction-finder, the following units are included: the panoramic digital receiver ARGAMAK-M N, the smartphone on the base of OS Android with the software packet SMO-ANDROMEDA [23], the handle with a holder for the smartphone and for fastening of changeable directional antennas; the power supply, the set of directional antennas, additionally, the notebook or the

Table 5.11 Technical characteristics of ARC-RP3M

Parameter	Value
Working frequencies range (base set)	9 kHz ... 3000 MHz
Working frequencies range (ARC-KNV3M)	9 kHz ... 8 GHz
Working frequencies range (ARC-KNV4M)	9 kHz ... 18 GHz
Maximal bandwidth of simultaneous processing	22 MHz
Spurious response selectivity, not less than	70 dB
Dynamic range on intermodulation of third and second orders, not less than	75 dB
Rate of panoramic spectral analysis (at spectrum resolution 6.25 kHz)	6500 MHz/s
Direction-finding sensitivity (20–3000 MHz)	3–15 μV/m
Instrumental accuracy of direction-finding (RMS error)	7–15°
Duration of autonomous operation	From 2 to 8 h
Range of working temperatures	−20 °C ... +45 °C

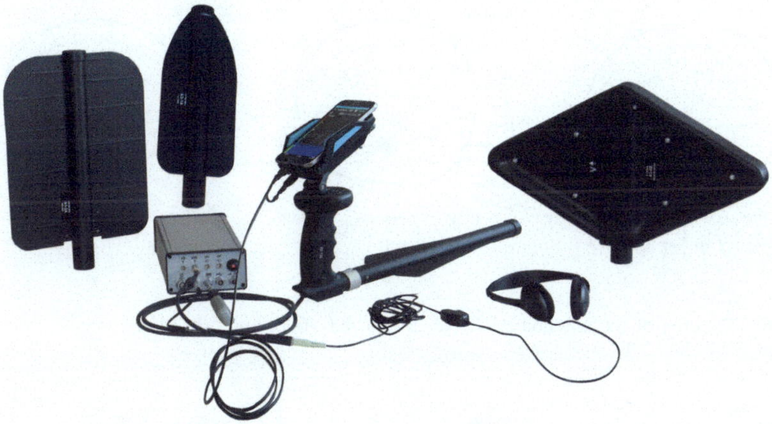

Fig. 5.60 Set of ARC-RP3M

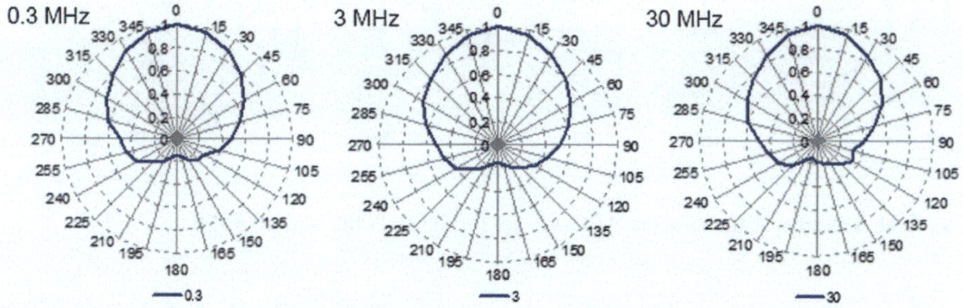

Fig. 5.61 Patterns of the antenna ARC-A3-KV for 0.3, 3, and 30 MHz

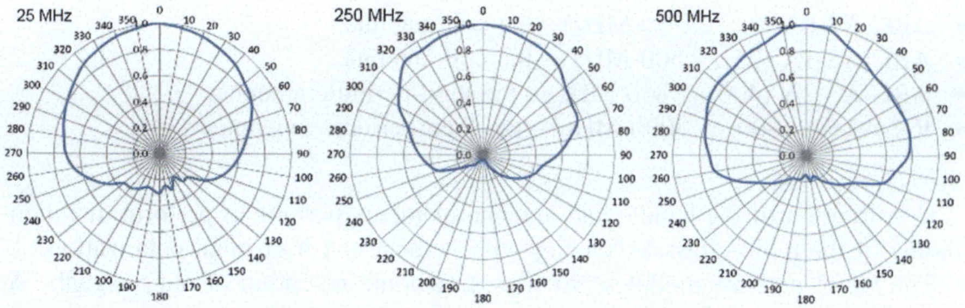

Fig. 5.62 Patterns of the antenna ARC-A3-1A for 25, 250, and 500 MHz

protected tablet computer may be included in the set. Main technical characteristics of the handheld direction-finder are listed in Table 5.11.

Antennas can be used in the active or passive mode. In the active mode antennas have high sensitivity, while in passive—the increased linearity, which make it

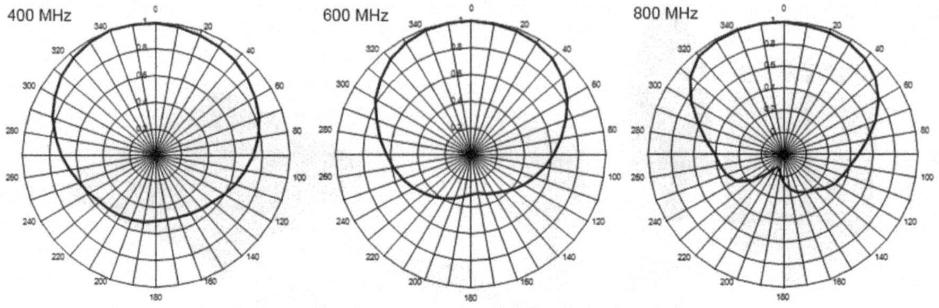

Fig. 5.63 Patterns of the antenna ARC-A3-2A for 400, 600, and 500 MHz

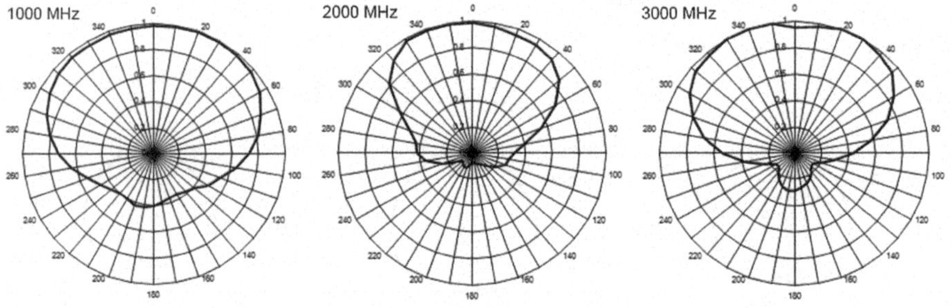

Fig. 5.64 Patterns of the antenna ARC-A3-3A for 1000, 2000, and 2000 MHz

possible to localize radio emission sources both in the far zone and in the direct nearness. The external view of antenna modules and the handle is presented in Fig. 5.60. Four active antennas are included in the delivery set:

- ARC-A3-KV (0.3 … 30 MHz)—the coil antenna;
- ARC-A3-1A (25 … 500 MHz)—the coil antenna;
- ARC-A3-2A (400 … 850 MHz)—the logo-periodic antenna;
- RK-A3-3A (800 … 3000 MHz)—the logo-periodic antenna.

Construction of the handle and antenna elements permits to perform direction-finding of the radio emission sources with vertical and horizontal polarization.

The set of antenna modules provides directional reception of radio signals. At that, the most actively used frequency range from 25 to 500 MHz is covered by the only antenna element ARC-A3-1A. Antenna patterns are shown in Figs. 5.61, 5.62, 5.63, and 5.64.

To expand range of working frequencies, the converters ARC-KNV3M (frequency range from 3 to 8 GHz) and ARC-KN4 M (frequency range from 3 to 18 GHz), each of which has the embedded directional antenna system, can be additionally included into the direction-finder set.

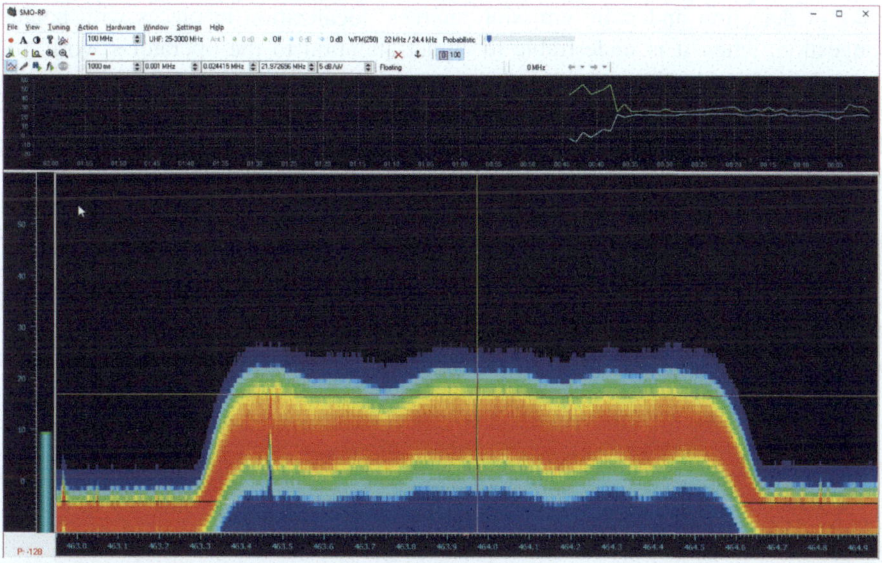

Fig. 5.65 Probabilistic spectrum of the CDMA signal

Changing of directional antenna modules is performed operatively without using special instruments. The antenna system (located in the bag) for concealed direction-finding is delivered optionally.

The handle of the handheld direction-finder is optimized for operator work in the field conditions. The weight of the handle with antenna does not exceed 700 g, and the center of gravity is below the operator's palm, which enables long operation without tiredness.

The signal analysis in the real time provided by the digital receiver ARGAMAK-M N with presentation of information in the form of the probabilistic spectrum enables reliable detection and distinguishing of the packet signals of various sources and performing of their direction-finding in one frequency band. At operation under control from the computer, the ARC-RP3M provides also decoding of service information of signals of digital communication systems GSM, IS-95, cdma2000, EV-DO, UMTS, LTE, TETRA, DECT , DMR, APCO P25, Wi-Fi, WiMAX, DVB-T/T2/H, etc.

As an example of spectrum representation in the real time, Fig. 5.65 shows the probabilistic spectrum of the CDMA signal, near left boundary of which we can see the narrowband interference.

To represent information and to control the direction-finder, the mobile device on the base of OS Android, usually the smartphone, and the special software SMO-ANDROMEDA are supplied. The control via USB 2.0 or wireless Bluetooth interface is used.

Application of the smartphone in the wireless control mode and the antenna system for concealed direction-finding make it possible to execute the operative tasks of detection and radio emission sources' localization in places with people congestion, when it is undesirable to attract attention to the operator's work, for example, at conduction of mass sport or public events. Figure 2.15 shows an example of open and concealed work on radio emission source localization. The directional antenna system and the digital receiver ARGAMAK-M are in the bag; the control is provided from the smartphone via the wireless Bluetooth interface.

Navigational possibilities of the smartphone enable monitoring of the location and orientation of the direction-finder. Using navigation data, the SMO-ANDROMEDA calculates and represents results of complex operation on the background of the local map: tracks of motion, signal levels depending on coordinates and antenna orientation, bearings, results of other measurement. The work with open formats of electronic maps, including Open Street Map, is possible.

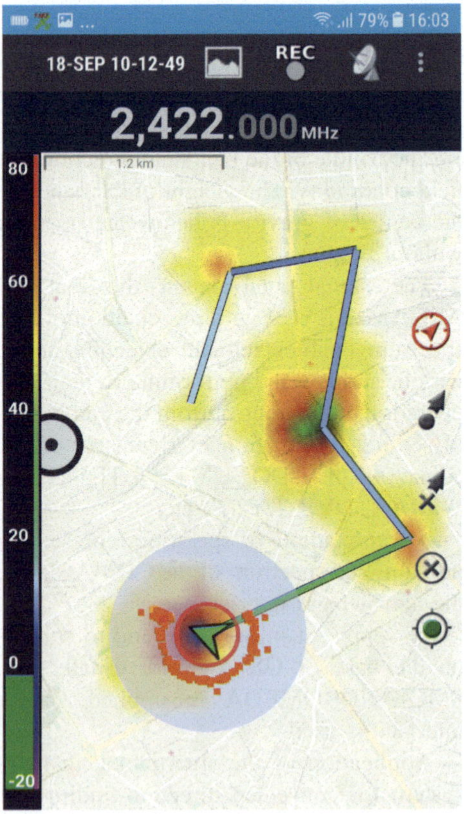

Fig. 5.66 Automatic determination of radio emission source location

5.23 Automatic Determination of Radio Emission Sources' Location

In the SMO-ANDROMEDA software, the automatic algorithm for calculation of the radio emission source location is realized, which essentially simplifies the operator's work. The operator with handheld direction-finder moves along the route and periodically rotates the antenna system in the azimuth plane along the circle, if possible, in the open places, for instance, in crossroads, eminences for estimation of signal direction of arrival. At that, the classical drive method, when operator moves toward to the signal increasing, is not required to be used, but one may move on the route that is remote enough from the source. On the smartphone display, the color zone is gradually formed, as shown in Fig. 5.66, more and more corresponding to the most probable location of the radio emission source.

The understandable and permanently updateable map of the Open Street Map format is used as basis. For orientation in space, the direction-finder sign, the motion track, the calculated matrix of the source location estimation, and a sign of detected radio emission source are imposed on the map image. During operation, the GLONASS/GPS data, the data from digital compass and the accelerometer of the smartphone are used, as well as data about signal level and direction of the antenna system. All obtained data are saved and transmitted to the module of automatic location calculation.

The algorithm of single-step coordinate calculation is used for bearing data processing. At antenna rotation, in conformity with the used model of additive noise, the rule of bearing estimation has a form

$$\widehat{\theta} = \arg\max_{\theta}(q0_{\theta}), \tag{5.77}$$

$$q0_{\theta} = \frac{\left(\sum_t S_t \cdot D(\theta - \psi_t)\right)^2}{\sum_t D^2(\theta - \psi_t)}, \tag{5.78}$$

where θ is the possible value of source bearing (in general case $-\pi \leq \theta \leq \pi$); S_t is the signal level; ψ_t is the bearing angle; D is the pattern of the receiving antenna.

At determination of coordinates, we use the function of source field strength with respect to the distance. In the case of reverse quadratic function and the log-normal measurement model, we have

$$(\widehat{x}, \widehat{y}) = \arg\min_{x,y}(q(x, y)), \tag{5.79}$$

$$q(x, y) = \Sigma_1(x, y) - \frac{1}{T}(\Sigma_2(x, y))^2, \tag{5.80}$$

$$\Sigma_1(x,y) = \sum_t (U_t - d_t(x,y))^2, \tag{5.81}$$

$$\Sigma_2(x,y) = \sum_t (U_t - d_t(x,y)), \tag{5.82}$$

where $U_t = 20\lg(S_t)$ is the result of taking the logarithm of amplitude, $d_t(x,y) = 20\lg\{D[\arg(\dot{r}_t(x,y)\cdot e^{-i\cdot\psi_t})]\} - 20\lg\{|\dot{r}_t(x,y)|^2\}$, $\dot{r}_t(x,y) = z(x,y) - Z_t$ is the complex distance from the direction-finder to the space point with complex coordinates $z(x,y) = y + jx$.

The resulting matrix is imposed to the digital locality map, the most probable location of radio emission source is designated by the separate marker in the form of a circle, and the direction-finder position is designated by arrow. During motion, the track of direction-finder motion along the route is constructed. During data accumulation, the zone of probable localization decreases, and the source position becomes more exact.

5.24 Address Direction-Finding

The spectrum analysis in the real time provided by the digital receiver ARGAMAK-M N with information representation in the probabilistic spectrum form allows detection and distinguishing of the packet signals of various sources overlapping in spectrum in one frequency band. Nevertheless, application of "classical" amplitude method based on the level or signal power estimation in the given bandwidth, which is received by the directional antenna for direction-finding and determination of localization of packet signal sources, which operate in the same bandwidth using time or code division, is not always effective. For operation with similar sources, it is necessary to know how to decode service identification signs transmitted in the digital packets and then to perform address direction-finding of the required signal extracting it with the background of others according to its identifier.

In Sect. 5.9 above, we considered algorithms of address direction-finding of GSM, Wi-Fi, and DECT signals. These algorithms provide identifier determination of the transmitting devices at standard and non-standard frequencies, extraction of the signal level corresponding to the given identifier, and then fulfillment of direction-finding and location determination using directional antennas. Described algorithms are implemented in the SMO-ANDROMEDA software.

In conformity with GSM signals, scanning of the working ranges and search for the mobile GSM stations are carried out. Figure 5.67 shows results, which correspond to the case of signal presence in the sixth-time slot, at that, the signal level is 71 dB μV, radio frequency number of the channel source ARFCN equals to 106.

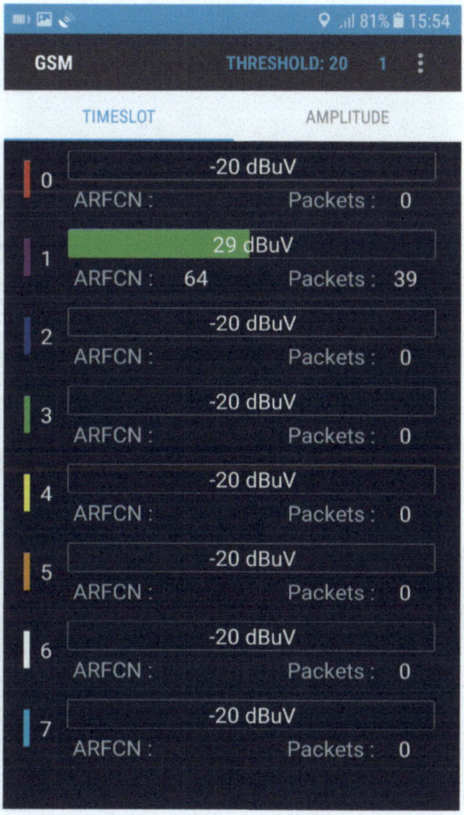

Fig. 5.67 Detection of GSM devices

The operator raising the detection threshold extracts the source with the largest signal level.

The history of levels for several sources working in one channel but in different time slots is shown in Fig. 5.68. The graph color corresponds to the number of the time slot. Level history graph allows operator to determine direction to the source (with the help of the complex amplitude direction-finding and source localization).

At localization of the access points and Wi-Fi devices, the search in the SMO-ANDROMEDA software is fulfilled both at the standard Wi-Fi frequencies in ranges 2.4 and 5 GHz and at arbitrary frequencies specified by the user. Representation of detected Wi-Fi sources is shown in Fig. 5.69. Similar to the GSM mode, there is a possibility of representation of the level history, an example is shown in Fig. 5.70.

For DECT sources, the implemented algorithm allows detection of sources both at the standard frequencies of DECT channels and at the specified manually non-standard frequencies. Results of DECT signal search are shown in Fig. 5.71. Similarly to Wi-Fi and GSM modes, the represented level history for the chosen RFPI value of the station is shown in Fig. 5.72.

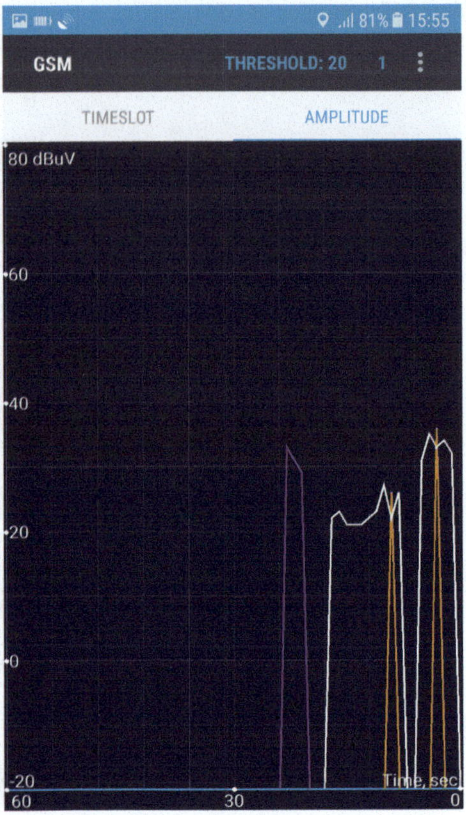

Fig. 5.68 Level history of GSM source

5.25 Manpack Measuring Complex ARC-NK5I

The difference of the manpack measuring complex ARC-NK5I from the radio direction-finder ARC-RP3M consists in the fact that it uses the measuring receiver ARGAMAK-M, which (compared to ARGAMAK-MN) has more accurate manufacture calibration and is attested as the measuring unit. Together with directional antennas for the handheld direction-finding, the measuring antenna is included into the set.

The high-stable external reference oscillator can be included in the complex set. The measuring complex provides not only detection, direction-finding, localization, and technical analysis of signals of radio emission sources. Measurement of the field strength level on the roof of the building from the cellular communication station is shown in Fig. 5.73.

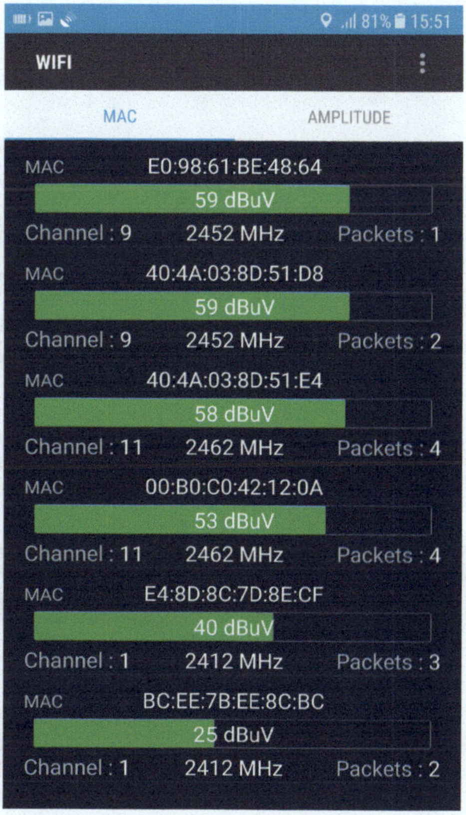

Fig. 5.69 Detected Wi-Fi source

5.26 Summary

The performed analysis and comparison of digital receivers of the super-heterodyne architecture and the direct conversion architecture (with frequency conversion to the "zero" frequency) confirmed that the super-heterodyne is preferable for measuring receivers with wide dynamic range and high measurement accuracy. The architecture with frequency conversion to the "zero" frequency allows implementation of compact economical devices with wide functional possibilities, which are able to solve various radio monitoring tasks. For example, the super-heterodyne receiver ARGAMAK-M N provides spurious response selectivity not less than 70 dB and dynamic range on intermodulation of third and second orders not less than 75 dB, the power consumption is 15 W. The digital radio receiving module with conversion to the "zero" frequency ARC-CPS3 has the power consumption less by half, however, the worse values of spurious response selectivity (equaled to 40 dB), the smaller dynamic range on intermodulation of third and second orders of near 60 dB, which may manifest itself in the negative manner on the receiver performance under conditions of complicate electromagnetic situation.

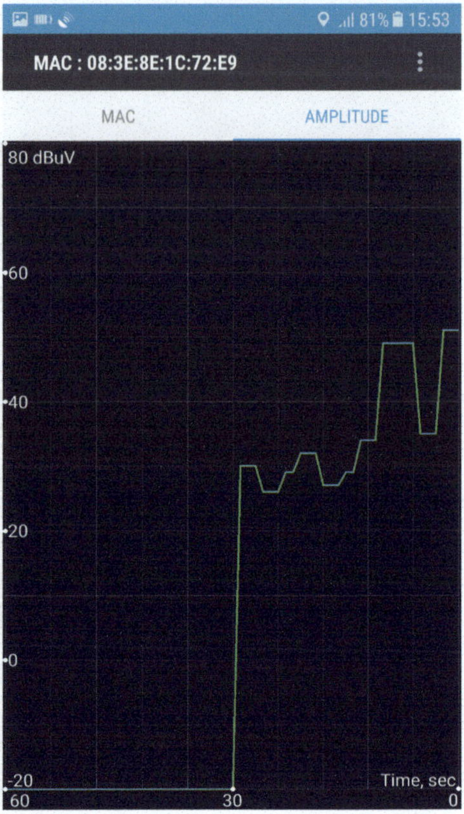

Fig. 5.70 Level history for the chosen Wi-Fi source

It is shown that application of integrated radio receivers has a series of technical advantages compared to traditional radio receiver accommodation outside of the antenna system, for which use the high-frequency lengthy feeder is necessary. We can attribute the following to technical advantages:

- Extension of the equipment dynamic range, approaching of its value to the width of the dynamic range of used receiver;
- Reduction or full elimination of the antenna effect influence, which, in turn, increases sensitivity, reduces errors of measurement and direction-finding of weak signals;
- Simplification of calibration of measuring unit;
- Application of cheap flexible coaxial cables of large length as the feeders (Ethernet and fiber-optic cables), which simplifies deployment and uninstalling of the antenna system, especially at mounting at high height.

Relationships for sampling frequency selection of the bandpass signals taking into consideration a possibility of non-informative spectrum components aliasing and also connecting spectrums of the initial discrete signal and the signal modulated

Fig. 5.71 Detected DECT sources

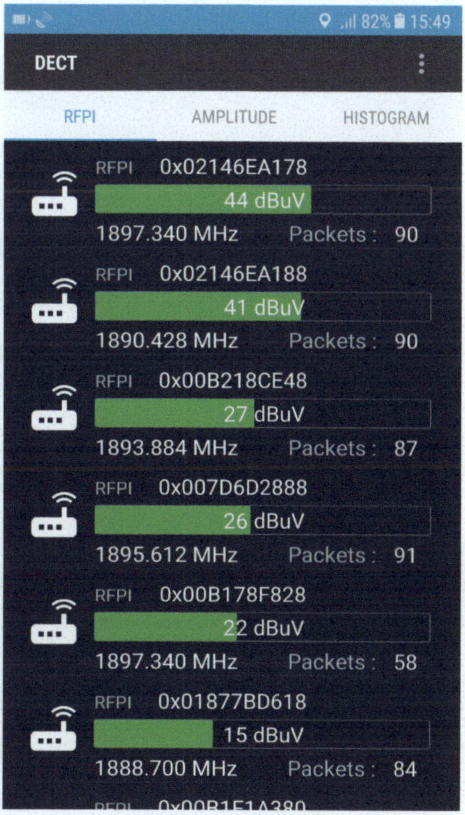

by the arbitrary periodic sequence are obtained in this chapter. The rule of spectrum transformation of the non-periodic discrete signal at its multiplication by the periodic sequence is formulated, which allows selection of the modulating sequence and provision of the required signal spectrum transformation, in particular, its conversion in the "zero" frequency and formation of quadrature components.

The typical algorithms of DSP are considered, which are applied at solution of radio monitoring tasks, including for useful signals filtering, changing of the sampling frequency, detection and direction-finding of small duration signals, as well as algorithms of address amplitude direction-finding of GSM, DECT , and Wi-Fi signals; their realization is shown in the software SMO-ANDROMEDA for the handheld direction-finder ARC-RP3M.

The presented family of digital radio receivers of the ARGAMAK family and radio direction-finders on its base provide solution of nearly all radio monitoring tasks including signal decoding of service information of digital communication systems GSM, IS-95, cdma2000, EV-DO, UMTS, LTE, TETRA, DECT , DMR, APCO P25, Wi-Fi, WiMAX, DVB-T/T2/H, etc.

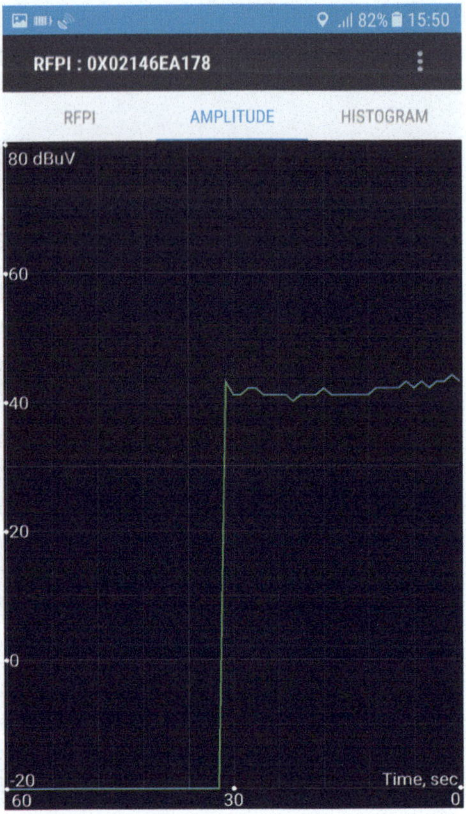

Fig. 5.72 Level history for chosen DECT source

The portable dual-channel radio receiving device ARGAMAK-2K is used in the manpack automatic radio direction-finders ARTIKUL-H1, as well as in the search systems for non-licensed radio emissions. The receiver has a case with the protection degree not worse than IP64, the wide working temperature range, the receiver weight does not exceed 5.5 kg, and the rate of the panoramic observation is up to 3.5 GHz/s.

Digital radio receivers ARGAMAK-M N and ARGAMAK-M, together with embedded rechargeable battery, have a weight not more than 1.5 kg. For the bandwidth of instantaneous observation of 24 MHz, the receivers provide signal processing in the real time without data missing; the rate of panoramic spectral analysis is up to 6.5 GHz/s. The receiver ARGAMAK-MN is used in the structure of the handheld radio direction-finder ARC-RP3M, and ARGAMAK-M in the structure of the portative measuring complex ARC-NK5I.

Progressive technical solutions are considered as the base of radio frequency modules ARGAMAK+, and the modern manufacture and adjustment technology provide high technical parameters in performance, linearity, path sensitivity, and characteristics stability, which allow development of the panoramic measuring

Fig. 5.73 Measurement of the field strength

receiver ARHGAMAK-IS, technical and metrological parameters of which correspond to ITU recommendations. The receiver provides all necessary types of measurements of radio signal parameters including electromagnetic field strength, frequency and bandwidth of radio emissions as well as many other parameters.

On the base of the radio frequency modules of the digital radio receiver ARGAMAK-IS, the digital radio receiver is created—the RF sensor ARGAMAK-RS intended for spectral and technical analysis of radio signals in the range 9 kHz...8000 MHz, measurement of the field strength and modulation parameters, identification of radio signal sources of digital technologies. At synchronization from GNSS, the receiver (in the radio monitoring network structure) provides determination of radio emission source location.

The family of automatic radio direction-finders ARTIKUL with the bandwidth of the simultaneous signal analysis up to 22 MHz, high performance and working frequency range 1.5 MHz … 8000 MHz consists of fixed, transportable, mobile, and manpack dual-channel correlation interferometers intended for signal direction-finding with arbitrary parameters: spectrum width, modulation type, time of broadcasting, etc. Radio direction-finders have comparatively small weight and dimensions; provide a possibility of fast deployment, installing and uninstalling, working in the wide temperature range.

Solution of radio monitoring tasks under conditions, which do not permit to use fixed and mobile complexes, is provided with help of manpack and transportable units such as the automatic radio direction-finder ARTIKUL-H1, the portable radio direction-finder ARTIKUL-MT, and the handheld radio direction-finder ARC-RP3M (ARC-NK5I), which have small weight-dimension indices in combination with high performance providing detection, direction-finding, and radio emission localization including short ones and working in one frequency band.

Radio monitoring stations ARCHA-INM and ARCHA-IT built on the base of measuring receiver ARGAMAK-IS with additional sync reception channel provide both direction-finding of radio emission sources and measurement of their parameters, as well as decoding and analysis of service information of digital communication networks.

The handheld radio direction-finder ARC-RP3M not only fulfills functions of radio emission source location as the equipment of the "last mile", but also provides secrecy at radio monitoring actions. It can be used in the modes of automatic determination of location and address direction-finding of radio emission sources.

For direction-finder control, the smartphone under control of the software SMO-ANDROMEDA is used, in which the automatic calculation method of radio emission source localization and algorithms of address direction-finding are realized. At source localization, it is not necessary to use the drive method, when operator should move in the direction of signal level growth. It is possible to move on the route at large remoteness from the place of source location; during displacement along the route, the color zone automatically forms on the electronic map of locality, which shows more and more accurately a position where radio emission source is situated. Address direction-finding and localization of radio signals of digital communication and data transmission systems of GSM, Wi-Fi, and DECT standards are carried out both at standard and non-standard frequencies, in the open locality and in premises.

Technical characteristics of digital radio receivers and radio direction-finders correspond to ITU requirements. They are used both under direct control by operators and in the structure of automated geographically distributed radio monitoring systems such as ARMADA, AREAL.

References

1. The Catalogue 2017 of IRCOS JSC. http://www.ircos.ru/zip/cat2017en.pdf. Accessed 28 Nov 2017
2. Rembovsky A, Ashikhmin A, Kozmin V, Smolskiy S (2009) Radio monitoring. In: Problems, methods and equipment. Lecture Notes in Electrical Engineering. Springer, 507 p
3. Ashikhmin AV, Kozmin VA, Myakinin IS, Radchenko DS, Spazhakin MI (2016, in Russian) Handheld direction finder for radio monitoring. Spetstehnika i Svyaz (4):101–105
4. Rembosky AM, Ashikhmin AV, Kozmin BA (2015, in Russian) Radiomonitoring: problems, methods, means. In: Rembosky AM (ed), 4th edn. Hot Line-Telecom Publ., Moscow, 640 p
5. Ashikhmin AV, Litvinov AI, Pershin PV, Polyakov AV, Sergienko AR, Spazhakin MI, Tokarev AB (2016, in Russian) Digital radio receivers of spectrum monitoring. Spetstehnika i Svyaz (4):90–97
6. Razavi B (1997) Design consideration for direct-conversion receivers. IEEE Transa Circ Syst-II Analog Digit Signal Proc 44(6):428–435

7. Tsui JB (2004) Digital techniques for wideband receivers, 2nd edn. SciTech Publishing Inc
8. Nezami MK (2003) RF architectures and digital signal processing aspects of digital wireless transceivers, 513 p
9. Kozmin VA (2016, in Russian) Application of integrated radio receivers to improve technical characteristics of monitoring equipment. Electrosvyaz (11):57–64
10. (2011) Handbook on spectrum monitoring. ITU-R, Geneva, 659 p
11. Volker J (2003/11) Better system sensitivity through preamplifiers/News from Rohde & Schwarz 43(178):41–45
12. Besser L, Gilmor R (2003) Practical RF circuit design for modern wireless systems. In: Passive circuits and systems, vol 1. Artech House Inc., 527 p
13. Goldenberg LM, Matyushkin BD, Polyak MN (1985, in Russian) Handbook of digital signal processing. Radio I Svyaz, Moscow, 312 p
14. Oppenheim AV, Schafer RW (2010) Discrete-time signal processing. Prentice-Hall, Englewood Cliffs, NJ, p 1108
15. Lyons R (2004) Understanding digital signal processing. Prentice Hall. Pearson Education Inc., Prentice Hall
16. Gu K (2005) RF system design of transceivers for wireless communications. Springer Science +Business Media, Inc.
17. Kozmin VA (2007, in Russian) Statistical modeling of radio engineering devices and systems (textbook), 2nd edn. Voronezh State Technical University, 258 p
18. Bityutsky VI, Kozmin VA (1987, in Russian) Periodic modulation of discrete signals and its use in digital processing. In: Methods and devices of digital and analog information processing. Voronezh, pp 20–23
19. Bityutsky VI, Kozmin VA (1986, in Russian) Using complex digital filtering to SSB signals receive. Radioelectron Commun Syst 29(4):48–53
20. Crochiere R, Rabiner L (1983) Multirate digital signal processing. Prentice-Hall Inc., Englewood Cliffs, New Jersey 07632, 411 p
21. Spazhakin MI, Repnikiv VD, Tokarev AB (2013, in Russian) Usage of farrow resampler in digital receiver for GSM direction finding. Bull Voronezh State Tech Univ 9(6–3):26–29
22. Spazhakin MI, Tokarev AB (2015) Digital receiver for addressed direction finding of modern communication standards. In: Control and Communications (SIBCON), 2015 International Siberian Conference OMSK 21–23 May 2015 Page(s):1–4 Print ISBN: 978-1-4799-7102-2. https://doi.org/10.1109/sibcon.2015.7147223
23. SMO-ANDROMEDA Software package for manual direction finding and radio monitoring for tablets and smartphones. http://www.ircos.ru/en/sw_andromeda.html. Accessed 28 Nov 2017

Chapter 6
Measurement of Radio Signals and Interferences Parameters

6.1 Introduction

Central frequency, spectrum width, and level of the observed signals are concerned, in particular, to a number of parameters to be estimated by the radio monitoring systems [1, 2]. Complicating requirements to measurements accuracy force continuous improvement of signal processing quality, as well as modifying of the used methods. The modern radio monitoring systems should provide measurements accuracy, which corresponds to the requirements of the normative documents, such as Russian State Standard P 53373-2009 [3], recommendations ITU-R SM.443 [4] and ITU-R SM. 377-4 [5].

Methods of parameters estimation essentially depend on types of the observed signals, on possibilities of the measuring equipment, and on requirements of standards in the area, to which the executed radio measurements are concerned. Usage of digital signal processing methods for measurements enables developers to improve separate characteristics of radio measuring systems significantly [2]. At the same time, it is necessary to take into consideration the interconnection and mutual influence of many indices of measuring systems; thereby, improvement of some characteristics can cause degradation of other characteristics directly or indirectly, sometimes leading to impossibility of providing some requirements of the normative documents. In this chapter, together with methods of measurement of radio signal parameters, such as central frequency, occupied bandwidth, etc., several similar "bottlenecks" in connection with measurements of radio interferences characteristics and analysis of radio situation as a whole are discussed.

A. M. Rembovsky et al., *Radio Monitoring*, Signals and Communication Technology, https://doi.org/10.1007/978-3-319-74277-9_6

6.2 Algorithms of Radio Signal Frequency Measurement

The key parameter of any narrowband radio signal is the central frequency. If the duration of observation is unlimited, then it can be measured with arbitrarily high accuracy. In practice, however, the duration of processed samples is limited. It may essentially affect the provided accuracy of the estimation [1].

The digital measurement methods described in literature, which are suitable for implementation in the automated hardware–software complexes of radio monitoring, use, as a rule, interpolation algorithms [6]. Investigation performed in [7] of different window functions influence on the frequency estimation accuracy does not reveal the significant qualitative variations of measurement accuracy as the response to the selected window function. Therefore, below we shall use the Hann window function for research of algorithms characteristics.

The recursive interpolation algorithm (RIA) can be represented as

$$f'_{m+1} = f'_m + \Delta f_m(x). \tag{6.1}$$

At initial iteration $\Delta f_0(x) = 0$, and at next steps, to obtain correction $\Delta f_m(x)$, we calculate:

$$\alpha = \sum_{n=0}^{N-1} x_n \cdot \exp\left(-j2\pi n\left(\frac{f'_m}{F_S} - \frac{1}{2N}\right)\right), \tag{6.2}$$

$$\beta = \sum_{n=0}^{N-1} x_n \cdot \exp\left(-j2\pi n\left(\frac{f'_m}{F_S} + \frac{1}{2N}\right)\right), \tag{6.3}$$

where x_n are the time samples, $n = 0, \ldots, N-1$ is an ordinal number of the time sample, and F_S is the sampling frequency.

These formulas are used for calculation of the frequency correction, which characterizes error of the frequency estimation if the real frequency varies within one interval between spectral samples [8]:

$$\Delta f_m(x) = \frac{1}{4 \cdot N} \cdot \frac{|\beta|^2 - |\alpha|^2}{|\beta|^2 + |\alpha|^2} \cdot F_S. \tag{6.4}$$

Investigation of this estimation accuracy by the mathematical modeling method shows that bias of the frequency estimation within the limits of the one interval between spectral samples behaves itself as shown in Figs. 6.1 and 6.2.

We see that already the first correction in expression (6.1) gives the improvement of the estimation accuracy. However, the high-accurate estimation is assured for arbitrary frequency of the processed signal only after third iteration. Thus, in practice, we should use $m = 3$ iterations according to the RIA algorithm, which coincides with recommendations of [9].

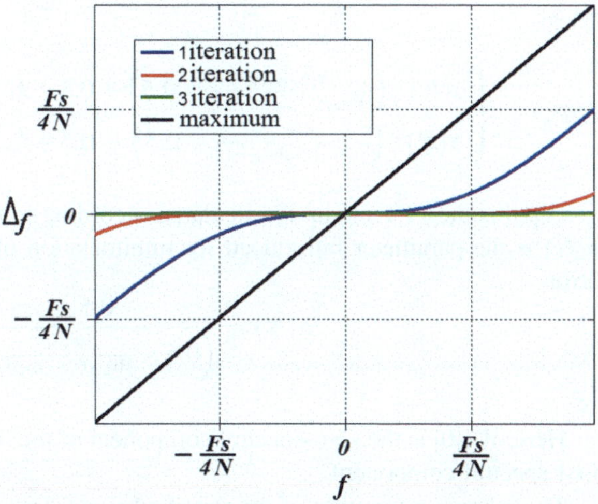

Fig. 6.1 Bias of the frequency estimation for different quantity of RIA iterations at high signal/noise ratio (SNR)

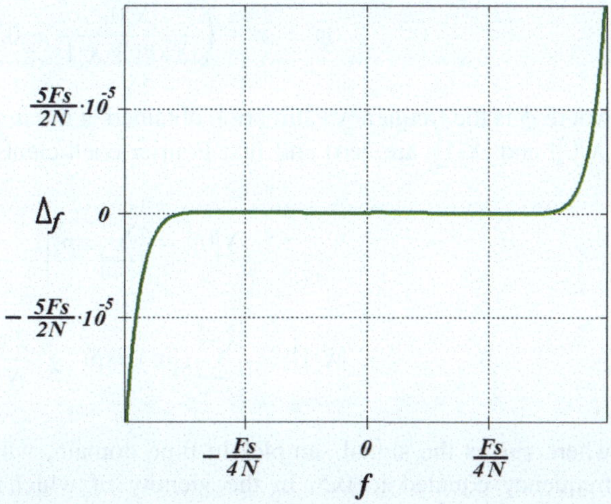

Fig. 6.2 Bias of the frequency estimation for the third RIA iteration at high SNR

The high-precision algorithm (HPA) is an improved version of the RIA algorithm in relation to calculation rate and estimation accuracy. This algorithm is described in [10] in detail.

The first stage of the HPA algorithm is calculation of estimation of the signal relative frequency according to the formula:

$$\phi = G(k_{\max}), \tag{6.5}$$

where

$$G(k_m) = \begin{cases} \alpha(N) \cdot \left(\frac{|X[k_m+1]|}{|X[k_m+1]| + |X[k_m]|} - 0.5 \right) + 0.5 + k_m, & |X[k_m+1]| \geq |X[k_m-1]| \\ \alpha(N) \cdot \left(\frac{|X[k_m]|}{|X[k_m]| + |X[k_m-1]|} - 0.5 \right) - 0.5 + k_m, & |X[k_m+1]| < |X[k_m-1]| \end{cases},$$

$|X[k]|$ is the FFT modulus calculated according to the weighted time signal, and $\alpha(N)$ is the parameter introduced for minimization of the frequency measurement error

$$\alpha(N) = \frac{0.5}{\left(0.5 - \frac{|W_N[1]|}{|W_N[0]| + |W_N[1]|}\right)}. \tag{6.6}$$

Here, $W_N[0]$ is the zero-spectral component of the time window, and $W_N[1]$ is the first spectral component.

The relative estimation of the specified signal frequency ϕ' is determined by the Fourier coefficients as:

$$\varphi' = \varphi + \left(\frac{|X[0]|}{|X[0]| + X[1]} - 0,5 \right), \tag{6.7}$$

where ϕ is the frequency estimation obtained at the first stage of measurements, and $|X[0]|$ and $|X[1]|$ are zero and first Fourier coefficients, relatively:

$$|X[0]| = \left| \sum_{n=0}^{N-1} z[n] \right|, \tag{6.8}$$

$$|X[1]| = \left| \sum_{n=0}^{N-1} z[n] \cdot \exp(-j \cdot \frac{2\pi}{N} \cdot n) \right|, \tag{6.9}$$

where $z[n]$ is the signal samples in time domain, which are displaced toward the frequency equaled to 0.5, in the vicinity of which the formula (6.5) gives the minimal error:

$$z[n] = x[n] \cdot \exp(-j \cdot \frac{2\pi}{N} \cdot (\phi - 0.5) \cdot n). \tag{6.10}$$

The required frequency is calculated as $f = \phi' \cdot F_S/N$.

Investigation results of the formed estimations accuracy at variations of the signal frequency within the interval between adjacent spectral samples are presented in Fig. 6.3.

Algorithms characteristics obtained with the help of statistical modeling, during which the sine signal with random uniformly distributed frequency in the range $[0 \ldots F_S]$ was subjected by processing, are presented in Fig. 6.4.

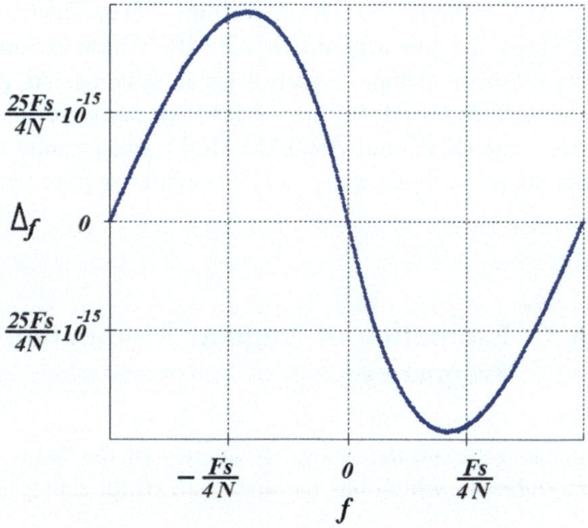

Fig. 6.3 Bias of the frequency estimation for the HPA algorithm at high SNR

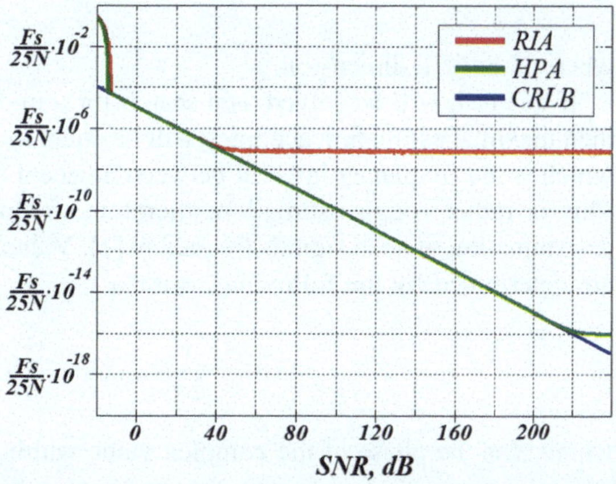

Fig. 6.4 Dependence of the mean-square error of the frequency estimation on the SNR

Investigation results show, that HPA and RIA algorithms are equally effective at low and medium values of SNR, and the error dispersion provided by them practically coincides with the Cramér–Rao lower bound, which can be written in accordance with [8] as

$$\sigma^2_{\text{CRLB}} = \frac{6}{(2\pi)^2 N(N^2-1)\cdot 10^{\frac{\text{SNR}}{20}}}, \tag{6.11}$$

where SNR is the signal/noise ratio. It should be noted that the RIA algorithm requires the computing effort higher, than the one for the HPA algorithm developed on its base for achievement of the same accuracy.

At high SNR, the HPA algorithm is also effective and, from theoretical position, it saves the linearity up to 220 dB, which is another advantage over the RIA algorithm at using in metrological systems. At present, the HPA algorithm is successfully used in the SMO-PAI software in the measuring systems with ARGAMAK-IS and ARGAMAK-M and permits execution of radio signal measurements with accuracy, which satisfies requirements of the normative documents [3, 5, 11].

6.3 Estimation of Angular Modulation Signal Central Frequency

Let us estimate the central frequency of the radio signal with frequency or phase modulation, which has the spectrum width satisfying the inequality

$$B_f \geq \frac{2F_S}{N}, \tag{6.12}$$

where N is FFT dimension.

If this task will be solved as a search for a middle of the radio signal spectrum, then maximal estimation accuracy will be equal to a half width of the frequency bin, which is the frequency interval between adjacent samples of the discrete spectrum. This is rather rough estimation; therefore, frequency detector is often used for determination of such signals frequency [1]. Values at the frequency detector output are determined by the following formula

$$X[i] = \frac{1}{2\pi} \cdot \angle(Z[i]) \cdot \overset{*}{Z}[i-1]), \tag{6.13}$$

where $\angle$ is the phase of the complex value within $(-\pi; \pi]$, $z[i]$ is the sample of the complex envelope of the signal, and $\overset{*}{z}[i-1]$ is the complex-conjugated value of the previous sample of the envelope.

To increase measurement accuracy, the estimation of received signal frequency should be calculated as the result of averaging of data, which arrived from the output of the frequency detector

$$F = \frac{F_S}{M \cdot N} \sum_{k=0}^{M-1} \sum_{i=0}^{N-1} W[i] \cdot X[i + kN/2] + F_0. \tag{6.14}$$

Here, F_0 is the receiver tuning frequency, N is the sample size, M is the amount of overlapping segments, k is the number of overlapping segment in the range from 0 to $(M-1)$, and $W[i]$ is the window function.

Figure 6.5 shows the dependence of the frequency measurement accuracy of the FM signal on the SNR, illustrating an advantage of frequency detector usage compared to estimation method according to the spectrum maximum. From this figure, we see that at SNR higher than 15 dB the method provides frequency estimation with accuracy not more than $F_S/(12N)$, which is several times less than the estimation error according to the spectrum maximum. At that, the high enough accuracy is provided for relatively short samples with the length starting from 16,000 samples.

6.4 Algorithm of the Digital Signals Central Frequency Estimation

Let us consider the spectral analysis method, with the help of which one can obtain frequency estimations with high accuracy of the amplitude-shift keying (ASK) signal, the frequency-shift keying (FSK and FSK-MS) signal and the phase-shift keying (PSK) signal in the real time [1].

Estimation of the central frequency of the ASK signal is not difficult at all because it has clearly expressed central spectral component, by which we can make estimation with the help of the HPA algorithm. At that, the estimation accuracy will be maximally closed to the Cramer–Rao boundary.

At frequency-shift keying, the definite frequencies of sine signals correspond to "0" and "1" values of the informational sequence, as shown in Fig. 6.6. Estimation of the central frequency of such a signal can be calculated as the arithmetic mean of the frequency estimations according to the spectrum maximum. Error of such estimation directly depends on a number of components of FSK signal, which are represented in the instantaneous spectrum in time interval under investigation.

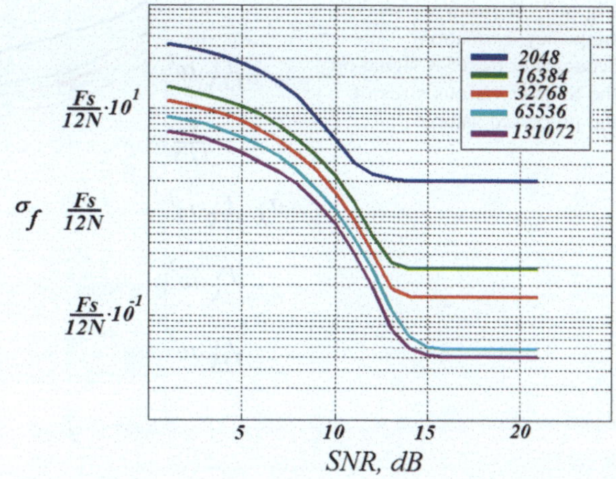

Fig. 6.5 Dependence of the frequency estimation accuracy of the FM signal on the SNR with the help of the frequency detector at various sizes of the processed sample

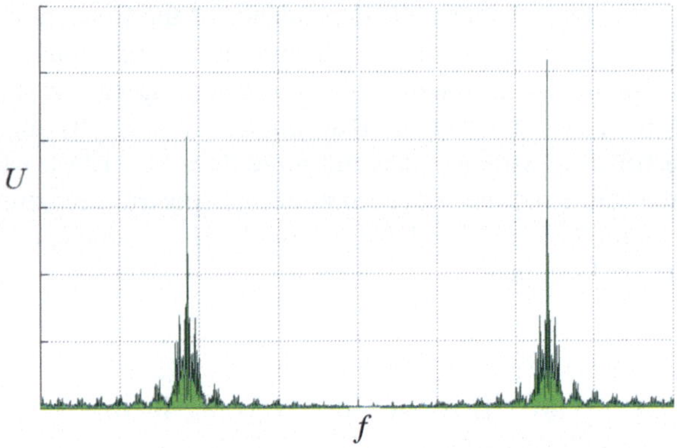

Fig. 6.6 FSK signal in the spectral domain

To increase accuracy of this estimation, we can use the HPA algorithm. However, it shows good results, when signal contains single maximal spectral component only. Hence, before execution of the calculations, it is necessary to determine the spectrum areas containing sine signals, and after this to carry out the estimation of each sine signal separately. Necessity of errorless extraction of the harmonic components of the signal does not enable to perform the frequency estimation of the FSK signal at low SNR in spite of effectiveness of the HPA algorithm.

The second more accurate method consists of the frequency detector usage (6.13). The array of instantaneous frequency values (two and more samples, depending on the quantity of shift keying positions) will be observed at output of the frequency detector. The mean value of the obtained array will give an estimation of the central frequency of the FSK signal at SNR >10 dB with accuracy from 1/2 to 1/8 of bin parts, which is confirmed experimentally. From Fig. 6.7, we see that at

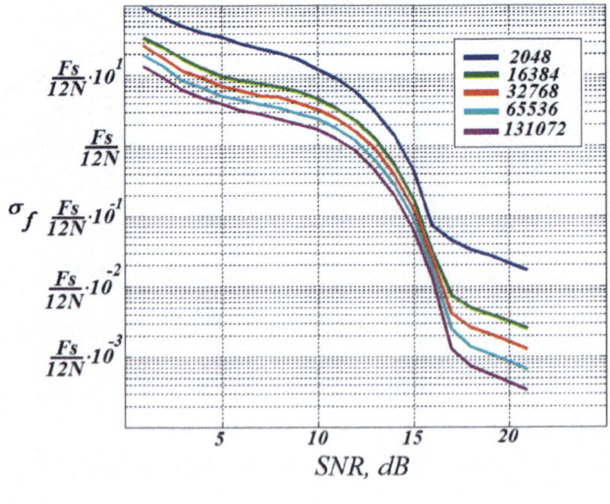

Fig. 6.7 Dependence of the frequency estimation accuracy of the FSK signal on the SNR at various sizes of the processed sample

SNR more than 15 dB, the error of the mean frequency method for FSK signal does not exceed hundredth part of the bin.

Minimal shift keying (MSK) manipulation is a specific case of FSK with the continuous phase and the minimal possible modulation index *m*, providing the orthogonality of signals, which transmit "0" and "1" of digital information. In other words, the MSK signal is the signal with minimal frequency shift between "0" and "1," at which these frequencies can be distinguished within the time interval T

$$\begin{cases} S_0 = A \cdot e^{j((\omega_0 - \omega_\partial)t)} \\ S_1 = A \cdot e^{j((\omega_0 + \omega_\partial)t)} \end{cases}, \; 0 \leq t \leq T. \tag{6.15}$$

To provide spectrum extraction of signals S_0 and S_1, we raise the signal to the n power. Taking into consideration the decrease of SNR at raising the signal to the power, to provide the distinguishing of spectral components, as a rule, it is expedient to set $n = 2, 4$:

$$\begin{cases} (S_0)^n = (A)^n \cdot (e^{j((\omega_0 - \omega_\partial)t)})^n \\ (S_1)^n = (A)^n \cdot (e^{j((\omega_0 + \omega_\partial)t)})^n \end{cases} \rightarrow \begin{cases} (S_0)^n = (A)^n \cdot e^{j((n\omega_0 - n\omega_\partial)\cdot t)} \\ (S_1)^n = (A)^n \cdot e^{j((n\omega_0 + n\omega_\partial)\cdot t)} \end{cases}, 0 \leq t \leq T.$$

Thus, at raising the MSK signal to the n power, both ω_0 frequency itself and the deviation ω_∂ of the signal frequency from ω_0 increase by n times. As a result, two clearly expressed maximums of the spectrum, which are distanced one from another, are formed. The spectrum transformation of MSK signal at rising to the power is illustrated in Fig. 6.8.

After this transformation, the central frequency estimation is expedient to be carried out with the help of the method based on the frequency detector, similarly with FSK signals. Results of the estimation error analysis obtained by this method are shown in Fig. 6.9. We see that at SNR not less than 15 dB, this method provides small error of MSK signal estimation even at small duration.

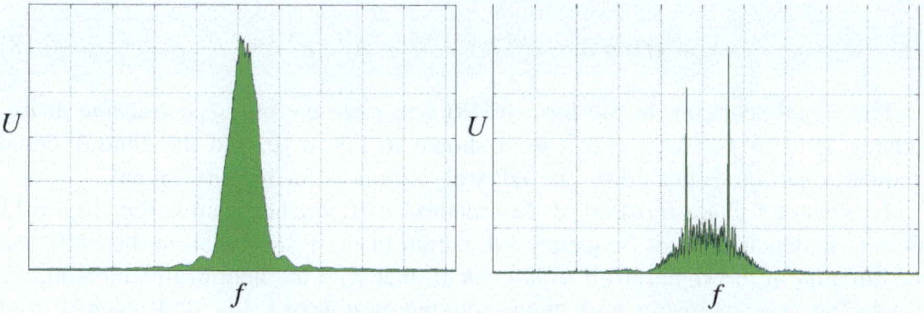

(a) The spectrum of the initial MSKsignal

(b) The spectrum of the MSK signal after transformation

Fig. 6.8 Initial and transformed spectrums of the MSK signal

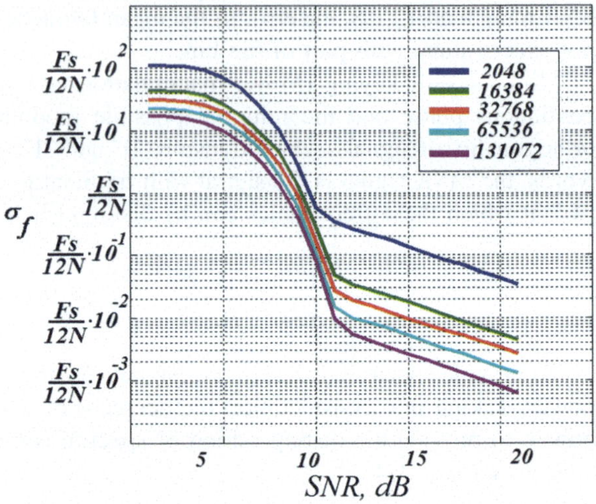

Fig. 6.9 Dependence of the frequency estimation accuracy of MSK signal on the SNR at various sizes of the processed sample

Phase-shift keying (PSK) signals can be written as [2]:

$$s_i(t) = A \cdot e^{j(\omega_0 t + \varphi_i)}, i = 0, \ldots (M-1), 0 \le t \le T, \tag{6.16}$$

where the phase φ_i may take M discrete values $\varphi_i = 2\pi i/M$, $i = 0, \ldots (M-1)$, ω_0 is the carrier frequency, and T is the duration of the measuring interval, $0 \le t \le T$.

At estimation of the PSK signal frequency, it is efficient to demodulate signal under investigation. It means fulfilling of the mathematical transformations sequence to enable extraction of the signal central component.

It can be written in general case as

$$s_i^M(t) = A^M \cdot e^{jM(\omega_0 t + \varphi_i)} = A^M \cdot e^{jM \cdot \omega_0 t + jM \cdot \varphi_i}. \tag{6.17}$$

Due to the fact, that $\varphi_i(t) = 2\pi i/M$, the expression (6.17) takes a form:

$$s_i^M(t) = A^M \cdot e^{jM \cdot \omega_0 t + jM \frac{2\pi i}{M}} = A^M \cdot e^{jM \cdot \omega_0 t}. \tag{6.18}$$

The signal spectrum in the form (6.18) will have the only spectral line at frequency $M\omega_0$ or $f = M\omega_0/(2\pi)$, as it shown in Fig. 6.10, and the central signal frequency can be estimated by the HPA algorithm as for the sine signal.

To illustrate characteristics of this method of frequency estimation, Fig. 6.11 shows the dependence of frequency estimation of the PSK signal on the SNR and the duration of the signal itself. It is evident, that with the help of this method, we can perform measurements with thousandth bin parts error at low SNR even at short duration of observation (at sample dimension not more than 2^{11} samples).

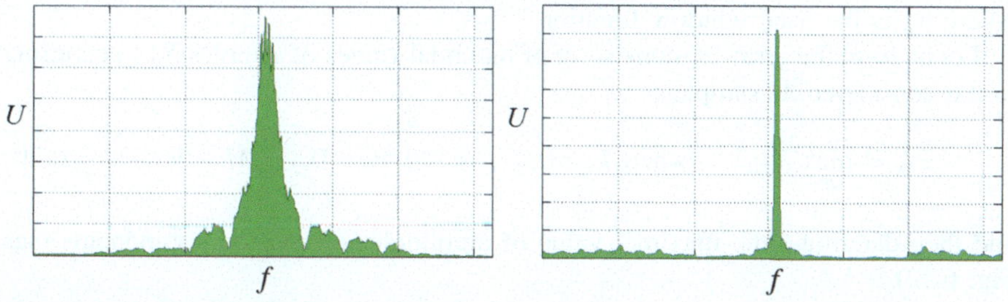

Fig. 6.10 Signal spectrum before and after rising to power

Fig. 6.11 Dependence of frequency estimation accuracy of the PSK signal on SNR at various sizes of the processed sample

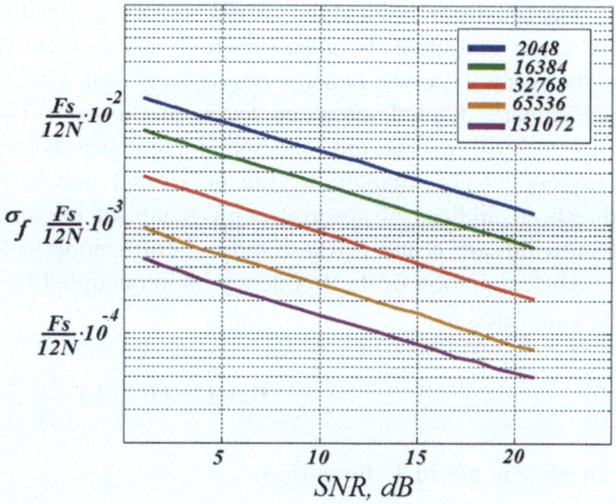

6.5 Measurement of the Occupied Bandwidth at the X dB Level

The spectrum width is one of the main parameters characterizing the signal. For estimation of this parameter, two methods are used. At that, the "X dB" method is less unprofitable.

Let the measurement of the radio signal spectrum width be carried out at the time interval with duration $T = T_0 \cdot M$, where M is the amount of samples, and T_0 is the duration of each separate sample with dimension N. At that, in the measurement interval with duration $T_0 = N \cdot \Delta t_0$, there are N complex samples x_n, taken with interval $\Delta t_0 = 1/F_S$. Then, the discrete spectrum of amplitudes for the sample sequence $x_{m,n}$ belonging to mth sample can be defined as

$$X_{m,k} = \left| \sum_{n=0}^{N-1} x_{m,n} \cdot w_n \cdot \exp(-j\frac{2 \cdot \pi}{N} kn) \right|, \quad k = 0,\ldots,N-1, \qquad (6.19)$$

where w_n is the used window function.

Let us form the array N_k consisting of maximal values of samples $X_{m,k}$ according to the considered M samples:

$$N_k = \max_m [X_{m,k}], \quad \text{where } k = 0, \ldots, N-1, m = 0, \ldots, M-1 \tag{6.20}$$

and then determine the maximal value of amplitudes of spectral components (see Fig. 6.12)

$$U_{\max} = \max_k [N_k]. \tag{6.21}$$

The spectrum width at the "X dB" level is defined as the frequency range beyond the lower and upper bounds, of which any discrete spectral component or continuous spectral power density has a level not less than by "X dB" lower, than preliminary defined reference level, in this case $U_{\max}$ [4]. Therefore, the value $U_{\max} - \text{X dB}$ will be the threshold for the search for extreme values of measured frequency band (marked by the horizontal line in Fig. 6.12). According to this threshold, indices of spectral components of beginning and ending of the signal bandwidth are L and R; these indices correspond to frequencies f_1 and f_2.

The estimation of the frequency band occupied by the signal under analysis can be calculated as

$$\text{Band} = (R - L) \cdot \frac{F_S}{N}, \tag{6.22}$$

where N is the FFT dimension.

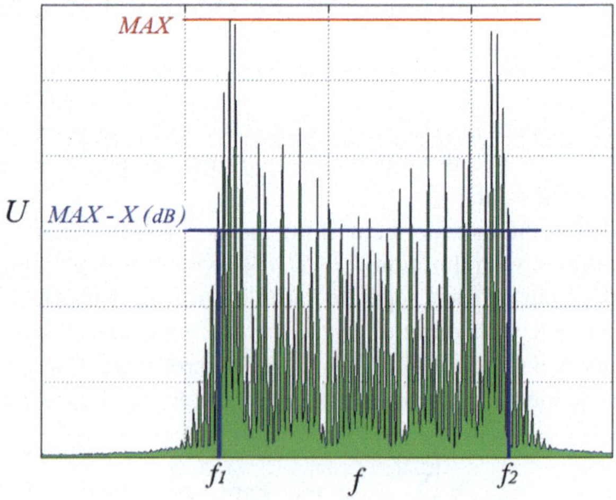

Fig. 6.12 Measurements of the radio signal bandwidth at the "X dB" level

To provide high accuracy of estimation, it is necessary for signal level to exceed the noise level not less, than by X dB. If this requirement is fulfilled with margins, the accuracy of bandwidth estimation will weakly depend on SNR. At the same time, since the bandwidth estimation (6.22) is based on samples of the discrete spectrum, the estimation accuracy directly depends on the bandwidth of the frequency bin.

6.6 Measurement of the Spectrum Width by the "Beta/2" Method

According to the "$\beta/2$" method, the spectrum width is understood as the frequency band beyond the lower and upper limits of which the emitted mean powers are equal (each) to $\beta/2$ percent of full mean power of radio emission

$$S_{(\mathrm{Spectr})} = \sum_{k=0}^{N-1} N_k^2. \tag{6.23}$$

It is the alternative to the "X dB" method.

To obtain the left boundary of the signal spectrum, beginning from the first sample, we calculate the sum of squared samples gradually increasing the sample index until we obtain the value closed to $S_{(\mathrm{Spectr})} \cdot \beta/2$. To obtain the right boundary of the signal spectrum, we can carry out the similar operation, but beginning from the last sample and moving toward the decreasing of the sample index:

$$\sum_{k=0}^{L} N_k^2 \approx \sum_{k=R}^{N-1} N_k^2 \approx S_{(\mathrm{Spectr})} \cdot \frac{\beta}{2}. \tag{6.24}$$

Two boundaries (indices L and R) corresponding to frequencies f_1 and f_2 shown in Fig. 6.13 are the calculation results.

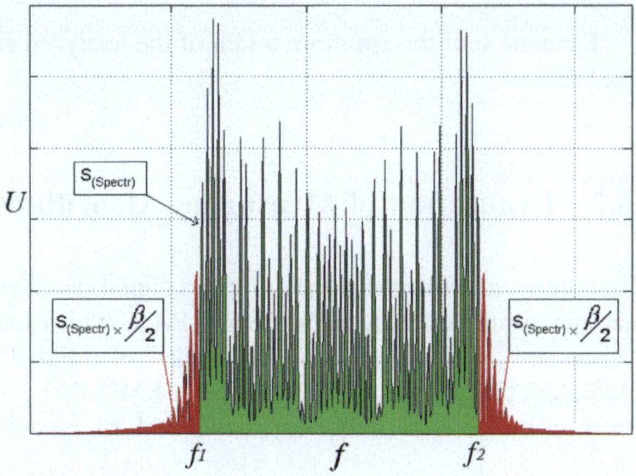

Fig. 6.13 Principle of signal bandwidth measurement according to the "$\beta/2$" method

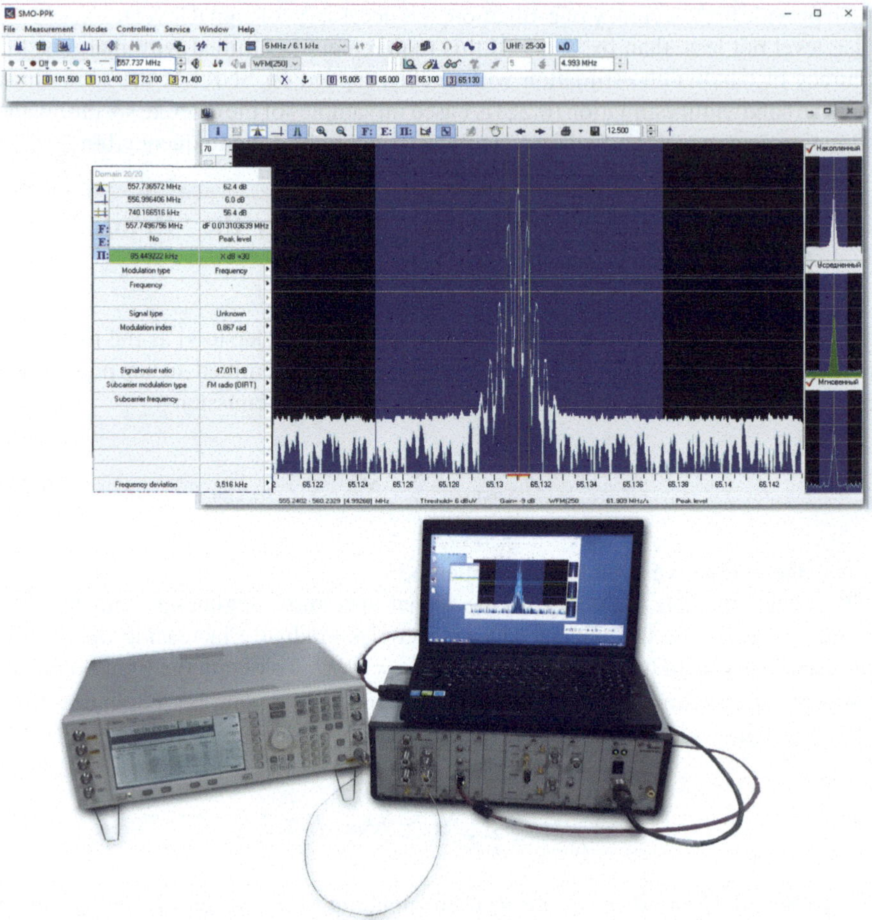

Fig. 6.14 Measurement of the AM signal frequency by the ARGAMAK-IS receiver

Estimation of the spectrum width of the analyzed signal is calculated by formula (6.22).

6.7 Realization of Measuring Algorithms

The above-described methods of radio signal parameters estimation are realized in the measuring software packet SMO-PA I, which has the state approach on measurements execution with the help of digital receivers ARGAMAK-IS, ARGAMAK-IM, and ARGAMAK-M. SMO-PAI provides functions of measurement of radio signal frequency, bandwidth of occupied frequencies, calculation of

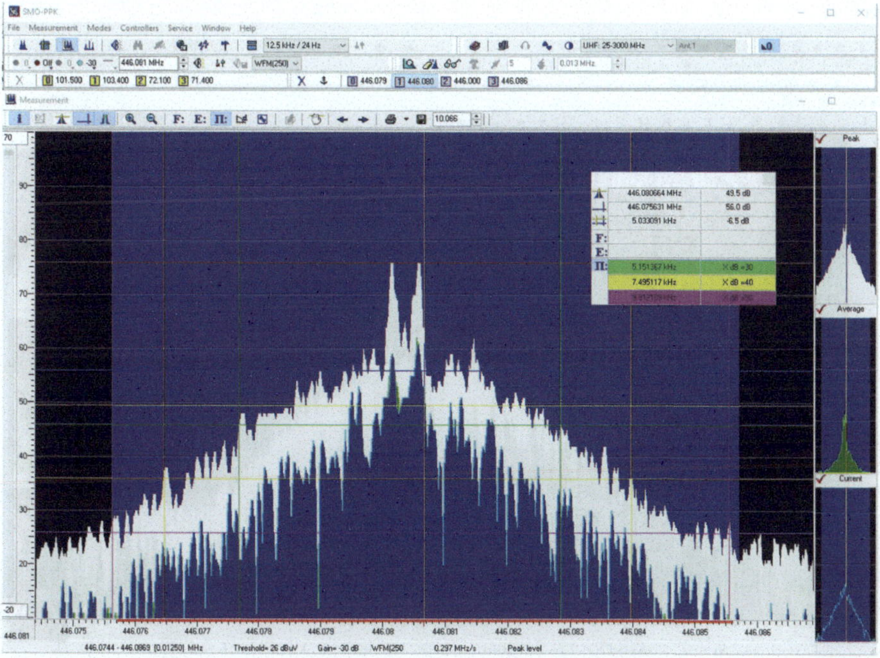

Fig. 6.15 Measurement of the occupied frequency band in SMO-PAI according to the "X dB" method

specific parameters of radio signals, such as frequency deviation, modulation coefficient, modulation frequency, satisfying the normative document [5]. The relative accuracy of frequency measurement for sine and AM signals is not worse than 1×10^{-9}, for signals with digital types of modulation not worse than 5×10^{-8}, and the estimation error of the signal spectrum width (according to the "X dB" method at 30, 40, 50 and 60 dB of X values) does not exceed 5%.

Figure 6.14 illustrates frequency and signal spectrum width measurement using the ARGAMAK-IS receiver and the SMO-PAI software.

Let us consider results of spectrum width measurement with the help of two above-mentioned methods. Measurements were executed at high SNR and frequency synchronization from the high-precision source of the synchronization VCH-311. The software windows corresponding to measurements by "X dB" and "$\beta/2$" methods are shown in Figs. 6.15 and 6.16, and the results are presented in Tables 6.1 and 6.2. Results prove that in the whole frequency range, the error of bandwidth measurement does not exceed 0.7%.

We note that growth of the measurement error at higher frequencies is not related to features of the processing algorithm, but is caused by increase of the phase noise level of radio receivers.

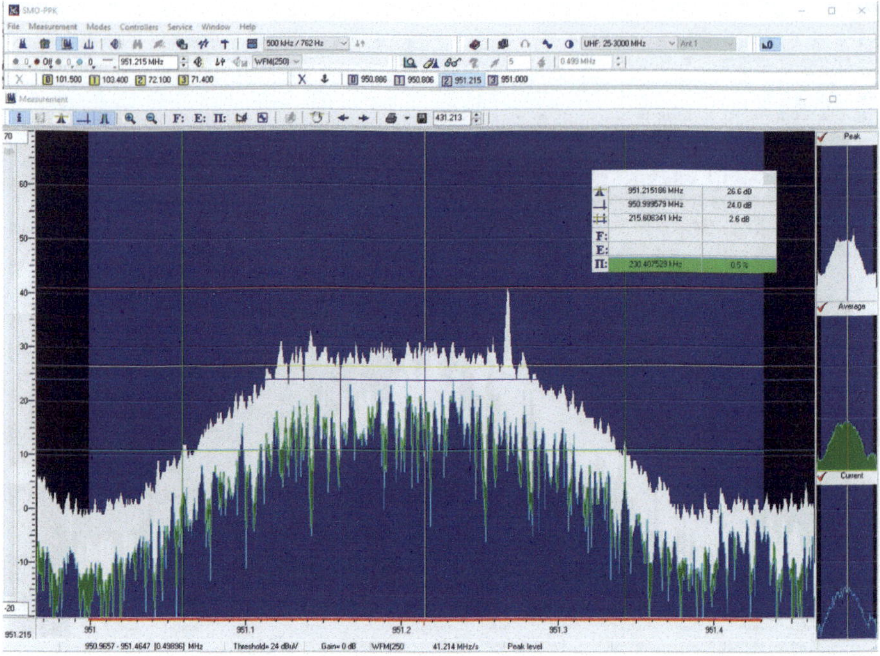

Fig. 6.16 Measurement of the occupied frequency band in SMO-PAI according to the "$\beta/2$" method

Table 6.1 Error of the spectrum width measurement according to the "X dB" method

f, MHz	X, dB			
	−30 (%)	−40 (%)	−50 (%)	−60 (%)
1	0.10	0.09	0.45	0.26
3	0.10	0.09	0.45	0.26
13	0.10	0.09	0.45	0.26
20	0.10	0.09	0.45	0.26
25	0.17	0.16	0.45	0.69
640	0.17	0.16	0.45	0.35
1000	0.17	0.16	0.45	0.69
2000	0.17	0.16	0.45	0.69
3000	0.17	0.16	0.45	0.69

6.8 Estimation of Interference Intensity

Requirements to the modern radio interference measurement devices are regulated by the set of normative documents including [12, 13] standards. These standards specify a set of control parameters, to which the measuring subsystem of the radio monitoring equipment should satisfy.

Let us focus on the problem of frequency selectivity in digital devices for interference measurement [14, 15]. Requirements to the bandwidth of measuring systems can be found in Fig. 2 of the standard [13]. Digital processing allows realization of filters with frequency response, which are rather close to the ideal filters, in the radio monitoring systems, and, having oriented to requirements to selectivity from the standard [13], it may seem to be expedient to use filters with high rectangular ratio of frequency-response characteristics in the digital systems for interference measurement. However, the more careful analysis shows that such a solution is not universal and, in particular, it leads to impossibility of providing standard requirements to the amplitude ratio for interference measuring system with the average detector.

In Sect. 6 of the standard [13], it is defined that average detector is used to measure mean value of the signal envelope, which passes the stages preceding the detector. The response of the interference measuring system with the average detector on the repeating pulses should be so, that at constant measurements of the measuring system, the ratio of amplitude and pulse repetition frequency at its input can be expressed by inverse proportionality [13, point 6.4.2]. In practice, the accuracy of these requirements fulfillment will depend on both the choice of sampling parameters and the filtering properties of the receiving path.

During testing of detector characteristics, the sequence of short radio pulses with the coherent high-frequency filling is applied to the input of the interference measuring system. The coherence of acting pulse filling provides the coherence of responses fillings on them of the intermediate frequency filter and, as the consequence, the direct summation of complex envelopes of these responses. As a result, the total complex signal applied to the input of the digital detector can be represented as

$$\dot{s}_{\text{вх. дет}}(t) = S_0\left[\sum_{k=0}^{\infty}\dot{A}\left(t-kT_n\right)\right]e^{j\psi_0}, \tag{6.25}$$

where T_n is the pulse repetition period; $\dot{A}(t)$ is the complex envelope of the receiving path pulse characteristic, which includes the final low path filter (LPF).

But when the repetition frequency of input probing pulses grows, the responses to them will not have time to fade and they will begin to overlap. Proportionality between amplitude and pulse repetition frequency will keep only under condition, that for each time moment t, the $\dot{A}(t - kT_n)$ components included in (6.25) will give complex values coinciding in phase, or (if the complex envelope of the pulse characteristics of the receiving path is represented by the only real values) under condition, that values of the complex envelope of the pulse characteristics of the receiving path will keep the sign for any t. For estimation of deviation of the specific LPF characteristic with regard to above-formulated requirement, we can use the index

Table 6.2 Error of the spectrum width measurement according to the "$\beta/2$" method

f, MHz	$\beta = 1\%$ (%)	$\beta = 6\%$ (%)	$\beta = 17\%$ (%)	$\beta = 24\%$ (%)
1	1.10	0.05	0.27	0.22
3	1.10	0.05	0.16	0.22
13	1.10	0.05	0.27	0.22
20	1.10	0.05	0.16	0.65
25	1.10	0.05	0.68	0.65
640	1.10	0.05	0.68	0.65
1000	1.10	0.05	0.68	0.65
2000	1.10	0.05	0.68	0.65
3000	1.10	0.05	0.68	0.22

$$\chi = \begin{cases} 20\lg\left(\dfrac{\int_{-\infty}^{\infty}|g(t)|dt}{\left|\int_{-\infty}^{\infty}g(t)dt\right|}\right) & \text{for analog filters;} \\ 20\lg\left(\dfrac{\sum_i |g(i)|}{\left|\sum_i g(i)\right|}\right) & \text{for digital filters;} \end{cases} \tag{6.26}$$

where $g(t)$ and $g(i)$ are the pulse characteristics of the analog and digital LPF. At strictly non-negative pulse characteristic, we have $\chi = 0\,\mathrm{dB}$, and with growth of the negative lobes area, the χ value increases.

Let us consider the Appendix A of the standard [13], in which there is a description of LPF providing the necessary system selectivity of filtering and having the filter transmission factor

$$\dot{K}_{\mathrm{CISPR}}(\omega) = K_0\left[\frac{(2\Omega^2)}{\left((\Omega + j\omega)^2 + \Omega^2\right)}\right]^2, \tag{6.27}$$

where the parameter Ω (rad/s) is calculated according to the rule:

$$\Omega = \frac{\pi B_f}{\sqrt{2}} \tag{6.28}$$

and provides the LPF with the bandwidth B_f (Hz) by the 6 dB level. Calculations show that the pulse characteristics of this filter have the weakly expressed alternate-sign character, due to which the index of distortion of the amplitude ratio may achieve $\chi_{\mathrm{CISPR}} \approx 1.1\,\mathrm{dB}$.

An attempt to use filter with almost rectangular amplitude–frequency characteristic, which has the alternating-sign sinc-like pulse characteristic, in the receiving path structure gives inadmissibly large value of the index (6.26) equaled to $\chi_{\text{DLPF}} \approx 5.6\,\text{dB}$.

For minimization of the χ index as the final LPF of the interference measuring system, we can recommend to use the finite impulse response (FIR) filter synthesized using the Hamming window at 10% predistortion of the pass bandwidth, which precedes the average detector

$$F_b = 0,45 \cdot TB_f. \tag{6.29}$$

At that, the filter order should be chosen in accordance with the rule

$$\text{FIR\_Len} = \text{ceil}(1,13/F_b), \tag{6.30}$$

where $\text{ceil}(\cdot)$ is the function of ceiling. The similar filter not only provides requirements to selectivity, but it has the pulse characteristic with index $\chi_{\text{FIR}} \approx 0\,\text{dB}$.

Errors in fulfillment of the amplitude ratio determined by the index

$$\chi(f_{\text{HF}}, F) = 20\ \text{lg}\left(\frac{s_{\text{out. det}}(f_{\text{HF}})}{s_{\text{out. det}}(F)}\frac{F}{f_{\text{HF}}}\right), \tag{6.31}$$

where f_{HF} and F are the high and low frequencies of the probing pulses, which pass to the input of interference measuring system; they are shown in Fig. 6.17.

These functions were obtained during full-scale experiments of measuring receivers of the ARGAMAK-family [2] for radio emissions belonging to C frequency bands (30…300 MHz) and D ones (300…1000 MHz), at the IF filter pass bandwidth B_f = 120 kHz by the 6 dB level. The FIR filter with the pulse characteristic length of 512 samples served as the analog of the ideal LPF. The two-sectional filter with the system function of the section

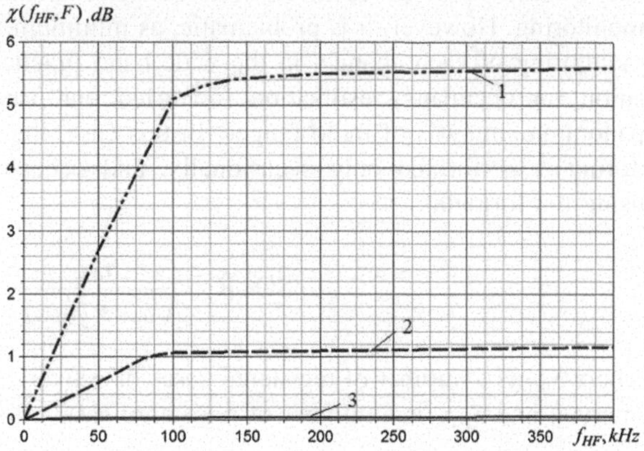

Fig. 6.17 Errors in the amplitude ratio $\chi(f_{\text{HF}}, F)$ at F = 5 kHz (1—is an analog of the ideal LPF; 2—CISPR-filter; 3—FIR filter)

$$H(z) = b\frac{1 + 2z^{-1} + z^{-2}}{1 - a_1 z^{-1} - a_2 z^{-2}}, \qquad (6.32)$$

and values of coefficients $b = 0.00778$; $a_1 = 1.7354$; $a_2 = -0.7665$ is the digital analog of the filter described in the standard [16]. Parameters of the special FIR filter synthesized using the "Intel Signal Processing Library" [17] are $F_b = 0.027$ FIR_Len = Ceil(1.13/0.027) = 42.

From Fig. 6.17, we see that improvement of the frequency selectivity at the expense of using the digital analog of ideal low pass filter is accompanied by inadmissible distortion of the amplitude ratio. Usage of the frequency-selective filter recommended in [13] is accompanied by formally admissible but quite essential deviation of this ratio from zero, and usage of FIR filter of low order with specially selected relatively small selectivity allows obtaining the "factor of safety" to check the amplitude ratio.

6.9 Estimation of Radio Frequency Spectrum Occupancy

The radio frequency occupancy, estimation of which is provided by the radio communication services of different countries with purpose to increase the effectiveness of the radio frequency resource usage, is the objective characteristic of the real usage degree of radio frequencies. Under radio spectrum occupancy, we understand a probability that in randomly chosen time moment the radio channel under investigation, the frequency band or another frequency resource will be used for information transmission. The necessity of the spectrum occupancy estimation is defined, in particular, by the governmental documents RD 45.193-2001 "Equipment of the radio monitoring stations. General technical requirements" and the National Standard of Russian Federation "Equipment of radio monitoring stations, automated. Technical requirements and methods of testing" [18].

Under the ideal conditions, estimation of occupancy requires round-the-clock and continuous observation of all frequencies included in the radio range under monitoring. However, it is problematic, as minimum, to implement the continuous monitoring of each channel in the wide band practically. Instead, the data acquisition for occupancy estimation is carried out usually with framework of the panoramic analysis of radio ranges. In this case, the separate channels state monitoring is performed only occasionally, and occupancy estimation is calculated using the formula

$$\text{SOCR} = \frac{S_{\text{act}}}{(S_{\text{act}} + S_{\text{pass}})}, \qquad (6.33)$$

where S_{act} is a number of registered cases of active channel states in the measuring interval, and S_{pass} is a number of registered cases of the passive states.

At occasional monitoring of states, we cannot exclude the situations, when in intervals between the moments of monitoring (control points) the radio pulses of small duration, or, oppositely, the small pauses in radio emission, which will not be registered and taken into account at occupancy calculation, may be situated. But even at absence of missed signals or pauses, impossibility of exact moment's determination of the channel switching from the active state to the passive one and vice versa may serve as a source of appreciable error of estimation [14]. Demonstration of this kind of error at occupancy measurement is shown in Fig. 6.18, where instead of the true value of occupancy SO = 50%, estimation according to formula (6.33) in the first case gives the value $SOCR_1 = 7/16 = 43.75\%$, while in the second case $SOCR_2 = 9/16 = 56.25\%$.

It is obvious, that with the growth of the control points amount $J_m = S_{act} + S_{pass}$, the potential disperse of measurement results will decrease, which provides insignificant occupancy estimation error at arbitrary choice of the time initial moment. The number J_m of the channel state control points at the occupancy measuring interval is defined by its duration T_m and value of the interval between moments of channel state control T_{rv} (which, in turn, is determined by radio monitoring equipment performance and by number of radio channels, in which occupancy measurement is carried out). However, it is impossible to increase the length T_m of the interval of occupancy measurement arbitrarily, because this kind of growth does not allow monitoring of the possible occupancy variation along the time axis. In accordance with requirements of international normative documents, intervals T_m must have fixed duration of 5 or 15 min [19, 20], and the quality of occupancy measurements at the similar intervals should satisfy the definite requirements in accuracy and reliability. Reliability P_{SO} of occupancy estimation is a probability that the SOCR estimation will differ from the true value SO not more than by the admissible absolute error Δ_{SO}

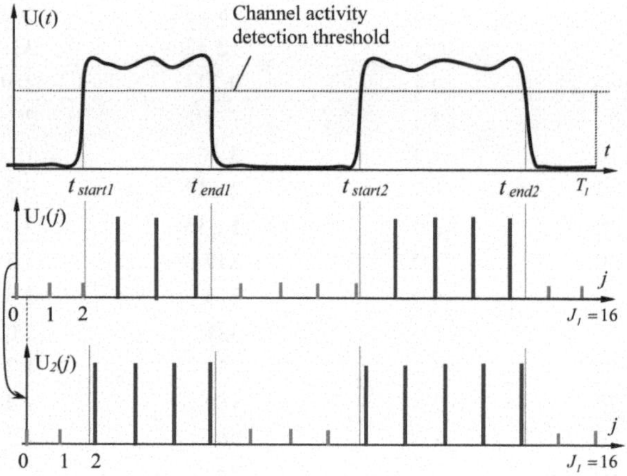

Fig. 6.18 Error origin at occupancy estimation

$$P_{SO} = P\{ |SOCR - SO| \leq \Delta_{SO}\}. \tag{6.34}$$

The admissible reliability values lay in the range from 90 to 99%, and the most frequently used value is P_{SO} = 95%.

The detailed information about features of occupancy measurement can be found in [21]; their usage enables optimization of the occupancy measurement process in various rather complicated conditions; however, the similar optimization is based on the a priori information about properties of the observed signals (especially, it concerns long signals with duration from several seconds) and on the detailed information about operation features of the used equipment. At the same time, at discussion of the occupancy estimation reliability, it is useful to rely on materials of the paper [22], defining the dependence of radio channel occupancy on the admissible error of its estimation and the sample volume necessary for achievement of reliability of P_{SO} = 95%. These dependences are represented in Table 6.3.

In the left part of Table 6.3, the data are presented, which are typical for international documents on occupancy measurement acting before 2012. The negative consequences of complying with these documents were the following restrictions of admissible relative estimation error:

Table 6.3 Number of samples required to achieve 10% maximum relative error δ_{SO} or 1% absolute error Δ_{SO} with 95% confidence level

Channel occupancy, %	Required relative error $\delta_{SO} = 10\%$		Required absolute error $\Delta_{SO} = 1\%$	
	Resulting magnitude of absolute error, %	Required number of independent samples	Resulting magnitude of relative error, %	Required number of independent samples
1	0.1	38 047	100.0	**380**
2	0.2	18 832	50.0	**753**
3	0.3	12 426	33.3	**118**
4	0.4	9 224	25.0	**1476**
5	0.5	7 302	20.0	**1 826**
10	1.0	**3 461**	10.0	**3 461**
15	1.5	**2 117**	6.7	4 900
20	2.0	**1 535**	5.0	6 149
30	3.0	**849**	3.3	8 071
40	4.0	**573**	2.5	9 224
50	5.0	**381**	2.0	9 608
60	6.0	**253**	1.7	9 224
70	7.0	**162**	1.4	8 071
80	8.0	**96**	1.3	6 149
90	9.0	**43**	1.1	3 459

Note We understand the quantity $\delta_{SO} = \Delta_{SO}/SO$ under error

(1) Necessity of obtaining a great number of samples when monitoring radio channels with almost no observed activity;
(2) Quite low quality of occupancy estimation in channels with high occupancy (for such channels, the admissible error could achieve ±10%).

In the right part of Table 6.3, the requirements to the procedure of occupancy measurement are presented, which were obtained at limitation of admissible absolute estimation error. At such an approach:

(1) At cost of sharp increasing of the relative measurement error at low occupancy levels, we can radically decrease requirements to the volume of the obtained data in weakly occupied channels;
(2) Requirements to the volume of the obtained data in radio channels with occupancy more than 30% become rathe strict.

Thus, an attempt of developers of normative documents to use one "simple and understandable solution" entailed high expenses of time and computing resources and not a very high estimation accuracy. And only after approval and acceptance of suggestion about possibility to use combined requirements to occupancy estimation accuracy [22], it became possible to make a decision balanced on engineering positions. Requirements to accuracy corresponding to this approach are shown in Table 6.3 by a bold type.

In the version of ITU Report "Spectrum occupancy measurements and evaluation" [20] renewed in 2016, the suggestion from the paper [22] is applied about admissibility to monitor the radio channels states for 3600 times in the occupancy measuring interval independently on the characters of the signals, which act in the frequency band under analysis, for qualitative estimation. For the measurement interval of 15 min, the above-mentioned requirements mean the necessity of channel state monitoring not rarely than 4 times per second, of which (for not too wide frequency ranges) the modern radio monitoring systems are quite capable. Even the twice lower temp (1800 points of monitoring in the measurement interval) is accepted as satisfying. The measurement accuracy provided at the expense of fulfillment of the similar requirements is presented in Table 6.4.

Values in Table 6.4 represent the fact that decreasing of number of control points J_m by K times leads to reliability decreasing or (for pulse signals in radio channels) proportional to $\sqrt{K}$ extension of the confidence interval, i.e., values in the right part of Table 6.4 (for $J_m = 1800$) exceeds values in its central part (for $J_m = 3600$) exactly by $\sqrt{2}$ times.

To predict occupancy measurement error in channels with pulse signals, we can rely on the below-presented functions from international Recommendations ITU-R SM.1880-1 [23, Fig. 1], which were initially published in [22]. The curves in Fig. 6.19 represent the dependence of relative measurement error δ_{SO} on the occupancy SO and a number of processed data samples J_m; the shaded prohibited area in the left upper part of this figure corresponds to the cases, when it is not recommended to estimate the occupancy, because small number of measurement points may lead to the inadmissibly high error in occupancy estimation.

Table 6.4 Occupancy measurement errors corresponding to 95% confidence level, achievable when estimating occupancy with exactly 3 600 and 1 800 data samples

Occupancy, %	Number of samples: 3 600		Number of samples: 1 800	
	Resulted absolute error, %	Resulted relative error, %	Resulted absolute error, %	Resulted relative error, %
1	**0.33**	32.5	**0.46**	46.0
2	**0.46**	22.9	**0.65**	32.3
3	**0.56**	18.6	**0.79**	26.3
4	**0.64**	16.0	**0.91**	22.6
5	**0.71**	14.2	**1.01**	20.1
10	**0.98**	**9.8**	**1.39**	**13.9**
15	1.17	**7.8**	1.65	**11.0**
20	1.31	**6.5**	1.85	**9.2**
30	1.50	**5.0**	2.12	**7.1**
40	1.60	**4.0**	2.26	**5.7**
50	1.63	**3.3**	2.31	**4.6**
60	1.60	**2.7**	2.26	**3.8**
70	1.50	**2.1**	2.12	**3.0**
80	1.31	**1.6**	1.85	**2.3**
90	0.98	**1.1**	1.39	**1.5**

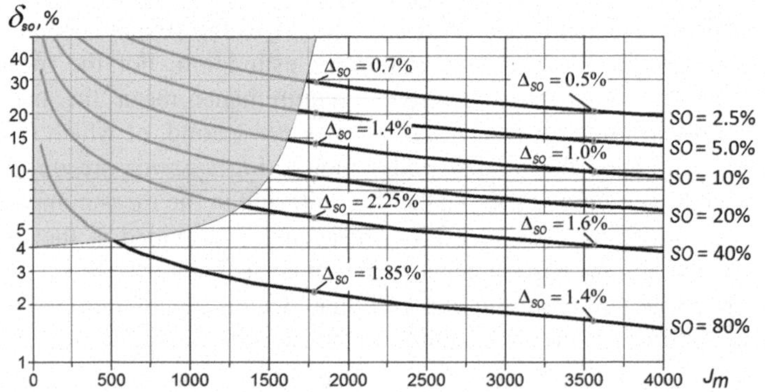

Fig. 6.19 Dependence of the relative error of occupation estimation δ_{SO} on the sample volume J_m at reliability 95%

At the same time, many modern radio monitoring systems at radio environment investigation of the frequency ranges, which are not too wide, enable not taking into account the providing of measurement reliability, because they have very high performance. So, the measuring receiver ARGAMAK-IS provides scanning performance up to several gigahertz per second in the mode of panoramic observation of the frequency range. It enables to fulfill requirements (on the safe side) on

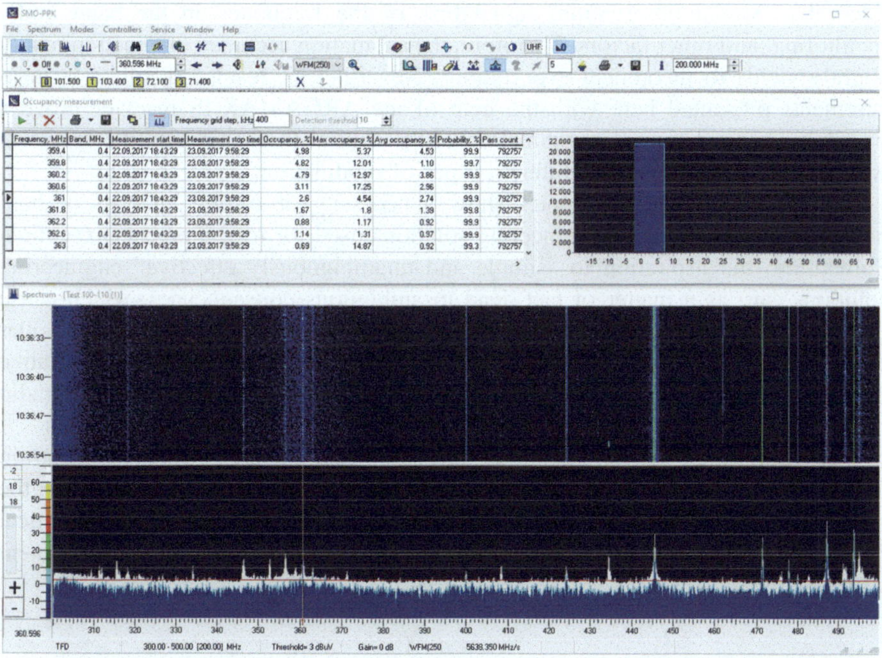

Fig. 6.20 An example of occupancy estimation

execution not less than 4 tests of radio channel condition per second. Thus, Fig. 6.20 obtained at frequency range from 200 to 600 MHz, shows that during the occupancy monitoring in the observation interval with 5-min duration, the radio channel state was monitored by receiver more than 21,000 times, which in several times exceeds the above-mentioned requirements and shows that receivers of the ARGAMAK-family permit to execute the reliable occupancy monitoring in the frequency bands wider than 1 GHz.

6.10 Summary

To provide high accuracy measurements of frequency parameters of modulated radio signals, it is necessary to take into consideration features of the used modulation types. Algorithms for frequency parameters measurements presented in this chapter are intended for central frequency measurement of the sine oscillations, signals with angular and digital types of modulation, as well as bandwidth of the occupied frequencies. At that, the estimation accuracy, which is achieved with these algorithms usage, can be closed to the Cramer–Rao bound.

At practical implementation of algorithms, the negative influence on the achieving estimation accuracy can be exerted by the low SNR, insufficiently long interval of signal observation, signal bias of the frequency in the band under monitoring, and other factors. Therefore, the estimation accuracy obtained in reality slightly yields to theoretical expectations. Nevertheless, results of full-scale measurements presented here with the help of the ARGAMAK-IS receiver prove a possibility of high-accuracy estimation of the central frequency and the spectrum width for radio signals with different types of modulation.

Considered practical examples show a necessity of careful analysis of both normative documents and engineering solutions used for measurements execution. Sometimes, the aspiration to "simple and unambiguously effective" engineering solutions can, as a matter of fact, lead to ambiguous consequences, which cause troubles in norms and standards' requirements providing. Anyway, all similar problems are surmountable, which is clearly proved by the experience of native measuring technique development and, in particular, the panoramic digital radio receivers ARGAMAK-IS and ARGAMAK-M.

The "combined" requirements to measurements of the frequency spectrum occupancy considered in this chapter, which are accepted by ITU and based on the relative and absolute estimation errors, allow obtaining of the acceptable practical recommendations to the number of samples necessary for occupancy measurement with the given error.

References

1. Gnezdilov DS, Kozmin VA, Kryzhko IB, Radchenko DS, Tokarev AB (2016) Estimation of frequency parameters of the modulated signals (in Russian). Spetstehnika i Svyaz (4):37–43
2. Rembosky AM, Ashikhmin AV, Kozmin BA (2015) Radiomonitoring: problems, methods, means (in Russian). In: Rembosky AM (ed), 4th edn. Hot Line-Telecom Publ., Moscow, 640 p
3. GOST (Russian State Standard) P 53373-2009. Technical requirements and test methods
4. Recommendation ITU-R SM.443 (2010) Bandwidth measurement at monitoring stations. ITU-R, Geneva, 9 p
5. Recommendation SM.377-4 (2007) Accuracy of frequency measurements at stations for international monitoring. ITU-R, Geneva, 2 p
6. Kayukov IV, Manelis VB (2006) Comparative analysis of different methods of signal frequency estimation (in Russian). Radioelectron Commun Syst 49(7):42–56
7. Gasior M (2006) Improving frequency resolution of discrete spectra. Krakow, pp 40–43
8. Reisenfeld S, Aboutanios E (2003) A new algorithm for the estimation of the frequency of a complex exponential in additive Gaussian noise. IEEE Commun Lett 7(11)
9. Bischl B, Ligges U, Weihs C (2009) Frequency estimation by DFT interpolation: a comparison of methods. Signal Process Mag
10. Antipov SA, Gnezdilov DS, Kozmin VA, Stopkin VM (2013) Digital interpolation algorithms of estimation the frequency of the harmonic signal, comparative analysis (in Russian). Radiotekhnika (3):42–46
11. SMO-PA/PAI/PPK panoramic analysis, measuring and direction finding software package. http://www.ircos.ru/en/sw_pa.html. Accessed 28 Nov 2017
12. IEC 60050-161—International electrotechnical vocabulary—Part 161: Electromagnetic compatibility

13. CISPR 16-1-1:2006 Specification for radio disturbance and immunity measuring apparatus and methods—Part 1-1: Radio disturbance and immunity measuring apparatus—Measuring apparatus
14. Kozmin VA, Radchenko DS, Tokarev AB (2016) Usage the modern radio monitoring systems for evaluation of intensity of interference and estimation of radio spectrum occupancy (in Russian). Spetstehnika i Svyaz (4):45–49
15. Kozmin VA, Tokarev AB (2012) Recommendations on the verification of the amplitude relationship for radio-disturbance measuring instruments. Meas Tech 55(1):79–82
16. GOST (Russian State Standard) P 51318.16.1.1-2007. Electromagnetic compatibility of technical equipment. Specification for radio disturbance and immunity measuring apparatus and methods. Part 1-1
17. Intel signal processing library. Reference manual. 630508-012
18. GOST (Russian State Standard) P 52536-2006. Automatic equipment for spectrum monitoring stations. Technical requirements and test methods)
19. Handbook on spectrum monitoring (2011) ITU-R, Geneva, 659 p
20. Report ITU-R SM.2256-1 (2016) Spectrum occupancy measurements and evaluation. ITU-R, Geneva, 52 p
21. Spectrum occupancy measurements in radio monitoring systems (2016) In: Tokarev AB (ed) Voronezh State Technical University, 237 pp
22. Kozmin VA, Pavlyuk AP, Tokarev AB (2014) Requirements to a radio-frequency spectrum occupancy evaluation. Electrosvyaz (6):47–50
23. Recommendation ITU-R SM.1880-1 (08/2015) (2015) Spectrum occupancy measurement and evaluation. ITU-R, Geneva,10 p

Chapter 7
Localization of Radio Emission Sources

7.1 Introduction

Localization or determination of radio emission source position is the complex problem due to multi-path radio wave propagation and absence of direct visibility of the radio emission source in most of the cases, small duration of emission and other factors. It is possible to execute measurements by fixed system RMPs network, one or a few mobile RMPs, at mutual usage of fixed and mobile depending on technical infrastructure. As a rule, the following main types of radio engineering measurements are used to solve the problem of localization: amplitude, goniometric, time difference of arrival (TDOA), frequency, temporal.

Many factors influence on the selection of the method to be used, such as number of active receiving positions and type of the used antenna systems, duration and spectrum width of the processed signals, range of signal-noise ratio (SNR), opportunities of equipment of radio monitoring.

In this chapter, we shall consider and analyze in general case the main methods of estimation of radio emission sources coordinates, which can be used at development of systems for RES localization. The special attention will be given to the TDOA method, to the procedures of correction of digital radio receivers reception paths to increase accuracy of RES localization by TDOA and amplitude methods [1–4].

7.2 General Formulation and Solution of the Problem

The general scheme of localization problem solution consists in the following. The measuring model is chosen in the form

A. M. Rembovsky et al., *Radio Monitoring*, Signals and Communication Technology, https://doi.org/10.1007/978-3-319-74277-9_7

$$z_i = F_i(P, l) + \xi_i, \tag{7.1}$$

where F is a function describing the dependence of measuring quantity on some set of parameters, P is coordinates of RES, l is the non-empty (in general case) set of additional unknown parameters necessary for calculation of F, i is a number of measurement in the sample, ξ is the measurement error.

Let us combine the unknown parameters P and l into the vector-column $\mathbf{x} = (P, l)^T$ and determine for the estimation $\hat{x}$ of the parameter set x the residual error at each measurement in the form

$$\Delta_i = z_i - F(\hat{x}). \tag{7.2}$$

For the (7.2) set, we can construct the functional, the exact view of which is determined by the solution method and in which minimum (maximum) is achieved in the exact solution. For the standard case of the least squares method (LSM) application, this functional has a form

$$J(P, l) = \Delta^T R^{-1} \Delta, \tag{7.3}$$

where R is a covariance matrix of measurement errors with elements $R_{ij} = M[\xi_i \xi_j]$. The set of parameters, which minimizes (7.3) functional, is its solution. In the case if measurements error ξ may be assumed as independent and of equal accuracy, the functional form is essentially simplified

$$J(x) = \sigma_\xi^{-2} \Delta^T \Delta = \sigma_\xi^{-2} \sum_{i=1}^{n} (z_i - F(x))^2. \tag{7.4}$$

Taking into consideration that the value of the minimum point of (7.4) does not depend on σ_ξ, the appropriate multiplier can be omitted and we can minimize the functional

$$J(x) = \Delta^T \Delta = \sum_{i=1}^{n} (z_i - F(x))^2. \tag{7.5}$$

Since in the general case, the functional (7.5) is the nonlinear function with several local minimums, its minimization is usually performed in two stages: at first stage, some initial approximations to parameters of interest are calculated; on the second stage, if necessary, these parameters are refined.

Owing to the above-mentioned nonlinearity of (7.5), the search for the initial approximation is usually carried out blindly by some grid of permissible values. This grid is selected with parameters providing detection and estimation of all local minimums, at that, parameters l need to be determined. In some cases, when unknown parameters l are included in (7.5) linearly, we can omit the blind search by these parameters providing their calculation at fixed coordinates P in according to evident formulas.

Refinement of solution is performed by one of the methods intended for the nonlinear task solution, for example, the Newton method.

In the case, when additional parameters l are absent, each measurement on the OXY plane corresponds to some line of possible positions, in which view is determined by the measurement type, for instance, for bearing this is a beam. Otherwise, the situation becomes complicate. In some cases, the transition to the difference measurements is possible or, to draw the line of possible positions, we should use several measurements. The manifold empiric or semiempirical matrix methods, based on the raster algorithm [5], are based on the line of positions. They assume formation of the grid, which covers operating area on the map. The grid step is selected depending on the operating area size and on the required accuracy; the grid is considered as a raster. Then, for each cell, we calculate either a number of position lines passed through it, or other similar quantities. As a result, we obtain the 3D-surface; it is the most probable that the source is situated within the bounds of that cell, which corresponds to the maximal value. For automatic determination of maximums, we use the algorithm of comparison with the preset threshold. As a result, we find the isolated areas, in which amplitude exceeds the threshold value. Then, the maximum coordinates are calculated within the bounds of each isolated area. As a result, we can reveal two or more radio emission sources operating at the same frequency. Implementation of such methods is rather simple, but usually their performance is poorer than the performance of the methods developed with account of distribution laws of measurement errors, in particular, LSM. But in some cases of location determination of several simultaneously operated RESs together with absence of information about the measurement error type, such methods may demonstrate good results.

In case, when several measurement types are simultaneously available, for example, bearings and amplitudes (typical for mobile stations), their mutual processing (integration) is implemented. At LSM application, we can write the functional in the form (7.3) or, assuming the independence of measurement errors of different types and measurement independence and equal accuracy inside the groups, in the form of the analog of (7.4):

$$J(P,l) = \sum_{k=1}^{K} \left(\sigma_{\xi_K}^{-2} \sum_{i=1}^{n} (F(x) - z_i)^2 \right), \tag{7.6}$$

where K is a number of measurement types, $\sigma_{\xi_K}^2$ is the measurement dispersion in each group.

The situation, when $\sigma_{\xi_K}^2$ values are a priori unknown (even approximately), is rather typical. In this case, at constructing of initial approximation, we can simultaneously perform the estimation of $\sigma_{\xi_K}^2$ dispersion values. For this, at the stage of initial approximation, we can use, instead of Eq. (7.6) functional, the multiplicative functional

$$J(x) = \prod_{k=1}^{K} J_k(x), \tag{7.7}$$

where $J_k(P, l)$ are the functional values calculated for each type of measurement independently according to formulas (7.5). At that, the dispersion estimation for measurements can be obtained by formulas

$$\hat{\sigma}^2_{\xi_K} = \frac{1}{n_K} \sum_{i=1}^{n_K} (F(\hat{x}) - z_i)^2, \tag{7.8}$$

where n_K is a number of measurements of kth type, $\hat{x}$ are parameter values, at which the minimum of (7.7) is achieved. Estimations (7.8) are used at the next stage of solution refinement.

In the case of LSM application, we may construct the accuracy estimation assuming the smallness of the error obtained. In this case, the covariance matrix of x parameter errors, at first approximation, takes a form

$$R_x = \left(\left(\frac{\partial F}{\partial x} \right)^T R^{-1} \frac{\partial F}{\partial x} \right)^{-1}. \tag{7.9}$$

For the case $R = \sigma^2_\xi I$, this formula takes a form

$$R_x = \sigma^2_\xi \left(\left(\frac{\partial F}{\partial x} \right)^T \frac{\partial F}{\partial x} \right)^{-1} \tag{7.10}$$

7.3 Application of Goniometric Measurements

Application of measurements of bearing to the source is based on account of its dependence on the mutual location of RMP and RES. The initial data are the bearing measurement sample as well as coordinates and orientation of the direction-finder at the moments of reception. The single measurement on the plane can be represented by beam or sector.

For illustration, Fig. 7.1 shows the configuration of three direction-finding RMPs R_1, R_2, R_3 and two RESs I_1, I_2. Dimensions of abscissa and ordinate axes are given in meters in this figure. It is evident (this is illustrated in the figure), that in the case of two RMPs location on one line with RES, the coordinates of such RES

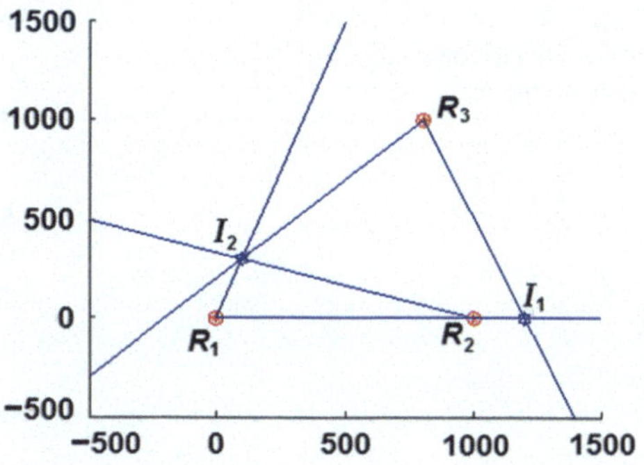

Fig. 7.1 An example of the position lines collection for goniometric measurements

according to bearings of these RMPs cannot be determined. To cover the whole plane, it is necessary to have three direction-finders.

Let the RMP be located in the point with coordinates Q, RES—with coordinates P, $r = P - Q = (\Delta x, \Delta y)^T$. Then, the bearing measurement may be presented as

$$z_B = \arctan\frac{\Delta x}{\Delta y} + \xi, \tag{7.11}$$

where ξ is the bearing error.

Let we have n measurements. The residual error of measurements has a form:

$$\Delta_i = z_i - \arctan\frac{\Delta x_i}{\Delta y_i}, \quad \Delta = (\Delta_1, \Delta_2, \ldots, \Delta_n)^T. \tag{7.12}$$

As the solution method, we choose the least squares method. Assuming that the measurement errors ξ are independent and of equal accuracy, we write the functional (7.5) for our task as

$$J_B(P) = \sum_{i=1}^{n}\left(z_i - \arctan\frac{\Delta x_i}{\Delta y_i}\right)^2 \tag{7.13}$$

The functional value (7.13) increases linearly together with growth of measurements amount. The mean residual error (not the total one) can be minimized. The formula for residual error can be represented in the following way:

$$J_B(P) = \sqrt{\frac{1}{n}\sum_{i=1}^{n}\left(z_i - \arctan\frac{\Delta x_i}{\Delta y_i}\right)^2}. \tag{7.14}$$

To illustrate potential accuracy characteristics of the bearings application to determine RES coordinates, the calculation field of expected errors determined

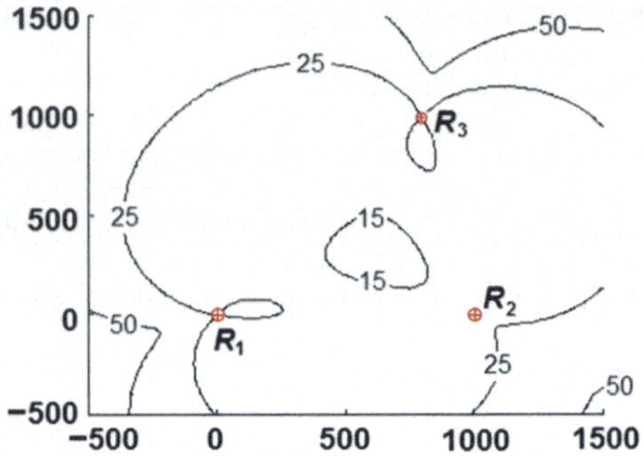

Fig. 7.2 Field of expected errors of goniometric measurements

according to the formula (7.14) at root-mean-square error (RMSE) of measurements of 1° is presented in Fig. 7.2.

7.4 Empirical Methods of Goniometric Measurements Processing

In the direction-finding equipment of the ARTIKUL-family, two empiric algorithms are used, which are traditionally named as matrix and cluster algorithms [6].

In the matrix algorithm, the bearings are calculated according to matrix cells by the raster algorithm and are summed by amplitude in the grid cells. At simultaneous operation of several radio stations at the same frequency, we may observe on this plane, which reflects the frequency of bearings registration in the grid cells, two or more maximums. The computation simplicity can be attributed to advantages of the matrix method, which gives a possibility of using the computer with low performance. The main drawback of this method, as any other methods using the bearing measurements only, is the impossibility of calculation of RES coordinates in the case when mobile station moves towards to the RES according to the driving method.

The cluster algorithm is the further modification of the matrix algorithm. The matrix grid is also used in this algorithm, which covers the operating area in the map, but in the grid cells, the intercepting points of the bearing lines (not the bearings lines themselves) are accumulated, which are the solutions of the triangular task. As a result, we obtain the 3D-surface formed by the intercepting points of bearings. Amplitude of this surface will be larger in those grid cells, where bearings are more often intercepted between each other. The cluster method advantages are the possibility of RES location determination according to less number of bearings than for the matrix algorithm. The main disadvantage of this method is high-computing effort. The cluster method as the matrix method does not enable RES coordinates calculation in the case of mobile station movement towards the source.

7.5 Application of Amplitude Measurements

Application of source field amplitude measurement is based on the account of its dependence on the distance to the source. The samples of amplitude and RMP coordinate measurements at the moments of reception are the initial data for calculation.

Let the RMP be located in the point with coordinates Q, the RES—in the point with coordinates P. Depending on the error types (additive or multiplicative), the model of signal power measurement can be represented accordingly in the form

$$z_P = \frac{D(\varphi)A}{r^p} + \xi \tag{7.15}$$

or

$$z_P = m \log \frac{D(\varphi)A}{r^p} + \xi, \tag{7.16}$$

where A is the signal amplitude in the point of emission, $D(\varphi)$ is the value of the receiving antenna directivity diagram in RES direction φ, $r = \|Q - P\|$ is the distance from RES to RMP, p is the amplitude decay factor depending on the distance, ξ are errors of measurement, m is the constant of transition to the logarithmic measurements (usually $m = 10$).

As we see from the (7.15) and (7.16) model's analysis, in addition to unknown coordinates, we have the unknown constant A, which is also a subject to be determined.

The principal form of measurement models (7.15) and (7.16) does not differ; therefore, we consider the model of additive measurements (7.15).

The single measurement on the plane could not be presented, and at measurement of signal intensity in several points for each pair of measurement, the circles serve as the lines of possible position of RES. Let us prove it. Let $z_1 = \frac{D(\varphi_1)A}{r_1} = a\frac{D(\varphi_2)A}{r_2} = az_2$. We assume that $a \geq 1$. Let us place the origin of the coordinate system in the point $P_1 = (x_1, y_1)^T$. Let the X-axis be directed towards P_2. In new coordinates system, $\widetilde{P_1} = (0,0)^T$, $\widetilde{P_2} = (r,0)^T$, where $r = \|P_1 - P_2\|$. Then, the measurement equation can be written in the form $\left(\frac{D(\varphi_1)}{D(\varphi_2)}a\right)^2(x^2 + y^2) = (r - x)^2 + y^2$. The last equation can be transformed to the form $\left(x + \frac{D^2(\varphi_2)r}{D^2(\varphi_1)a^2 - D^2(\varphi_2)}\right)^2 + y^2 = \frac{D^2(\varphi_1)D^2(\varphi_2)a^2r^2}{(a^2D^2(\varphi_1) - D^2(\varphi_2))^2}$. This is the equation of the circumference with the center in $\left(\frac{-D_2^2 r}{D_1^2 a^2 - D_2^2}, 0\right)^T$ and of radius $\frac{D(\varphi_1)D(\varphi_2)ar}{a^2D^2(\varphi_1) - D^2(\varphi_2)}$. The case $a = \frac{D(\varphi_2)}{D(\varphi_1)}$ is the singular one; in this case, the circumference is degenerated into the straight line.

At amplitude measurements, the minimal necessary number of RMPs (points of measurement by the mobile RMP) is three; however (excluding the degenerated case of equal-remoteness from all three RMP), three RMPs correspond to two interception points of position lines, and therefore, four RMPs are necessary to provide a possibility to make the unambiguous solutions in the whole operating area, which is explained in Fig. 7.3.

The residual error in measurements has a form:

$$\Delta_i = z_i - \frac{D(\varphi_i)A}{r^p}, \ \Delta = (\Delta_1, \Delta_2, \ldots, \Delta_n)^T. \tag{7.17}$$

We choose the LSM as the solution method. Assuming measurement errors ξ independent of and equal accuracy, we write the functional (7.5) for our task in the form

$$\begin{aligned} J_{1P}(A,P) &= \Delta^T\Delta = \sum_{i=1}^{n}\left(z_i - \frac{D(\varphi_i)A}{r_i^p}\right)^2 \\ &= A^2\sum_{i=1}^{n}\frac{D^2(\varphi_i)}{r_i^{2p}} - 2A\sum_{i=1}^{n}\frac{D(\varphi_i)z_i}{r_i^p} + \sum_{i=1}^{n} z_i^2. \end{aligned} \tag{7.18}$$

Since the functional (7.18) additionally depends on the A parameter, for implementation of the matrix algorithm, we construct it for the case of known P. At fixed P, Eq. (7.18) transforms into the quadratic equation for A The minimum at fixed P is achieved for

$$\hat{A} = \left(\sum_{i=1}^{n}\frac{D(\varphi_i)z_i}{r_i^p}\right)\left(\sum_{i=1}^{n}\frac{D^2(\varphi_i)}{r_i^{2p}}\right)^{-1}. \tag{7.19}$$

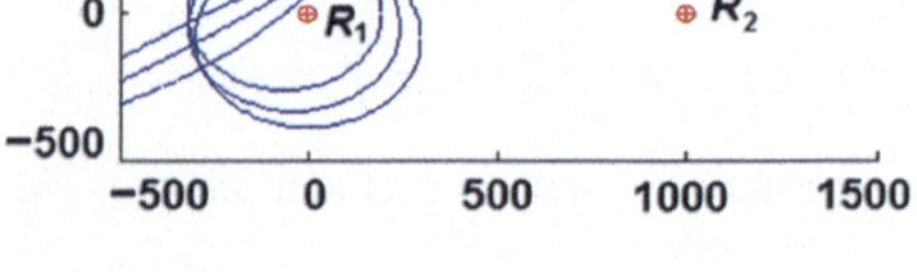

Fig. 7.3 An example of position lines collection for the amplitude measurements

After substituting this estimation into (7.18), we have:

$$J_{1P}(A,P) = \left(\sum_{i=1}^{n} \frac{D(\varphi_i)z_i}{r_i^p}\right)^2 \left(\sum_{i=1}^{n} \frac{D^2(\varphi_i)}{r_i^{2p}}\right)^{-2} \sum_{i=1}^{n} \frac{D^2(\varphi_i)}{r_i^{2p}}$$
$$- 2\left(\sum_{i=1}^{n} \frac{D(\varphi_i)z_i}{r_i^p}\right)\left(\sum_{i=1}^{n} \frac{D^2(\varphi_i)}{r_i^{2p}}\right)^{-1} \sum_{i=1}^{n} \frac{D(\varphi_i)z_i}{r_i^p} + \sum_{i=1}^{n} z_i^2. \quad (7.20)$$

and after simple manipulations, we get

$$J_{1P}(A,P) = \sum_{i=1}^{n} z_i^2 - \left(\sum_{i=1}^{n} \frac{D(\varphi_i)z_i}{r_i^p}\right)^2 \left(\sum_{i=1}^{n} \frac{D^2(\varphi_i)}{r_i^{2p}}\right)^{-1}. \quad (7.21)$$

The functional (7.21) value grows linearly with increase of the measurements amount; therefore, it is more convenient to minimize not the total residual error, but the mean one. The formula for the functional in this case takes a form:

$$J_P(A,P) = \sqrt{\frac{J_{1P}(A,P)}{n}} = \sqrt{\frac{1}{n}\left(\sum_{i=1}^{n} z_i^2 - \left(\sum_{i=1}^{n} \frac{D(\varphi_i)z_i}{r_i^p}\right)^2 \left(n\sum_{i=1}^{n} \frac{D^2(\varphi_i)}{r_i^{2p}}\right)^{-1}\right)}. \quad (7.22)$$

Thus, in connection of amplitude measurements, the algorithm for determination of RES coordinates consists of minimization of the functional (7.22) over coordinates of radio emission source P in nodes of the operation zone grid.

Application of directed and non-directed (having the unitary pattern $D(\varphi) \equiv 1$) antennas is defined by the type of the used equipment, the class of tasks, and other parameters. As a rule, amplitude measurements are fulfilled as additional to the main ones; for example, at fulfillment of goniometric measurements (direction-finding), at that, omnidirectional antennas are used. The application of amplitude measurement in manual direction-finders is an example of amplitude measurements with directed antennas.

To illustrate the potential accuracy characteristics of amplitude measurements usage by the non-directed antenna for RES coordinate determination, Fig. 7.4 shows the field of expected errors calculated using the formula (7.10) for the unitary decaying exponent in the formula (7.15). The noise level was considered as constant for all coordinates. The amplitude of emitted signal was chosen so that at RES location in the system center (in point with coordinates (500, 500)), the SNR in reception points location was about 20 dB.

We should note that such factors, which are difficult to be taken into consideration, as type of underlying surface, relief of radio wave propagation path, housing density, etc., may affect the effectiveness of localization systems functioning, which use the amplitude measurements.

Fig. 7.4 Field of expected error at amplitude measurements

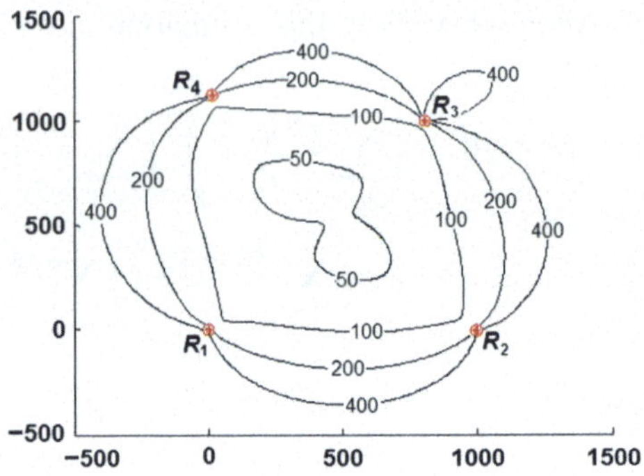

7.6 Application of Time Difference of Arrival Measurements

Difference in distance from RES to different reception points of the geographically distributed system causes the difference in delays of signals arrival from RES to the different RMP. At high-accurate RMP timescale synchronization, the mutual delays of signal arrivals registered by these RMPs can be recalculated into the distance differences. The similar samples of distance differences from RES to various RMPs together with coordinates of RMP themselves serves as initial data for estimation of RES coordinates at TDOA measurements.

The single measurement on the plane can be represented by hyperbola or hyperbola with vicinity (by two hyperbolas).

The minimal number of RMPs is three, as it is shown in Fig. 7.5, but at that, the appearance of two interception points of position lines is possible, and to provide a possibility to make the final unambiguous decisions in all operating area, we need four RMPs [7].

Let the RMPs with numbers i and j, included in the mth pair, be located in points with coordinates Q_i and Q_j, and RES be located in the P point. Then, the

Fig. 7.5 An example of position lines collection for TDOA measurements

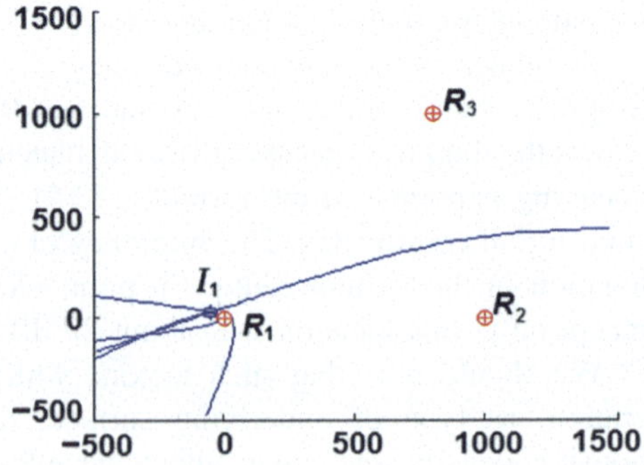

interconnection between these coordinates and the measured mutual delay can be presented as

$$z_m = t_{ij} \cdot c = \|Q_i - P\| - \|Q_j - P\| + \xi_m, \quad (7.23)$$

when c is the electromagnetic wave velocity, ξ_m is the measurement error for mth pair of RMP.

Assuming that ξ_m errors are distributed in accordance with normal law, the solution of RES localization problem based on the LSM for measurements (7.23) can be reduced to the task of functional minimization

$$J(P) = \mathbf{w}^T R^{-1} \mathbf{w}, \quad (7.24)$$

where $\mathbf{w}$ is the vector-column consisting of elements $w_m = z_m - \|Q_i - P\| - \|Q_j - P\|$, and R is a covariance matrix of measurement errors with elements $R_{ij} = M[\xi_i \xi_j]$.

If measurements (7.23) can be considered as independent and of equal accuracy, the R matrix has a simple form $R = \sigma_\xi^2 I$, where I is the unitary matrix (the diagonal matrix, in which diagonal elements equal to unit), and the functional (7.24) essentially simplifies

$$J(P) = \sigma_\xi^{-2} \mathbf{w}^T \mathbf{w} = \sigma_\xi^{-2} \sum_k w_k^2. \quad (7.25)$$

The functional (7.25) is nonlinear in the general case, so that in the permissible zone of P variation, the several local extremes may appear, due to which the minimization task is solved in two stages. At first stage, the search for approximations to the local minimums is performed, after which, at the second stage, their refinement by the iteration Newton method is executed.

Figure 7.6 shows the RMSE field of RES localization at assuming that the TDOA system includes three RMPs, for which the measurement errors of mutual delays are independent and distributed according to the normal law with zero mean and the RMS deviation of 100 ns, in its structure. The locations of RMPs are marked with circles. From Fig. 7.6, we see that the least localization error is expected inside the triangle, restricted by points of RMP location, and at continuation of this triangle sides the error is significant and increases sharply at RES removal from the points of RMP location.

There is a variety of factors influencing on the quality and effectiveness of the TDOA system functioning. We can attribute the following to their number:

(1) Mutual location of RES and antennas of RMP system;
(2) Identity of reception paths of used radio receivers;
(3) Synchronization accuracy of spaced RMPs;
(4) Performance of commands and data exchange between RMPs, etc.

Fig. 7.6 Field of the expected error of TDOA measurements

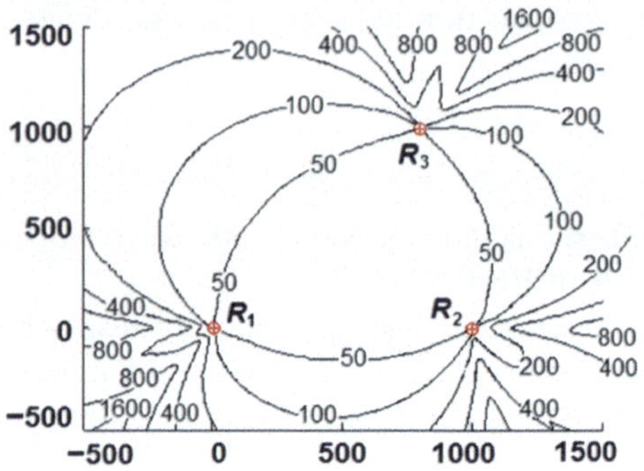

Non-identity of receiving paths manifests itself in the form of dependence of registered mutual delays on the frequency band occupied by the signal under processing. Even at use of single-type RMPs, the concrete specimens of receivers have differences in paths calibration and high-selective filters included in its structure differ in properties. As a result, the simple variation of signal frequency emitted by the same RES may lead to variation of registered (in RMPs) mutual delays up to several tens of nanoseconds.

An example of deviations of mutual delays from their true value is presented in Fig. 7.7. This diagram is obtained for receivers ARGAMAK-IC with the bandwidth of 22 MHz, it happens to be frequency-dependent and "attached" to the position of the processed signal in the receiver bandwidth. From the presented data we see that without additional processing, the measurement error of mutual delays can achieve ±50 ns for signals with the bandwidth $B_f \ll 1$ MHz.

The local timescales synchronization of the RMPs, which are included into the TDOA system, can be carried out, for instance, with the reference to PPS-signals, which are formed by timing navigation receivers. Investigation results presented in [2] show that adjustment of time local scales by means of PPS-signal processing by Kalman filters allows synchronization of time scales of RMP with mismatch, which

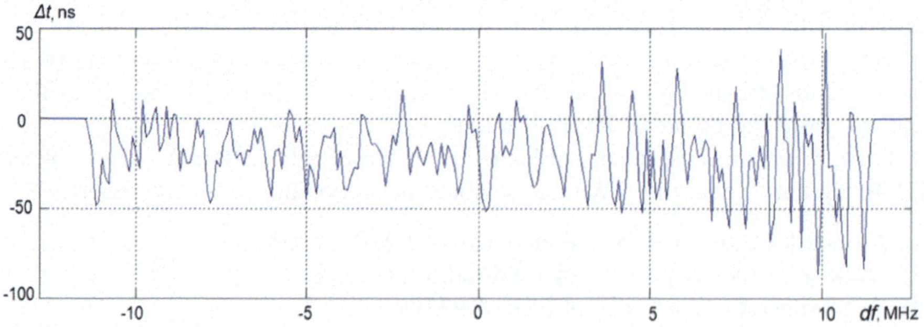

Fig. 7.7 An example of deviations of signal mutual delay from the received signal frequency with respect to the bandwidth center of the receiver

does not exceed ±30 ns. At absence of other error sources, the similar synchronization quality allows obtaining errors of the 10 m order inside the RMP polygon. This corresponds to the needs of radio monitoring services (on the safe side) and other consumers of RES localization results.

At last, importance of urgent exchange of commands and data relates to the fact that at geographical diversity of RMP and RES the signal of the specific RES may be observed at any RMP with high SNR, but at other RMPs, it will be practically unnoticeable on the noise background. In this situation, the procedure of data acquisition for localization at RMP with low SNR may be activated by the outside command only. However, the command exchange between RMPs cannot be fulfilled in a moment, and therefore, it is necessary to arrange buffering allowing recovery of the recently observed fragments of data samples from the buffer at the external request arrival, at each RMP. At low rate of command and data exchange, requests on data obtaining will proceed to the remote RMPs with significant delays, and to provide the system efficiency, we must require buffer with very high volume. Thus, for effective functioning of distributed TDOA systems, we must use the high-rate links for command and data exchange. It complicates the practical application of TDOA measurements at the mobile RMPs [8].

7.7 Application of Frequency Measurements

The high-accurate measurement of the received signal frequency may become useful at utilization of the mobile RMPs. At mutual motion of transmitter and receiver, the frequencies of emitted and received signals do not coincide. Let the fixed RMP be located in the point with coordinates Q, and the moving RES—in the point with coordinates P, $r = P - Q = (\Delta x, \Delta y)^T$, $v = \frac{\mathrm{dP}}{\mathrm{dt}} = (v_x, v_y)^T$ is the RMP velocity. The signal frequency can be calculated with enough accuracy according to the formula

$$z_f = f_0\left(1 + \frac{\Delta v}{c}\right) + \xi, \tag{7.26}$$

where f_0 is the frequency of emitted signal, c is the radio wave propagation speed (speed of light), Δv is the speed of mutual approach of the emitter and the receiver, ξ are errors.

Let RES be fixed. Then measurements of the received signal frequency can be written as

$$z_f(t) = f_0\left(1 + \frac{v^T r}{\mathrm{c}||r||}\right) + \xi(t), \tag{7.27}$$

where «T» is the transposition symbol.

For illustration, we consider the variant of circular movement. Assuming that RES removal significantly exceeds the motion circumference diameter, the model of frequency measurement can be presented in the form:

$$z_f(\varphi_i) = f_0\left(1 + \frac{v}{c}\sin(\varphi_i - \psi)\right) + \xi, \tag{7.28}$$

where ψ is the RES azimuth, φ_i is the azimuth of RMP trajectory points. With assumption that errors ξ are independent, we obtain the accuracy estimation of ψ determination at using LSM. The covariance matrix of errors for determination of the unknown pair ψ, f_0 has a form

$$\begin{aligned} R &= \frac{v^2}{c^2}\begin{pmatrix} \sum_{i=1}^{n}\sin^2(\varphi_i - \psi) & \sum_{i=1}^{n}\sin(\varphi_i - \psi)\cos(\varphi_i - \psi) \\ \sum_{i=1}^{n}\sin(\varphi_i - \psi)\cos(\varphi_i - \psi) & f_0^2\sum_{i=1}^{n}\cos^2(\varphi_i - \psi) \end{pmatrix} \\ &\approx \frac{v^2}{c^2}\begin{pmatrix} \frac{\pi}{\Delta\varphi} & 0 \\ 0 & f_0^2\frac{\pi}{\Delta\varphi} \end{pmatrix} = \frac{v^2}{c^2}\begin{pmatrix} \frac{\pi}{\Delta\varphi} & 0 \\ 0 & f_0^2\frac{n}{2} \end{pmatrix}, \end{aligned} \tag{7.29}$$

so that

$$\sigma_\psi \approx \sqrt{\frac{2}{n}\frac{c}{v}\frac{\sigma_\xi}{f_0}}. \tag{7.30}$$

The obtained result can be interpreted as follows: Having fulfilled the circular motions in several points during the movement, we obtain some equivalent of bearings in these points in result, the accuracy estimation of which can be calculated according to (7.30). At values $v = 20$ m/s, $\sigma_\xi = 2$ Hz, $f_0 = 300$ MHz, $n = 32$, we obtain the value of $\sigma_\psi \approx \frac{1}{40}$ rad $\approx 1.5°$.

Figure 7.8 shows the RMSE field of RES localization assuming that the RMP moving along the circumference with the speed 20 m/s carries out eight

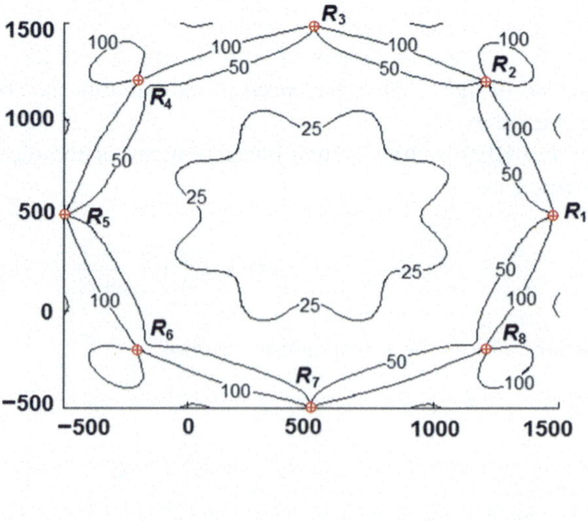

Fig. 7.8 Field of the localization error distribution at frequency measurements

measurements of frequency with the nominal value of 300 MHz and in which errors are independent and distributed according to the normal law with zero mean and the RMS deviation equaled to 1 Hz. RMP location at the measurement moments is marked by red points.

7.8 Application of Time Marker Measurements

We assume that there is a possibility to fix set of sync pulses generated by RES, which moments of emission are known (can be predicted) with accuracy up to some constant shift, at the moving RMP. Then, subtracting the moments of sync pulse emission from the moments of their reception, we obtain measurements

$$z_t(T, P) = T + \frac{\|Q - P\|}{c} + \xi, \tag{7.31}$$

where T is the displacement of RES and RMP timescales, c is the radio wave propagation speed, ξ are errors.

At fixed P, the T value estimation can be obtained by the formula

$$\hat{T} = \frac{1}{n}\sum_{i=1}^{n}\left(z_t(T, P) - \frac{\|Q - P\|}{c}\right). \tag{7.32}$$

In the case when ξ errors are strongly correlated, we may proceed to the following difference measurements

$$\tilde{z}_t(T, P) = \frac{\|Q_i - P\|}{c} - \frac{\|Q_{i-1} - P\|}{c} + (\xi_i - \xi_j). \tag{7.33}$$

To illustrate the potential accuracy characteristics at using the time marker measurements in order to determine RES coordinates, the calculated field of expected errors of measurement 100 ns is represented in Fig. 7.9.

7.9 TDOA System on the Base of ARGAMAK-IS

In connection with wideband signals and radio emissions of small duration, the manufacturers of radio monitoring equipment offer to use the time difference of arrival method, which have got the name TDOA, increasingly over the last years [9]. In technical literature, the positive tests of TDOA systems are presented for systems realized in the radio receiving devices [7, 10].

As an example, we examine the localization system on the base of TDOA measurements being fulfilled by the network of fixed RMPs [4]. To widen covering

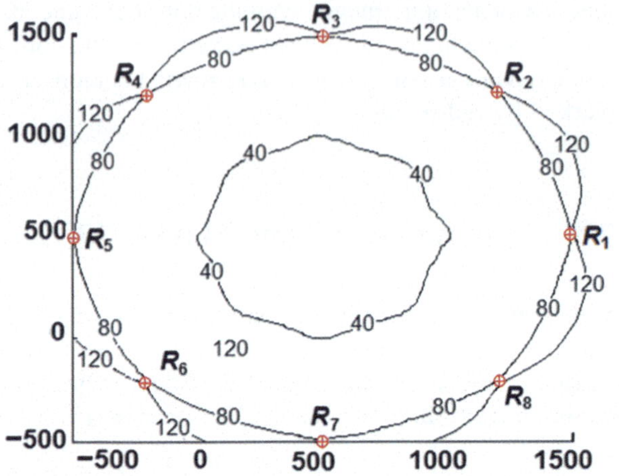

Fig. 7.9 Field of the localization error distribution over the collection of temporal measurements

zone, it is expedient to deploy the RMP of such a system on the highly located points of locality or on the roofs of high buildings; however, due to physical or administrative reasons, we cannot realize these recommendations far from always. So, at carrying out the comparable testing of TDOA localization methods, the choice of RMP location points was defined by the simplicity of approach to them, the minimization of interference from outside transport units going by, as well as the availability radius of signals, which are generated by the test-generator.

Measurement of difference of the radio signal arrival time from RES to several spaced receiving points is the basis of TDOA system functioning for estimation of RES location [8]. If signals under processing contain the prolonged fragments (e.g., a preamble) of the exactly known form, then their time position estimation can be fulfilled at each radio receiver independently, using the correlation receivers adjusted to the given signal type. To estimate the mutual delays of signal, received by two receivers with arbitrary forms, the mutual processing of two data samples is required. To minimize the data volume transmitted between radio receivers and also to simplify the frequency filtering of processed signals against other emissions, which reside in the receiver bandwidth, we used the spectral method at development of the data processing algorithm, and the calculation of signal mutual delays was provided on the base of the totality of spectral samples. Data acquisition for estimation of mutual delays anticipates the spectrum calculation at each of the used radio receivers:

$$\dot{S}[n] = \mathrm{fft}\{s[k]\} = \sum_{k=0}^{N-1} s[k] \cdot w[k] \cdot \exp\left\{-j \cdot 2\pi \cdot \frac{n \cdot k}{N}\right\}, \tag{7.34}$$

where $s[k]$ is a sequence of data samples at the IF filter output, $w[k]$ is a weighting function. Then, from this spectrum, the subset $n_b \leq n \leq n_e$ of samples corresponding to the processed signal is excluded; they are transmitted between the receivers.

We should also take into account, that from the engineering point of view, it is rather difficult to provide the strict synchronism of data sample obtained at different receivers, and therefore, besides the spectral samples $\dot{S}[n]$, we should register information about high-accurate timing of the receiver local timescales for calculation of Δt_{sc} difference of "time affixment" of data samples, by which spectrums are calculated at the various system points, to determine the mutual delays.

The coordinate of modulus maximum of cross-correlation function serves the rough estimation of the signal mutual delay:

$$k_m = \arg\ \max\left\{\left|\dot{R}_{ij}[k]\right|\right\}, \tag{7.35}$$

which is calculated by means of inverse transform of the complex mutual signal spectrum from the receivers with numbers i and j

$$\dot{R}_{ij}[k] = \text{ifft}\left\{\dot{S}_j[n] \cdot S_i^*[n]\right\}. \tag{7.36}$$

Here $\dot{S}_i[n]$, $\dot{S}_j[n]$ are the samples of the complex signal spectrums observed in different TDOA points, $\{\cdot\}^*$ is the sign of the complex-conjugated quantity, ifft$\{\cdot\}$ is the operation of calculation of the inverse discrete Fourier transform.

To correct the peak position, we use the parabolic approximation of the CCF top. The appropriate correction is calculated as

$$\text{dk} = \frac{R\_\text{pl} - R\_\text{mi}}{2 \cdot (R\_\text{pl} + R\_\text{mi} - 2 \cdot R\_\text{max})}, \tag{7.37}$$

where $R\_\text{max} = \left| \dot{R}_{ij}[k_m] \right|$ is the modulus of the maximal CCF sample, R_pl and R_mi are the absolute values of two adjacent samples.

The total mutual delay for the current receiver pair is

$$t_{ij} = \frac{k_m - dk}{F_{Д}} + \Delta t_{sc}, \tag{7.38}$$

where F_s is the sampling frequency of data samples at IF filter output, Δ_{sc} is the correction reflecting the difference of "time affixment" of data samples, on which the spectrums at different receivers were calculated.

Figure 7.10 shows the RMSE field for RES location estimation under assumption that TDOA system includes three receivers, for which measurement errors of mutual delays are independent (for each triplet of receivers, two independent measurements were chosen only) and distributed according to the normal law with zero mean and the RMS deviation equaled to 100 ns. Points in the figure designate the receivers' location in the angles of the triangle with sides 4.9, 6.1, and 7.8 km, which corresponds to the receiver location under condition of urban locality at fulfillment of the full-scale experiment. Marks in the form "x" show the RES location, and signs on the level lines correspond to the expected RMS error of location determination in meters.

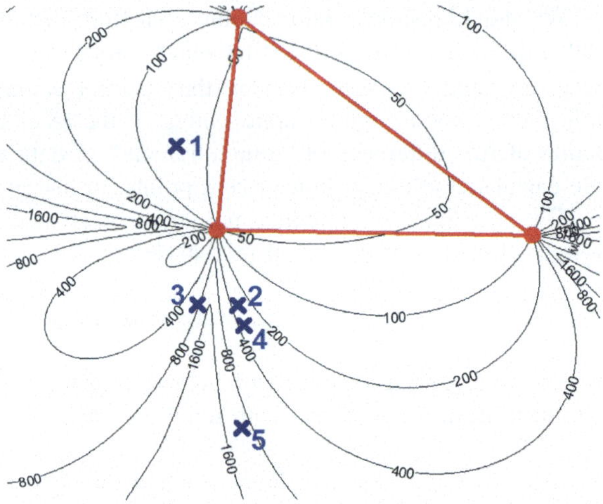

Fig. 7.10 Field of RMSE distribution of location estimation

From Fig. 7.10, we see that the least error is expected inside the triangle restricted by points of receivers' location, and on the triangle sides' continuation, the error is significant and increases sharply at RES removal from the points of receivers' location. This fact should be taken into consideration at results estimation of full-scale experiment, which will be discussed later.

According to recommendations of the ITU report [7], the less strict requirements are made to selection of platforms of TDOA receiving antennas installation than to the platforms of the antenna system location for goniometric direction-finders. However, the practice shows that interference effects influence on the determination accuracy of signal mutual delays in the very negative manner in connection with signals with the $B_f < 1$ MHz bandwidth. If receiving antennas cannot be installed on masts and at distance from the possible reflectors, the signal distortions caused by interference effects may lead to the bias of the mutual delay estimations by tens and sometimes hundreds of nanoseconds, which is much more than errors related to receiver's path non-identity. As a consequence, at selecting positions for receiving antenna localization, we need to take reasonable efforts in order not to have any interference sources near them.

Fulfillment of the full-scale experiments was provided with usage of three panoramic digital measuring receivers ARGAMAK-IS operating under control of the SMO-ARMADA software [9]. An example of used antenna system consisting of remote RF sensor unit active antennas and the receiving GLONASS/GPS antenna is presented in Fig. 7.11. The construction and technical characteristics of the ARGAMAK-IS measuring receiver were described in detail in Chap. 2.

We would like to note that the above-mentioned recommendations concerning antenna systems location cannot often be implemented in full scale in practice, because a number of tall buildings available for receivers for installation in towns are always limited. Thus, at preparation to the full-scale system testing, we cannot find places, operating in which the receivers may cover totally all possible points of

Fig. 7.11 Atenna system for ARGAMAK-IS

RES location. The scheme of mutual location of receivers and RES is shown in Fig. 7.12.

The procedure of location determination begins from a selection (by the operator) of the frequency and the reception bandwidth of the signal of interest, and it finishes by appearance of results, which contain RES coordinates estimation, in the SMO-ARMADA database. The software windows, corresponding to the task of the digital TV transmitting station localization and to the obtained results of location determination, are presented in Figs. 7.13 and 7.14.

Results presented in Fig. 7.14 show that at presence of three receivers in the system structure, the ambiguity of RES location determination can occur since intersections of hyperbolas, which correspond to the delay differences in signal reception for several measurements under condition of noise action, is concentrated in two various areas. To eliminate the ambiguity, we should have not less than four receivers in the system structure, which not only eliminates the ambiguity in location determination, but also reduces RMSE of location determination. In the

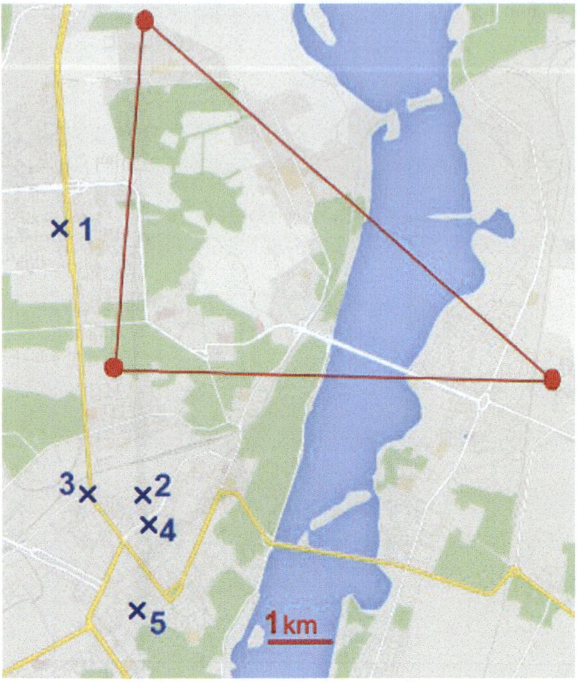

Fig. 7.12 Scheme of mutual location of receivers and RES during the full-scale experiment

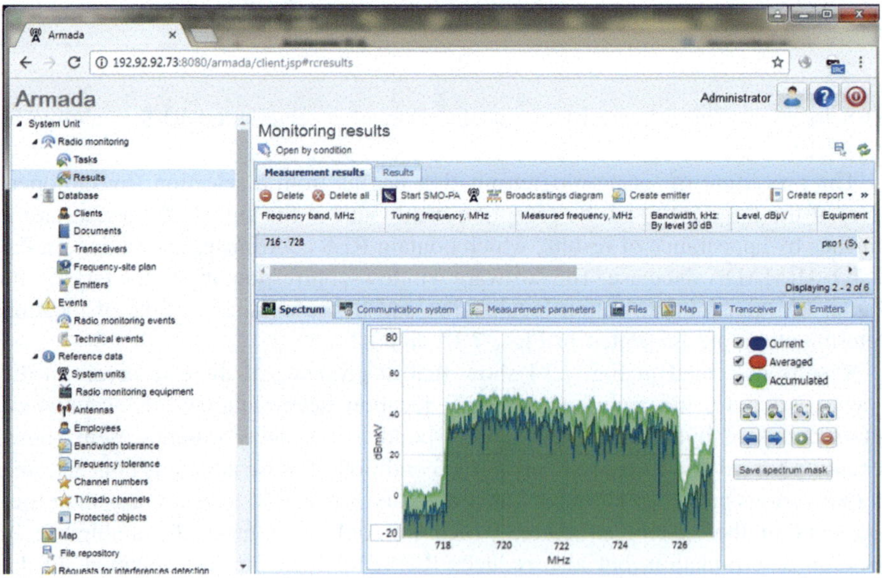

Fig. 7.13 Signal of digital TV (RES No.2)

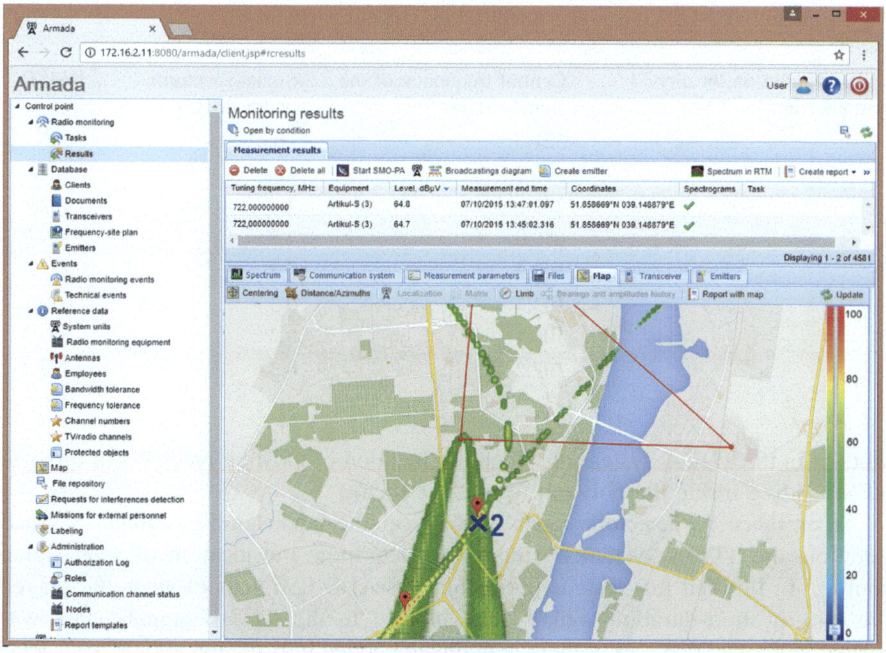

Fig. 7.14 Determination of digital TV transmitter location

measurement fulfilled, the best (true) estimation had an error of the RES location about 100 m.

The peak of cross-correlation function for narrowband signals is much flatter than for digital TV signals; therefore, measurement of mutual delays of UHF radio station signals is accompanied by the higher error. To increase accuracy in such a situation, we must use multiple measurements; during the full-scale experiment for UHF radio stations the eight data samples were sequentially registered, by which the RES coordinates were calculated. Nevertheless, RMSE of location determination was essentially higher here than for wideband TV signals. Results of fulfilled tests are presented in Table 7.1.

Results of the full-scale tests are well-agreed with theoretical curves of the RMSE presented in Fig. 7.10. In particular, the first two lines of Table 7.1 reflects the fact that RES No. 1 is located in more convenient zone than the RES No. 2 in respect to the receiver, and therefore, processing of the uniform signal samples of analog TV is accompanied in the first case by the noticeably less error of RES location determination. The final part of Table 7.1 demonstrates the difference in the RES location determination accuracy on UHF radio stations signals. In spite of uniformity of processed signals, the accuracy of location determination for various RESs differs by several times due to the fact that RESs with large numbers are located farther from the receiver triangle in the zone of high error. To increase accuracy of similar RESs location determination, we must increase the receivers'

Table 7.1 Results of the full-scale experiment of TDOA system for RES with continuous wave signals

RES location on locality (number in Fig. 7.12)	Central frequency of the signal (MHz)	Signal spectrum width (MHz)	RMSE (m)
1	521.3	<6	125.7
2	519.7	<6	167.0
2	722.0	7.8	132.1
2	650.0	7.8	109.9
2	100.3	0.3	411.5
3	103.4	0.3	669.2
4	106.1	0.3	593.3
5	107.2	0.3	1202.4

amount in the TDOA system or widen the location of existing receivers in order to RES residing inside the polygon formed by them.

At presence of detectors capable to reveal the packet radio signals of small duration, the TDOA system is suitable to determine the location of such signal sources. In fulfilled full-scale experiments, the ADS-B signals of air traffic played the role of short-duration signals. The interest to the ADS-B signals class was related to the fact that part of them contains the coded information about "true" RES coordinates, determined with the help of the high-accuracy radio navigation. Detection of air-traffic control signals was performed by the special filter matched with the ADS-B signal preamble. Fulfilled tests demonstrated the TDOA system capability to determine the target coordinates on the base or radio pulses with duration about 100 μs, emitted by targets.

It is quite natural that with target removal from the receiver polygon, the accuracy of location determination decreased. For example, the error of airplane location determination, which were in the 10 km-zone from the TDOA center, as a rule, did not exceed 1 km. For targets located at distance of 30 … 50 km from the TDOA center, the RMSE of location determination increased up to 2 … 3 km. At attempt to determine airplane location, which flied at distance 70 … 100 km, the observed error increased up to 5 … 10 km. For air-traffic control, the similar error is too high. For better results in location determination, it is necessary to have greater mutual receiver's separation; however, the direction of "merged" hyperbolas factually shows the direction of signal arrival from the airplane, and it can be used as bearing (Fig. 7.15).

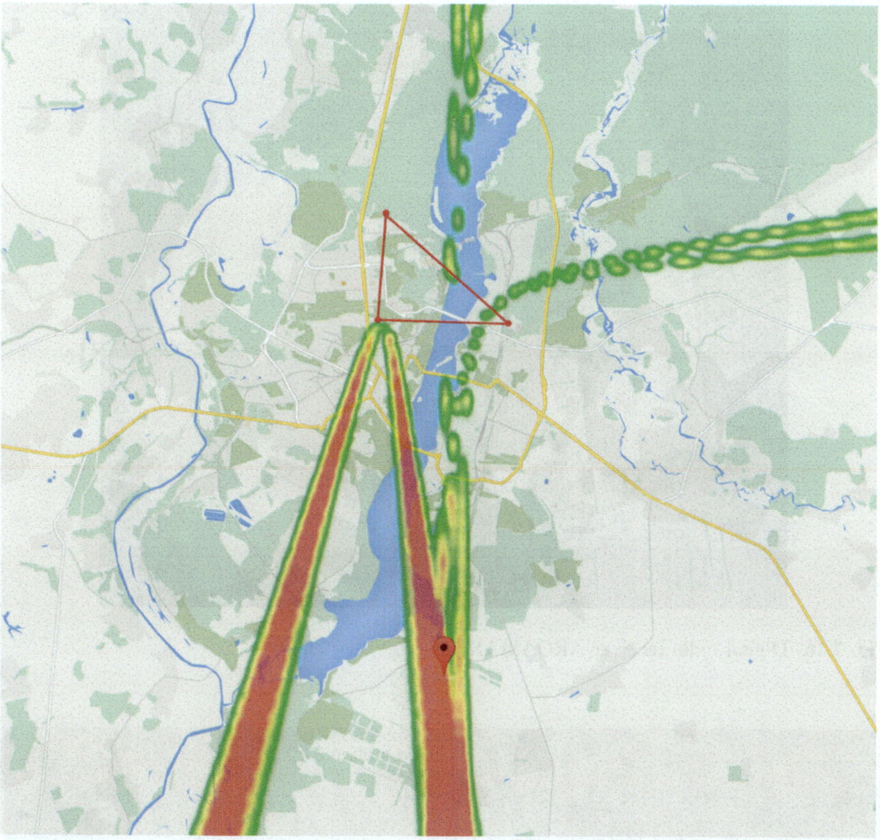

Fig. 7.15 An example of airplane location determination according to the ADS-B signal

7.10 System of Location Determination on the Base of ARGAMAK-RS

As the second example, we consider the localization system on the base of TDOA and amplitude measurements fulfilled by the network of fixed compact digital radio receivers ARGAMAK-RS [3].

The localization system consisted of three ARGAMAK-RS receivers operating under control of the SMO-PPK software [11, 12]. The digital radio receiver ARGAMAK-RS is presented in Fig. 7.16, and location of fixed station in locality is presented in Fig. 7.17. Signals, which were used for localization, were the frequency-modulated oscillations with the spectrum width about 500 kHz.

In Figs. 7.18, 7.19, 7.20, and 7.21 presented later, besides the calculated estimations of RES localization, the raster matrices are drawn above the locality photograph. These matrices characterize a possibility of RES localization in the given point of the space in accordance with measurement fulfilled; in areas with small point density, the RES localization in unlikely, and point concentrations

Fig. 7.16 Digital radio receiver ARGAMAK-RS

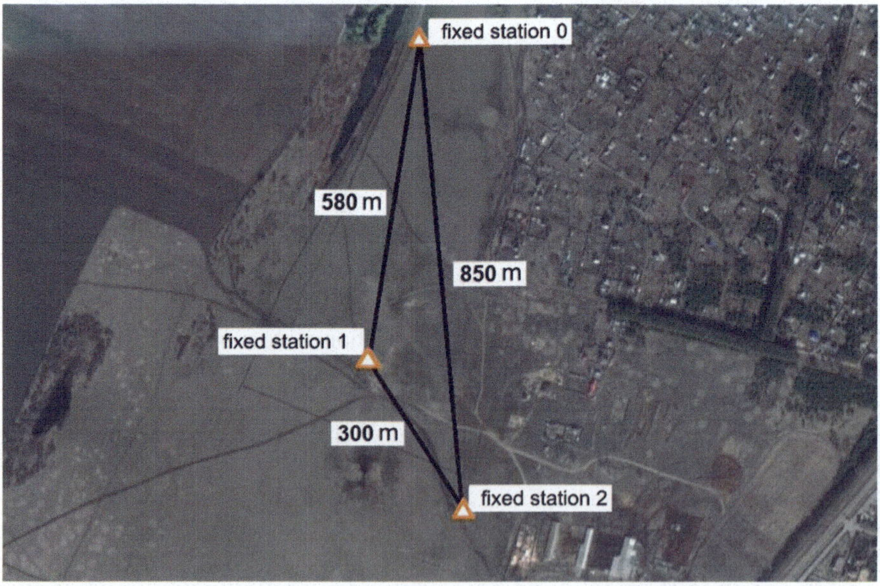

Fig. 7.17 Scheme of mutual localization of RMPs and RES at full-scale testing

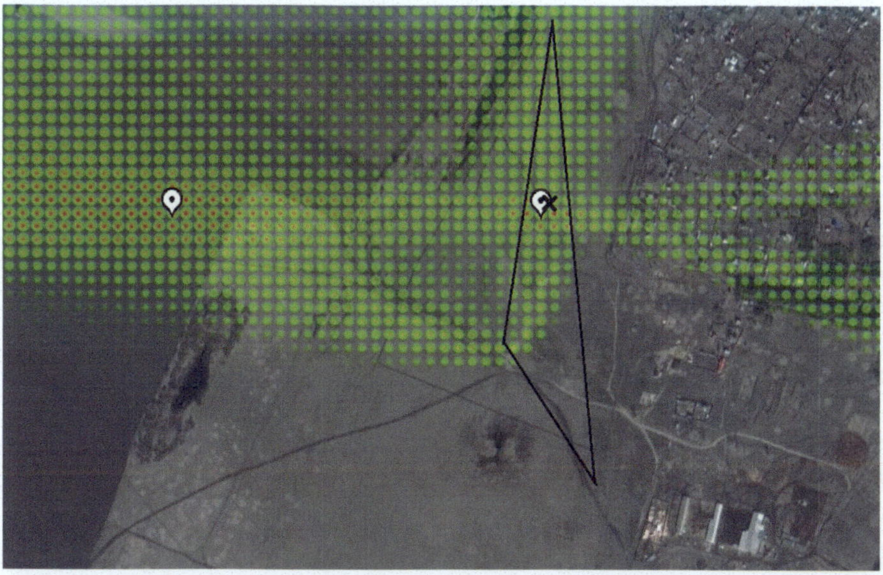

Fig. 7.18 Results of RES localization, which is in the control point No. 1, on the base of TDOA measurements. Error of the nearest solution is 22 m

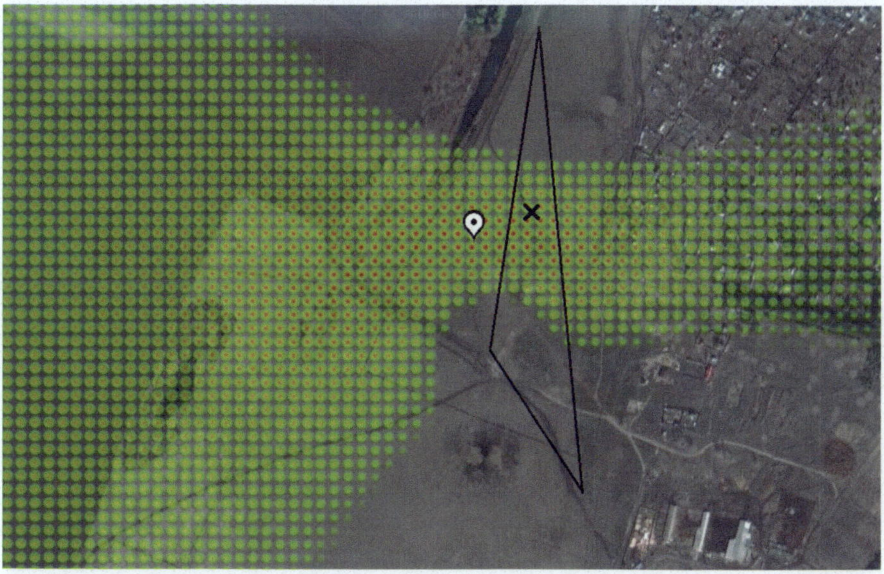

Fig. 7.19 Results of RES localization, which is in the control point No. 1, on the base of amplitude measurements. Error of the nearest solution is 120 m

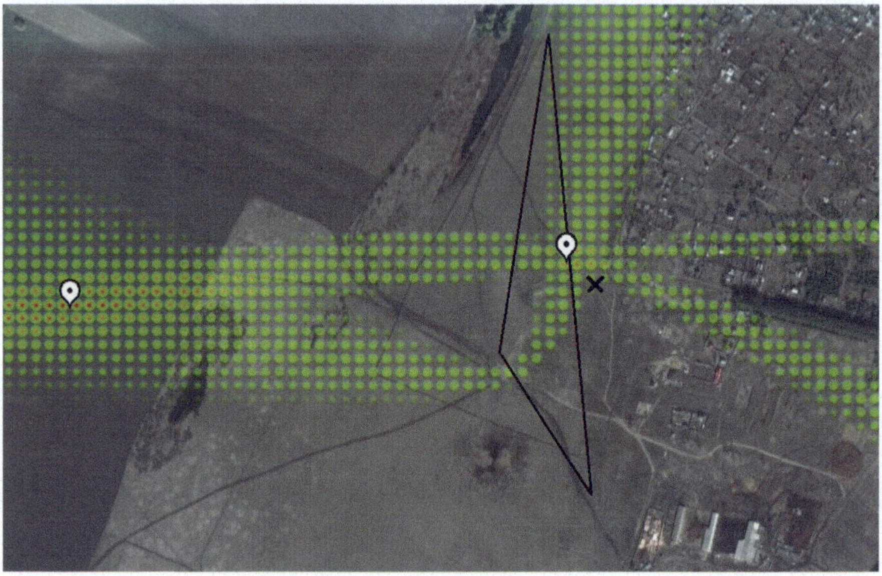

Fig. 7.20 Results of RES localization, which is in the control point No. 2, on the base of TDOA measurements. The error of the nearest solution is 58 m

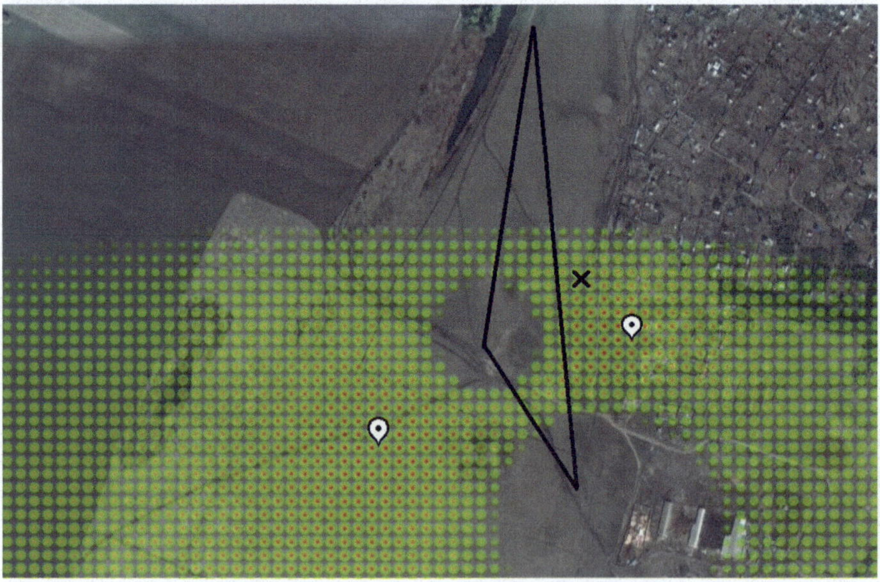

Fig. 7.21 Results of RES localization, which is in the control point No. 2, on the base of amplitude measurements. The error of the nearest solution is 124 m

correspond to areas with the maximal probability of RES localization. In several figures, there are, as minimum, two areas of concentration of raster matrices, which indicates on presence of two possible solutions of the localization task. Mentioned ambiguity is caused by small number of RMPs in the system structure.

Information presented in Figs. 7.18 and 7.19 corresponds to the case, when the RES being localized is placed inside the triangle formed by RMPs. For similar RES, the localization errors are, at average, minimally possible, which is well agreed with the results obtained on the base of TDOA measurements. The error of localization estimation here is 22 m, while the usage of amplitude measurements only was accompanied in the same situation by the error exceeding 100 m.

Localization errors are something higher for the cases presented in Figs. 7.20 and 7.21 due to the fact that the RES under localization is placed now beyond the triangle, forming by RMPs. This leads to noticeable growth of errors of RES coordinates determination with the help of TDOA method, although it does not seriously influence on accuracy of RES coordinates estimation by the amplitude method. The error of localization estimation for TDOA system happened to be 58 m, while at amplitude measurement it was 124 m.

7.11 Formation of Monitoring Receiver's Frequency-Time Scale

Formation of the high-accuracy frequency-time scale of the monitoring receivers is necessary both for determination of RES localization by network of receivers and for measurement of radio signal parameters with necessary accuracy. In particular, the documents [13, 14] indicate that maximal permissible value of carrier frequency deviation in the single-frequency networks of digital TV (SFN) is not more than 1 Hz for the DVB-T/H standard and not more than 0.5 Hzfor the DVB-T2 standard. Accordingly, measuring units should have the measurement error not worse than 0.05 Hz. Taking the 1 GHz frequency as the calculation basis for the digital communication networks signal carrier frequency, we obtain, that relative instability of the reference oscillator of the measuring unit should be not more than 5×10^{-11}.

Thus, the problem of high-accuracy estimation of radio emission source localization and measuring of signal parameters is inseparably connected with the problem of formation of the local high-accuracy frequency-timescale in the monitoring receiver [1, 2].

7.11.1 Formation of Local Timescale in the Receiver with Usage of GNSS

Synchronization of reference oscillator (REF) operation of the receiver with international universal time coordinates (UTC) can facilitate to growth of accuracy of oscillation frequency measurements made by the monitoring receiver. We can realize the synchronization with UTC on the base of "pulse-per-second" (PPS) signal processing, which are formed by the GNSS navigation receiver at the beginning of each second by the UTC scale. However, it should be taken into consideration that PPS signals are the synchronization units only, but they themselves do not form the timescale. At first, it is impossible to ensure stability of the PPS-signal observation; a possibility of their generation essentially depends on the location of the satellite navigation grouping with respect to the point of reception, and at insufficient number of available satellites, the PPS signals disappear. Secondly, the radio wave propagation features and actions of interference lead to error in navigation signals reception, due to which the sample-pulses may appear with deviation from the etalon time grid by tens nanoseconds at the navigation receiver output. Thus, high-accuracy local time–frequency scale (LTFS) of the receiver can be formed only on the base of its internal high-stable reference oscillator, which is adjusted in accordance with the PPS-signals totality. At that, it is not often necessary to provide the physical adjustment of the reference oscillator itself; LTFS can be obtained by means of usage of calculated corrections to the sample sequence, determined on the base of the PPS signals; these corrections are formed by this oscillator itself.

If we could guarantee the ideal frequency stability of the monitoring receiver REF and the ideal reception of navigation signals, then appearance of sample-pulses at the navigation receiver output will always be after the fixed number of REF cycles. In reality, the REF frequency instability leads to LTFS deviation from the UTC scale, due to which the interval k_i between the moments of sample-pulse observation, which is measured in the REF cycles, smoothly changes in random manner. If to consider the fact of sample-pulses jitter usage additionally, then mathematical model of LTFS functioning can be written in the form

$$\begin{aligned} z_i &= p_i + \xi_i, \\ p_i &= p_{i-1} + k. \end{aligned} \tag{7.39}$$

Here p_i is the real LTFS shift from the UTC scale on ith second of the observation interval, z_i is the observed LTFS shift distorted by the random correction ξ_i caused by the jitter, k is a coefficient of incessant mutual offset of scales caused by the REF current frequency shift from its nominal value. In reality, due to instability of REF frequency, the k coefficient itself demonstrates the definite random drift; however, at utilization of high-stable oscillator, the rate of this drift is low enough to allow regarding k as unknown constant on intervals with several hours duration and considering errors caused by the drift as a consequence of inaccuracy of k measurement and uncertainty of correction ξ_i values. Values of ξ_i can be considered as the sequence of independent random quantities with acceptable accuracy for practice. At absence of reliable information about ξ distribution law, we shall, when it is possible, assume values as distributed by the double exponential Laplace law, but in the cases, when this distribution will lead to very complex algorithms, we shall use normal distribution.

7.11.2 The Algorithm of Local Time-Frequency Scale Formation

At start stage of LTFS formation, the value p of the LTFS shift from the UTC scale and the scale offset coefficient k corresponding to the REF current frequency are unknown. And if to the moment of LTFS formation the value array z_i of enough volume was available, then we would be able to obtain the primary estimations $\tilde{p}$ and $\tilde{k}$, for instance, on the base of the least squares method (LSM).

Let n sequential measurements be included in the observation array z_i. Assuming that the moment of the last measurement is the origin for counting, we obtain that previous measurement will correspond to "minus first" second, and the "oldest" element of this array corresponds to the point with coordinate (1–n). Then, we integrate measurement results z_i into the vector $\mathbf{z} = (z_{1-n}, z_{2-n}, \ldots, z_{-1}, z_0)^T$, and unknown parameters of LTFS—into the vector $\mathbf{x} = (p, k)^T$. Then, the system of linear equations, which unified them, can be written as

$$\mathbf{H}_n \cdot \mathbf{x} = \mathbf{z} + \boldsymbol{\xi}, \tag{7.40}$$

where $\xi = (\xi_{1-n}, \xi_{2-n}, \ldots \xi_{-1}, \xi_0)^T$ is the vector of random corrections with the correlation matrix $\mathbf{R} = M\{\boldsymbol{\xi}\boldsymbol{\xi}^T\}$, and the $\mathbf{H}_n$ matrix has a form

$$\mathbf{H}_n = \begin{pmatrix} 1 & (1-n) \\ 1 & (2-n) \\ \ldots & \ldots \\ 1 & -1 \\ 2 & 0 \end{pmatrix}. \tag{7.41}$$

The LSM estimation of the parameter vector $\mathbf{x}$ can be calculated according to the rule

$$\tilde{\mathbf{x}} = \left(\mathbf{H}_n^T \mathbf{R}^{-1} \mathbf{H}_n\right)^{-1} \mathbf{H}_n^T \mathbf{R}^{-1} \mathbf{z}. \tag{7.42}$$

At independent measurements with equal accuracy, the correlation matrix of corrections $\boldsymbol{\xi}$ can be written in the form $\mathbf{R} = \sigma_\xi^2 \mathbf{I}$. Values $\mathbf{H}_n^T \mathbf{R}^{-1} \mathbf{H}_n$ do not depend on the specific sample; results of their preliminary calculation can be presented as

$$\begin{aligned} \mathbf{H}_n^T \mathbf{R}^{-1} \mathbf{H}_n &= \frac{1}{\sigma_\xi^2} \begin{pmatrix} n & \sum_i (i-n) \\ \sum_i (i-n) & \sum_i (i-n)^2 \end{pmatrix} \\ &= \frac{1}{\sigma_\xi^2} \begin{pmatrix} n & -n(n-1)/2 \\ -n(n-1)/2 & (2n^3 + 3n^2 + n)/6 \end{pmatrix}. \end{aligned} \tag{7.43}$$

from which

$$\left(\mathbf{H}_n^T\mathbf{R}^{-1}\mathbf{H}_n\right)^{-1} = \sigma_\xi^2\begin{pmatrix} (2n^3+3n^2+n)/6 & n(n-1)/2 \\ n(n-1)/2 & n \end{pmatrix}\Big/((n^4+12n^3-n^2)/12). \tag{7.44}$$

At $n \gg 1$, the expression (7.44) can be written in first approximation as

$$\left(\mathbf{H}_n^T\mathbf{R}^{-1}\mathbf{H}_n\right)^{-1}\Big|_{n\gg 1} \approx \sigma_\xi^2\begin{pmatrix} 4/n & 6/n^2 \\ 6/n^2 & 12/n^3 \end{pmatrix}. \tag{7.45}$$

Returning to LTFS formation, we take into account that to obtain the measurement array z_i in reality, it is necessary to organize acquisition of the appropriate data, for which rather large time interval is required, because each new measurement can be obtained only after a second compared to the previous one. As a consequence, it is more reasonable to take as the primary estimation of the offset coefficient—$\tilde{k}_0 = 0$, and as the shift $\tilde{p}_0$—the value obtained on the base of the just first sample-pulse from the navigation receiver, and after that to correct these values in the tracking mode, not making long preliminary data accumulation.

7.11.3 The Tracking Mode and the Local Time-Frequency Scale Adjustment

For correction of estimations $\tilde{p}$ and $\tilde{k}$, we use the Kalman filter to form the estimation

$$\tilde{\mathbf{x}}_i^+ = \tilde{\mathbf{x}}_i^- + \mathbf{P}_i^-\mathbf{h}^T(\mathbf{h}\mathbf{P}_i^-\mathbf{h}^T + \mathbf{R}_i)^{-1}(z_i - F(\tilde{\mathbf{x}}_i^-)), \tag{7.46}$$

where $\tilde{\mathbf{x}}_i = (\tilde{p}_i, \tilde{k}_i)^T$ is the vector of estimations corresponding to the ith estimation step, z_i is the observed at this step LTFS shift from the UTC scale, determined by the observation moment of the sample-pulse from the navigation receiver, $F(\tilde{\mathbf{x}}_i) = (\tilde{p}_i + \tilde{k}_i)$ is the predicted value of shift, $\mathbf{h} = (1 \;\; 0)$ is the matrix of partial derivatives for measuring parameters (the matrix $\mathbf{H}_n$ reduced to the only last measurement), $\mathbf{P}_i = (\mathbf{H}^T\mathbf{R}^{-1}\mathbf{H})^{-1}$ is the covariance matrix of solution error, being a subject to estimation at each ith step in accordance with the rule

$$\mathbf{P}_i^+ = \mathbf{P}_i^- - \mathbf{P}_i^-\mathbf{h}^T(\mathbf{h}\mathbf{P}_i^-\mathbf{h}^T + \mathbf{R}_i)^{-1}\mathbf{h}\mathbf{P}_i^-, \tag{7.47}$$

where $\mathbf{R}_i$ is the correlation matrix of random corrections, which is transformed into the σ_ξ^2 constant for the case of the single measurement. Assuming that the number of earlier fulfilled measurements $n \gg 1$, we shall consider the usage of approximation (7.45) being admissible instead of the iterative rule (7.47) for the matrix $\mathbf{P}_i$. Then,

$$\begin{aligned} \mathbf{P}_i^- \mathbf{h}^T &\approx \sigma_\xi^2 \begin{pmatrix} 4/n & 6/n^2 \\ 6/n^2 & 12/n^3 \end{pmatrix} \cdot \begin{pmatrix} 1 \\ 0 \end{pmatrix} = \sigma_\xi^2 \begin{pmatrix} 4/n \\ 6/n^2 \end{pmatrix} \text{ and} \\ \mathbf{h}\mathbf{P}_i^- \mathbf{h}^T &\approx (1 \;\; 0) \cdot \sigma_\xi^2 \begin{pmatrix} 4/n \\ 6/n^2 \end{pmatrix} = \frac{4\sigma_\xi^2}{n}. \end{aligned} \tag{7.48}$$

At $n \gg 1$, the term $\mathbf{h}\mathbf{P}_i^- \mathbf{h}^T$ in (7.46) will be significantly less than $\mathbf{R}_i = \sigma_\xi^2$, which allows rewriting of (7.46) in much more compact form

$$\tilde{\mathbf{x}}_i^+ = \tilde{\mathbf{x}}_i^- + \begin{pmatrix} 4/n \\ 6/n^2 \end{pmatrix} (z_i - (\tilde{p}_i + \tilde{k}_i)). \tag{7.49}$$

If we take into consideration that the LTFS offset coefficient k is not a constant, but it is the subject of the random drift due to the REF frequency instability, then it is impossible to rely on the infinitely lengthy sequence, and the estimation rule of the LTFS parameters must be modernized. For this, it is enough to use the composition filter: Before processing of n_F measurement, it will work as the Kalman filter, and after as the filter with constant coefficients

$$\begin{pmatrix} \tilde{p} \\ \tilde{k} \end{pmatrix}^+ = \begin{pmatrix} \tilde{p} \\ \tilde{k} \end{pmatrix}^- + \begin{pmatrix} 4/n \\ 6/n^2 \end{pmatrix} (z_i - (\tilde{p}_i + \tilde{k}_i)), \tag{7.50}$$

where $n = \min(\max(10, i),\ n_F)$.

Here n_F is the characteristic time of REF parameter variation (in seconds) determined by the product data sheet or via experiment. This parameter defines the convergence rate (the transient length is typically 2–4n_F) and also influences on the filter operation accuracy. Under conditions of good coincidence of the real event with the linear model (absence of parabolic and other components of higher order), the more n_F is, the higher filter accuracy is. If the condition of parabolic and other component absence is not satisfied, the excessive increase of n_F will lead to accuracy loss.

7.11.4 Modified Tracking Algorithm

Algorithms developed under assumption about the normal distribution law of errors are instable to violation of this assumption and to abnormal overshoots. The algorithms developed in accordance with assumption about error distribution on the base of the Laplace law are more stable, although such algorithms, as a rule, are more complicate in implementation and converge slower. However, taking into consideration [15], we may construct simple enough filter, which is quasi-optimal for the case of errors distributed according to the Laplace law.

Let us set out some constant max_err, exceeding of which by the error modulus of z_i is observed rather rarely (statistically unlikely). Usually, as such a constant, one takes doubled or tripled a priori estimation of measurement error.

Modification of the (7.49) algorithm consists of z_i correction according to the rule

$$\widetilde{z}_i = z_i \frac{\text{max\_err}}{\max(|z_i|, \text{max\_err})}. \tag{7.51}$$

The physical sense of (7.51) consists of the fact that if $|z_i| > \text{max\_err}$, the measurement is normalized in such a manner that the residual error sign is saved, and the modulus reduces to max_err.

7.11.5 Results of Full-Scale Tests

To check the synchronization accuracy provided by the above-mentioned algorithm, the experiment was performed, during which the wideband signal from the output of the Agilent E4438c generator was applied simultaneously to the input of two ARGAMAK-IS radio receivers via the splitter. Each receiver registered the signal time of arrival according to its own LTFS, and then the registered signal observation moments were compared between themselves. Comparison results are shown in Fig. 7.22, where horizontal axis represents time in seconds passed from the moment of experiment beginning, and on vertical axis, the difference of registered moments of signal arrival is shown, which characterizes the mismatch of two LTFSs formed in accordance with (7.49)–(7.51). As we see from Fig. 7.22, at lengthy enough interval, the deviation of two scales practically did not exceed 10 ns, i.e., the root-mean-square error of each scale affixment was units of nanoseconds.

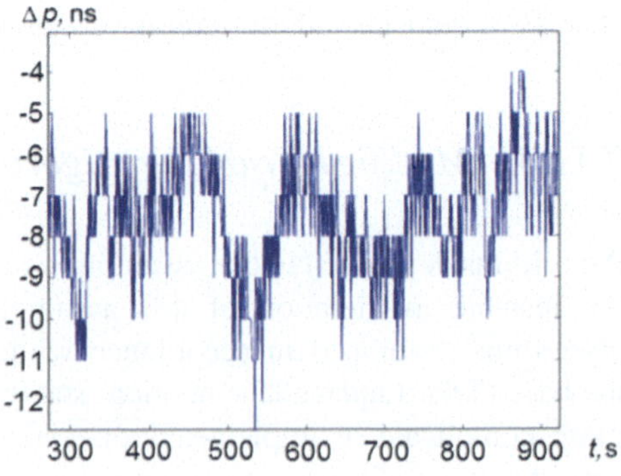

Fig. 7.22 Time-variation of LTFS mistiming of two monitoring receivers

Above-mentioned LTFS offset from the UTC scale arises due to shift of the real REF frequency equaled to f from its stated frequency f_p, and the offset value is determined by relative frequency detuning

$$k = -\frac{\Delta f}{f_p}, \tag{7.52}$$

where the absolute detuning $\Delta f = f - f_p$.

The estimation of $\tilde{k}$ of the mutual offset of timescales obtained at the expense of the tuning algorithm (7.50)–(7.51) application enables consideration that the real REF frequency can be calculated according to the rule (on the f_p base):

$$\tilde{f} = f_p + \Delta\tilde{f} = f_p(1 - \tilde{k}). \tag{7.53}$$

After similar correction, the estimation error of the external signal frequency caused by REF instability will be determined by the value

$$\frac{\Delta\tilde{f} - \Delta f}{f_p} = k - \tilde{k}. \tag{7.54}$$

For practical verification of signal frequency estimation accuracy, the receiver-synchronizer VCH-311 was used, which is characterized (according to its product data sheet) by instability of 1×10^{-13} order. The sine signal, in which frequency was considered as exactly known, was applied to the input of the ARGAMAK-IS receiver from the output of the oscillator clocked by the reference signal, which was formed by VCH-311. The receiver performed the periodic frequency measurements. In parallel with measurement, it formed LTFS according to rules (7.50)–(7.51) and using the estimation $\tilde{k}$ obtained for the specification of the REF frequency according to (7.53). The influence of such LTFS tuning to UTC is presented in Fig. 7.23. Line 1 shows results of external signal frequency measurement by the monitoring receiver without correction usage. Line 2 characterizes the relative error of the frequency estimation after correction. As it follows from this figure, errors of frequency estimations are reduced by 3 … 4 times already after 5 … 10 min of tuning.

7.12 Correction of Radio Receiving Paths Non-identity in TDOA Systems

TDOA systems are used for RES location estimation and have the set of geographically distributed receivers in its structure. TDOA system functioning is based on measurement of the mutual time delay between moments of radio signal arrival from RES to receivers included in the system structure. The mutual signal delay

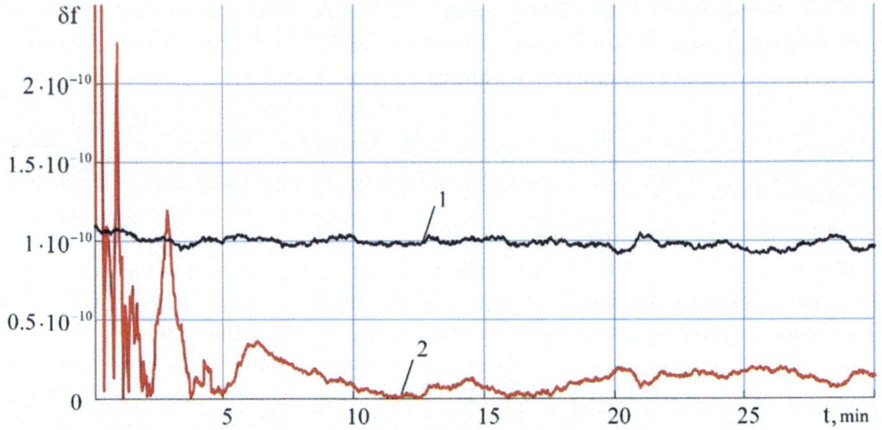

Fig. 7.23 Influence of LTFS correction on the relative error of frequency measurement of the sine oscillation by the monitoring receiver: 1—without correction, 2—with correction

should be measured with high accuracy up to several nanoseconds, which makes the requirements to the receiving equipment to be strict enough.

The goal of this section is to describe the receiver correction approach for measuring the mutual delay values with required accuracy [2].

7.12.1 Revelation of Non-identity of the Receiving Paths

The signal delays registered by radio receivers and processed by TDOA system depend not only on the distance between RES and the receiver, but on the signal delays after their transformations by the receiving paths. Extraction and processing of the specific relatively narrowband signal anticipates, in particular, its frequency filtering. The narrowband filtering, as a rule, mates with the significant response delay at filter output with respect to the time of the signal application to the filter input. Since RES location estimation is based on measurement of difference between the signal times of arrival to the receiver, the absolute delay value, which accompanies the signal processing, does not play any principal role, if receivers are of the same type and delay for different receivers becomes coincidental. However, at high demands to the estimation accuracy of RES localization, the usage of even the same type of receivers cannot guarantee equality of the existing delays with necessary accuracy. Due to this, there is a need in providing of specialized equipment calibration.

Necessity of receiving path calibration can be determined, for example, with the help of the following experiment. Let us consider the unit, in which structural scheme is presented in Fig. 7.24. At identical processing paths, the same connecting cable length, and at ideal receiver synchronization, the observed mutual signal

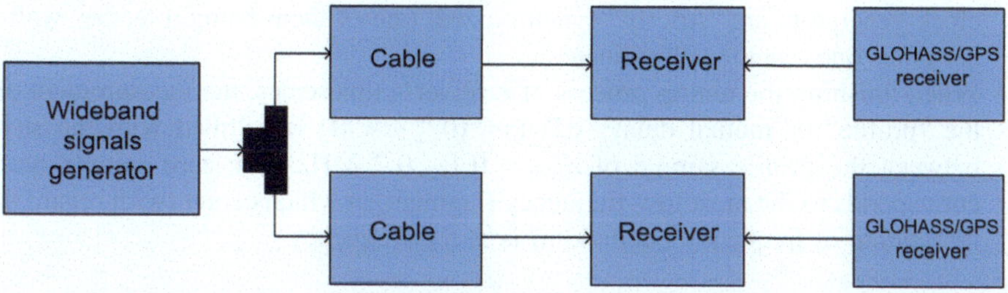

Fig. 7.24 Structural scheme of the unit for identity testing of the receiver analog paths

delay obtained by two receivers from a splitter, should be absent. Owing to the tuning error of local time scales of receivers, performing on the base of PPS-signals from GLOHASS/GPS receivers [2], the delay may be nonzero, but at coincidence of receiving paths characteristics, it should represent the constant, which is insignificant in value for all signals of the processed frequency range. Unfortunately, this assumption was not confirmed.

To investigate the influence of delays of non-identical receiving paths on the measurement accuracy, the following experiment was fulfilled. The wideband signal of digital TV proceeded to the ARGAMAK-IS receivers and was divided into the frequency fragments of 100 … 500-kHz width, and estimation of signal mutual delays was carried out separately for each fragment. The obtained variation diagram of the observed mutual delays turned out to be frequency-dependent, "attached" to the frequency deviation of the processed signal fragment from the receiver tuning frequency. At variation of the receiver tuning frequency, the "profile" of mutual delays variations was shifted following the tuning frequency, remaining unchanged with respect to the processed signal offset from the central frequency of IF path of the receiver. This means that mutual delay measurement error relates to the non-identity of IF filters of the used receivers. From Fig. 7.7, we see that for receivers ARGAMAK-IS the mutual delay measurement error caused by non-identity of IF path may achieve ±50 ns.

The revealed "profile" of variations of the observed mutual delays was saved unchanged at repeated fulfillment of the experiment in different days and for different carrier frequencies of radio signal processed to the splitter input.

7.12.2 *Approach for Correction of Receiver Paths*

The profile constancy of the observed mutual delays allows using of the following procedure of receiving paths mutual correction:

1. Unit to examine the profile of mutual delays using radio receivers to be corrected should be installed. Its scheme is shown in Fig. 7.24. At that, the receiver

clock generators are "strictly" synchronized; one of them being a leader, while the other one—being subordinate.

2. When finishing the tuning process of receiver's timescales, the measurement of the "profile" of mutual delays $\mathrm{dTau}[m]$ $(0 \leq m < M)$ is fulfilled with the step between the profile sample of $df_{\mathrm{clb}} = 0.1\ldots0.2$ MHz. The zero sample here corresponds to the most low-frequency fragment, in which center (with regard to the middle of frequency bandwidth) is at the frequency

$$f_{0\,\mathrm{clb}} = \quad -0,5 \cdot (F_s - df_{\mathrm{clb}}), \tag{7.55}$$

where F_s is the sampling frequency of data sampling at the IF filter output.

3. According to the measured profile, the samples of the calibration data array are calculated for usage in further calculation of FFT of size N

$$\mathrm{corr}[n] = \exp\left\{-j \cdot 2\pi \cdot df \cdot \sum_{i=0}^{n} \Delta_\tau[i]\right\} \tag{7.56}$$

where the zero sample corresponds to the left boundary of IF filter bandwidth

$$f_0 = \quad -0,5 \cdot (F_S - df), \tag{7.57}$$

The frequency interval between FFT samples is equal to

$$df = \frac{F_S}{N}, \tag{7.58}$$

and corrections $\Delta_\tau[i]$ corresponding to the frequency $f_i = f_0 + i \cdot df$ are determined by the rule

$$\Delta_\tau[i] = \frac{df}{df_{\mathrm{clb}}} \cdot \left\{\mathrm{dTau}\left[1 + \mathrm{floor}\left(\frac{f_i}{df_{\mathrm{clb}}}\right)\right] - \mathrm{dTau}\left[\mathrm{floor}\left(\frac{f_i}{df_{\mathrm{clb}}}\right)\right]\right\}, \tag{7.59}$$

where $\mathrm{floor}(\cdot)$ is the operation of roundup downwards. The totality of samples $\mathrm{corr}[n]$, determined by (7.56), is sufficient to perform the correction procedure.

To calculate the correction array $\mathrm{corr}[n]$, the estimation of signal mutual delays is carried out in the spectral domain with cross-correlation function computation according to the algorithm examined in Sect. 7.9.

Results of correction procedure application are presented in Fig. 7.25. The blue line corresponding to the correction results shows, that after correction the delay dispersion reduces to ± 10 ns.

We now take into consideration that mismatch of the local time scales is usually more than ± 10 ns [2], and interference distortions caused by ineffectual location of the receiving antennas may be more in order. As a consequence, possibilities of

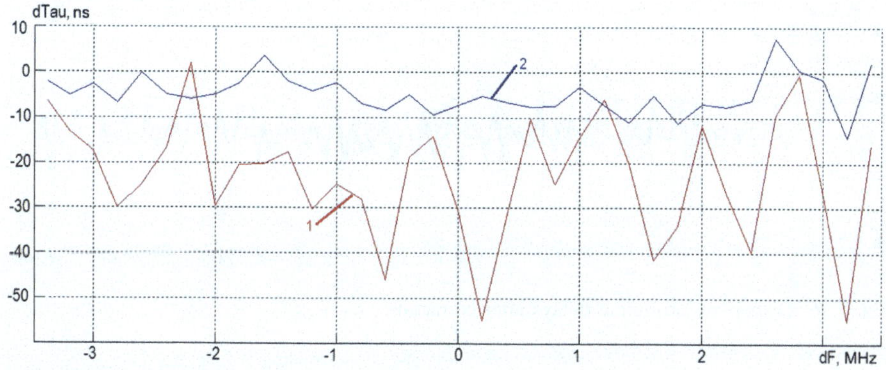

Fig. 7.25 An example of estimation correction of signal mutual delays

receiving paths corrections providing by the above-described procedure, in the most of cases meet the practical requirements.

The fact, that the described approach cannot be realized "remotely", can be attributed to its disadvantage; to measure the "profile" of mutual delays dTau$[m]$ all receivers included in the TDOA system should be gathered together and calibrated being connected in pairs in the installation structure shown in Fig. 7.24.

7.12.3 Practical Results

We should take into consideration that correction results of measured delays in Fig. 7.25 demonstrate the error reduction exactly due to the non-identity of IF filters of the receivers. Measurement errors observed at practical application of calibrations will be larger because of non-identity of receiving paths, which proceeds to the IF path, as well as the mutual drift of the local time scales of receivers. Let us examine, for instance, results of the full-scale tests presented in Fig. 7.26, during which measurements were performed, with the help of calibrated receivers' pair, consequently, with 10–15-min interval, for digital TV signal mutual delays for different variants of the receiver central frequency. As we see, in practice, the observed dispersion of mutual delays obtained after correction (curve 2) increases up to ±20 ns, however, as earlier, remains in 2–4 times less than without correction (curve 1).

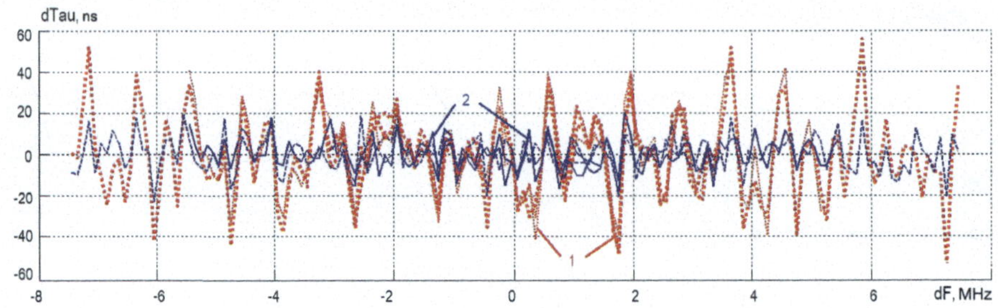

Fig. 7.26 Dispersion of mutual delays after correction

7.13 Summary

In this chapter, the general approach for solution of RES localization problem is described on the base of the main types of radio engineering measurements: amplitude, goniometric (angular), TDOA, frequency and time. Irrespective to the measurements type, the sequence of steps to solve the localization problem remains unchanged:

- Construction of the mathematical model of measurements;
- Choice of the functional;
- Minimization of the functional;
- Creation of initial approximation;
- Linearization of the initial task—reducing it to the task "in the small";
- Solution of the task "in the small" with obtained solution accuracy estimation.

The choice of the specific scheme of RES localization depends on the radio signal type and parameters, in which source should be localized, and on the technical equipment of the radio monitoring services. The algorithmic support of digital radio receivers and radio direction-finders enables implementation of the localization scheme, which maximally completely takes into account the RES characteristics and opportunities of existing RMPs.

Presented results can be also used at planning of the radio monitoring network. Possibilities for fixed RMPs arranging, the required characteristics of accuracy and performance of different types of RESs localization, which are taken into consideration together with presented characteristics of accuracy of localization methods, will permit to choose the expedient structure of the measuring equipment.

The mentioned examples of RES localization systems on the base of digital radio receivers ARGAMAK-IC and ARGAMAK-RS have demonstrated the efficiency of the offered approach in connection with the TDOA and amplitude measurements. Characteristics of the localization estimation system depend on the number of radio receivers in the TDOA system structure, the mutual positions of the receiver and required RES, as well as on the interference distortions, which are defined by the mounting platform of the receiving antennas. The measuring receivers

ARGAMAK-IC and ARGAMAK-RS can be replaced by less expensive models, for example, ARGAMAK-MN or ARC-CPS3, if they are equipped with high-accuracy synchronization system.

Offered algorithms of high-accuracy frequency-time scale formation of the radio monitoring receiver are distinguished by simplicity, reliability, and high enough rate of synchronization with the UTC scale. At stable reception of navigation signals, which depends on the current configuration of the satellite grouping and usually is provided in practice, the sufficient accuracy of the frequency-time scale is usually achieved after 5–10-min adjustment. After transient finishing, the RMSE of the timing is units of nanoseconds, and the guaranteed frequency instability is less than 5×10^{-11}.

The developed approach of mutual delay correction (used in the TDOA systems) caused by the receiver path non-identity enables several times reduction of estimation systematic errors. The importance of practical application of the offered correction procedure is the highest for TDOA systems consisting of the small number of RMPs, and it slightly decreases at using of large number of receivers for measurements, when negative effects from non-identity of receiving paths are compensated by the significant volume of acquired and processed data.

References

1. Gnezdilov DS, Kozmin VA, Kryzhko IB, Tokarev AB, Fateev AA (2016, in Russian) The formation of local high-precision frequency-time scale for radio monitoring receivers. Spectehnika i Svyaz (4):85–89
2. Kalinin YE, Kozmin VA, Kryzhko IB, Poliakov AV, Salikov AA, Tokarev AB (2014, in Russian) Synchronization subsystems of TDOA direction-finding system points. Radiotekhnika (3):51–54
3. Kozmin VA, Kryzhko IB, Tokarev AB, Fateev AA (2016, in Russian) Locating of radiation sources. Spectehnika i Svyaz (4):60–68
4. Kozmin VA, Kryzhko IB, Provotorov AS, Salikov AA, Tokarev AB (2016, in Russian) Assessment of accuracy of radio emission sources location made by TDOA system. Electrosvyaz (1):35–40
5. Rogers D, Adams A (1976) Mathematical elements for computer graphics, 2nd edn. McRaw-Hill Publishing Company
6. Rembosky AM, Ashikhmin AV, Kozmin BA (2015, in Russian) Radio monitoring: problems, methods, means. In: Rembosky AM (ed), 4th edn. Hot Line-Telecom Publ., Moscow, 640 p
7. Report ITU-R SM.2211-1(06/2014) (2015) Comparison of time-difference-of-arrival and angle-of-arrival methods of signal geolocation. SM Series. Spectrum management, 32 p
8. Kozmin VA, Pavlyuk AP, Tokarev AB (2014, in Russian) Comparison of spectrum monitoring coverage features of AOA and TDOA geolocation methods. Electrosvyaz (2): 37–40
9. SMO-ARMADA software package of automated spectrum monitoring systems. http://www.ircos.ru/en/sw_armada.html. Accessed 28 Nov 2017
10. Agilent (2014) N6841A RF sensor for signal monitoring network. Data Sheet. Agilent Technologies, Inc. 2012–2014. Published in USA, April 24, 2014. 5990-3839EN
11. The catalogue 2017 of IRCOS JSC. http://www.ircos.ru/zip/cat2017en.pdf. Accessed 28 Nov 2017

12. SMO-PA/PAI/PPK Panoramic analysis, measuring and direction finding software package. http://www.ircos.ru/en/sw_pa.html. Accessed 28 Nov 2017
13. GOST (Russian State Standard) P 55696–2013 Digital video broadcasting. Transmission equipment for digital terrestrial television broadcasting DVB-T/T2. Technical requirements. General parameters. Measurement methods (in Russian)
14. Norm of State Committee on Raduo Frequencies 17–13 (2015, in Russian) "Radio Transmitters of all categories of civil application. Requirements on the permissible frequency deviations" with modifications dated July 2015
15. Mudrov VI, Kushko VL (1983, in Russian) Methods of measurement processing: quasi-likely estimations, 2nd edn. Radio I Svyaz Publishing, Moscow, 304 p

Chapter 8
Television and Radio Broadcasting Monitoring

8.1 Introduction

Television (TV) and radio broadcasting (RB) can be attributed to the electronic units of mass media information, which are capable to carry information to the very wide population stratum almost immediately. It influences on the modern society greatly. Government services are obliged to check TV-RB companies to fulfill requirements foreseen by laws, to perform monitoring of content and technical quality of broadcasted programs, to take correcting measures in cases of violations revelation.

Under conditions of permanent increase of TV channels number and widening of the territory covering, it is expedient for radio monitoring effectiveness growth to use the automated distributed-in-territory system, which is controlled by the unified center. This chapter is devoted to consideration of the DVB-T2 analyzer and the multi-channel radio monitoring complex ARC-TVRV for TV and RB monitoring intended both for operation in ARMADA ARMS and outside of this system [1, 2].

8.2 DVB-T2 Radio Signals Analysis

During last decade, the on-air digital TV broadcasting has got a wide distribution. Digital TV technologies have a series of advantages compared to analog broadcasting: a possibility to transmit larger number of TV programs in one frequency channel, robustness to the multi-path signal transmission, higher noise immunity, flexibility to variations of TV signal parameters, possibility to add new services, etc. At present, more effective DVB-T2 systems, which are intended to both fixed and mobile subscribes, displace first DVB-T/H systems.

For planning and exploitation of TV and RB systems, checking of the conformity of requirements to transmitter parameters, and estimation of the covering zone,

A. M. Rembovsky et al., *Radio Monitoring*, Signals and Communication Technology, https://doi.org/10.1007/978-3-319-74277-9_8

it is necessary to execute analysis of radio signal parameters regularly. Equipment for such analysis is manufactured by well-known manufacturers of the radio monitoring equipment Rohde&Schwarz, Keysight, etc.

The analyzer of DVB-T/H digital TV signals is described in [3, 4], which is implemented on the panoramic digital receivers of the ARGAMAK family. In this chapter, we present analyzer of DVB-T2 signals, which is implemented on the DRR of the same family.

8.3 DVB-T2 Technology Features

A great number of mechanisms to increase noise immunity and to reduce non-informative part of the frequency–time resource are anticipated in the DVB-T2 system. This allows the system to exceed the digital TV system of the first generation by 30–50% in capacity. In the DVB-T2 system, the several independent data streams can be transferred, at that, these streams may differ in modulation, encoding, coverage zone, and other parameters, which provide flexibility in different service presentation to various types of subscribers. For instance, one of the streams may be intended for stationary subscribers, while another—for mobile ones. At present, this system is under wide deployment in many regions of Russia.

The Orthogonal Frequency Division Multiplexing (OFDM) transmission is the base of digital TV technology [5, 6]. At that way of transmission, the data stream is divided into several low-speed streams, which are transmitted at different orthogonal sub-carriers. It allows usage of wide signal bandwidth and, accordingly, high rate of data transmission without reducing of the symbol duration and saving inter-symbol interference on the acceptably low level. OFDM systems also have additional advantages compared to traditional single-frequency systems: robustness to multi-path radio signal propagation, simplicity of digital implementation, etc.

When we use sub-carriers, the inverse fast Fourier transform (IFFT) is carried out for each transmitted block of N modulated symbols. As a result, the N-dimensional vector of samples is formed in the time domain. To suppress inter-symbol interference, we add a guard interval (a prefix), which represent the last N_{CP} samples of this vector, to beginning of the obtained vector. The guard interval length is chosen of bigger value, than the multi-path delay spread. Since the signal consists of a lot of modulated sub-carriers, each OFDM symbol can be considered as a totality of cells, each of which corresponds to the modulated sub-carrier of the single-symbol interval. The OFDM symbols formed in such a manner are transmitted into the channel. On the receiving side, for each OFDM symbol, the guard interval samples are suppressed, and the last samples are undergone to the fast Fourier transform (FFT) with N dimension. As a result, the signal in frequency–time domain is formed, which is undergone to further processing.

The DVB-T2 standard [7] allows utilization of a great number of modes and transmission parameters. The following values of the signal bandwidth are acceptable: 10 MHz, 8 MHz, 7 MHz, 6 MHz, 5 MHz, 1.7 MHz. Six values of FFT length are possible: $N = 32,768$ (32 K), $N = 16,384$ (16 K), $N = 8192$ (8 K), $N = 4096$ (4 K), $N = 2048$ (2 K), and $N = 1024$ (1 K). Not all N sub-carriers are used for transmission; a part of sub-carrier (at ends of the signal spectrum) forms the protective frequency bands. In order to increase capacity for large FFT (8 K, 16 K, 32 K), we may apply the mode of increased number of used sub-carriers. In the general case, seven relative values of the prefix length are anticipated: $N_{CP}/N = 1/4;\ 19/128;\ 1/8;\ 19/256;\ 1/16;\ 1/32;\ 1/128.$ To increase noise immunity in DVB-T2 signals, the cascade error correction coding is used: LDPC and BCH. The following coding rates are possible: 1/4, 1/2, 3/5, 2/3, 3/4, 4/5, 5/6.

The DVB-T2 signal contains both data streams and signaling necessary for reception. There are three types of signaling: the preamble P1 signaling, the L1-pre signaling, and the L1-post signalling. The BPSK modulation is always used for the L1-pre signaling, the following types of modulation can be used for the L1-post signalling: BPSK, QPSK, 16-QAM, 64-QAM, and for information streams we use: QPSK, 16-QAM, 64-QAM, 256-QAM. For the last ones, we may use the rotated symbol constellations, which real and imaginary parts are transmitted in different cells. In some cases, this allows growth of the reception noise immunity.

To struggle with deep fading, together with the traditional single-antenna transmission, the DVB-T2 standard anticipates a possibility of simultaneous signal transmission from two spaced antennas using spatial-frequency coding technology [8].

There are specific mechanisms to reduce a peak-factor of OFDM signal transmitted with a goal to decrease the cost of the analog equipment part.

Depending on conditions of signal propagation, the operator may choose one from eight possible scattered pilot patterns, which differ in pilot cell density. These pilot cells are used on the reception side for estimation of the channel frequency response, which is necessary at signal demodulation. A presence of a number of pilot pattern variants allows minimization of the non-informative part of the resource ensuring the required accuracy of the channel estimation.

In order to improve the decoding quality under fading conditions, the different types of interleaving are used including bit and symbol interleaving, interleaving in time and frequency domains.

Choice of DVB-T2 signal parameters is performed by the network operator depending on the required coverage zone, the district character, the propagation conditions, the capacity required, and so on.

The possible sets of DVB-T2 parameters are grouped in the two profiles: BASE and LITE. Utilization of the last one poses essentially less requirements to the receiving equipment, which is particularly important for compact mobile receivers.

8.4 The Structure of DVB-T2 Signal

Let us consider the structure of DVB-T2 signal in detail. The diagram of DVB-T2 signal formation is presented in Fig. 8.1.

The initial data for DVB-T2 signal formation are the one or several logical transport streams (TS) [9] and/or one or several logical generalized streams of one of the following formats: Generic Encapsulated Stream (GSE), Generic Continuous Stream (GCS), Generic Fixed-length Packetized Stream (GFPS) [10]. In general case, the logical stream contains several services (TV-radio programs, etc.) and control data, which detail the stream content. Each of these streams is transferred in its own channel of physical level (PLP), which is characterized by the modulation type and encoding parameters. Single-type logical streams can be combined into groups. The mutual stream, which is transferred by the mutual PLP and formed from the same packets of the logical streams of the group, may be added to the group consisting of identical logical TSs with the same control data. These packets are transmitted in the mutual stream only, which endures reduction of the total volume of the control data.

Each PLP is structured into a sequence of logical baseband (BB) frames, which formation, in the general case, includes extraction of required number bits of the stream, addition of check-up bits, header of BB-frame and scrambling. There are two modes of BB-frame formation: the normal mode and the high-efficiency mode, in which the control data volume is reduced.

Logical BB-frames of PLP passes to the encoding, interleaving and modulation unit, the structure of which is presented in Fig. 8.2.

At first, BB-frames are exposed to BCH encoding with $(N_{\text{bch}}, K_{\text{bch}})$ parameters. BB-frames of $N_{\text{bch}} = K_{\text{ldpc}}$ in length, which are expanded by the check-up BCH bits, pass to the LDPC encoder with $(N_{\text{ldpc}}, K_{\text{ldpc}})$ parameters. As a result, the coded blocks of N_{ldpc} in length are formed, which are schematically presented in Fig. 8.3.

Two values of the coded block are anticipated by the standard: $N_{\text{ldpc}} = 64,800$ and $N_{\text{ldpc}} = 16,200$. The other coding parameters are defined by the N_{ldpc} value and the chosen coding rate (1/2, 3/5, 2/3, 3/4, 4/5, 5/6).

Bits of each coded block are interleaved. Parameters of bit interleaving depend on the coded block length and the used modulation.

After interleaving, bits of coded blocks are united in words consisting of η_{mod} bits. For QPSK $\eta_{\text{mod}} = 2$, for 16-QAM $\eta_{\text{mod}} = 4$, for 64-QAM $\eta_{\text{mod}} = 6$, for 256-QAM $\eta_{\text{mod}} = 8$. Besides, bits of each word (for QPSK; 256-QAM,

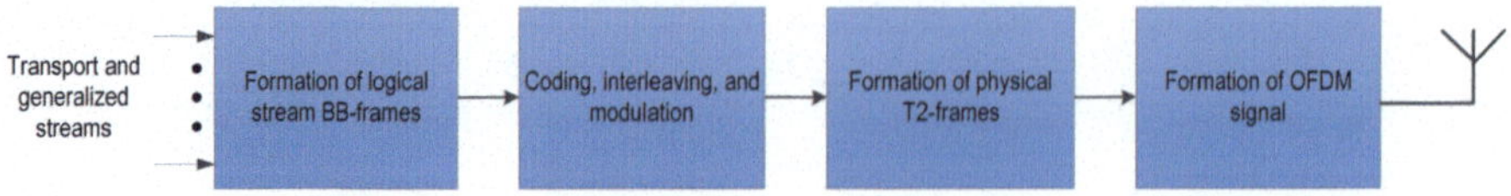

Fig. 8.1 Diagram of digital DVB-T2 signal formation

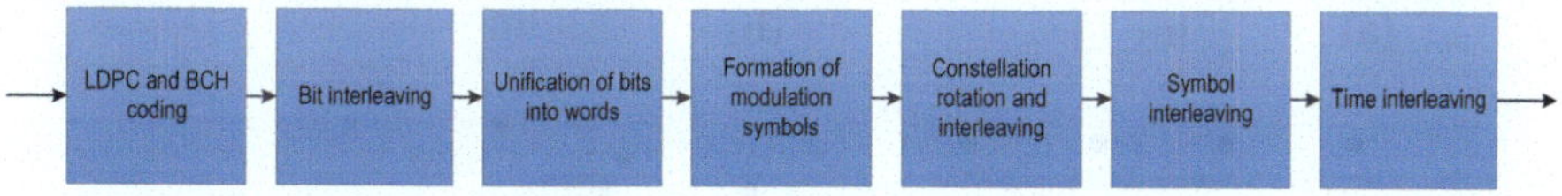

Fig. 8.2 Diagram of encoding, interleaving, and modulation of the one PLP

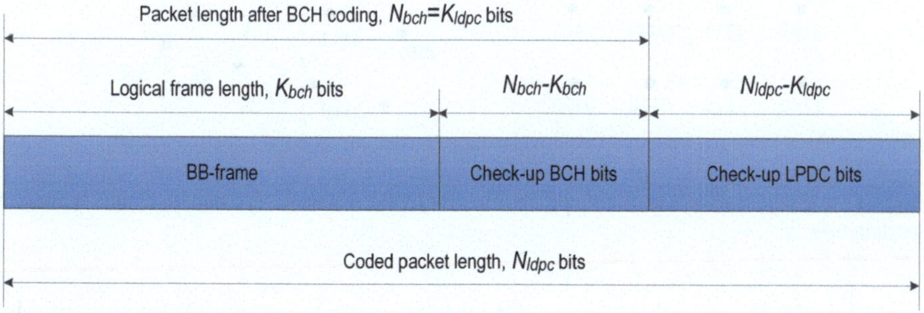

Fig. 8.3 Structure of the coded block

$N_{\text{ldpc}} = 16,200$) or a pair of words (for 16-QAM; 64-QAM; 256-QAM, $N_{\text{ldpc}} = 64,800$) are mixed between themselves.

Then, words of coded blocks are reflected on appropriate symbols of the modulation constellation. In the case of rotated constellation application, the obtained complex symbols are multiplied by the $\exp(j2\pi\Phi/360)$ phase factor, for QPSK $\Phi = 29°$, for 16-QAM $\Phi = 16.8°$, for 64-QAM $\Phi = 8.6°$, for 256-QAM $\Phi = \arctan(1/16)$. Besides, interleaving of the symbol imaginary parts within the limits of each coded block is carried out, for which their cyclic shift on one symbol is fulfilled. As a result, the imaginary part of the symbol turns into the adjacent symbol, and the imaginary part of the last symbol turns into the first symbol. An example of the normal and rotated modulation constellations together with gray bit mapping for 16-QAM is presented in Fig. 8.4.

Obtained modulation symbols are exposed to symbol interleaving. Interleaving is carried out within the limits of coded blocks according to a rule depending on the block number and the number of signals in the block.

Then, to noise immunity increasing for signal reception by the mobile subscribers, the temporal symbol interleaving is carried out in the time interleaving block interval. Several interleaving blocks are grouped into the time interleaving frame. A number of coded blocks in the interleaving frame may vary from 1 to 1023.

Interleaving frames of PLP streams are further mapped onto OFDM symbols, which are structured in T2-frames. We can distinguish three types of PLP: general PLP, PLP of type 1, and PLP of type 2. General PLP and PLP of type 1 have one fragment in the T2-frame, PLP of type 2 has from 2 to 6480 fragments. Several T2-frames are united into super-frames. Each T2-frame consists of one special

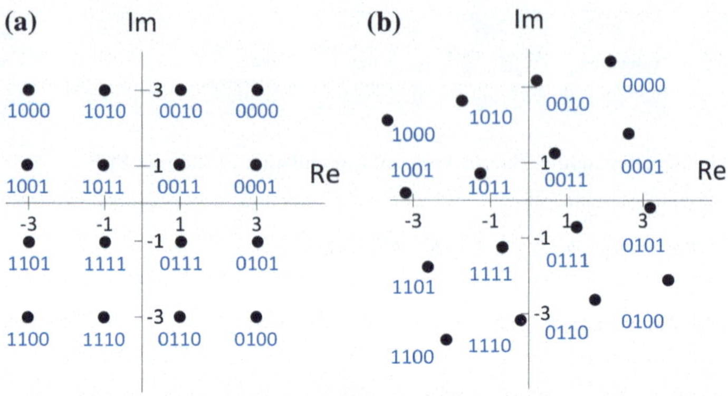

Fig. 8.4 A constellation and bit mapping for 16-QAM, **a** normal constellation and **b** rotated constellation

symbol of the P1 preamble, several OFDM P2 symbols, and OFDM data symbols. To ensure reception of PLP streams in each T2-frame, the signaling is transmitted. It consists of three parts: the P1 preamble signaling, L1-pre signaling and L1-post signalling transmitted in P2 symbols, as shown in Fig. 8.5.

The P1 signaling includes L1-pre signaling and L1-post signalling formats (the transmission profile is BASE or LITE, a number of transmitting antennas is one or two), FFT size (1 K, 2 K, 4 K, 8 K, 16 K, 32 K).

In the L1-pre signaling, the parameters necessary for reception and decoding of the L1-post signalling and other following information is transmitted:

- Type of the transmitted stream (TS, GSE, GFPS, GCS, and their combinations);
- Flag of usage of the expanded sub-carriers set;
- Length of the guard interval;
- Presence and mechanism of PAPR reduction;

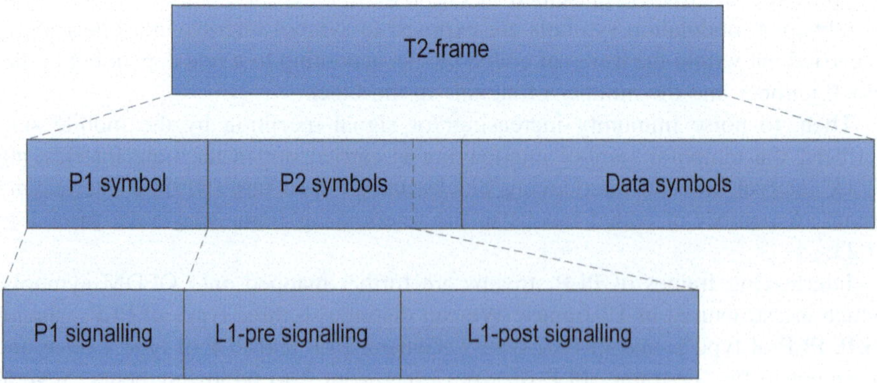

Fig. 8.5 Structure of the T2-frame and signaling

- Parameters of L1-postdata transmission:
- Modulation,
- Type and rate of encoding,
- Size of the symbol block,
- Size of the information block,
- Flag of expansion field presence,
- Flag of scrambling presence;
- Plot pattern type (PP1–PP8);
- Identification parameters (may be not transmitted):
- Cell identifier CELL_ID,
- DVB network identifier NETWORK_ID,
- T2 system identifier in DVB network T2_SYSTEM_ID;
- Number of T2-frames in the super-frame;
- Number of OFDM data symbols in the T2-frame;
- Number of carrier frequencies in the T2 system;
- Number of the current carrier frequency;
- Version of the T2 system.

The L1-post signalling contains parameters necessary for reception of PLP streams of the T2-frame. It consists of two parts: static, which parameters are constant in the super-frame interval, and dynamic, which parameters may change in each T2-frame. The static parameters of the L1-post signalling are the following:

- Number of fragments of all PLP in one T2-frame;
- Number of PLP streams in the current super-frame;
- Index and central frequency of T2 system carriers;
- For each PLP stream:
  - Identifier,
  - Type,
  - Payload type of transmitted data,
  - Initial T2-frame in the super-frame,
  - Coding rate,
  - Modulation,
  - Flag of modulation constellation rotation usage,
  - Size of the LDPC coding block,
  - Period of fragments (in T2-frames),
  - Type and parameter of time interleaving,
  - Formation mode of BB-frames.

Dynamic parameters of the L1-post are the following:

- Number of T2-frame in the super-frame;
- Distance between fragments of one PLP;
- Beginning of the first PLP fragment of the type 2;
- For each PLP stream:

- Identifier,
- Initial symbol in the frame,
- Number of coded blocks in the interleaving block.

The L1-pre signaling of 200 bits in length is exposed to encoding and modulation. After signaling expanding, the BCH encoding with parameters (3240, 3072) is carried out, and then the LDPC encoding with parameters (16,200, 3240) is executed. The checking LDPC bits are exposed to decimation (11,488 bits are discarded). Expansion bits are also discarded. As a result, the coded block of the L1-pre signaling with length of 1840 bits is formed. These bits are mapped on the appropriate BPSK modulation symbols, which then are mapped on the logical cells of P2 symbol according to the established rule.

The L1-post signalling may have different lengths; it also is exposed to encoding, interleaving and modulation. If the length of the L1-post signalling is more than 7032 bits, it is divided into several fragments, which length is less than 7032 bits. After expansion of each fragment, the BCH encoding with parameters (7200, 7032) is carried out, and then the LDPC encoding with parameters (16,200, 7200) is executed. The check-up LDPC bits are exposed to decimation. The expansion bits are also discarded. As a result, one or several coded blocks of the same length are formed. Bits of each coded block are exposed to interleaving, which parameters depend on the modulation type. When using BPSK or QPSK modulations, bit interleaving is not used. After interleaving, bits of coded blocks are united into words consisting of η_{mod} bits. In addition, when using 16-QAM, the 64-QAM bits of word pairs of the coded block additionally get mixed up. Then, the coded block words are mapped to the appropriate symbols of the used modulation constellation. Formed modulation symbols of the L1-post signalling are mapped to OFDM P2 symbols.

Data and signaling mapping of formed modulation symbols to the logical cells of the T2-frame is carried out in the following order: At first, symbols of the L1-pre signaling are mapped to cells of the P2 symbols, then, the L1-post signalling symbols are mapped. Remained P2 symbol cells and all OFDM data symbols are filled by modulation symbols of general PLP, then PLP of type 1, and further PLP of type 2. The signal frequency–time structure of the T2-frame is presented in Fig. 8.6.

Modulation symbols of sub-carriers of each OFDM symbol of the T2-frame are exposed to frequency interleaving, which rule depends on the FFT size. After that, they are applied to the unit of OFDM signal formation, which diagram is shown in Fig. 8.7.

At application of double-antenna transmission, the spatial-frequency encoding of transmitted symbol blocks is performed. Each block consists of two symbols, which are simultaneously transmitted to the adjacent sub-carriers from two different antennas. The encoding scheme for two transmitting antennas 0 and 1 is shown in Fig. 8.8. Further signal processing is carried out for two antennas in the same manner.

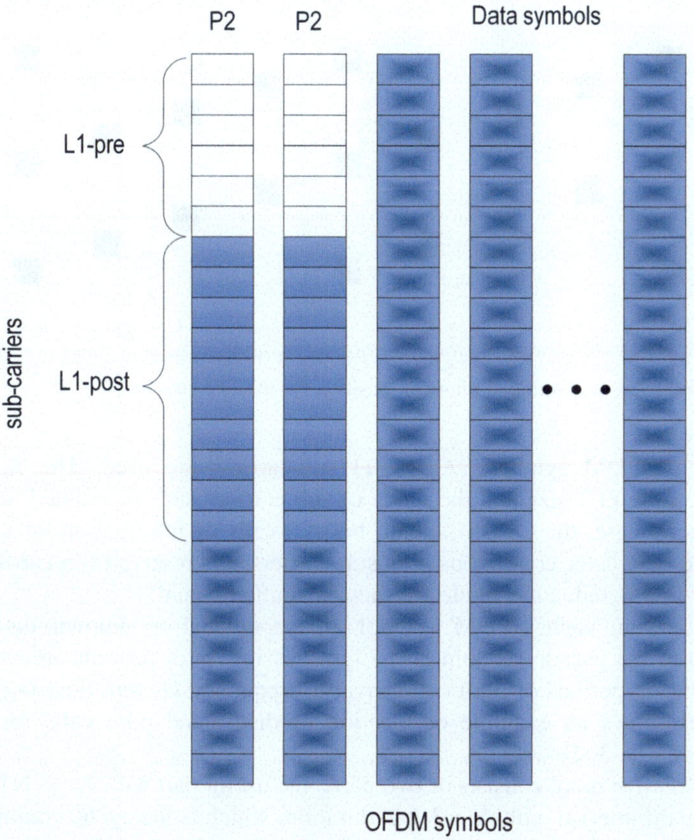

Fig. 8.6 Frequency–time structure of the T2-frame

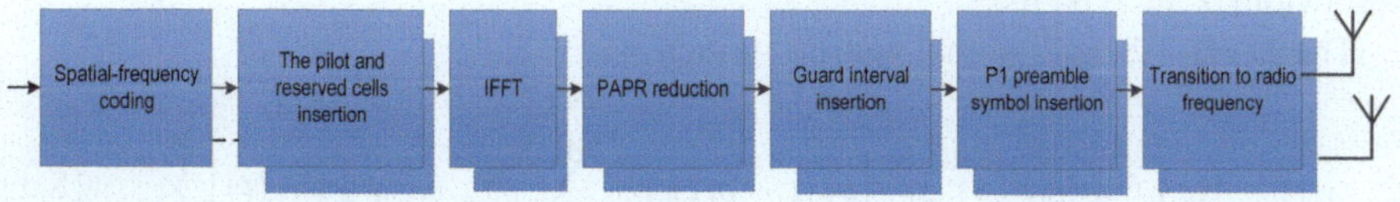

Fig. 8.7 Diagram of OFDM signal formation

Fig. 8.8 Scheme of spatial-frequency encoding

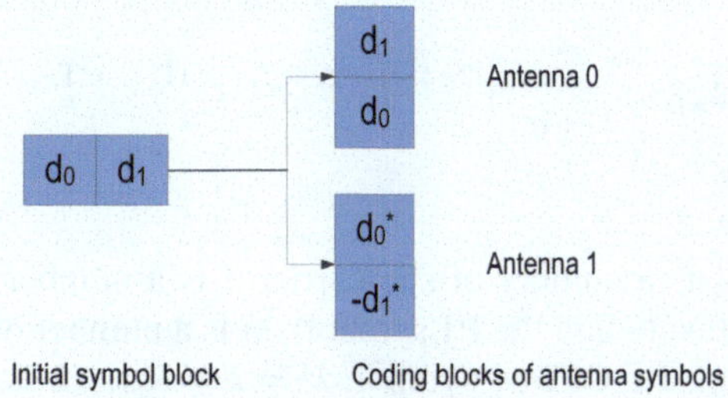

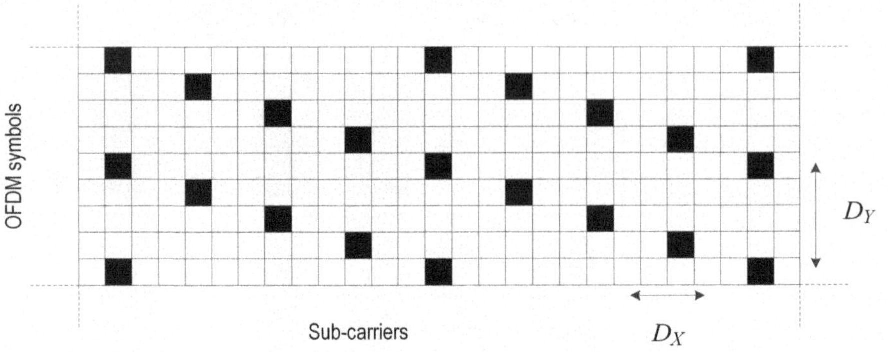

Fig. 8.9 Fragment of distributed pilot cell position for the PP1 structure

In each OFDM symbol, $(K_{\max}+1)$ sub-carriers are used. The $K_{\max}$ value depends on the FFT size and the mode of sub-carrier usage (expanded or normal).

In general case, the OFDM signal, besides cells of information data, contains distributed pilot cells, continuous pilot sub-carriers, and reserved sub-carriers (when using the PAPR reduction mode for the transmitted signal).

Pilot cells in each OFDM symbol are modulated by known data formed according to the pseudo-random law. Eight possible pilot patterns are anticipated, which differ in periods of pilot cell arrival to frequency D_X and time D_Y domains. Figure 8.9 shows an example of location of distributed pilot cells for the PP1 structure $(D_X = 3,\ D_Y = 4)$.

The OFDM symbol consists of two parts: the useful part with $T_U = \mathrm{NT}$ duration and the guard interval with $\Delta = N_{\mathrm{CP}}T$ duration, which is the cyclic continuation of the useful part and is inserted in front of it. Distance between adjacent sub-carriers is $1/T_U$. Duration of the elementary period T is the main basic parameter of the OFDM signal and is determined by its bandwidth. For instance, for 8 MHz bandwidth $T = 7/64\,\mu\mathrm{s}$.

The DVB-T2 radio signal can be written as

$$s(t) = \mathrm{Re}\left\{\exp(j2\pi f_c t)\sum_{m=0}\left[p_1(t-mT_F) + \sum_{l=0}^{L_F-1}\sum_{k=0}^{K_{\max}} c_{m,l,k}\,\psi_{m,l,k}(t)\right]\right\}, \tag{8.1}$$

where

$$\psi_{m,l,k}(t) = \begin{cases} \exp\left[j2\pi\frac{k'}{T_U}(t-\Delta-T_{P1}-\mathrm{l}T_S-\mathrm{m}T_F)\right], & t\in[\mathrm{m}T_F+T_{P1}+\mathrm{l}T_S,\ \mathrm{m}T_F+T_{P1}+(l+1)\,T_S] \\ 0, & t\notin[\mathrm{m}T_F+T_{P1}+\mathrm{l}T_S,\ \mathrm{m}T_F+T_{P1}+(l+1)\,T_S] \end{cases}, \tag{8.2}$$

k is a number of sub-carrier, l is a number of the OFDM symbol in the T2-frame (excluding the P1 symbol), m is a number of the T2-frame, $T_s = T_U + \Delta$ is duration of the OFDM symbol, $k' = k - K_{\max}/2$ is the sub-carrier index with regard to a number of the central sub-carrier, $c_{m,l,k}$ is the normalized modulation symbol, f_c is the central radio signal frequency, L_F is a number of the OFDM symbols in the T2-frame, $T_{P1} = 2048T$ is duration of the P1 symbol, $T_F = L_F T_s + T_{P1}$ is duration of the T2-frame, $p_1(t)$ is the time signal of the P1 symbol.

To reduce the peak-factor of the transmitted OFDM signal, two mechanisms are anticipated. The first one consists of reservation of some part of sub-carriers for non-informative signal transmission. It can be calculated for each OFDM symbol by the iterative algorithm in such a manner that signal power would not exceed the given threshold value. In accordance with the second method, limitation of signal power of the OFDM symbol is provided by expansion of the end points of the used modulation constellation for informative sub-carriers. This procedure allows reduction of the transmitted OFDM signal peak-factor without system capacity decrease. Both mechanisms can be used mutually and separately.

The symbol of the P1 preamble is intended for fast DVB-T2 signal detection, reception of base parameters of the P1 signaling and initial frequency–time synchronization.

Symbol structure of the P1 preamble is presented in Fig. 8.10. It represents the 1K OFDM symbol with two guard intervals. The useful part of the symbol "A" constitutes 1024 samples (elementary symbols). The first guard interval "C" is situated ahead and represents the first $M_C = 542$ samples of the useful part shifted by the $f_{SH} = 1/1024\,T$ frequency, equaled to the frequency difference of adjacent sub-carries of the P1 symbol. The second guard interval "B" is situated behind and represents the last $M_B = 482$ samples of the signal useful part, which are also shifted by the f_{SH} frequency. During formation of the symbol useful part, only 384 sub-carriers are modulated, which use the DBPSK modulation and are distributed in the signal spectrum using pseudo-random approach. Such a location of informative sub-carriers allows frequency-shift estimation up to 500 kHz (at 8 MHz signal bandwidth). Informative sub-carriers transfer two modulating field sequences of S1 and S2, in which the P1 signaling is coded.

Thus, the formed OFDM signal is transferred onto radio frequency, amplified and emitted into the air.

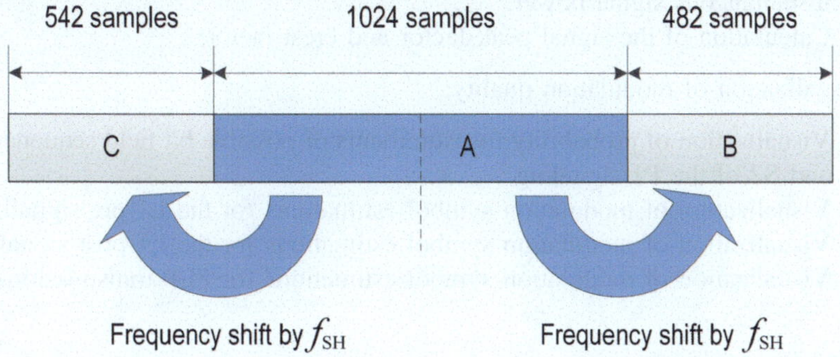

Fig. 8.10 Symbol structure of the P1 preamble

8.5 Analyzer Functional Possibilities

The analyzer of the DVB-T2 signal is made in accordance with requirements to the DVB signal measurements standard execution [11, 12]. It allows obtaining of the signals of digital TV in the given spectral range, receiving of the control data, obtaining of the informational streams (PLP), extraction of sub-streams of different programs and performing a large measurement volume of physical and transport levels.

The analyzer consists of the hardware part—the digital panoramic measuring receiver of the ARGAMAK family [4, 13], and the software part—programs of system mathematical support. The last part can be installed on the personal computer using the unified package together with other software, for example, performing the analysis of DVB-T/H signals and signals of base stations of different cellular communication systems.

The following functions are realized in the analyzer:

- Detection of the DVB-T2 signal of given bandwidth at the given frequency;
- Reception of control data including P1, L1-pre-, L1-post signalling;
- Reception of PLP streams;
- Reception of service information tables PSI/SI of the transport streams;
- Extraction of separate program streams;
- Measurement of the full bitrate of the transport stream, the bitrate of each program, and the bitrate of each type of packets (PID);
- Spectral measurements:
  - High-accuracy estimation of the frequency shift,
  - Visualization of the DVB-T2 signal spectrum,
  - Detection of narrowband interference in the signal bandwidth;
- Power measurements:
  - Estimation of the signal mean power,
  - Formation of the distribution function of the instantaneous signal power,
  - Formation of the complementary cumulative distribution function of the instantaneous signal power,
  - Calculation of the signal peak-factor and crest-factor;
- Visualization of modulation quality:
  - Visualization of probability measurements of possible bit field sequences S1 and S2 of the P1 signaling,
  - Visualization of modulation symbol estimations for the L1-pre signaling,
  - Visualization of modulation symbol estimations for the L1-post signalling,
  - Visualization of modulation symbol estimations for PLP transmissions;

- Measurement of modulation quality of the L1-pre, L1-post signalling and various PLP including:
  - Estimation of MER and EVM,
  - Estimation of the system target error (STE),
  - Estimation of the amplitude imbalance (AI),
  - Estimation of the quadrature error (QE),
  - Estimation of the phase jitter (PJ);
- Channel measurements:
  - Estimation of SNR,
  - Estimation of the BER before the LDPC decoder,
  - Estimation of the BER before the BCH decoder,
  - Estimation of the FER for BB-frames,
  - Estimation of the PER for user packet (UP) of**** transmitted PLP (only for Normal Mode, GFPS, or TS),
  - Estimation of the channel frequency response (amplitude, phase, and delay),
  - Estimation of the channel impulse response,
  - Estimation of the power delay (multi-path) profile and the RMS delay spread (effective channel length);
- Analysis of the transport streams including measurements of the first, second, and third priorities, etc;
- Analyzer allows according to executed measurements:
  - Realization of the covering zone of the DVB-T2 network taking into account the district digital map data,
  - Determination of detected transmitting station localization;
- Formation of the analysis reports in different formats: XML, MS-Word, Excel. The flexible choice of the outlet information and forms of its representation.

8.6 Reception of the DVB-T2 Signal

The analyzer input radio signal is amplified, down-converted into the baseband, filtered to suppress higher harmonics and to transform the initial signal band to the DVB-T2 signal band. Then, the analog-to-digital conversion and its oversampling are carried out. As a result, the discrete baseband signal $x_i,\ i = 0, 1, 2, \ldots$ is formed with the sample frequency of $1/T$ corresponding to one sample per elementary interval. The processing sequence of the obtained baseband signal is presented in Fig. 8.11.

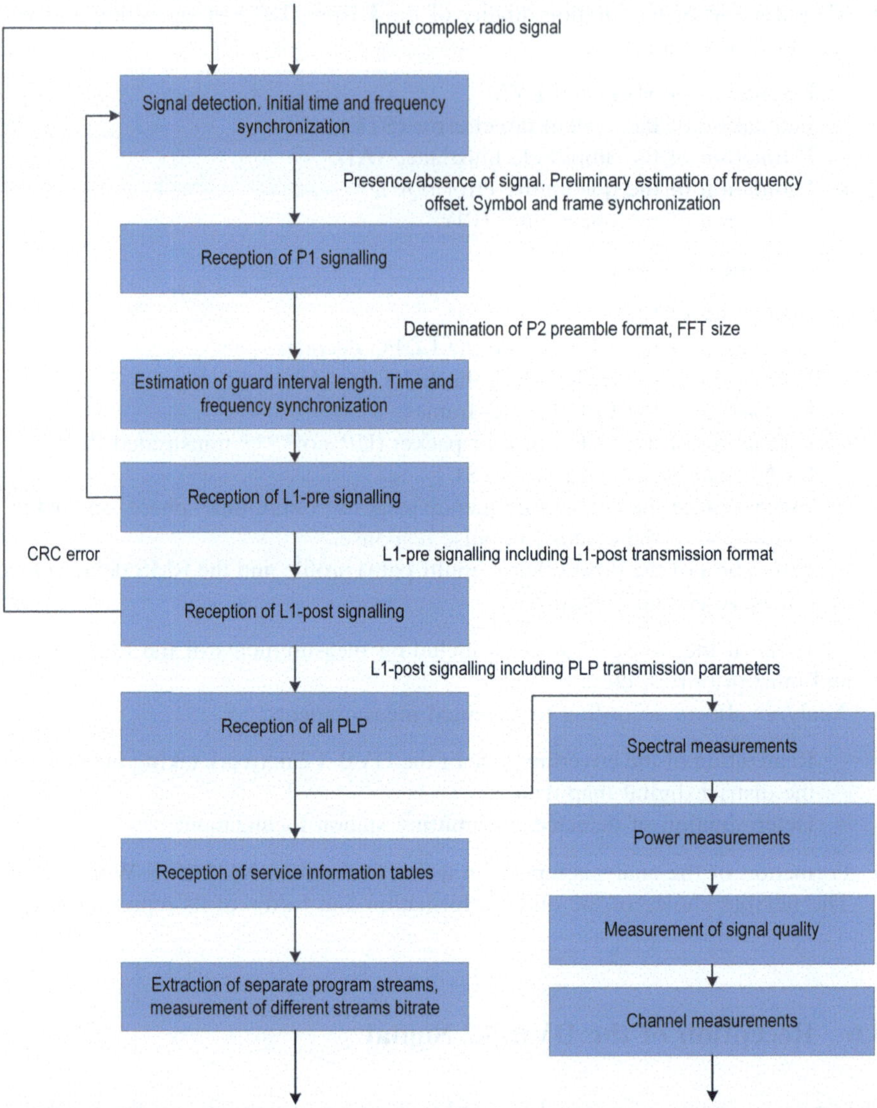

Fig. 8.11 Processing sequence of the DVB-T2 signal

8.6.1 Detection of the DVB-T2 Signal and Reception of Signaling Data

Detection of the DVB-T2 signal and initial frequency–time synchronization are carried out according to the P1 symbol located in the beginning of the each T2 frame. For this, the output discrete signals of two correlators are formed, which correspond to the parts of the "C" and "B" of preamble

$$Q_n^{(C)} = \sum_{i=0}^{M_C-1} x_{n+i} x^*_{n+i-M_C} \exp[j2\pi(i+n-M_C)\mathrm{T}f_{\mathrm{SH}}], \quad (8.3)$$

$$Q_n^{(B)} = \sum_{i=0}^{M_B-1} x_{n+i} x^*_{n+i-M_B} \exp[-j2\pi(i+n)\mathrm{T}f_{\mathrm{SH}}], \quad (8.4)$$

and decision function for the preamble search

$$R_n = \left|Q_n^{(C)}\right|^2 + \left|Q^{(B)}_{n+M_C+M_B}\right|^2. \quad (8.5)$$

Then we determine the maximal value of (8.4) $R_{\max} = \max R_n$, the position of this maximum $n_{\max} = \arg\max R_n$, as well as the mean value of the interval under investigation $\overline{R} = \langle R_n \rangle$. In the case, when the inequality $R_{\max} > h\,\overline{R}$ is satisfied, we make a preliminary decision about the presence of the DVB-T2 signal in the observed realization, and $n_{\max}$ is considered as the first sample of the useful part of the P1 symbol, h is the detection threshold. Otherwise, after given number of failure attempts, the decision is made about absence of the DVB-T2 signal in the analyzed frequency channel.

In the case of the DVB-T2 signal detection, the initial frequency synchronization is carried out. The frequency offset can be represented in the form of $F = F_1 + F_2$, where $F_2 = m_F f_{\mathrm{SH}}$ is a value divisible by f_{SH}, and $|F_1| < f_{\mathrm{SH}}$. Taking into consideration the time structure of the P1 symbol, according to the phase-difference method [14], we find estimation of the first frequency offset

$$F_1 = \frac{1}{2\pi(M_C+M_B)T} \arg\left\{Q^{(C)}_{n_{\max}} \, Q^{(B)}_{n_{\max}+M_C+M_B}\right\}. \quad (8.6)$$

Baseband signal samples are exposed to phase correction according to (8.6) estimation. The value m_F estimation of the frequency offset requires the P1 symbol analysis in the frequency domain. For this, we perform the FFT of the sample block with 1024 in length starting from the $n_{\max}$ sample

$$X_k = \sum_{i=0}^{1023} x_{n_{\max}+i} \exp\left[\frac{-j2\pi i(k-426)}{1024}\right], \quad (8.7)$$

where $k = \overline{0.1023}$ is the sub-carrier number. The m_F value is estimated according to the function maximum position

$$Y_{m_F} = \sum_{k=0}^{1023} \left| X_{(k+m_F)\bmod 1024} \right|^2 \gamma_k, \tag{8.8}$$

taking into account the known position of the active sub-carriers of the P1 symbol, $\gamma_k = \begin{cases} 1, & k \in \mathrm{KP1} \\ 0, & k \notin \mathrm{KP1} \end{cases}$, $KP1$ is an array of active sub-carrier numbers.

To receive the P1 signaling from (8.7), we extract the active sub-carrier array, then its de-scrambling and demodulation are performed. As a result, the soft solutions of the bit information fields S1 and S2 are formed. According to them, we calculate the probability measures for possible bit sequences of these fields. The most likelihood sequences determine the informational bits of the P1 symbol.

After reception of the P1 signaling, the FFT size N and the possible set of the prefix size are known. The N_{CP} estimation from this set is carried out in the similar manner as [3].

The time tracking, at reception of the control signaling and data, is carried out once per the T2-frame by the pilot cells of the P2 symbols. For time estimation of the FFT block start, we form the function

$$G_j = \sum_{m=0}^{N_{P2}-1} \left| \sum_{i=0}^{N-1} x_{j+i+m(N+N_{\mathrm{CP}})} s1_i^* \right|^2, \tag{8.9}$$

where the reference signal in the time domain $s1_i$, $i = \overline{0,\ N-1}$ is matched with the pilot signal of the OFDM P2 symbols, and it can be calculated a priori, N_{P2} is a number of P2 symbols depending on the FFT size. The $G_{j_{\max}}$ maximum, its position $j_{\max}$, and the threshold $h_S = v\,G_{j_{\max}}$ of the decision function (8.9) are determined, v is the relative threshold. The argument minimum of the decision function (8.9) j, which satisfies the condition $G_j > h_S$, is considered as the first sample of the useful part of the OFDM P2 symbol, and it determines the FFT block start of all symbols of the frame $j_m^{(\mathrm{FFT})}$, $m = \overline{0,\ L_F - 1}$.

FFT is performed with each block corresponding to the useful part of OFDM symbols. The residual frequency offset is estimated according to the pilot sub-carriers of the P2 symbols by the phase-difference method. After its compensation, the estimation of the channel frequency response by the method of the linear interpolation of the reference pilot sub-carriers is performed.

The soft decisions of the informative cells of the P2 symbols are

$$Z_{m,k} = \frac{X_{m,k}}{A_{m,k}}, \tag{8.10}$$

where $X_{m,k}$ is the FFT result, $A_{m,k}$ is an estimation of the channel frequency response, m is a number of the OFDM symbol, k is the sub-carrier number. From (8.10), one can extract the soft decision array of informative cells of the OFDM P2 symbols by means of elimination of pilot sub-carriers, reserved and two pairs of edge sub-carriers.

Then, for each OFDM P2 symbol, we perform the frequency de-interleaving of the soft decision of the cell symbols and the channel estimation according to the known rule depending on the FFT size.

8.6.2 Reception of the L1-Pre Signaling

To receive the L1-pre signaling from the array obtained after frequency de-interleaving, we can extract a vector of the L1-pre signaling $Z_i^{(\mathrm{pre})}$, $i = \overline{0,\ 1839}$. Similarly, the cell power vector $P_i^{(\mathrm{pre})}$, $i = \overline{0,\ 1839}$ is formed. For these, the soft decisions of the bit L1-pre-message are formed

$$\mu_i = P_i^{(pre)} \mathrm{Re} Z_i^{(\mathrm{pre})},\ i = \overline{0,\ 1839}. \tag{8.11}$$

The soft decision vector (8.11) consists of three parts: first 200 elements correspond to message informational bits, the next 168 elements—to check-up BCH bits, and the last 1472 elements—to check-up LDPC bits. The vector (8.11) expands in special manner up to the length of the coding block $N_{\mathrm{ldpc}} = 16,200$ for compensation of bits rejected after the LDPC encoder in the transmitter. The obtained block passes to the LDPC decoder and then to the BCH decoder. After decoding, the first 200 bits are extracted, which represent the informative block of the L1-pre signaling. According to its bits, the check-up bits are formed in conformity with the CRC. Formed and received check-up bits are compared. If all appropriate bits coincide, we make a decision about faultless reception of the L1-pre signaling, from the fields of which we can extract necessary parameters. Otherwise, the detection procedure and further operations are anew performed. If the L1-pre signaling is received with error for a given number of cases, the operation with this frequency channel is interrupted due to poor reception conditions.

8.6.3 Reception of the L1-Post Signalling

In the L1-pre signaling, a series of parameters is transmitted, which are used for reception of the L1-post signalling: a number of logical cells $N_{\mathrm{MOD\_total}}$ in the signaling, the type of modulation, a number of informative bits in the signaling, an indicator of scrambling presence, etc.

To receive the L1-post signalling from obtained after the array frequency de-interleaving of soft decisions of informative P2 symbols, we extract the vector of the L1-post signalling and the appropriate vector of cell powers $\left\{Z_i^{(\mathrm{post})}\right\}, \left\{P_i^{(\mathrm{post})}\right\}$, $i = \overline{0,\ N_{\mathrm{MOD\_total}} - 1}$. Using elements of these vectors, the soft decisions of L1-postbits signaling are formed in conformity with the used modulation. For instance, for QPSK, the soft decisions of coded bits of the ith cell of the L1-post signalling are

$$\mu_{i,0} = P_i^{(\text{post})} \operatorname{Re} Z_i^{(\text{post})}, \; \mu_{i,1} = P_i^{(\text{post})} \operatorname{Im} Z_i^{(\text{post})} \tag{8.12}$$

Then, multiplexing of bit soft decisions is performed according to the rule depending on the modulation type. As a result, the vector of bit soft decisions of the L1-post signalling is formed. It divides into calculated number of coding blocks of signaling.

At using of 16-QAM or 64-QAM modulation, we apply the bit interleaving of the coding blocks. For this, recording of elements of each coding block into the matrix is carried out, sequentially in rows. The output vector is formed by means of reading of matrix elements in columns. The matrix parameters depend on the length of the coding block and on the used modulation. For BPSK and QPSK modulation, the bit interleaving is not performed.

The vector obtained after de-interleaving expands in the special manner up to the length of the coding block $N_{\text{ldpc}} = 16\,200$ for compensation of bits rejected after the LDPC decoder in the transmitter. The expanded block passes to the LDPC decoder and then to the BCH decoder. From blocks obtained after decoding, we extract bits of the L1-post signalling. These bits are exposed to de-scrambling. Then, de-scrambled bits of all coding blocks are united, and as a result, the bit vector of the L1-post signalling is formed. Using its bits, the check-up bits are formed in conformity with the CRC. Formed and received check-up bits are compared. If all appropriate bits coincide, the decision about faultless reception of the L1-post signalling is made, from fields of which the necessary parameters are extracted. Otherwise, the reception procedure for the L1-post signalling is performed for the next frame. In the case of unsuccessful L1-postreception during the given number of frames, the operation with this frequency channel is interrupted due to poor reception conditions.

8.6.4 PLP Reception

A sequence of operations at PLP reception is presented in Fig. 8.12.

The frequency tracking is performed once per frame using its permanent pilot sub-carriers. A number and a position of the continual pilot sub-carrier set (Ω_{CP}) is known and depends on the FFT size and on the scheme of the pilot pattern PP1–PP8. Estimation of the frequency offset can be fulfilled by the phase-difference method:

$$F_2 = \frac{1}{2\pi(N + N_{\text{CP}})T} \arg\left[\sum_{k\in\Omega_{\text{CP}}} \sum_{q=N_{P2}}^{L_F-2} X_{q+1,k} c_{q+1,k} X^*_{q,k} c_{q,k}\right], \tag{8.13}$$

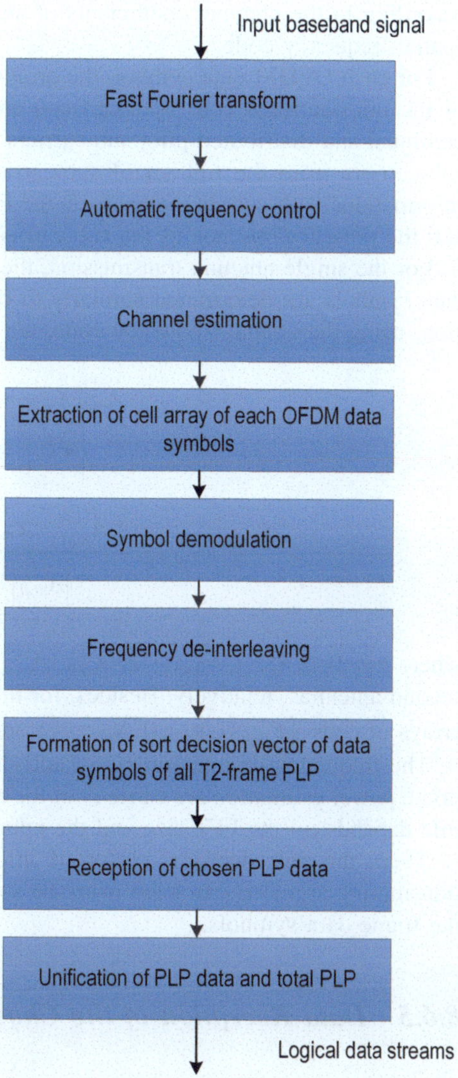

Fig. 8.12 Operation sequence at PLP reception

where $X_{q,k}$ is the FFT result, $c_{q,k}$ is the transmitted real symbol in the pilot cell of the qth OFDM symbol of the kth sub-carrier.

For channel estimation, at first, we form the reference channel coefficients of distributed pilot cells of the frame as the ratios of the FFT result to the transmitted symbols of these cells. OFDM data symbols of the frame are divided into blocks with the length of D_Y symbols. For the single-antenna transmission, the estimation of the channel frequency response is determined by the linear interpolation method for reference channel block coefficients. For double-antenna transmission, the estimations of the channel of the first and the second antennas are determined according to the channel coefficients of adjacent pilot cells of the block and their spatial-frequency code.

For each OFDM data symbol, the arrays of FFT result and channel estimations of the informational cells (sub-carriers) are formed by means of rejection of the continual and distributed pilot sub-carriers, the edge sub-carriers, and the reserved sub-carriers from the full set of sub-carriers $k \in [0,\ K_{\max}]$. The number N_{data} of informational sub-carriers depends on the FFT size, the scheme of the pilot pattern, and the presence/absence of the sub-carrier reservation mode.

For the single-antenna transmission, the soft decisions of informational cells of data symbols are determined similarly to (8.10). For the double-antenna transmission, using the spatial-frequency code, we obtain

$$Z_{m,p} = \frac{A_{m,p}^{(1)*} X_{m,p} + A_{m,p+1}^{(2)} X_{m,p+1}^{*}}{\left|A_{m,p}^{(1)}\right|^2 + \left|A_{m,p+1}^{(2)}\right|^2}, \tag{8.14}$$

$$Z_{m,p+1} = \frac{A_{m,p+1}^{(1)*} X_{m,p+1} - A_{m,p}^{(2)} X_{m,p}^{*}}{\left|A_{m,p}^{(1)}\right|^2 + \left|A_{m,p+1}^{(2)}\right|^2}, \tag{8.15}$$

where $p = 0, 2, 4, \ldots, N_{\text{data}} - 2$, $A_{m,p}^{(1)}$, $A_{m,p}^{(2)}$ are channel estimations of the first and second antennas, relatively. Besides, for informational cells, the power estimations arrays of cells $\{P_{m,p}\}$, $p = \overline{0, N_{\text{data}} - 1}$ are formed for the decoding procedure.

The frequency de-interleaving of soft decisions of cell symbols and the appropriate power estimation are carried out for each OFDM symbol. The de-interleaving rule depends on the FFT size and the number of the OFDM symbol in the frame.

Then, the soft decision vector of informational symbols of all PLP of the transmitting signal is extracted from the soft decision array of the P2 symbols and the frame data symbols.

8.6.5 Data Reception of the Chosen PLP

For data reception of chosen PLP, we can use parameters of the L1 signaling: the stream type, the position of stream beginning, the number of coding blocks and T2 frames in the interleaving frame, the modulation, the coding block size and the coding rate, a number and the fragment period in the frame for PLP type 2, a number of frames in the super-frame, a period of T2-frames used, and so on.

The operation sequence for logical data stream obtaining of chosen PLP is presented in Fig. 8.13.

According to parameters of the L1 signaling, we determine the number of frames in the super-frame, in which the chosen PLP is transmitted, and the soft decisions of informative symbols of this stream frame are extracted. Then the number of interleaving frames in the super-frame is determined, and arrays of soft decisions

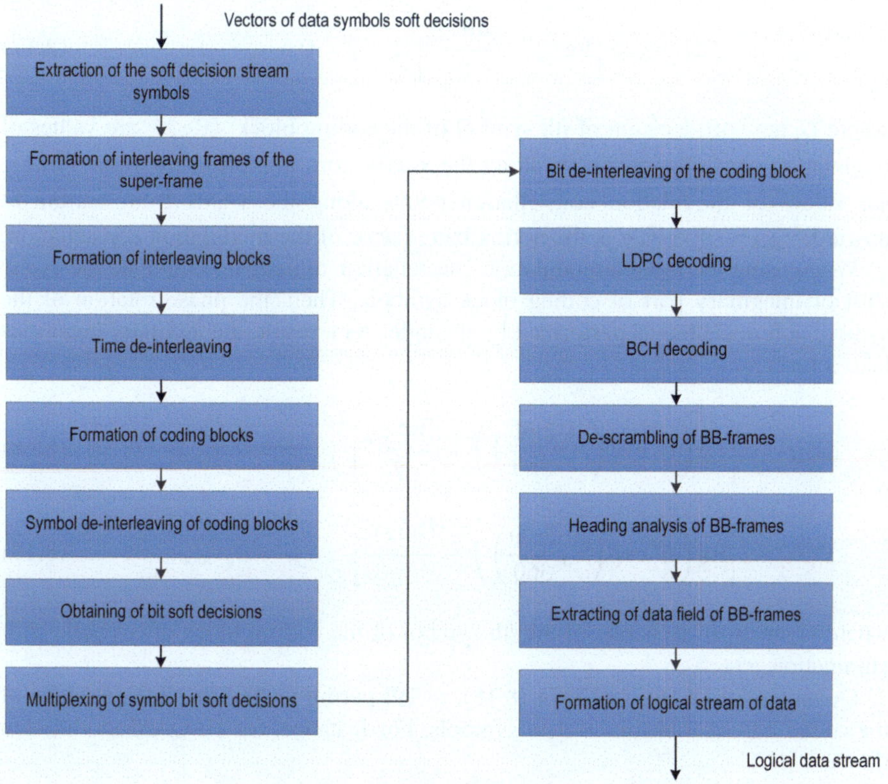

Fig. 8.13 Operation sequence for logical data stream obtaining of chosen PLP

of informational symbols are formed for each interleaving frame. More than one interleaving block may be in the interleaving frame. In this case, the soft decisions of their symbols are extracted. For each interleaving block, the time de-interleaving of symbol soft decisions is carried out by means of recording of its elements into a matrix on rows and reading on columns. A number of rows and columns are determined by the modulation type and the coding block size. Each interleaving block obtained after time de-interleaving is divided into several coding blocks. Then, for each coding block, the symbol de-interleaving is carried out by the known rule depending on a number of the coding block in the interleaving block and its length.

To obtain soft decision of coding block bits when using the usual modulation constellation, at first, we calculate the following auxiliary quantities

$$\Omega_q^x(i) = \left[\operatorname{Re} G_i - \frac{(\operatorname{Re} z)_q}{\gamma}\right]^2, \quad q = \overline{0, v_{\text{mod}} - 1}; \tag{8.16}$$

$$\Omega_q^y(i) = \left[\operatorname{Im} G_i - \frac{(\operatorname{Im} z)_q}{\gamma}\right]^2, \quad q = \overline{0, v_{\text{mod}} - 1}, \tag{8.17}$$

where G_i are soft decision of ith symbol of the coding block, $(\operatorname{Re} z)_q$ are values of modulation constellation points along the x-axis from the left to the right, $(\operatorname{Im} z)_q$ are values of modulation constellation points along the y-axis from bottom to upward, $v_{\text{mod}} = 2^{\eta_{\text{mod}}/2}$, γ is the normalizing factor of the modulation constellation.

When using the rotated modulation constellation, at first, we eliminate the cyclic shift of imaginary part of coding block symbols. Then, the phase rotation of the symbol soft decisions is performed by Φ angle. As a result, the auxiliary quantities take a form

$$\Omega_q^x(i) = \left[\operatorname{Re}\left(G_i' \exp\left(-j\frac{2\pi\Phi}{360}\right)\right) - \frac{(\operatorname{Re} z)_q}{\gamma}\right]^2, \quad q = \overline{0, v_{\text{mod}} - 1}; \tag{8.18}$$

$$\Omega_q^y(i) = \left[\operatorname{Im}\left(G_i' \exp\left(-j\frac{2\pi\Phi}{360}\right)\right) - \frac{(\operatorname{Im} z)_q}{\gamma}\right]^2, \quad q = \overline{0, v_{\text{mod}} - 1}, \tag{8.19}$$

where G_i' are soft decisions of the ith symbol of the coding block after cyclic shift elimination.

Quantities of (8.16), (8.17) or (8.18), (8.19) permit to find the soft decisions of the coded bits of transmitted data symbols. For instance, for 16-QAM modulation they are

$$\mu_{i,0} = P_i\left[\min\left(\Omega_0^x(i), \Omega_1^x(i)\right) - \min\left(\Omega_2^x(i), \Omega_3^x(i)\right)\right], \tag{8.20}$$

$$\mu_{i,1} = P_i\left[\min\left(\Omega_0^y(i), \Omega_1^y(i)\right) - \min\left(\Omega_2^y(i), \Omega_3^y(i)\right)\right], \tag{8.21}$$

$$\mu_{i,2} = P_i\left[\min\left(\Omega_1^x(i), \Omega_2^x(i)\right) - \min\left(\Omega_0^x(i), \Omega_3^x(i)\right)\right], \tag{8.22}$$

$$\mu_{i,3} = P_i\left[\min\left(\Omega_1^y(i), \Omega_2^y(i)\right) - \min\left(\Omega_0^y(i), \Omega_3^y(i)\right)\right]. \tag{8.23}$$

Soft decisions of each symbol bits of the coding block or the symbol pair of the coding block (depending on modulation used and the block size) are interleaved. After that, bit soft decisions of the coding block are multiplexed in one vector according to a known rule.

Then, the bit de-interleaving of the coding block is performed. In general case, this is fulfilled sequentially by two de-interleaving devices. At first de-interleaving, the input vector elements are written on rows into the matrix, which parameters are defined by the modulation type and the coding block size. The output vector elements of the first de-interleaving device are read with a shift on the matrix columns. The second de-interleaving device interleaves the check-up LDPC bits of the coding block according to a known rule. Its parameters depend on a coding rate.

The obtained block of bit soft decisions passes to the LDPC decoder and then to the BCH decoder. Decoding parameters are known from the L1 signaling.

After decoding, we obtain the bit block—BB-frame. Its de-scrambling is carried out. The BB-frame header is analyzed, which results in determination of the BB-frame formation mode (normal or high efficiency), the stream type in PLP, a length in bits of the BB-frame data field, the beginning position of the first used UP packet in the BB-frame, a presence of the stream synchronization and the elimination field of zero packets. According to the header parameters, the data field bits of each BB-frame are extracted. The logical data PLP stream is formed, in the general case, by means of modified fragments of the sequential BB-frames multiplexing. Modification depends on a combination of the stream type and the BB-frame formation mode. In particular, for TS and normal mode, the bit stream of expanded UP is formed from the data field stream of BB-frames. Then we reject the check-up bits of each UP, and also, if necessary, the synchronization fields of the input stream (ISSY) and a counter of zero packets. Then, the synchro-bytes are inserted, if they had been rejected, and zero packets, if they had been eliminated at transmission.

At presence of the mutual PLP, its multiplexing with PLP data is performed. This is necessary for programs data reproduction included in PLP. The multiplexed stream is formed by means of replacement of the zero PLP packets by the corresponding nonzero packet of the mutual PLP. If packets do not satisfy this condition, the packet of PLP data remains.

8.7 Example of DVB-T2 Signal Analyzer Operation

Let us discuss some results of DVB-T2 analyzer operation, which is implemented on the hardware base of the panoramic digital measuring receiver ARGAMAK-IS, as shown in Fig. 8.14. The other digital receivers of ARGAMAK family, for instance, ARGAMAK-M, ARC-D11 [13] can be also used as the hardware base.

The measuring receiver ARGAMAK-IS has functions of synchronization according to signals of global navigational satellite systems GPS and GLONASS, which allows reduction of frequency instability of the reference oscillator from 10^{-10} to 2×10^{-12} and realization of high-accuracy measurement of the receiving signal frequency.

Algorithm of frequency measurement of the digital TV signal is based on the phase-difference method, which functions in the presence of AFC at reception of the DVB-T2 signal and does not require the specific operation mode. An error of the signal carrier measurement for single-frequency broadcasting network at synchronization from GNSS does not exceed $\pm 5 \times 10^{-11}$. At that, error of level measurement is not more than ± 1 dB, and error of the check-up frequency band is $\pm 0.1\%$. Magnitudes of measurement errors satisfy requirements of the Russian

Fig. 8.14 Radio receiver ARGAMAK-IS (below) with the wideband receiving antenna ARC-A7A-3 (on the right) and the navigation GNSS antenna (on the left)

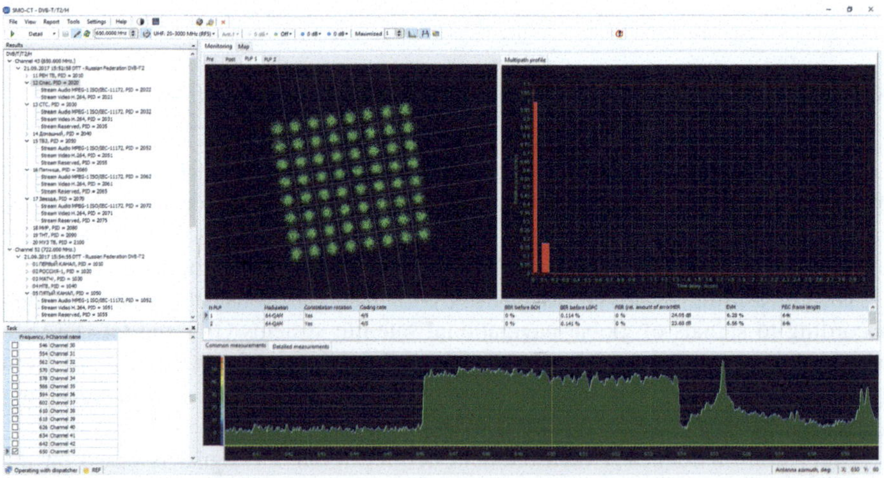

Fig. 8.15 Interface window of the analyzer

standards 55696–2013, 5939–2014, and acting Russian General Radio Frequency Centre norms [15].

Figure 8.15 shows the variant of the main window picture of the analyzer interface. In the left window part, the treelike structure of the received programs of the TV signal transport streams at the 722-MHz frequency is present. In the right part, the signal constellation is shown, which gives a representation of the modulation type and the quality of the received signal, the power delay (multi-path) profile indicating a number and the delay patches (effective channel lengths), symbol estimations of the second transport PLP 2 stream, as well as the signal spectrum. The interface window has the flexible adjustments for a choice of shown elements, graphs, tables, and their content.

Possible version of the report about analyzer operation results is presented in Table 8.1.

8.8 High-Accuracy Frequency Estimation of Digital TV Carrier

In the digital TV systems, the strict requirements are made to the accuracy of the signal carrier frequency. This relates to the fact that, in the general case, a receiver may simultaneously receive signals from different transmitters working in the single-frequency channel (single-frequency network). In this case, for effective TV signal reception, the maximal admissible deviation of the carrier frequency should be not more than 1 Hz for the DVB-T/H standard and not more than 0.5 Hz for the DVB-T2 standard [15, 16].

In this connection, the measuring equipment should have the frequency measurement error not worse than 0.1–0.05 Hz.

High-accuracy estimation of the digital TV signal carrier frequency includes the following procedures [17]:

- Search for the digital TV signal;
- Initial time and frequency synchronizations;
- Estimation of the FFT size and the guard interval size;
- Fast discrete Fourier transform;
- Time and frequency tracking;
- High-accuracy estimation of the carrier frequency.

The block diagram of high-accuracy estimation of the TV signal carrier frequency is presented in Fig. 8.16.

To obtain the input digital baseband signal, the received radio signal is amplified, and the analog-to-digital conversion, down-conversion into baseband, filtering, and oversampling are carried out.

The DVB-T/H signal detection and estimation of N, N_{CP} are based on the prefix presence in each OFDM symbol. The correlation functions of repeated segments are formed for all possible sets of N, N_{CP}. The absolute maximum of these functions is determined, which is compared to a threshold. In case of its exceeding, the decision is made about presence of the TV signal. The pair N, N_{CP} corresponding to this maximum is an estimation of the FFT size and the guard interval size, which unambiguously define the totality of pilot cells.

Table 8.1 DVB-T2 analyzer operation results

	Parameter of TV signal
Date	08.07.2016
Time	19:15:36
Signal type	DVB-T2
Channel	Канал 52
ID of transmitting station	13604
NID	13583
ONID	8835
Network name	DTT-Russian Federation
Frequency, MHz	722
Frequency offset, Hz	0.03
Measured frequency, MHz	722.00000003
Bandwidth, MHz	7.61875
Level, dB μW	62.3
SINR, dB	46
Calibration factor, dB	18.2
Field strength, dB μV /m	80.5
Number and mode of sub-carrier usage	32k
Guard interval	1/16
Modulation type of L1 post	QPSK
Pilot pattern	PP4
Number of transmitting antennas	1
Number of transport stream	3
--------Transport stream1--------	
TSID	1
Modulation	64-QAM
Constellation rotation	Yes
Coding rate	4/5
FEC frame length	64k
BER before LDPC	0.313 %
BER before BCH	0 %
FER	0 %
MER	20.08 dB
EVM	9.98 %
Transport stream 2	
TSID	2
Modulation	64-QAM
Constellation rotation	Yes
Coding rate	4/5
FEC frame length	64k
Constellation rotation	Yes
Coding rate	4/5
FEC frame length	64k
BER before LDPC	0.284 %
BER before BCH	0 %
FER	0 %
MER	21.59 dB
EVM	5.47 %

Transport stream data

PID of program	Russian program	Stream	PID of stream
1010	01 FIRST CHANNEL	Video H.264	1011
		Audio MPEG-1 ISO/IEC-11172	1012
		Teletext	1014
		Reserved	1015
1020	02 RUSSIA-1	Video H.264	1021
		Audio MPEG-1 ISO/IEC-11172	1022
		Audio MPEG-1 ISO/IEC-11172	1023
		Teletext	1024
		Reserved	1025
1030	03 MATCH!	Video H.264	1031
		Audio MPEG-1 ISO/IEC-11172	1032
		Teletext	1034
		Reserved	1035
1040	04 NTV	Video H.264	1041
		Audio MPEG-1 ISO/IEC-11172	1042
		Teletext	1044
		Reserved	1045
1050	05 FIFTH CHANNEL	Video H.264	1051
		Audio MPEG-1 ISO/IEC-11172	1052
		Reserved	1055
1060	06 RUSSIA-K	Video H.264	1061
		Audio MPEG-1 ISO/IEC-11172	1062
		Teletext	1064
1070	07 RUSSIA-24	Video H.264	1071

Table 8.1 (continued)

BER before LDPC	0.308 %
BER before BCH	0 %
FER	0 %
MER	19.76 dB
EVM	10.32 %
Transport stream 3	
TSID	3
Modulation	64-QAM

		Audio MPEG-1 ISO/IEC-11172	1072
1080	08 CARUSEL	Video H.264	1081
		Audio MPEG-1 ISO/IEC-11172	1082
		Teletext	1084
1090	09 OTR	Video H.264	1091
		Audio MPEG-1 ISO/IEC-11172	1092
1100	10 TV CENTR	Video H.264	1101
		Audio MPEG-1 ISO/IEC-11172	1102
		Teletext	1104
		Reserved	1105
1110	VESTI FM	Audio MPEG-1 ISO/IEC-11172	1112
1120	MAYAK	Audio MPEG-1 ISO/IEC-11172	1122
1130	Russian Radio	Audio MPEG-1 ISO/IEC-11172	1132

Initial time synchronization is necessary for accurate determination of OFDM symbols beginning. It is based on the analysis of the filter output signal matched with the pilot component of the OFDM signal.

Initial estimation of the frequency offset may be represented in the form $F = F_1 + F_2$, where F_2 is a quantity divisible by the frequency difference of adjacent sub-carriers $1/\mathrm{NT}$, and $|F_1| < 1/\mathrm{NT}$.

Estimation of the first frequency offset F_1 is determined in accordance with the phase difference between repeated parts of the OFDM symbols. The frequency shift F_2 is obtained according to the maximum of the correlation function of assumed pilot sub-carriers of two adjacent OFDM symbols. Baseband signal samples are corrected in conformity with estimations of F_1 and F_2. Processing algorithms for DVB-T/H are presented in detail in [3, 4].

Signal detection and initial frequency–time synchronization for DVB-T2 are carried out according to the P1 signal located in the beginning of the each T2-frame. For this, we use the specific P1 symbol structure.

Then, we perform the reception of the P1 symbol signaling, which contains the FFT with size N and the possible set of the prefix size values. Estimation of N_{CP} from this set is carried out in the same manner as for DVB-T/H [3].

Time tracking for DVB-T/H and DVB-T2 is provided continuously, step-by-step, according to the signal pilot component in the similar manner as the initial time synchronization for DVB-T/H. The difference is in the fact that it is expedient to use the estimation of the first sample position of the OFDM symbol's useful part of the previous step of the time tracking as the a priori interval center.

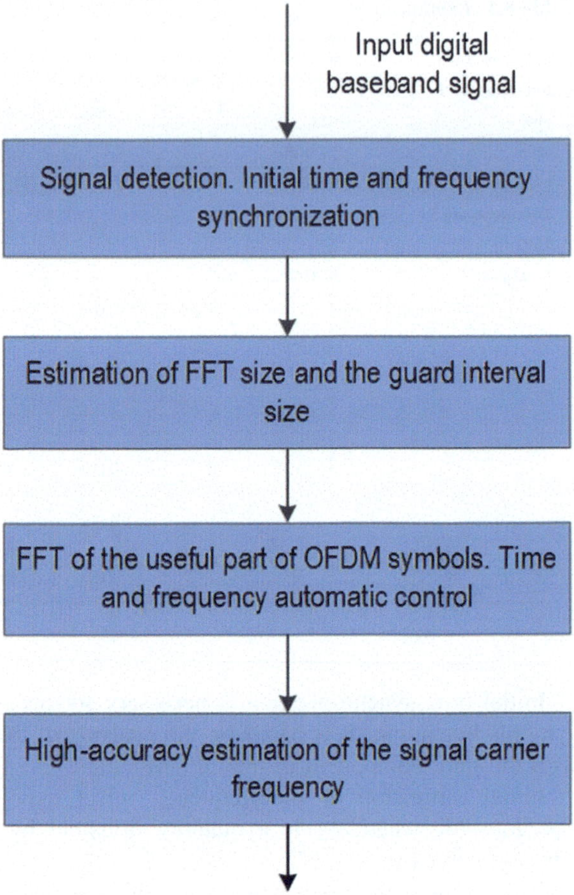

Fig. 8.16 Diagram of high-accuracy estimation of the carrier frequency

For frequency tracking over each block corresponding to the useful part of the OFDM symbol, the discrete fast Fourier transform is carried out.

The automatic frequency tracking is carried out for DVB-T/H and DVB-T2 according to continual signal pilot sub-carries. Estimation of the frequency shift is pronounced periodically (with a period of I frames) by means of the phase-difference method [14]. Frequency offset estimation of the pth stage of the frequency control has a form:

$$f_p = \frac{1}{2\pi T_{\mathrm{fr}}} \arg(U_p), \tag{8.24}$$

$$U_p = \sum_{i=\mathrm{pI}+1}^{\mathrm{pI}+I-1} \sum_{k\in\Omega_{\mathrm{CP}}} \sum_{q} X_{q,k}^{(i)} X_{q,k}^{(i-1)*}, \tag{8.25}$$

where $X_{q,k}^{(i)}$ is the FFT result, q is the number of the OFDM symbol in the frame, k is a number of the sub-carrier, i is the frame number, T_f is the frame duration, I is a number of frames at each stage of the automatic frequency control.

The detected offset is compensated by the baseband signal phase correction.

The high-accuracy estimation $\widetilde{F}$ of the frequency offset between central frequencies of the receiver and the receiving digital TV signal is performed by addition to the preliminary estimation of the frequency offset F of the additional calculating quantity

$$F_p = \frac{1}{2\pi T_{\mathrm{fr}}} \arg\left[\sum_{m=1}^{p} U_m \exp\left(j2\pi T_{\mathrm{fr}} \sum_{n=1}^{m-1} f_n\right)\right]. \tag{8.26}$$

The procedure (8.26) realizes the phase-difference method over whole time interval under analysis. Utilization of distributed pilot cells in this procedure does not result in the noticeable growth of estimation accuracy, so the phase-difference procedure is carried out according to pilot sub-carriers only. It takes into consideration the periodic frequency control carried out by the receiver-measuring complex at signal reception. Since quantities U_p,f_p are calculated by the receiver complex at estimation of frequency offset, one cannot practically require additional computing expenses for calculation of the frequency amendment (8.26). For estimation of high-accuracy determination procedure inaccuracy, we can use the quantity δ, which is equal to the maximal modulus of the three last differences

$$\delta = \max(|F_M - F_{M-1}|, |F_{M-1} - F_{M-2}|, |F_{M-2} - F_{M-3}|), \tag{8.27}$$

where M is a number of the last stage of frequency control. The δ quantity characterizes the estimation accuracy and particularly important for signals received under the complicate channel and interference conditions.

Results of the frequency-shift estimation for DVB-T/H and DVB-T2 systems are presented in Tables 8.2 and 8.3 in the format $(\widetilde{F} - F_0) \pm \delta$, where F_0 is the true frequency shift. For obtaining these results, the TV signal source and a receiver were synchronized by the mutual high-stable reference oscillator (REF).

We see that for the same FFT size, the estimation accuracy for the DVB-T/H signal is slightly higher than for DVB-T2. This can be explained by the fact that the number of pilot sub-carriers is more in the DVB-T/H signal. For the standard DVB-T2 with the FFT size of 32 K, which is mostly widespread in Russia, we have the relative error of the 10^{-11} order.

Figure 8.17 shows a practical example of high-accuracy frequency-shift convergence versus time for the DVB-T2 system, the FFT size is 32 K. The true value of the frequency shift in Fig. 8.17 is 1 Hz. We see that just after 0.2 s, the measurement result differs from true one less than by 0.05 Hz, and total fluctuations of frequency estimations obtained at observation duration about 2 s were of order of 0.01 Hz in the given example.

Table 8.2 High-accuracy estimation error of the frequency shift for DVB-T/H systems

Frequency shift (Hz)	FFT size		
	2 K	4 K	8 K
0	0.0119 ± 0.0130	−0.0028 ± 0.0069	−0.0077 ± 0.0051
1	0.0113 ± 0.0037	0.0029 ± 0.008	0.0039 ± 0.0056
10	0.0041 ± 0.0098	0.0121 ± 0.0112	0.0069 ± 0.0120

Table 8.3 High-accuracy estimation error of the frequency shift for DVB-T2 systems

Frequency shift (Hz)	FFT size					
	1 K	2 K	4 K	8 K	16 K	32 K
0	−0.0345	−0.0710	−0.1257	−0.0125	−0.02594	0.0038
	±0.0004	±0.00084	±0.0014	±0.00079	±0.00037	±0.0021
1	0.0032	0.0265	−0.0565	−0.0535	−0.0028	−0.0019
	±0.00084	±0.0003	±0.0012	±0.00051	±0.0036	±0.0022
10	−0.0753	−0.0922	−0.1340	−0.0268	−0.0108	−0.0009
	±0.00055	±0.00082	±0.0018	±0.0017	±0.0022	±0.002

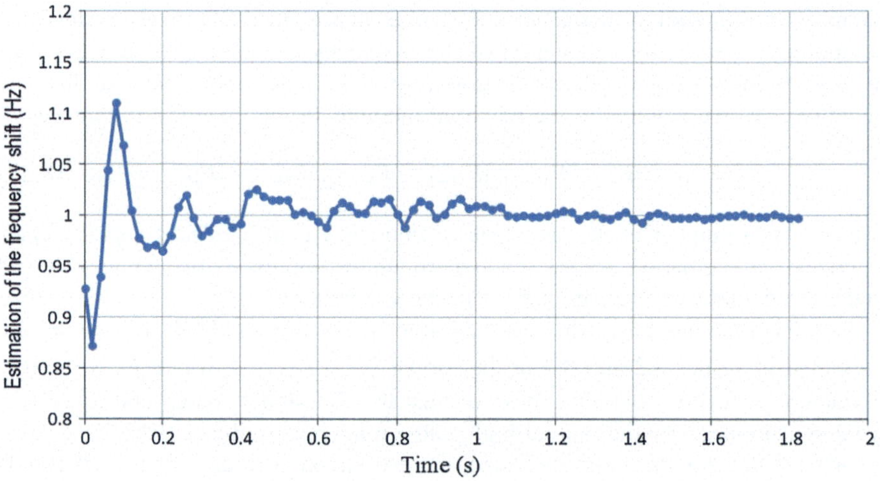

Fig. 8.17 Convergence example of the frequency-shift estimation, $F_0 = 1$

The procedure of high-accuracy estimation of the frequency shift is included into the SMO-CT software part of the DVB-T/H, DVB-T2 signals analyzer [1, 3, 18].

To check the analyzer operation, the measurements of estimation accuracy of the TV signal carrier frequency were performed. Measurements were carried out in accordance with the block diagram presented in Fig. 8.18. To reduce the effect of the REF frequency instability of the measuring radio receiver ARGAMAK-IS influencing on the measurement accuracy, measurements were accompanied by the continuous check of the REF frequency deviation from its nominal value. For this,

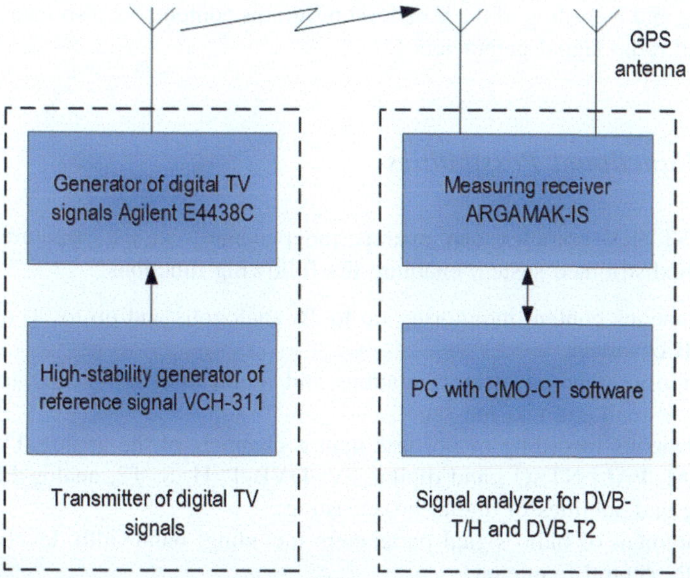

Fig. 8.18 Diagram of experiment conducting

Table 8.4 Estimation of frequency offset for the DVB-T2 signal (in Hz)

Carrier frequency (MHz)	FFT size			
	4 K	8 K	16 K	32 K
96	0.03	0.03	0.02	0.01
514	0.02	0.02	0.01	0.01
858	0.02	0.02	0.03	0.01

the high-accuracy local scale of frequency–time synchronized by time marks of the satellite navigation systems and insured the relative instability not more than 5×10^{-11}, was formed. The generator Agilent E4438C synchronized by the high-stable external oscillator VCH-311 was used as the DVB-T/H, DVB-T2 signal source. Its relative instability is about 1×10^{-13}. The Agilent E4438C generator allows obtaining the test signal with different variations of TV signal parameters. Results of carrier frequency estimations are given in Table 8.4.

8.9 Automated Multi-channel Monitoring of TV and RB Signals

A goal of this section is examination of construction peculiarities and functioning principles of the ARC-TVRV complex developed on the base of receivers of Russian implementation and manufacture, which is intended for operation in the structure of automated radio monitoring system providing not only multi-channel monitoring and recording of the broadcast programs content, but also measurements of received radio signal parameters [2].

8.9.1 Functional Possibilities

The ARC-TVRV complex can operate independently in the structure of geographically distributed system enabling the following functions:

- Simultaneous content monitoring up to 16 analogous and up to 20 digital TV and RB channels;
- Registration of broadcasting violations and notification using operator's interface by e-mail and text message;
- Simultaneous recording of not less than 8 channels of the analog TV signals: SECAM, PAL, NTSC , and digital TV: DVB-T, H, S, T2, analog FM broadcasting and subtitles of digital broadcasting;
- Measurement of radio signal parameters including: bandwidth, level and field strength, central frequency;
- Estimation of radio signal modulation parameters;
- Execution of tasks according to schedule without continual control channel;
- Equipment self-diagnostics, support of guarding signaling sensors.

8.9.2 Peculiarities of Construction and Functioning

The ARC-TVRV complex is manufactured in transportable and fixed variants of implementation shown in Figs. 8.19 and 8.20.

In the transportable variant, the electronic equipment is located in the thermostabilized and damp-proof case for outdoor use all year round. The fixed variant has no protective case; it is intended for indoor allocation. Structural diagram of ARC-TVRV is shown in Figs. 8.21 and 8.22 shows the internal composition of the transportable variant.

The signal from the TV antenna passes into the active splitter and then to receiver's blocks for analog and digital radio signals. The analog signal receiver receives simultaneously 16 informational channels; the capture card transforms them into the digital code. The digital signal receiver embedded into the industrial computer has two independent receiving channels and allows simultaneously decoding of two PLP streams, ensuring reception of up to 20 digital TV channels. The industrial computer is intended for signal processing and recording, control of external devices, and it realizes the local control server for the ARMS ARMADA, which supports the user interface, the execution of tasks according to the schedule, the electrical power control, the autonomous operation and functioning in the radio monitoring network, the equipment self-diagnostics, etc.

Fig. 8.19 Transportable variant

Embedded power supply ensures about one hour of autonomous operation; this time can be increased when connecting to additional accumulators. Remote control is executed via the main Ethernet channel or via the reserved wireless channel supplying 2G, 3G, 4G technologies. Transfer to the reserved channel occurs automatically at main channel disappearing. This complex allows remote control from the workstation, at which one does not require utilization of additional software, and all control is carried out via the Web browser. The self-diagnostic system allows in-time detection and notification about abnormal situations and equipment faults. Besides, there is a guarding system ensuring the case openness check and a tilt of the case and the knock to it. Self-diagnostic and guarding systems use the standard SNMP protocol.

The free volume for additional equipment mounting may serve for additional communication equipment, for measuring equipment, accumulators, storages for information recording. The digital panoramic measuring receiver ARGAMAK-IS [4, 13], which provide panoramic and spectral analysis and signal parameter measurement, service identifier and parameters' analysis for TV and RB systems, can be included into the complex structure. The signal measurement errors for digital TV systems meet requirements of Russian standards 55696–2013, 55939–2014, and of acting Russian General Radio Frequency Centre norms:

Fig. 8.20 Stationary variant

- At measurement of the signal level: ±1 dB;
- At measurement of the signal central frequency of the multi-frequency broadcasting network at operation from embedded REF: $\pm 1 \times 10^{-9}$;
- At measurement of the signal central frequency of the single-frequency broadcasting network for synchronization from the embedded REF according to GLONASS/GPS signals of exact time: $\pm 5 \times 10^{-11}$;
- At measurement of the signal bandwidth: ±0.1%.

As an example, Fig. 8.23 shows the SMO-ARMADA [19] software window with results of analyzer operation for the DVB-T2 digital TV signal, which was considered earlier in the first part of this chapter.

The measuring receiver ARGAMAK-IS included in the measuring complex ensures measurement of radio signal parameters with analog and digital modulations, including: frequencies of radio signals with AM, FM and digital modulation types, bandwidths, frequency deviations of FM signals, frequency channel occupancy, as well as digital TV signal parameters.

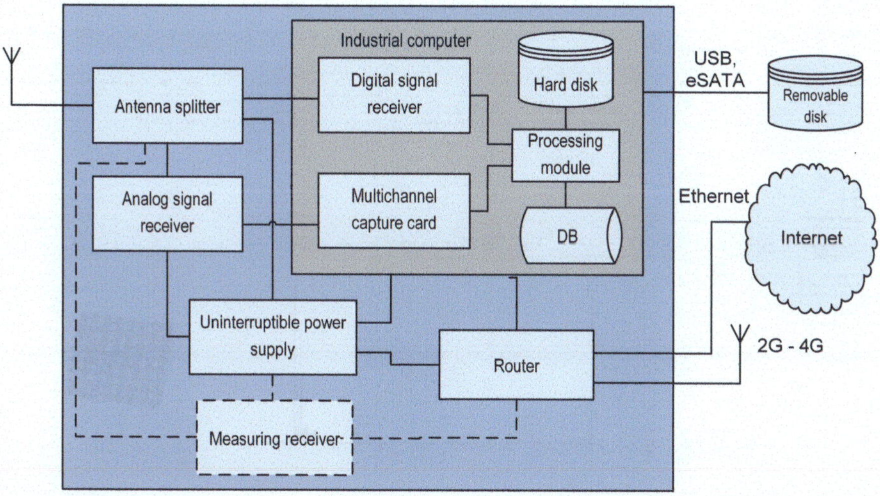

Fig. 8.21 Structural diagram of ARC-TVRV

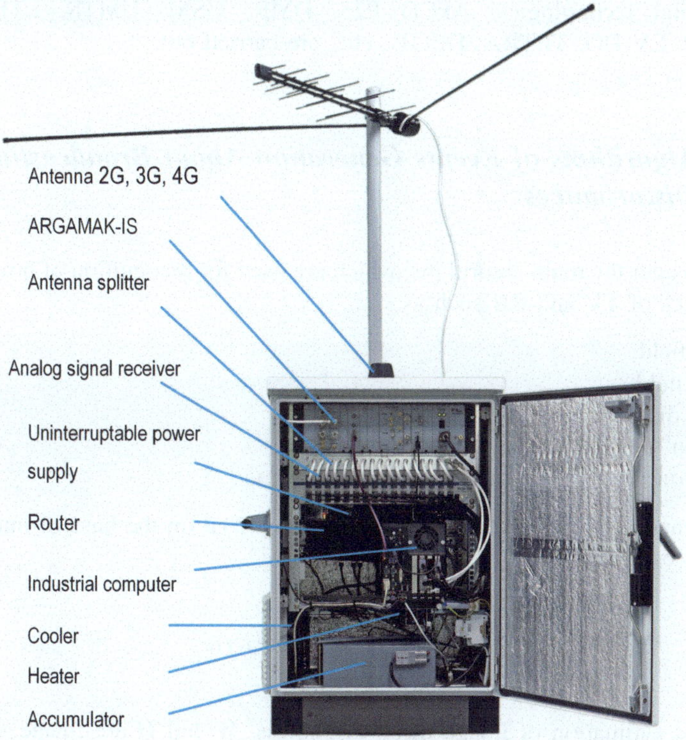

Fig. 8.22 Internal composition of the ARC-TVRV transportable complex

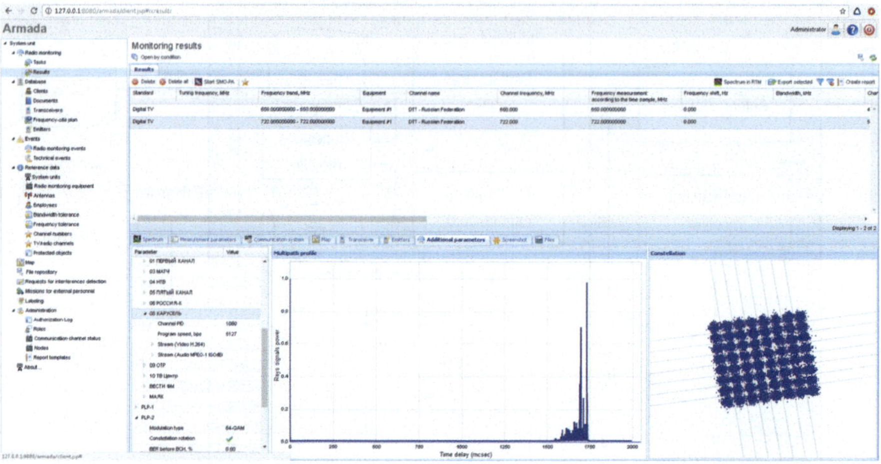

Fig. 8.23 DVB-T2 signal measurement

Besides, the identifier determination and radio signal parameter estimation of other digital technologies: APCO P25, DMR, GSM, UMTS, LTE, IS-95, cdma2000, EV-DO, TETRA, DECT , etc., are carried out.

8.9.3 Algorithms of Events Generation About Broadcasting Disturbances

Let us consider the main algorithms, which are used for recognition of broadcasting disturbances of TV and RB such as:

- Black field;
- White field;
- Signal disappearing;
- “Frozen” frame;
- Variation of the sound accompaniment volume.

Appearance of black and white fields is recognized on the base of image mean brightness estimation

$$B = \frac{1}{\mathrm{WH}} \sum_{i=1}^{W} \sum_{j=1}^{H} \mathrm{pxl}(i,j), \tag{8.28}$$

where B is estimation of image mean brightness, W and H are image width and height expressed in pixels, elements $\mathrm{pxl}(i,j)$ contain information about brightness of separate image pixels. If H_{black} is the threshold value for the black field, H_{white} is the threshold value of the white field, then under fulfillment of the condition $B < H_{\mathrm{black}}$ the decision is making that the black field was indicated in the display, while for $B > H_{\mathrm{white}}$ the decision is making that the white field was indicated in the display.

For recognition of cases of initial signal disappearing, when transmitted "image" has noisy character, the variance proximity of the image spectrum to the variance of white noise spectrum is estimated. For this, according to 2D image spectrum,

$$F(m,n) = \sum_{i=1}^{W}\sum_{j=1}^{H} \mathrm{pxl}(i,j) \cdot e^{-j2\pi\frac{(m-1)(i-1)}{W} + \frac{(n-1)(j-1)}{H}}. \tag{8.29}$$

the normal variance of spectral component intensity is estimated

$$D_F = \frac{1}{\mathrm{WH}}\frac{1}{M_F^2}\sum_{m=1}^{W}\sum_{n=1}^{H} (|F(m,n)| - M_F)^2, \tag{8.30}$$

where $M_F = \frac{1}{\mathrm{WH}}\sum_{m=1}^{W}\sum_{n=1}^{H} F(m,n)$ is the mathematical expectation of the image spectrum samples' modulus. The sign of the noisy character of the image frame analyzed is the condition $D_F < H_{\mathrm{white\_noise}}$ fulfillment, where $H_{\mathrm{white\_noise}}$ is the threshold value for "noisy" image.

To track the cases of image fading, we should estimate the maximum of image mutual correlation on the base of their complex spectrums F_1 and F_2:

$$F_1(m,n) = \sum_{i=1}^{W}\sum_{j=1}^{H} \mathrm{pxl}_1(i,j) e^{-j2\pi\left[\frac{(m-1)(i-1)}{W} + \frac{(n-1)(j-1)}{H}\right]}; \tag{8.31}$$

$$F_2(m,n) = \sum_{i=1}^{W}\sum_{j=1}^{H} \mathrm{pxl}_2(i,j) e^{-j2\pi\left[\frac{(m-1)(i-1)}{W} + \frac{(n-1)(j-1)}{H}\right]}; \tag{8.32}$$

Then, the conjugated spectrum of the second image is determined

$$F_{2c} = \mathrm{Re}[F_2(m,n)] - j\mathrm{Im}[F_2(m,n)] \tag{8.33}$$

and the product of the complex spectrum:

$$M(m,n) = F_1(m,n) \cdot F_{2c}(m,n). \tag{8.34}$$

The correlation function is calculated by means of inverse discrete Fourier transform application to the product of complex spectrum

$$R_{12}(m,n) = \frac{1}{\mathrm{WH}}\left|\sum_{i=1}^{W}\sum_{j=1}^{H} M(i,j)e^{j2\pi\left[\frac{(m-1)(i-1)}{W}+\frac{(n-1)(j-1)}{H}\right]}\right|. \tag{8.35}$$

Maximum of the mutual image correlation $R_{12\max} = \max(R_{12})$ is compared with minimal autocorrelation peaks $R_{11\max}, R_{22\max}$ of each images

$$R_{A\min} - \min(R_{11\max}, R_{22\max}). \tag{8.36}$$

If the offset of correlation function peak with respect to image center does not exceed the given value S_{XY} and the inequality fulfills

$$\frac{R_{12\max}}{R_{A\min}} > H_R, \tag{8.37}$$

the frames are considered as the same, and the decision about image fading is made. We can adjust the algorithm sensitivity at the expense of S_{XY} and H_R parameters.

For volume variation determination, we can estimate the intensity level of the sound signal according to the approach presented in [20]. If the volume level variation exceeds the given threshold H_L, then events about volume level disturbance are generated.

Task execution on check of the informational TV component is accompanied by representation of a mosaic from images obtained for different channels, as shown in Fig. 8.24.

There is a service field in each image frame, in which the channel name, the frequency value, etc., are displayed. The order, image frame repetition frequency, their size and content are adjusted by the user. At that, one or several channels (as operator chooses) may be displayed in real time in the separate window. The eventual software mechanism generates events about broadcasting violations including image saving at the moment of the event arising.

The status of TV and RB channels under monitoring is integrally mapped in the form of a diagram consisting of color bands, each of which shown in Fig. 8.25. The band color informs about channel status in the appropriate time moment.

Notifications about happened events are stored in the database and sent by e-mail to the required address or by text messages to given phone number. It is possible to adjust the notification way depending on the round-the-clock time.

Besides detection of broadcast disturbances, the complex provides simultaneous multi-channel signal recording of digital and analog TV and RB signals. At that, geographic coordinates of ARC-TVRV, date and time of recording, and also subtitles of digital broadcasting are registered. Recording is carried out to the embedded, additional, or network storages. Dividing of record into separate fragments with the given size or duration is anticipated. Volume of the data recorded on the storage depends on the used resolution, the digital stream speed, and record duration. Table 8.5 shows a dependence of necessary size on the storage upon the record parameters.

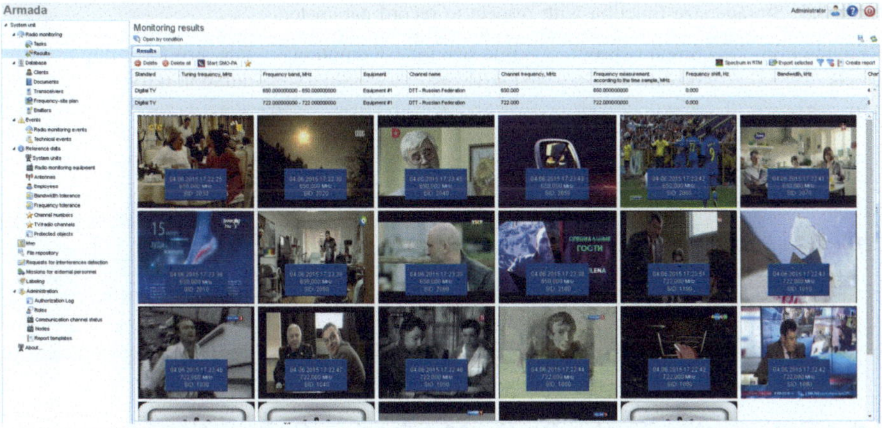

Fig. 8.24 Visualization of content

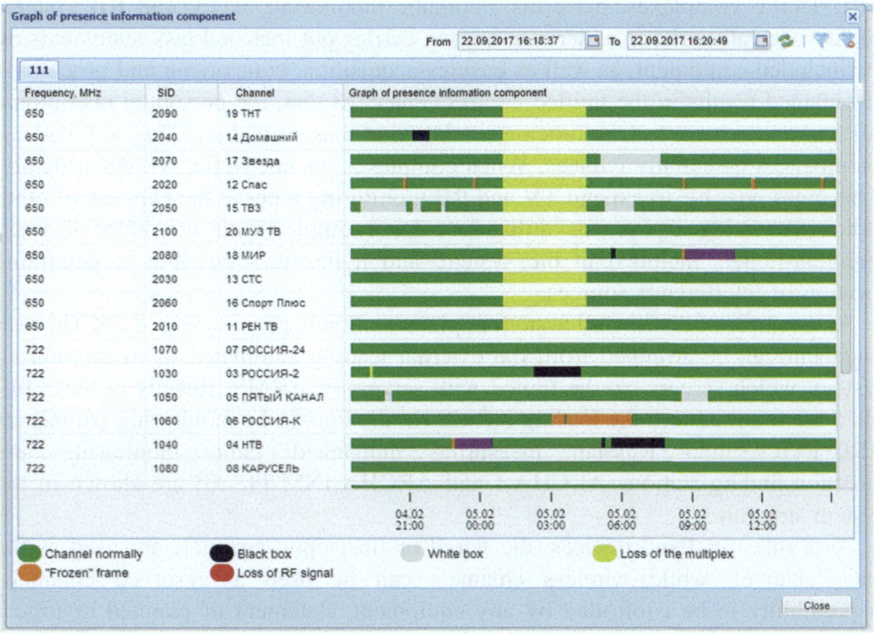

Fig. 8.25 Informational component presence

At displaying resolution of 320 × 256 and the stream speed of 200 kb/s, the 1 TB storage will provide simultaneous recording of the eight TV channels during 60 days. When free size will be finished, the old files are automatically replaced with the new ones. At that, it is possible to protect those files, which must be saved instead of re-recording.

Table 8.5 Storage size function with respect to record parameters

Resolution in pixels	Stream speed (kb/s)	File size for 1-h duration (MB)	File size for 7-day duration (GB)
320 × 256	80	35	6
	200	88	15
	250	109	18
	1020	436	73
720 × 576	670	284	49
	1660	713	120
	2050	883	148
	8230	3534	594

8.9.4 Operation in the Radio Monitoring System Structure

The ARMS distributed in the territory provides the centralized control of ARC-TVRV complexes, performs automatic monitoring of TV and RB systems quality over all territory under monitoring, carries out matched task statements on the included equipment, as well as ensures acquisition, comparison and processing of obtained results in the unified control center. At that, the personnel operational efficiency increases, and the time on revelation of reasons and rejecting of broadcast disturbances essentially reduces. When complexes operate in the ARMS structure, it becomes possible to expand TV and RB monitoring types at the expense of other radio monitoring equipment utilization, for example, with the help of radio direction-finders included in the system, and it becomes possible to determine location of interference sources.

An example of distributed system implementation is given in Fig. 8.26. The task statement can be arranged from the external leading automated radio monitoring system, which servers can be united with servers of ARMS. Results of tasks fulfilled are returned into the leading ARMS. In the figure, the monitoring complexes ARC-TVRV and Russian measuring unmanned radio monitoring and direction-finding stations ARCHA-I and ARCHA-INM [4, 13] are shown in the system structure.

As a rule, for fixed devices, the wired or fiber-optic channels are used as the main channel, while wireless channels can be used as reserved channels. A possibility to be controlled by any equipment, statement of planned or urgent tasks by the particular ARMS operator is defined by adjustment of the access rights and the policy of system safety. So, control of ARC-TVRV, operating on the district level, can be realized from both ARMSs located in the same district, and ARMSs located at the region level. We can assign tasks from one ARMS for different radio monitoring equipment, which allows effective solution of radio monitoring tasks for TV and RB systems.

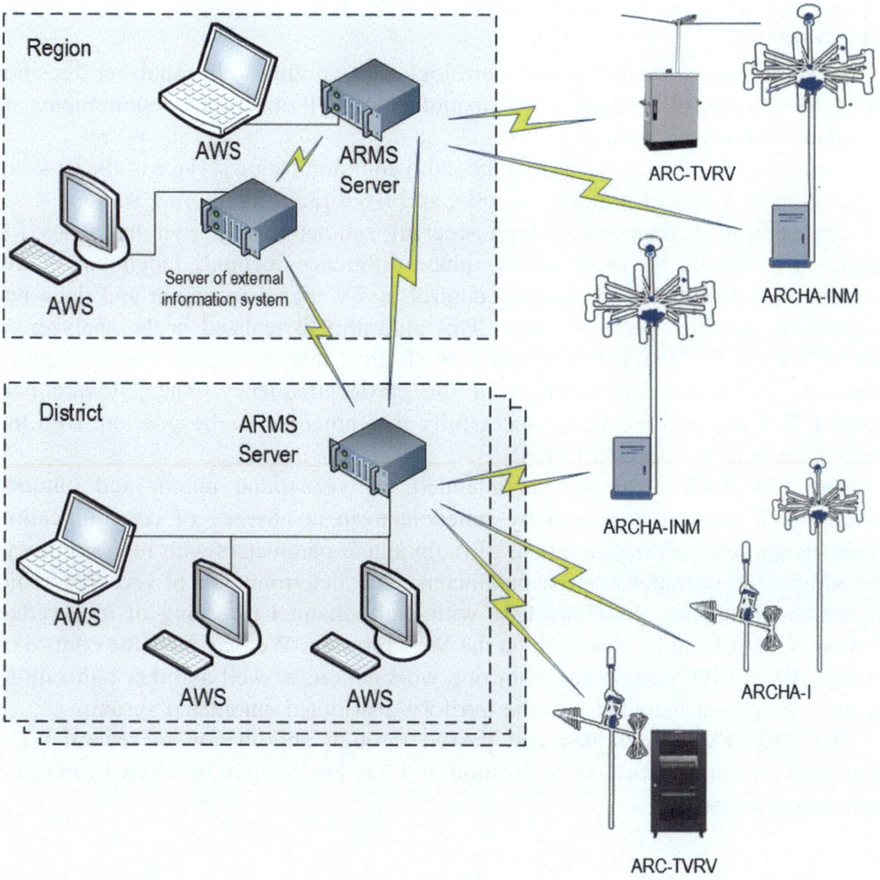

Fig. 8.26 Structure of the radio monitoring network

8.10 Summary

The DVB-T2 analyzer described in this chapter is intended for search for digital TV station signals, reception of service data, obtaining of informational streams, extraction of sub-streams of different programs, fulfillment of large volume of measurements of physical and transport levels. It can be used for planning and operation of digital TV systems, measurement of TV transmitter parameters, estimation of the coverage zone.

This analyzer consists of the hardware part—the measuring receiver of ARGAMAK family, and the software part—the application of system mathematical support. Utilization of the universal measuring receiver as equipment enables (together with the DVB-T2 signal analysis) to solve other radio monitoring tasks, for instance, to perform investigations of DVB-T/H signals, as well as signals of various communication systems: GSM, UMTS, LTE, cdma2000, EVDO, DMR, APCO P25 etc.

According to its functional and metrological possibilities, this analyzer does not yield to systems of leading world manufacturers. It meets the requirements of Russian normative documents.

Currently, the analyzer is used in the radio communication service of the Russian Federation as a part of portable, mobile, and fixed radio monitoring stations.

The developed algorithm of high-accuracy estimation of carrier frequency for digital TV signals is based on the phase-difference method, functions on the background of automatic frequency control at TV signal reception and does not require the specific operation mode. This algorithm is realized in the analyzer of DVB-T/H, DVB-T2 signals and meets (with great margin) the normative requirements to the estimation accuracy of the carrier frequency. The software part SMO-CT of analyzer has been successfully recommended in the structure with the measuring receiver ARGAMAK-IS.

The ARC-TVRV complex is intended for year-round indoor and outdoor operation. It ensures operation by schedule even at absence of communication channels and enables measurement of radio signal parameters with high accuracy, decoding of transmitter service parameters, and determination of received radio signal characteristics simultaneously with multi-channel recording of transmitted content. Control can be arranged via the Web browser. We may have the control of many ARC-TVRV complexes from one workstation, as well as other radio monitoring equipment included into the territory-distributed automated system.

The ARC-TVRV complex had passed through approbation in several radio frequency centers of Russian Federation and has got positive opinions from radio monitoring experts.

References

1. Alexeev PA, Ashikhmin AV, Kayukov IV, Kozmin VA, Manelis VB (2016, in Russian) DVB-T2 signals analyzer. Spetstehnika i Svyaz (4):15–28
2. Kozmin VA, Korochin SV, Bocharov DN (2016, in Russian) Automated multi-channel television and radio broadcasting monitoring. Electrosvyaz (10):62–68
3. Ashikhmin AV, Kozmin VA, Kayukov IB, Manelis VB (2012, in Russian) Digital TV radio signal analyzer. Radiotekhnika (2):75–91
4. Rembosky AM, Ashikhmin AV, Kozmin BA (2015, in Russian) Radiomonitoring: problems, methods, means. In: Rembosky AM, 4th edn. Hot Line-Telecom Publ., Moscow, 640 p
5. Cimini Jr LJ (1985) Analysis and simulation of a digital mobile channel using orthogonal frequency division multiplexing. IEEE Trans Commun COMM–33:665–675
6. van Nee R, Prasad R (2000) OFDM wireless multimedia communications. Artech House, Boston-London
7. ETSI EN 302 755 V1.3.1 (2012–04) Digital Video Broadcasting (DVB); Frame structure channel coding and modulation for a second generation digital terrestrial television broadcasting system (DVB-T2)

8. Alamouti S (1998) A simple transmit diversity technique for wireless communications. IEEE J Sel Areas Commun 16(8):1451–1458
9. ISO/IEC 13818-1: Information technology—Generic coding of moving pictures and associated audio information: Systems
10. ETSI TS 102 606: Digital Video Broadcasting (DVB); Generic Stream Encapsulation (GSE) Protocol
11. A14-2 Digital Video Broadcasting (DVB); Measurement guidelines for DVB systems; Amendment for DVB-T2 System, July 2012
12. ETSI TR 101 290 V1.2.1 Digital Video Broadcasting (DVB); Measurement guidelines for DVB systems (2001–05)
13. The catalogue 2017 of IRCOS JSC. http://www.ircos.ru/zip/cat2017en.pdf. Accessed 28 Nov 2017
14. Kayukov IV, Manelis VB (2006, in Russian) Comparative analysis of different methods of signal frequency estimation. Radioelectron Commun Syst 49(7):42–56
15. Norm of State Committee on Raduo Frequencies 17–13 "Radio Transmitters of all categories of civil application. Requirements on the permissible frequency deviations" with modifications dated July 2015 (in Russian)
16. GOST (Russian State Standard), pp 55696–2013. Television broadcasting digital. Transmitting equipment for digital ground-based TV broadcasting DVB-T/T2. Technical requirements. Main parameters. Methods of measurement (in Russian)
17. Bespalov OV, Kayukov IV, Kozmin VA, Manelis VB (2016, in Russian) High-precision estimation of digital television carrier frequency. Spetstehnika i Svyaz (4):97–100
18. SMO-CT Software package for digital TV signal analysis. http://www.ircos.ru/en/sw_ct.html. Accessed 28 Nov 2017
19. SMO-ARMADA software package of automated spectrum monitoring systems. http://www.ircos.ru/en/sw_armada.html. Accessed 28 Nov 2017
20. Recommendation ITU-R BS.1770-3 (2012) Algorithms to measure audio programme loudness and true-peak audio level, Geneva, p 22

Chapter 9
Detection and Identification of Digital Radio Sources

9.1 Introduction

For planning and exploitation of digital radio communication and data transmission networks, for checking of correspondence of radio transmitter parameters to the frequency-spatial plan (FSP), for investigation of the conflict situations and for search for interference sources, it is necessary to identify the transmitting stations and to decode of their service parameters.

In this chapter, we consider the family of signal analyzers for digital technologies implemented of the base of digital receivers of the ARGAMAK-family including signal analyzers of GSM, UMTS, LTE, IS-95, cdma2000, EV-DO, TETRA, DECT base stations, DMR, APCO P25 stations and, at the end of the chapter, the wireless Wi-Fi networks. All analyzers have the similar hardware structure presented in Fig. 9.1 and differ in software only.

One of the digital radio receivers of the ARGAMAK-family ARGAMAK-M, ARGAMAK-RS, ARGAMAK-IS, ARCHA-IN, etc., can be used as the radio monitoring unit. The receiver forms I/Q components of the digital signal, which bandwidth and sampling frequency correspond to the system being analyzed, and further, the signal is processed by the computer. The software running on the computer realizes the receiving–measuring algorithms of signal processing and an interface for the corresponding analyzer.

The specific feature of considered analyzers is a possibility of search and decoding of digital signals, which can be performed not only at the standard frequencies used in the specific digital technology, but in the whole range of operating frequencies of digital radio receivers.

A. M. Rembovsky et al., *Radio Monitoring*, Signals and Communication Technology, https://doi.org/10.1007/978-3-319-74277-9_9

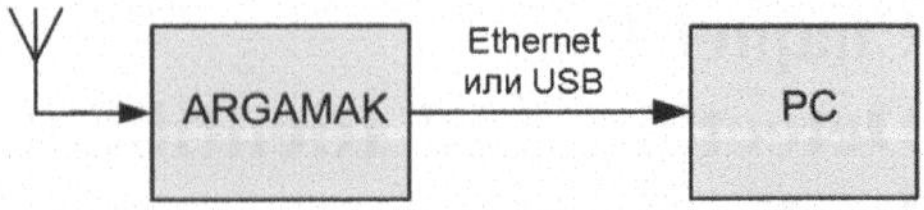

Fig. 9.1 Structure of the digital signal analyzer

9.2 Analyzer of GSM, UMTS, LTE Base Stations Signals

Cellular networks of GSM, UMTS, LTE European digital standards are widely distributed in Russia and in many other countries. Networks of the second generation of GSM standard are based on technologies of frequency and time division multiple access (TDMA). The first commercial GSM network was deployed in 1992. These networks now remain popular owing to their reliability, functionality, cheapness, and developed infrastructure. Amount of GSM networks' subscribers exceeds two billion.

In recent decades, the technologies of cellular communication systems are developed intensively due to necessity of increasing data transmission rate and providing new services with improved quality for subscribers. In addition to the second-generation systems, the wideband systems of the third (UMTS) and fourth (LTE) generation appeared. They are focused on the packet data transmission, which allows providing of such new services to subscribers as high-speed Internet, video conferences, mobile TV, etc.

The third-generation systems are based on the technology of code division multiple access (CDMA). Compared to GSM, they have higher spectral efficiency and much higher capacity.

The fourth-generation systems are based on usage of orthogonal sub-carrier signals. They have higher data transmission rate (up to 300 Mbit/s in the 20 MHz bandwidth), stability to multi-path propagation, flexibility in signal parameters choice, and other advantages compared to technologies of frequency, time, and code division of previous generations.

For planning and exploitation of GSM, UMTS, and LTE networks, verification of the requirements' correspondence to the transmitters' parameters, and analysis of the coverage area, it is necessary to perform identification of active base stations and analysis of their radio signals regularly.

The equipment for analysis of the cellular systems of second, third, and fourth generations is produced by the leading worldwide manufacturers of radio monitoring equipment: companies Rohde & Schwarz (TSMQ, CMW-500), Tektronix (YBT250, K2Air), Agilent (E7495A/B, N9080A LTE FDD, N9082A LTE TDD).

In this section, we consider the Russian analyzer of base station signals of GSM, UMTS, LTE networks, which does not yield to analyzers of the leading worldwide manufacturers in functional opportunities. This analyzer is intended for search and reception of base station signals, their identification, parameters estimation and localization. One of the radio monitoring equipment on the base of the digital radio receivers of the ARGAMAK-family ARGAMAK-M, ARGAMAK-RS, ARGAMAK-IS, ARCHA-IN, etc., can be its hardware base [1, 2]. The receiver forms the digital signal, which bandwidth and sampling frequency correspond to the system being analyzed. The analyzer includes also three software modules realizing the reception-measuring algorithmic complexes for signal processing of appropriate systems, and the general interface application.

Peculiarities of the base station signal structure will be briefly described below for each cellular communication system under investigation. Moreover, we present the general block scheme of signal processing, the functional possibilities of the analyzer, and examples of analysis results [3].

9.2.1 GSM Signals Analysis

In general case, all base stations of the GSM network have several cells in their composition providing independent transmission and reception within frameworks of the own angular sector. Some quantity of frequency channels from a set assigned to the network operator is assigned to each cell. One radio frequency channel from this set is used for transmission of the broadcast data.

In the GSM standard, the multi-station access with TDMA is used in accordance with which the signal transmitted by the base station represents the structured sequence of frames [4]. In the sequence period, each frame has its own serial number *FN* from 0 to 2715647. The frame is divided into eight-time slots with 576.9 mks length each. For broadcast channel, the sequence of 51 frames forms one multi-frame. The digital information stream represents the sequence of symbol bursts inserted in the time slots. The GMSK modulation is used. Information is transmitted with the rate of 270.833 kbit/s. This means that one-time slot includes 156.25 symbols.

The base station signal contains different channels: synchronization channel (SCH), frequency correction channel (FCCH), broadcast control channel (BCCH), as well as traffic and other control channels. The analyzer processes SCH, FCCH, BCCH only; the traffic channels are not analyzed.

Figure 9.2 shows multi-frame structure of the broadcast channels. Frames of the multi-frame are grouped into 5 groups of 10 frames each, at that, one frame remains free. The first frame of each group contains frequency correction channel; the second frame of each group contains synchronization channel. The other 8 frames in each group form two blocks of four frames. The first block of the first group is intended for the broadcast control channel, whereas other 9 blocks of the multi-frame are used for transmission of other logic channels.

Information symbols of SCH are interpreted in the appropriate fields of sync bursts, which are transmitted in the slot 0 of frames 1, 11, 21, 31, 41 of each multi-frame. The structure of the sync burst is shown in Fig. 9.3a.

The bursts of frequency correction are transmitted in FCCH, which do not carry any information symbols and are situated in the slot 0 of frames 0, 10, 20, 30, 40 of each multi-frame. The structure of the frequency correction burst is shown in Fig. 9.3b.

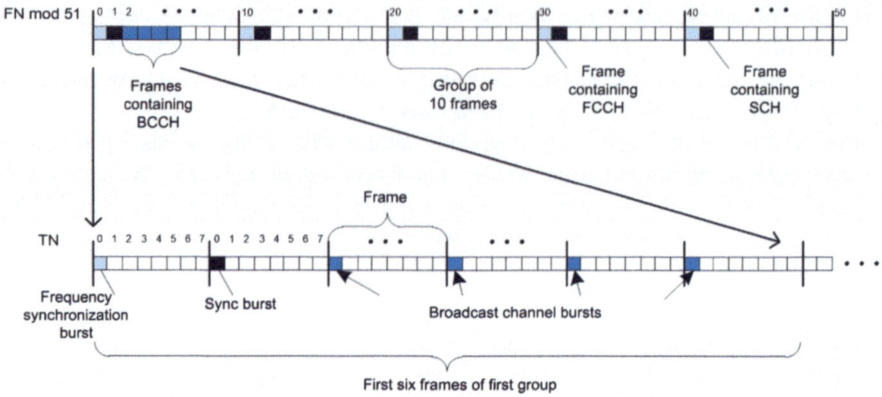

Fig. 9.2 Multi-frame structure of the GSM signal

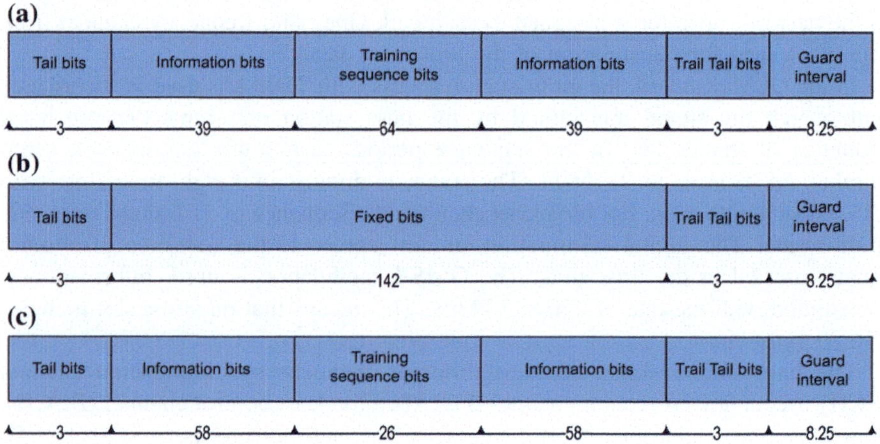

Fig. 9.3 Structure of the GSM bursts: **a** the sync burst, **b** the frequency correction burst, **c** the normal burst

Information symbols of BCCH are represented in the appropriate field of the normal bursts, which are always transmitted in the slot 0 of frames 2, 3, 4, 5 of each multi-frame, and also in some modes in other positions. The structure of the normal burst is shown in Fig. 9.3c.

The search for the cells signals, their identification, and parameters measurement is carried out by the analyzer for all possible frequency channels of the GSM standard from the given range.

For each frequency channel, the baseband signal of the necessary bandwidth and the sampling frequency is formed, the procedures of search for base station signal, time–frequency synchronization, and reception of sync and broadcast channels messages are executed, from which the identification data of the cell are extracted.

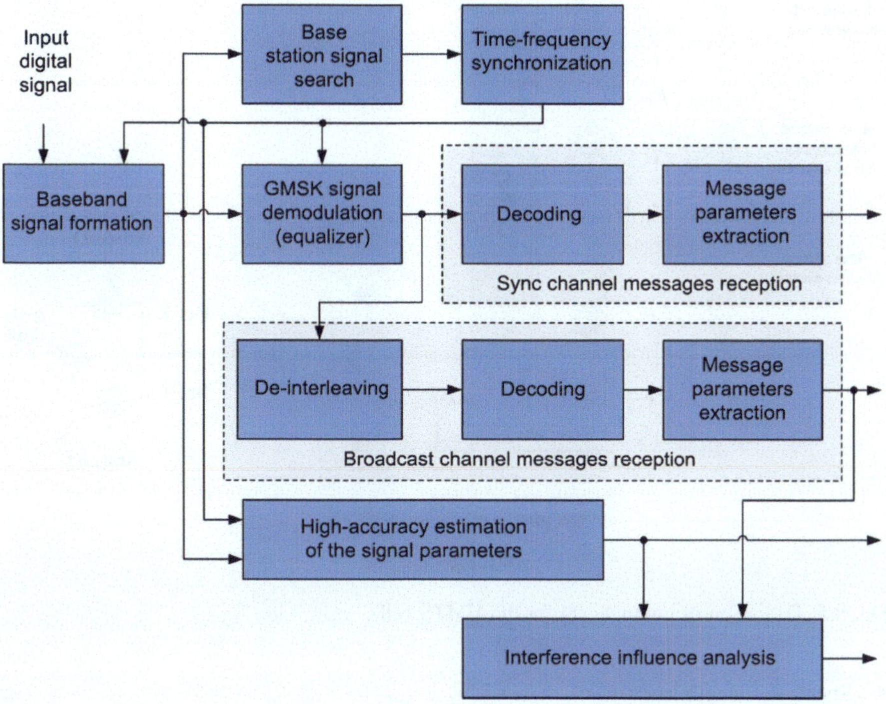

Fig. 9.4 Block diagram of GSM signal processing

The structural scheme of digital processing of the GSM signal is presented in Fig. 9.4. The operation of the receiving–measuring algorithmic GSM complex is described in [2] in detail.

The analyzer of base station signals of GSM networks enables to find out the base station broadcasting in the required spectral range and to receive their following broadcast data:

- Base station identification code (BSIC):
  - Network color code (NCC),
  - Base station color code (BCC);
- Local area identifier (LAI):
  - Mobile country code (MCC),
  - Mobile network code (MNC),
  - Local area code (LAC);
- Cell identifier;
- Frequency channels assigned to the given cell (CA);
- Frequency channels of broadcast adjacent cells channels (BA).
- The analyzer performs the following high-accurate measurements:

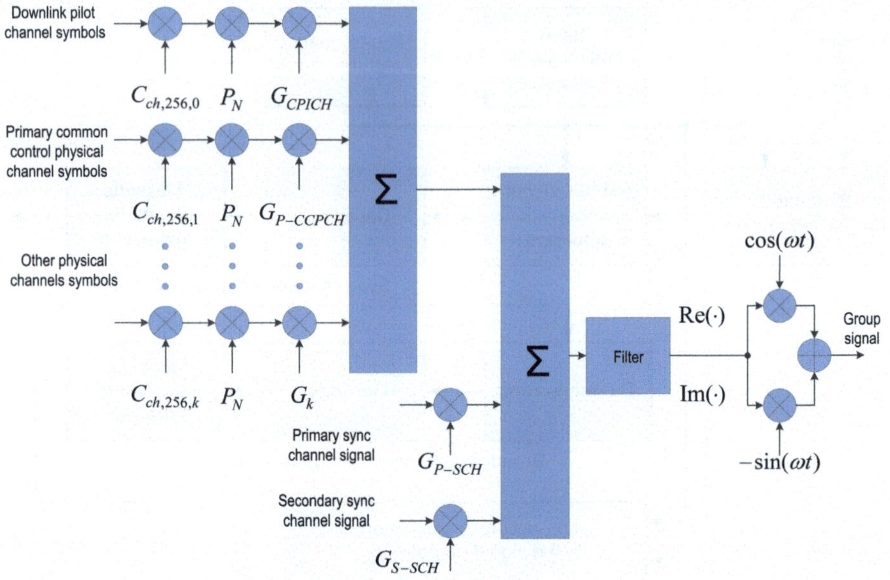

Fig. 9.5 Formation of group signal for the UMTS cell

- Spectral measurements:
  - Frequency shift,
  - Bandwidth (99% of power);
- Power measurements:
  - Mean power of the signal,
  - Power of signal symbols;
- Signal-quality measurement:
  - Phase error,
  - Level of carrier suppression,
  - Burst power level versus time;
- Channel measurements:
  - Power delay profile and RMS delay spread,
  - Signal/(interference + noise) ratio,
  - Carrier/noise ratio,
  - Probabilities of bit and packet errors.

On the base of fulfilled measurements, the analyzer enables:

- To calculate location of the detected base stations;
- To execute analysis of inside- and inter-network interference influences of base stations from one to another;

- To synthesize total channel quality coefficient according to the given criterion for effective monitoring of communication;
- To display network coverage area taking into account the digital maps of locality.

9.2.2 *UMTS Signals Analysis*

A principle of code division multiple access [5] is the base of the UMTS technology. Signals from various network cells and various channels of each cell are transmitted simultaneously in one spectral range. Each cell has its own scrambling code which enables the receiving side to extract the signal of this exact cell. For division of different physical channels of the cell, the orthogonal channel codes are used.

The UMTS standard allows usage of two duplex modes: the frequency division duplex (FDD) and the time division duplex (TDD). In the FDD mode, the downlink and uplink channels (from the base station to the subscriber one and vice versa) have separate frequency bands. In the TDD mode, the division of the downlink and uplink channels is executed at the expense of using various time intervals in one frequency band.

The UMTS signal of the cell includes the totality of different physical channels: the primary common pilot channel, the primary/secondary synchronization channel, the primary common control physical channel, as well as other service and informational (traffic) channels. Each physical channel is structured in time into frames with duration of 10 ms. Frame consists of $N_{sl} = 15$ slots of $N_{ch} = 2560$ elements (chips) in each.

The pilot signal is continuously transmitted and used by the subscriber stations for synchronization and estimation of the propagation channel. On the base of sync channels, the search for cell is provided, the scrambling cell sequence is determined, and the slot and frame synchronization are performed. In the primary common control physical channel, the broadcast information including cell identification parameters is transmitted.

Figure 9.5 shows the structural diagram of the cell group signal formation.

The complex modulation symbols of each physical channel (besides sync channels) expand in spectrum with appropriate channel code $C_{SF,k}$. Then, the chip sequence of obtained symbols is multiplied element-wisely by chips of the scrambling cell code P_N. The frequency of chip sequence is 3.84 MHz. The signal samples of the physical channel are multiplied by the G_k coefficient, which defines the channel relative power. Then signals of all physical channels are unified by means of element-wise complex summation. The obtained signal is filtered and transferred to the carrier frequency. After filtering, the group signal formed in such a manner proceeds to the transmitter analog path, where it is amplified and then broadcasted.

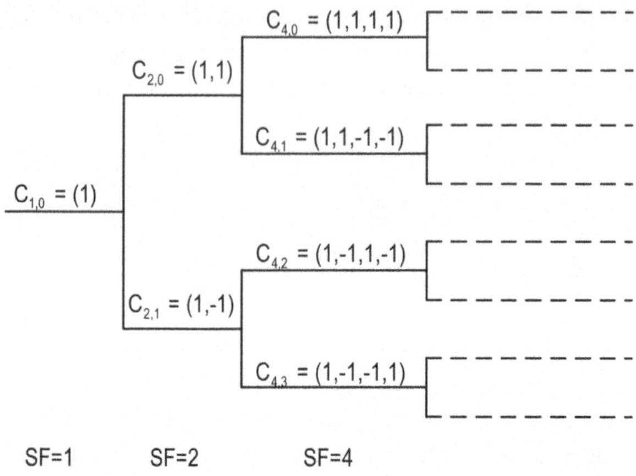

Fig. 9.6 Code tree for OVSF codes formation

In the UMTS system, 512 primary scrambling codes can be used. Each primary scrambling code produces 15 secondary scrambling codes. The set of primary scrambling codes is divided into 64 coding groups, each of which consists of 8 primary scrambling codes.

The scrambling code represents the complex pseudo-random sequence with length of 38400 chips, which is periodically repeated in each frame. The real and imaginary components of the scrambling code are the different segments of the Gold sequence.

To assure orthogonality between various physical cell channels, the channel orthogonal variable spreading factor (OVSF) codes are used in UMTS. OVSF codes are formed by the code tree, as shown in Fig. 9.6, where they are designated as $C_{SF,n}$, SF is the code spreading factor,[1] n is the code number, $n = \overline{0,\ SF - 1}$. The full tree consists of 8 levels, which correspond to $SF = 256$. The ensemble of OVSF codes depends on the used quantity of the spreading factor. For common pilot channel, the channel code $C_{256,0}$ is always used, for the primary common control channel—the code $C_{256,1}$. Channel codes of other physical channels can be various.

The UMTS standard admits a possibility of transmission diversity in the downlink channel. The time-switched transmit diversity (TSTD) mode for antenna switching can be applied for sync channels only. In this mode, even slots are transmitted via the first antenna, while odd slots—via the second antenna.

The space time transmit diversity (STTD) mode can be used for all informational and control channels. In this mode, data transmission is performed by blocks with 4 bits. The STTD coding scheme is presented in Fig. 9.7. The UMTS base stations may not support the mode of transmit diversity. However, for subscriber stations, and hence, for the analyzer, the support of transmit diversity is mandatory.

The time structure of sync channels is presented in Fig. 9.8.

[1]The code spreading factor is the chip quantity of the pseudo-random sequence in one symbol.

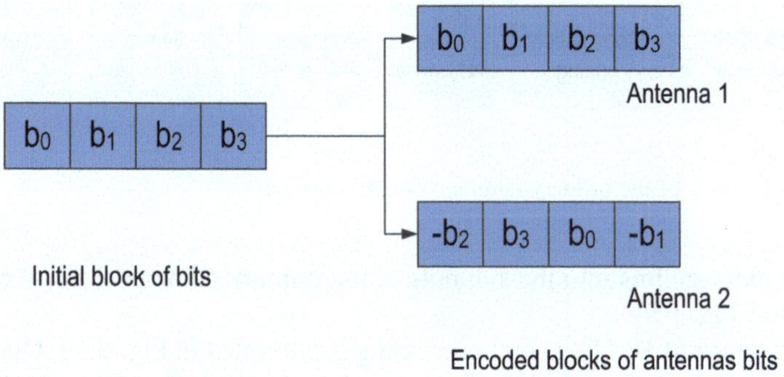

Fig. 9.7 Coding scheme STTD

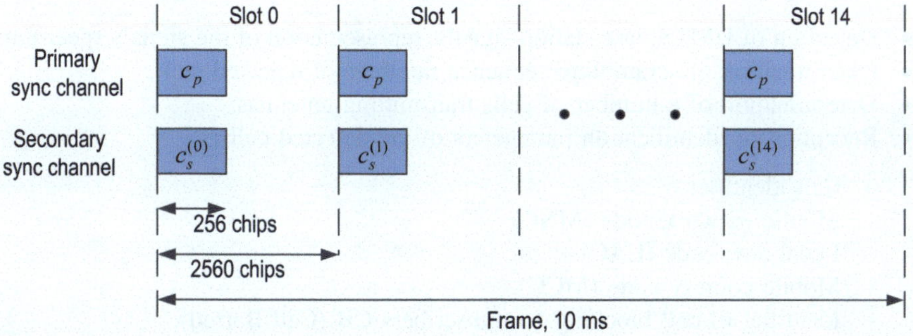

Fig. 9.8 Structure of sync channels

The primary sync signal for each frame represents the periodically transmitted (in the beginning of each slot) complex sync code c_p with length of 256 chips. The primary sync signal is the same for all network cells.

The secondary sync signal for each frame consists of the sequence of various 15 codes $c_s^{(m)}$, $m = \overline{0, 14}$ with length of 256 chips transmitted in the slot beginning, m is a slot number. In each slot, one of the 15 secondary sync codes is possible to be used. They represent the known complex orthogonal pseudo-random sequences with the same real and imaginary parts. Combination of secondary sync codes in the frame defines the coding group, to which the cell scrambling code belongs.

The pilot channel is continuously transmitted and carries the known symbolic sequence. The number of chips in the symbol equals to 256, so a number of symbols in the slot equals to 10. In the transmit diversity mode, the pilot channel is transmitted from both antennas using the same channel and scrambling codes. In this case, the known symbolic sequence differs for first and second antennas.

The primary common control channel is continuously transmitted except the first 256 chips of each slot. The spreading factor for this channel $SF = 256$, so that a number of symbols in each slot equal to 9. The scheme of transformation of

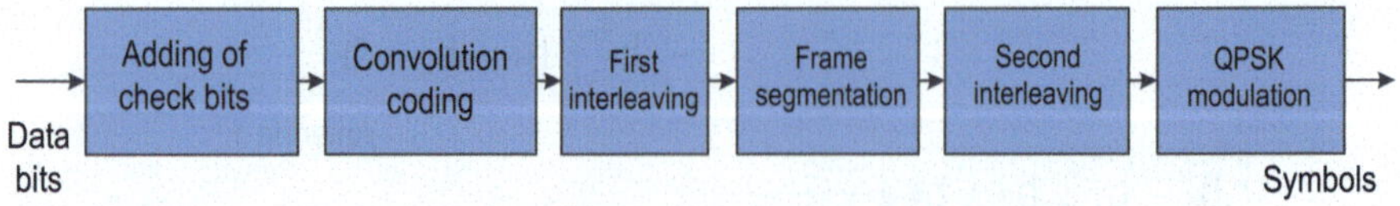

Fig. 9.9 Formation of the primary common control channel

broadcast message bits into the symbols of the primary common control channel is presented in Fig. 9.9.

The sequence of UMTS signal processing is presented in Fig. 9.10. Operation of the receiving–measuring algorithmic complex is described in [2] in detail.

The analyzer of UMTS base station signals enables to detect cell signals in the given spectral range, to receive their identification data and to perform a great volume of physical level measurements. It performs the following functions:

- Detection of UMTS base station signals, representation of the signals spectrum;
- Determination of scrambling sequence numbers of detected cells;
- Determination of a number of cells transmitting antennas;
- Reception of identification parameters of the detected cells:
  - Cell identifier,
  - Mobile network code (MNC),
  - Local area code (LAC),
  - Mobile country code (MCC),
  - Identifier of cell blocking for subscribers CB (Cell Barred),
  - Estimation of frequency shift;
- Power measurements:
  - Estimation of mean power of the UMTS signal,
  - Estimation of pilot signal power, primary sync signal power, secondary sync signal power, and primary common control channel power,
  - Formation of codogram—signal power of all code channels,
  - Calculation of utilization coefficient of code channels,
  - Determination of peak-factor and crest-factor of the UMTS signal;
- Signal-quality measurement:
  - Visualization of modulation symbols estimation of the primary common control channel,
  - Estimation of EVM of pilot signal and primary common control channels,
  - Signal/(interference + noise) ratio for the pilot signal,
  - Estimation of noise level,
  - Estimation of carrier suppression level,
  - Estimation of bit and packet errors probability;
- Measurements of the propagation channel—determination of the multi-path profile and effective channel length;

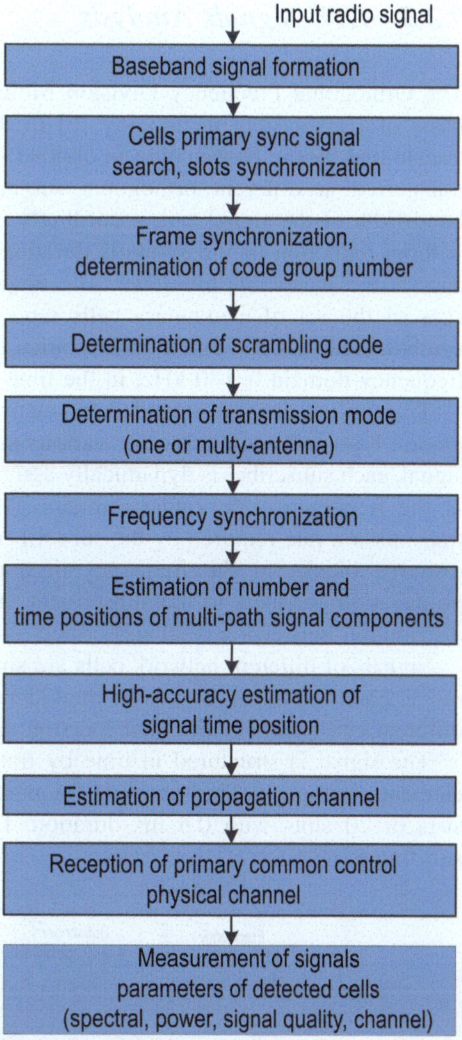

Fig. 9.10 Sequence of UMTS signal processing

- According to measurements fulfilled, the analyzer permits:
  - To provide formation of the coverage area of the UMTS network taking into account data from the digital maps of locality;
  - To calculate location of detected base stations.

9.2.3 LTE Signals Analysis

The Orthogonal Frequency Division Multiple Access (OFDMA) approach is the basis of LTE system technology [6] in the downlink channel. The data stream transmitted by the base station is divided into several low-rate streams, which are transferred at different orthogonal sub-carriers. This permits to use the wide bandwidth of the signal and, accordingly, to obtain high rate of data transmission without reduction of the symbols duration and keeping the inter-symbol interference at the acceptable low level. The frequency–time resource of the system consists of the set of elementary cells representing one sub-carrier of one OFDM symbol. Cells are combined into resource blocks. Size of the resource block in the frequency domain is 180 kHz, in the time domain −0.5 ms.

Transmitted data are intended for many subscribers. In general case, the modulation type and coding rate for various subscribers are different. In the OFDMA signal, each subscriber is dynamically assigned by some quantity of resource blocks in the frequency–time signal domain. This quantity is determined by the data transmission rate required by the subscriber and by the cell occupancy. Position of resource blocks in the frequency–time domain depends on characteristics of propagation channels to the subscribers. An example of frequency–time resource distribution between several subscribers is shown in Fig. 9.11.

Signals of different network cells are simultaneously transmitted in one spectral range. Each cell has its own individual identifier, which defines signal parameters of different cell channels and their location in the frequency–time domain.

The signal is structured in time by frames with duration of 10 ms or 307200 elements (samples). The sampling frequency $f_{LTE} = 30,72$ MHz. One frame consists of 20 slots with 0.5 ms duration. Each consecutive pair of slots forms a sub-frame.

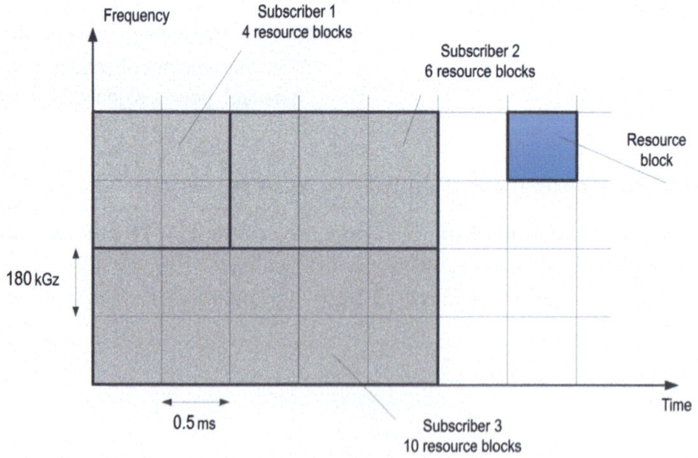

Fig. 9.11 An example of frequency–time resource distribution

In LTE two types of guard interval (prefix) are possible for OFDM symbols: standard and extended. At typical interval between sub-carriers of 15 kHz the number of OFDM symbols in the slot equals to 7 for the standard prefix and for extended one equals to 6. At that, the length of the standard prefix equals to 160 elements for the initial slot symbol and to 144 elements for the other slot symbol. The length of extended prefix is 512 for all slot symbols.

The LTE standard anticipates two modes of duplex: FDD and TDD. For the FDD mode, all sub-frames are used for transmission both in downlink and uplink channels at different carriers. For the TDD mode, some sub-frames are used for signal transmission in the downlink channel, while other sub-frames—in the uplink channel. Seven different frame configurations are possible, which differ in number of sub-frames assigned to downlink and uplink channels and by their distribution in the frame.

The bandwidth of the LTE system can take one from the following values: 1.4; 3; 5; 10; 15; 20 MHz.

The signal of the LTE cell includes a totality of various physical channels: the primary/secondary synchronization signal, the reference signal, the physical broadcast channel (PBCH), the physical Hybrid ARQ indicator channel (PHICH), the physical control format indicator channel (PCFICH), the physical downlink control channel (PDCCH), the physical downlink shared channel (PDSCH). Each channel is transmitted in the definite frequency–time elementary signal cells. The scheme of cell signal formation is presented in Fig. 9.12.

Search for cells, slot and frame synchronization, determination of the cell identifier and duplex mode (FDD/TDD) are provided according to sync signals.

The reference signal is used by the subscriber stations for estimation of a number of transmitting antennas, frequency response of the propagation channel and for frequency tracking.

In the broadcast channel, the MIB message is transmitted, which contains the frame number, the signal bandwidth value, and PHICH parameters, which are necessary for localization of symbols of this and other channels. In the channel of

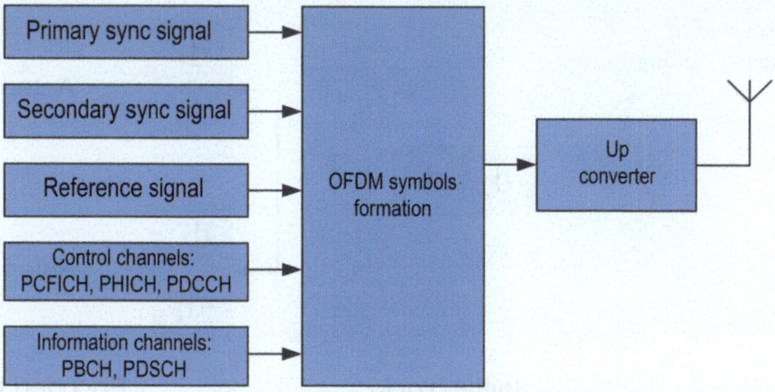

Fig. 9.12 Formation scheme of the LTE cell signal

the control channel indicator (PCFICH) the number of OFDM symbols in sub-frames used by the control channels PHICH is indicated.

In the control channels (PDCCH), parameters necessary for reception of the common information channel PDSCH are transmitted. In the information stream of the PDSCH the SIB messages are transmitted, in which identification and other data are contained.

The LTE standard leaves open a possibility of multi-antenna transmission in the downlink channel. The mode of spatial multiplexing can be used for information channels only. In this mode, the various data streams are transmitted from different antennas. This transmission method provides maximal capacity of the communication system; however, it requires utilization of several antennas not only on the transmitting side, but on the receiving side too (MIMO technology). The mode of spatial-frequency coding can be used for informational and control channels. It is possible to use two or four transmitting antennas. In this mode, the data transmission is performed by pairs of symbols transmitted simultaneously at the adjacent sub-carriers from two different antennas. The coding scheme for two transmitting antennas 0 and 1 is presented in Fig. 9.13.

For the case of four transmitting antennas, the spatial-frequency coding is performed in the similar manner, alternately, for different pairs of antennas: 0, 2 and 1, 3. LTE base stations may not support mode of diversity transmission; however, for subscriber stations, and hence, for the analyzer the support of diversity transmission is mandatory.

The primary sync signal represents the complex elements of the Zadoff-Chu sequence with one of three values of the parameter, which is defined by the base station sector. Usage of the Zadoff-Chu sequence is caused by good correlation properties of this sequence both in frequency and time domains. The primary sync signal is the same for all network cells with appropriate sector number. For FDD, it is transmitted in the last OFDM symbols of slots 0 and 10, for TDD—in third OFDM symbols of sub-frames 1 and 6 of each frame. In the frequency domain, the primary sync signal is transmitted in the central part of the signal spectrum.

Fig. 9.13 Scheme of spatial-frequency coding

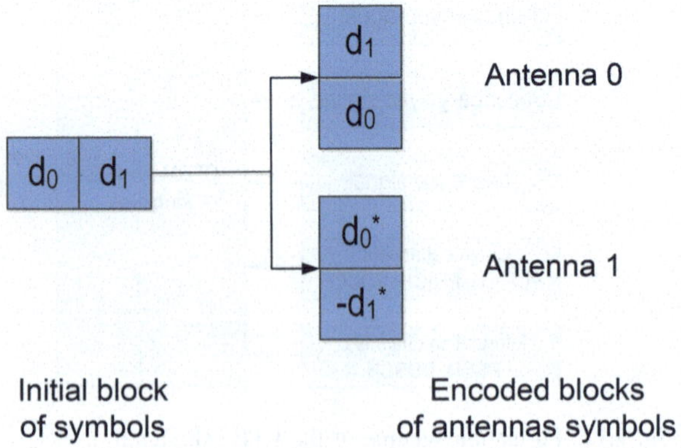

Symbols of secondary sync signal are transmitted in sub-frames 0, 5 and depend on the group identifier. The last, together with the sector identifier, defines the cell identifier of physical level.

To transmit the secondary sync signal, the same antenna is used as for transmission of the primary sync signal. For FDD, the secondary sync signal is transmitted in the third OFDM symbol from the end of slots 0 and 10 of each frame. For TDD—in the second OFDM symbol from the end of slots 1 and 11 of each frame. In the frequency domain, the transmission is carried out at the same sub-carriers as for the primary sync signal.

Symbols of the reference complex signal are determined by the Gold sequence with length of 31, and also by the slot number in the frame, the number of OFDM symbol in the slot, the prefix length and the cell identifier of physical level. The number of transmitted reference symbols depends on the bandwidth of the LTE signal.

The reference signal is transmitted from all antennas used for transmission. In the time domain, reference symbols for antennas 0 and 1 are transmitted in zero and also in fourth (standard prefix) or in third (extended prefix) OFDM symbols, and for antennas 2, 3—in the first OFDM symbol of each slot. In the frequency domain, the sub-carriers, which numbers depend on the number of OFDM symbol in the slot, the number of slot in the frame, the antenna number, and cell identifier of physical level, are used. At multi-antenna transmission, cells occupied by the reference signal are not used by other transmitting antennas. An example of reference cells position for different transmitting antennas is presented in Fig. 9.14 in the case of extended prefix.

The data packets representing the MIB messages serve as input data to form symbols of the broadcasting channel. They are transmitted with period of 40 ms. Each message occupies the interval of four frames. The broadcast channel is transmitted at 72 central sub-carriers of the first four OFDM symbols of the slot 1 of each frame. The scheme of symbol formation of the broadcasting channel is presented in Fig. 9.15.

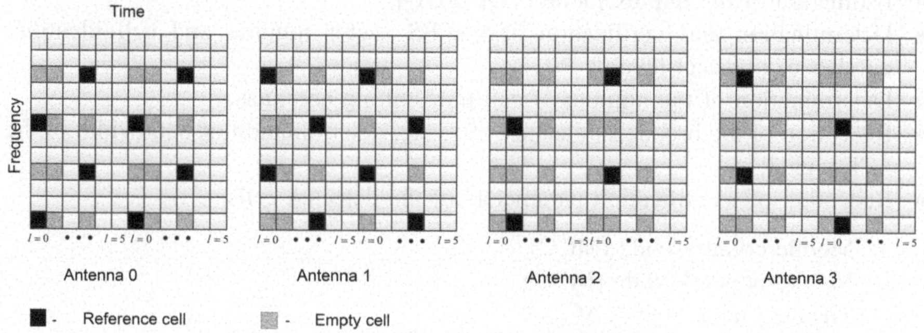

Fig. 9.14 Position of reference cells in the frequency–time domain

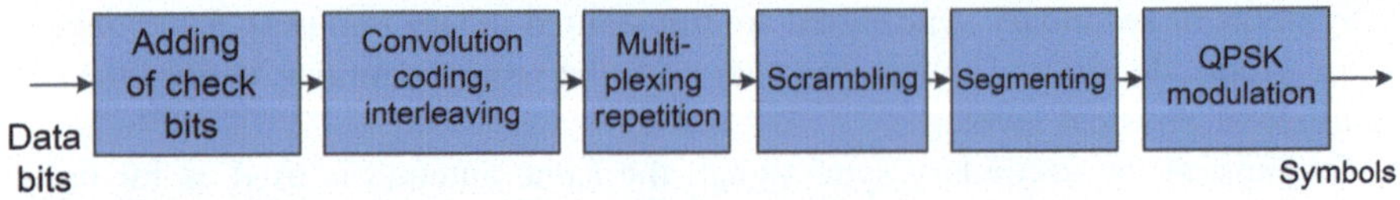

Fig. 9.15 Formation of the broadcasting channel

In the PCFICH, each value of the symbol number is coded in the form of appropriate sequence of bits. After scrambling, bits are mapped into symbols of QPSK modulation. These symbols are placed in 16 cells of the slot 10 in the first OFDM symbol of each even frame by four quadruplets (4 symbols each).

The control channel consists of one or several control channel elements. Each element includes 9 groups of cells. One group consists of the one quadruplet of symbols. Several control channels can be transmitted in the sub-frame. In connection with the SIB1 message, each control channel can contain 4 or 8 control channel elements. Procedures of check bits addition, convolution encoding, interleaving, repetition of obtained bits, multiplexing, scrambling, modulation, cyclic shift, and interleaving of symbol quadruplets are performed for bits of control channel message. Parameters of some procedures depend on the format of the transmitted message.

In PDSCH, the broadcast SIB1 messages are transmitted with a period of 80 ms. In contrast to the control channel, in PDSCH the procedure of cyclic shift and interleaving of quadruplets is absent, instead of convolution encoding the turbo-encoding is used, and representation of formed symbols is fulfilled onto the resource-blocks.

The sequence of LTE signal processing is presented in Fig. 9.16.

The LTE base station signal analyzer allows detecting of the cell signals, receiving its identification parameters and performing a great volume of physical measurements in the given spectral range.

Developed LTE BS analyzer executes the following functions:

- Detection of broadcasting LTE BS signals, representation of signal spectrum;
- Estimation of prefix length of the LTE signal;
- Estimation of the duplex mode (TDD/FDD);
- Determination and verification of the BS sector number and cell identifier number of physical level;
- Determination of the number of cell transmitting antennas;
- Reception of the broadcast channel message, determination of bandwidth of the LTE signal;
- Reception of identification parameters of the detected cells:
  - Mobile country code (MCC),
  - Mobile network code (MNC),
  - Tracking area code (TAC),
  - Identifier of BS in the eNB network (ID),

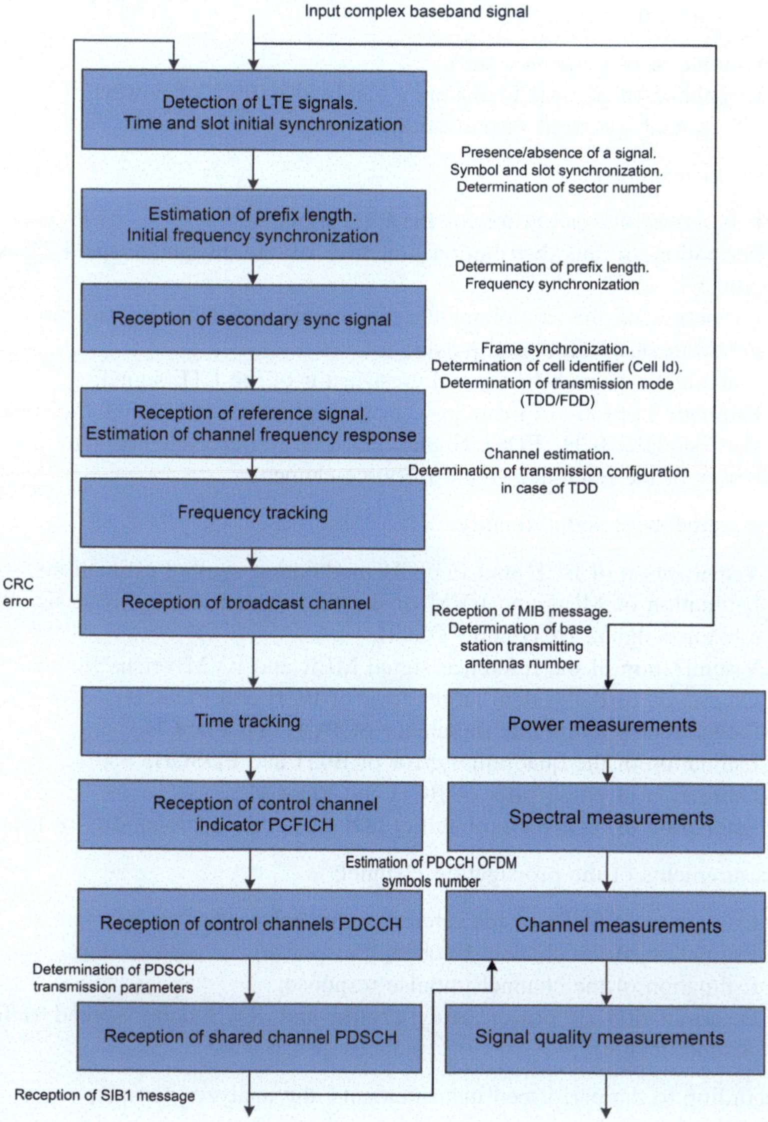

Fig. 9.16 Sequence of LTE signal processing

- Identifier of BS cell (Cell ID),
- E-UTRAN cell global identifier (ECGI),
- Identifier of Cell Barred for subscribers (Cell Barred);

- Spectral measurements:
  - Estimation of frequency shift,
  - Estimation of occupied frequency bandwidth (99% of power),
  - LTE signal spectrum visualization;
- Power measurements:
  - Estimation of mean power of the LTE signal,
  - Formation of the distribution function of the instantaneous LTE signal power,
  - Formation of the complementary cumulative distribution function of the instantaneous LTE signal power,
  - Calculation of peak-factor and crest-factor of the LTE signal,
  - Estimation of ratio of mean power of the primary sync signal, the secondary sync signal, BCH, PDCCH and PDSCH resource-elements to the mean power of the reference signal resource-element;
- Measurements of signal quality:
  - Visualization of BCH and PDSCH modulation symbol estimations,
  - Estimation of MER and EVM of the primary, secondary sync signals, the reference signal, BCH and PDSCH,
  - Visualization of the reference signal MER and EVM versus frequency,
  - Estimation of the system target error of BCH and PDSCH,
  - Estimation of amplitude imbalance of BCH and PDSCH,
  - Estimation of the quadrature error of BCH and PDSCH,
  - Estimation of phase jitter of BCH and PDSCH,
  - Estimation of bit error probability and packet error probability of the BCH;
- Measurements of the propagation channel:
  - Estimation of SNR for the reference channel,
  - Estimation of the channel frequency response,
  - Estimation of the channel impulse response,
  - Determination of power delay profile and RMS delay spread (effective channel length).
- According to the performed measurements, the analyzer permits:
  - To execute the formation of the LTE network coverage area taking into account the data of digital maps of the locality;
  - To localize the detected base stations.

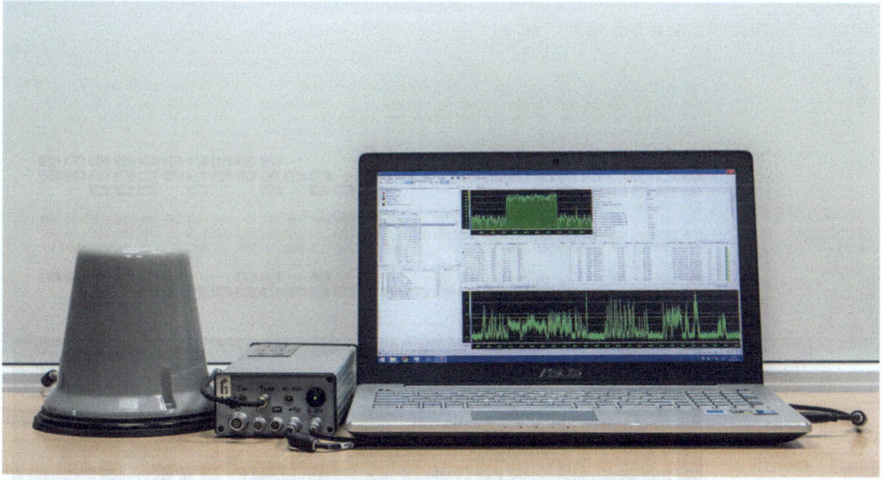

Fig. 9.17 Hardware–software analyzer of base station signals

9.2.4 Examples of Analyzer Operation

Figure 9.17 shows the hardware–software analyzer of base station signals of GSM, UMTS, LTE networks of cellular communication. Its hardware part represents the portable measuring receiver ARGAMAK-M. The SMO-BS 3GPP software part of the analyzer includes modules realizing receiver-measuring algorithmic complexes of GSM, UMTS, LTE signal processing and an interface [7].

Figures 9.18, 9.19, 9.20, 9.21, and 9.22 show examples of SMO-BS 3GPP software windows. Signal reception was executed in the fixed point of the industrial center. In the left upper part of the window, we see the tree of detected cells, which are grouped according to the network operators. In the left lower part, there is a list of tasks, in which standards, network operators and range of analyzed frequencies can be selected. In the lower part of the window, the spectrum of the received signal is represented in the preset frequency band.

In the center of the GSM window in Fig. 9.18, we see numbers of operator channels, in which cell signals are detected. Below, there is a table of detected cells including detection time and main identification parameters. In the right upper part of the window, there is extended list of received and measured signal parameters for the chosen cell.

In the center of the UMTS window in Fig. 9.19, the output signal of the primary sync signal correlator is presented in the graph for the chosen frequency channel. This signal characterizes general signal-interference situation in the reception point, in particular, allows estimation of relative signal powers for detected cells. The horizontal line corresponds to the search threshold. We see that in the frequency channel under investigation five cells are detected, signals from which are powerful enough.

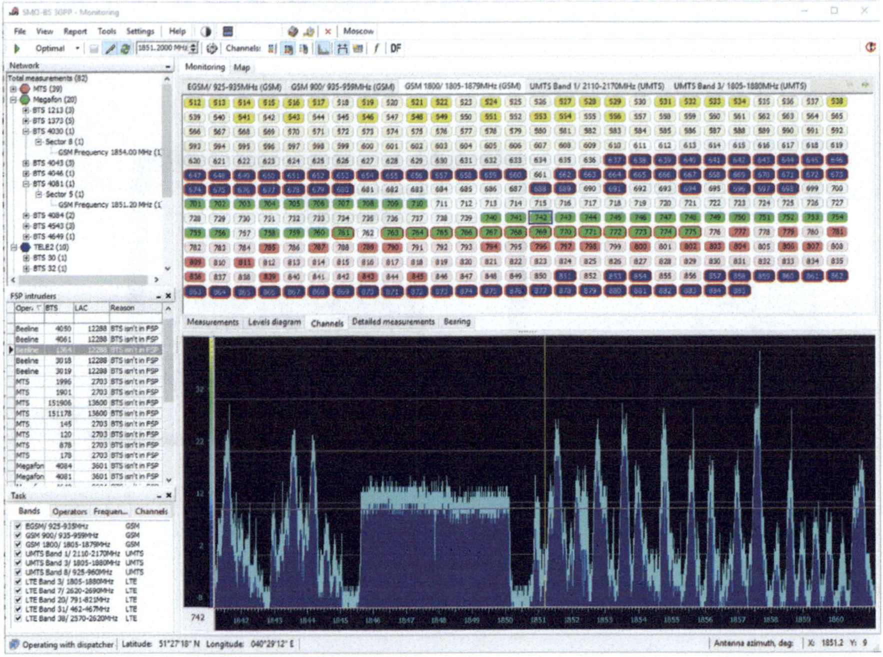

Fig. 9.18 Window of the GSM signals analysis software

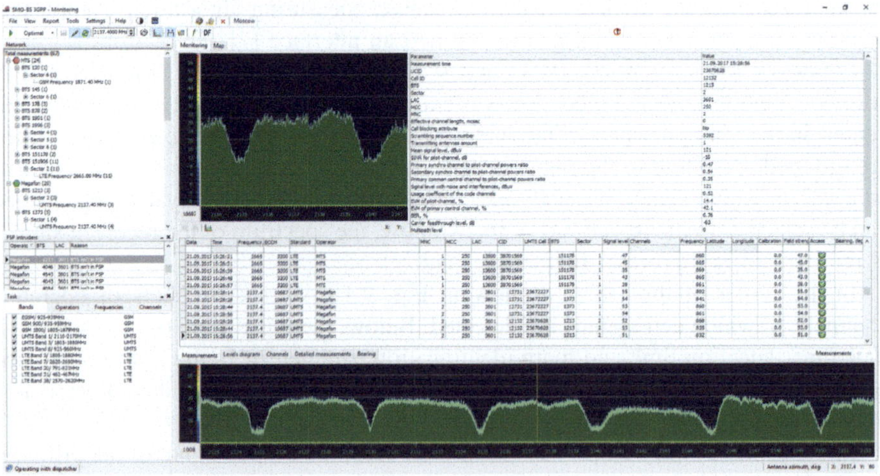

Fig. 9.19 Window of UMTS signals analysis software, the "Measurements" mode

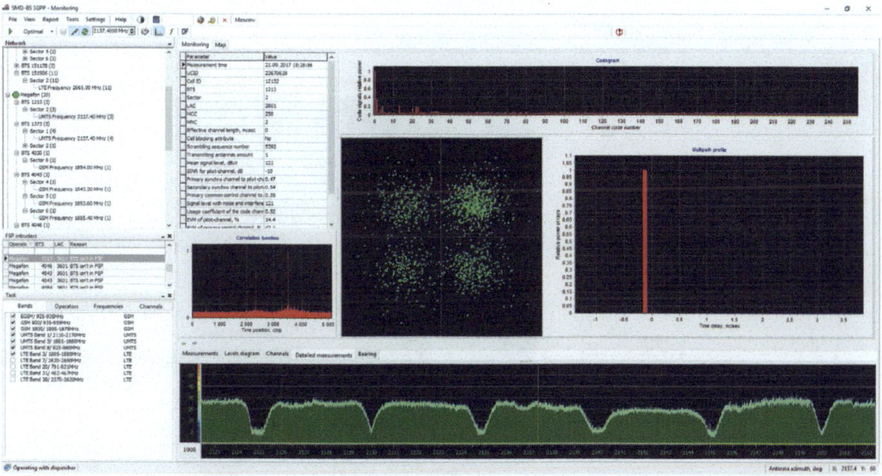

Fig. 9.20 Window of UMTS signals analysis software, the "Detail measurements" mode

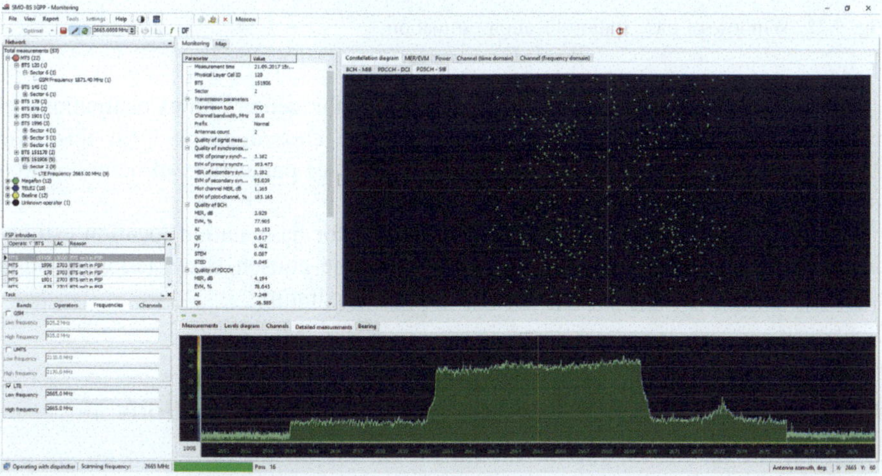

Fig. 9.21 Software window of the LTE signals analysis

In the UMTS window in Fig. 9.20, three graphs for the chosen cell are shown. In the upper part, the codogram is shown—the function of signal power estimation versus a number of its code. The channel code with number 0 corresponds to the pilot signal, while the channel code with number 1 corresponds to the primary common control channel. For visual estimation of demodulation quality, below on the left, there are estimations of modulation symbols of the primary common control channel. On the right, estimation of multi-path profile of the propagation channel is presented. Relative signal powers of the paths are 1, 0.32, 0.26.

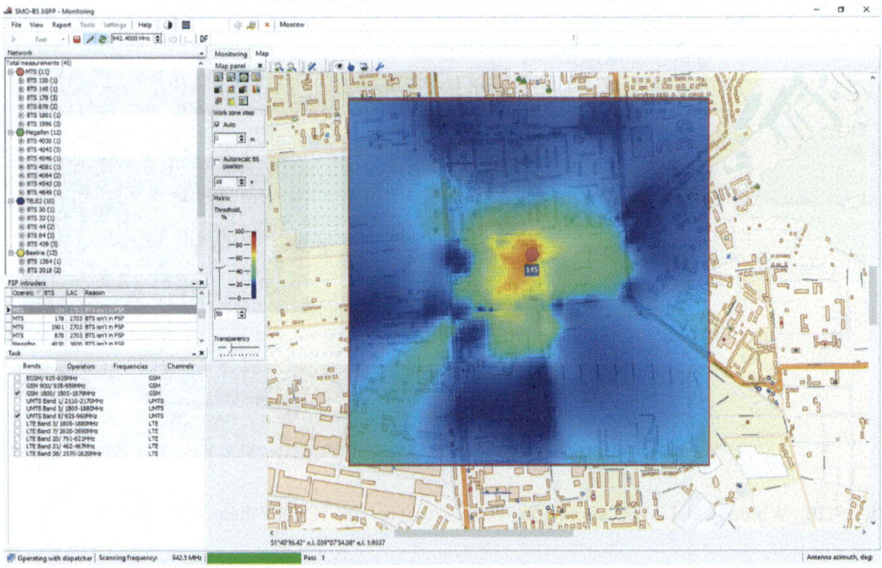

Fig. 9.22 Window of base station location estimation

In the LTE window in Fig. 9.21, the graph in the center shows estimations of modulation symbols (signal constellation) of the chosen channel (in this case PDSCH) for the given cell. On the left from it, we see parameters of demodulation quality.

Figure 9.22 shows an example of the window for base station location estimation. The two-dimensional decision function of the chosen base station location estimation was made according to the function maximum.

9.3 Analyzer of IS-95/cdma2000/EV-DO Base Stations Signals

The technology of Code Division Multiple Access (CDMA) allows all subscribers to operate in one frequency range. In the cellular communication, the CDMA method appeared in 1995. Before it, the method of code division was widely used in military communication systems.

Cellular communication standards based on the CDMA technology have evolved from IS-95 standards (2G) to cdma2000 standard (2.5G), and then—to EV-DO standard (3G).

The principle of code division is used to extract signals of different BS in all these systems. To divide signals in IS-95 and cdma2000 systems, the CDMA principle is used and in EV-DO systems—the TDMA principle.

During improvement of CDMA technology, the portability of technological solutions has been saved, and maximally, the reverse technical and frequency compatibility have been realized.

9.3.1 IS-95/cdma2000 Signal Structure

Let us consider the IS-95/cdma2000 base station signal structure [2]. The central frequency of the base station transmitted signal is defined by a number of the frequency channel. For synchronization, in accordance with a standard, two fixed frequency channels—primary and secondary, are anticipated. Each base station transmits at least one pilot channel and one sync channel at the frequency of either primary or secondary channels. Numbers of operator frequency channels of the CDMA cellular communications are usually known.

The CDMA signal includes the set of parallel channels: the pilot channel, up to one sync channel, up to seven paging channels, and some quantity of traffic channels. For cdma2000 system, depending on configuration, the base station signal may contain other control, pilot, and traffic channels. The pilot signal is transmitted continuously and is used by mobile stations for initial synchronization. Identification parameters, parameters intended for frame synchronization in mobile stations, and parameters of the Long code permitting the scrambling and data reception of the paging channel are transmitted in the sync channel messages. Paging channels are used for service information transmission to the mobile stations. In the paging channel, the various types of messages including broadcast ones are periodically transmitted.

Channel signals are expanded in spectrum by the orthogonal Walsh functions and by the general pseudo-random sequence (PRS). For base station signals of IS-95 networks, the Walsh codes of 64 chips in length are used. For base station signals of cdma2000 networks, the Walsh codes of various lengths (128, 64, 32, and 16 chips) are used. For both standards, the pilot channel uses the Walsh function W_0^{64}, the sync channel uses the Walsh function W_{32}^{64}, and the paging channel uses the Walsh function W_1^{64}. In designation of the Walsh function W_k^n, the lower index k is the length of the Walsh function, and the upper index n is its number. Other orthogonal Walsh functions are distributed between traffic, control, and other active channels of the system.

The time structure of various channels is shown in Fig. 9.23.

The pilot signal represents the periodically repeating PRS with duration of 26.67 ms. The chip (PRS elementary symbols) frequency is $f_{rep} = 1.2288$ MHz. The beginning of the appropriate period (nearest to the time moment of beginning of even seconds) of the pilot signal is shifted with respect to the start of even seconds by PILOT_PN × 64 chips. The value of the shift index PILOT_PN is distinctive for each base station. All pilot signal bits equal to 0.

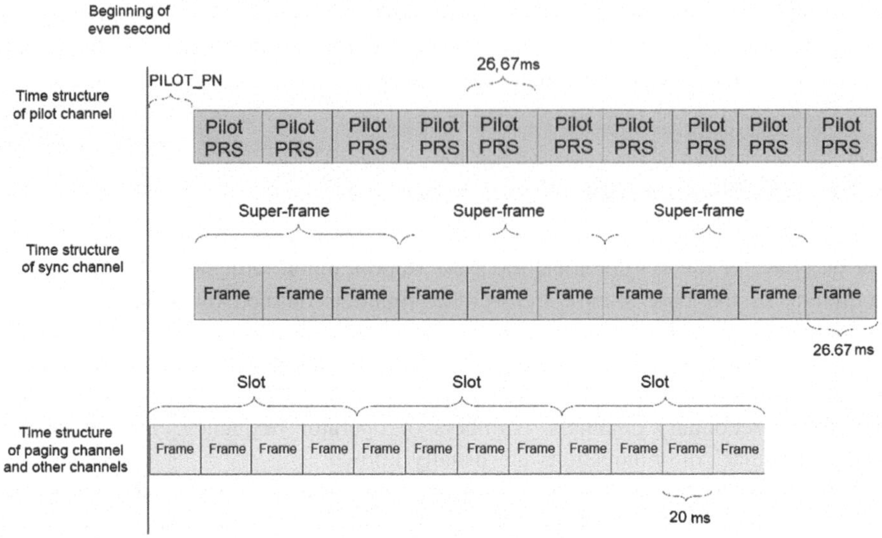

Fig. 9.23 Time structure of various channels of the CDMA signal

The sync channel is structured in the form of super-frame sequence with 80 ms duration, each of which is divided into frames with 26.67 ms duration. At that, intervals of sync channel frames coincide with appropriate intervals of PRS. The sync channel super-frame begins in the moment of even second plus shift of the pilot PRS. In the sync channel, the continuous transmission of its messages is executed. Transmission rate of sync channel data is 1200 bit/s. The sync channel message consists of bits of the message body, where message parameters are encoded, bits of message length and check bits. Bits of sync channel message are added by zero bits so that the formed message capsule (Fig. 9.24) of the sync channel would occupy three super-frames of the sync channel. The special bits (SOM—Start of Message), which are the first bits of each frame and indicate beginning of the sync channel message capsule, are periodically inserted into the capsule block. The capsule bits are transmitted in the sync channel frames after convolution encoding, repetition, and interleaving.

The signal of paging channel represents the sequence of slots with 80 ms duration. Slots of the paging channel are divided into frames with 20 ms duration which are divided into half-frames with 10 ms duration. Beginning of each 25th slot

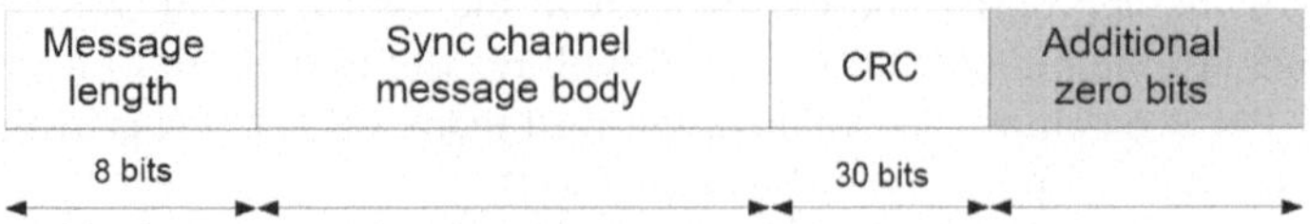

Fig. 9.24 Structure of sync channel message capsule

of the paging channel coincides with beginning of even second independently on the value of the shift index PILOT_PN. The rate of data transmission of the paging channel is 4800 bit/s or 9600 bit/s. The paging channel message consists of bits of the message body, where paging message parameters are encoded, bits of message length, and check bits. To form a capsule, the message bits of the paging channels are added by zero bits. Their number depends on the message type, data transmission rate in the paging channel, and on the fact whether transmitted messages are synchronized (starting in the beginning of the half-frame) or not. The special Synchronized Capsule Indicator (SCI) bits, which are the first bits of each half-frame and indicate the beginning of the message capsule of the paging channel for synchronized messages, are periodically inserted into the capsule block. The capsule bits, after convolution encoding, repetition, interleaving, and scrambling (superposition of the Long code), are transmitted in the frames of the paging channel.

From the large number of various messages of the paging channel, the analyzer receives data of two types of messages: the system parameters message containing the base station identifier and its coordinates, and the message about cell channels (CDMA Channel List Message). Other messages of the paging channel are not analyzed.

The time structure of other channels is similar to the paging channel structure, i.e., time intervals of these channel frames are the same.

The bit streams of all channels of the CDMA signal, description of which is discussed above, are the initial data for formation of the CDMA modulated baseband signal. Transformation of bit values (0; 1) to the set (1; −1), channel weighting, orthogonal channel expansion, multiplexing of symbol channels streams, signal spreading by the pseudo-random pilot sequence, and band-pass filtering is carried out. The general scheme of baseband signal formation for the cdma2000 standard is presented in Fig. 9.25. For IS-95A, IS-95B standards, the similar scheme is used.

The data stream passing with the rate of 28.8 kbit/s, after interleaving at the subscriber side, is divided into 6-bit portions. One of 64 Walsh sequences ($2^6 = 64$) is associated with each fragment. After this procedure, the 307.2 kbit/s stream is transformed with the help of the Long code sequence, which is similar to the sequence at the base station. Each element of the Walsh sequence is presented by 4 elements of the pseudo-random sequence. After that, the stream is divided between channels, additionally transformed with the help of the short code sequence, delayed by half-symbol in the quadrature channel, as it is required at PSK with a shift, and is filtered and passed to the modulator.

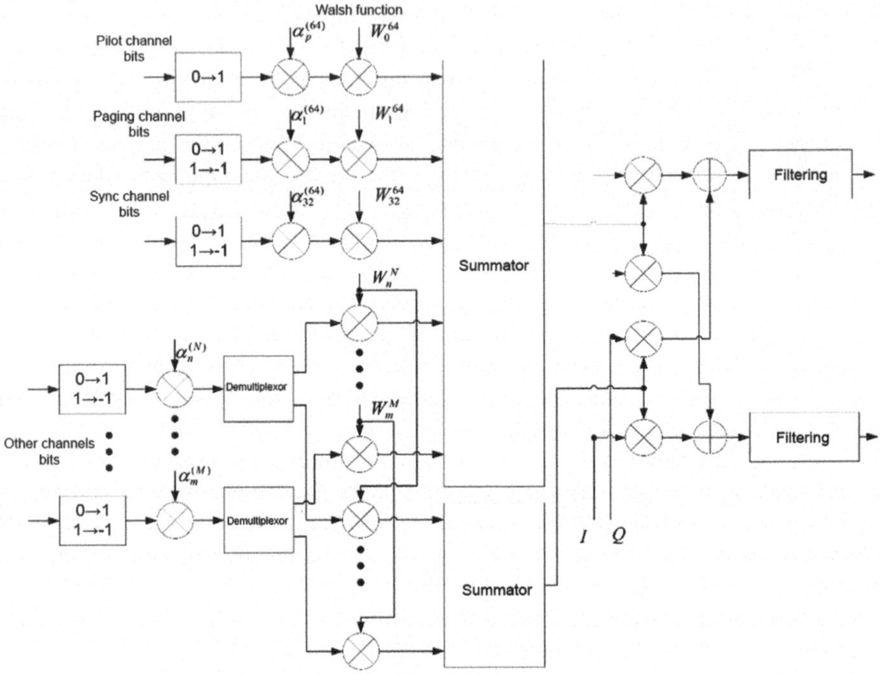

Fig. 9.25 Formation of baseband signal for the cdma2000 standard

9.3.2 The Analyzer of IS-95/cdma2000/Base Stations Parameters

This analyzer is intended for identification and parameters measurements of CDMA base station networks radio signals, which operate according to IS-95A, IS-95B, cdma2000 (IMT MC-450) standards. The analyzer can also operate with signals of last generations of CDMA networks, such as 1xEV-DO, 1xEV-DV. It uses only broadcast data and ignores information of the subscriber traffic channels. Since the analyzed channels are transmitted in the bandwidth of 1.25 MHz, the analyzer can successfully operate not only in 1x network but also in networks 2x, 3x, etc. The structure of CDMA signal was considered in Sect. 9.3.1.

As the hardware–software complex for signals of GSM, UMTS, LTE base stations, the CDMA analyzer consists of the hardware part—the measuring DRR of the ARGAMAK-family and the software part—the SMO-BS CDMA system mathematical support software [7]. The last can be installed on the computer using the unified packet together with other software. The window of SMO-BS CDMA in the mode of base stations monitoring is shown in Fig. 9.26.

The analyzer provides obtaining of the following identification broadcast data of the base station signals:

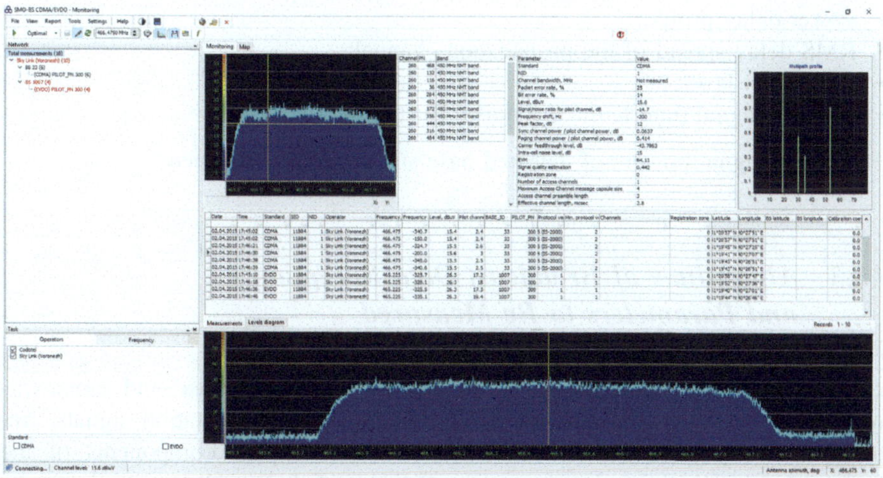

Fig. 9.26 Window of SMO-BS CDMA software

- System identifier,
- Network identifier,
- Level of protocol (standard),
- Minimal level of the mobile station signal, which allows functioning with this base station,
- Shift index of the pilot PRS,
- Identifier of the base station,
- Longitude and latitude of base station position,
- Registration zone,
- Numbers of the network frequency channels.

At that, the algorithms of signal parameters estimation of CDMA base stations are realized, including:

- Frequency shift;
- Bandwidth;
- Sum signal power;
- Ratio of the peak power to the mean power of the sum signal;
- Pilot signal power;
- Codogram—signals power of all code channels;
- Utilization factor of code channels;
- Ratio of the paging channel power to the power of the pilot channel;
- Ratio of the sync channel power to the pilot channel power;
- Level of inter-cell interference;
- Level of carrier suppression;
- Vector of errors EVM;
- Quality parameter ρ of the signal;

- Power delay profile;
- RMS delay spread of the propagation channel;
- Multi-path level;
- Ratio of signal/(interference + noise) for the pilot signal.

Besides, signal powers of detected base stations are determined.

9.3.3 Algorithms of Broadcast Data Reception and Parameters to Be Measured

Reception of broadcast information supposes procedures of baseband signal formation with necessary sampling frequency, search for base station signals, frequency–time synchronization, determination of multi-path components of the detected base station signals, estimation of the channel, as well as directly procedure of sync and paging channels messages reception. Under channel estimation, we understand estimation of complex envelopes of path signals. The structure of the CDMA base station signals processing realized in the SMO-BS CDMA software is presented in Fig. 9.27.

The digital signal at the intermediate frequency with the bandwidth of 2 MHz with sampling frequency $F_s = 6,4$ MHz, from which the baseband signal is formed, is applied to the input of the software part. For this, the mean value of the input sample stream is determined, which is subtracted from these samples values with the aim of compensation of ADC mean value of the receiver. Then transfer to the baseband is executed. After frequency transfer, the signals are filtered to suppress higher harmonics and to change initial signal bandwidth to the band of 1.25 MHz. Then, signal oversampling is performed with the aim to transfer the sampling frequency to $F_s = 6,4$ value.

To search for the pilot signal, the decision function is formed for time positions from the a priori interval with 26.67 ms duration located after ½ of the chip. The decision function is defined by the output signal of the filter matched with the pilot PRS. Formed values of the decision function are compared with the adaptive threshold. If these values do not exceed the threshold over all a priori interval, the decision is made about pilot signal absence in this frequency channel, and the further operation with this frequency channel is interrupted. If for some time positions the threshold is exceeded, the decision is made about pilot signal presence and appropriate time positions are assumed as approximate estimations of time positions of the base station signals. Initial time shift of different base stations signals is divisible by 64 chips, which allows distinction of multi-path components of one base station from the signal of other base stations.

Estimation of frequency offset between generators of transmitting and receiving sides is executed according to the pilot signal of detected base stations. If necessary, the frequency correction of the input baseband signal is provided.

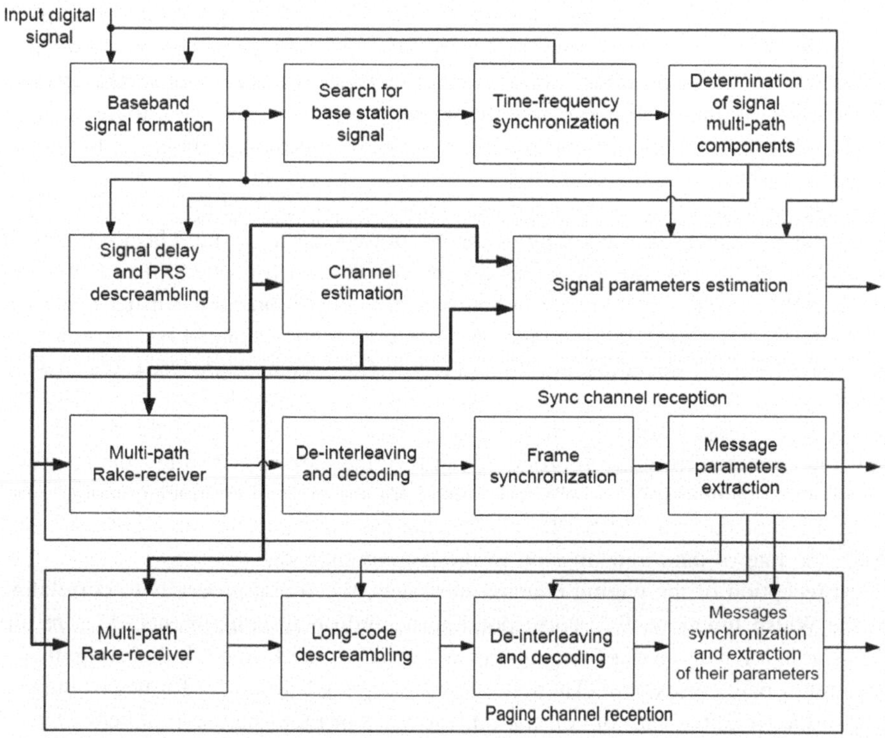

Fig. 9.27 Structure of the processing algorithms of the CDMA signal

Further, the multi-path components of the base stations signals are determined. At that, the decision function values taken each ¼ of chips in some vicinity of earlier-obtained time positions of the detected signal are analyzed. These estimations are used then for synchronization of single-beam receivers of the Rake receiver at estimation of the channel and data demodulation.

For multi-path signal components of each detected base station, the signal is decimated with coefficient of 1/4. As a result, for each single-beam receiver, the samples stream with 1 sample per chip is formed. Then pilot PRS de-scrambling is performed. For this, multiplication of decimated signal samples by values of reference complex pseudo-random pilot sequences is executed. Resulting signal samples proceed to the block of channel estimation and blocks of message reception of the synchronization channel and the paging channel.

The channel estimation is formed using the sliding window method for each single-path receiver. These estimations are used for demodulation of received symbols of the paging channel and the sync channel.

At reception of the synchronization channel messages, the signal proceeds to the correlator of the Walsh function W_{32}^{64} corresponding to multi-path components. As a result, the correlation responses of the sync channel symbols for each multi-path component are formed at their outputs. On their base, decisions about symbols in the form of the weighted sum real part of multi-path components correlation responses are formed, where weights represent the complex-conjugated channel estimations of these components (the Rake receiver). Then, de-interleaving of symbol blocks and multiplexing of the same (repeated) symbols solutions are performed. Then Viterbi decoding is executed. Beginning of the sync channel message is determined in the decoded symbol stream according to the SOM bit. Verification of message decoding is carried out according to the check-up bits. If sync channel message decoding was executed with error for several times one after another, the decision is made about bad reception conditions, and further operation with signal of this base station in the considered frequency channel is interrupted. In the case of correct decoding, the main parameters, which characterize the investigated base station—the system identifier SID, the network identifier NID, the protocol level (standard), the minimal supported protocol level of the mobile station, are extracted from the appropriate fields of the message. Besides, the auxiliary parameters are determined, which are used for the paging channel reception: the shift index of the pilot PRS PILOT_PN, the system time, the status of the Long code, the rate of data transmission in the paging channel.

At reception of the paging channel messages, the signal proceeds to correlators of the Walsh functions W_1^{64} corresponding to multi-path components. As a result, the correlation responses of paging channel symbols are formed at their outputs for each multi-path component. Then, decisions of symbols (bits) are formed as the real part of the weighted sum of correlation responses of multi-path components, where weights represent complex-conjugated estimations of the channel of appropriate components. After that, de-interleaving of symbol blocks and multiplexing of soft solutions of similar symbols in the case, when there is a repetition in the paging channel, are performed. Than Viterbi decoding is executed. The type of messages in the decoded symbols stream is performed using SCI bits, the type of messages (synchronized or non-synchronized) is determined, and also beginning and end of the stream messages are defined. The verification of decoding of paging messages according to check-up bits is fulfilled. If paging channel messages are decoded with error for several times one after another, the decision is made about bad reception conditions, and further operation with paging channel of this base station is interrupted. In the case of correct decoding, the necessary messages are selected, from the appropriated fields of which the parameters are extracted characterizing the receiving base station: base station identifier BASE_ID, longitude and latitude of base station position, registration zone, frequency channels of the network.

Broadcast data received by the analyzer and main parameters of base station signals of CDMA networks being estimated are given in Table 9.1.

The frequency offset ΔF as deviation of mean frequency of the base station signal from the average frequency of assigned channel is estimated for detected base stations. Estimation of the frequency offset is performed according to the pilot signal and fulfilled in accordance with interpolation algorithm [8], where characteristics of this algorithm are calculated, and it is shown that estimation of the

Table 9.1 Received broadcast data and parameters to be measured

Received data	Sync channel data: • System identifier, •Network identifier, •Protocol level (standard), •Minimal protocol level of mobile station, •Shift index of pilot PRS; Paging channel data: • Base station identifier, • Longitude and latitude of base station position, • Registration zone, • Network frequency channels.
Spectral measurements	Frequency offset Frequency bandwidth Spectrogram
Power measurements	Power of sum signal Ratio of peak power to mean power of sum signal Relative signal powers of detected stations
Measurements in the code domain	Power of pilot signal Codogram—signals powers of all code channels Utilization coefficient of code channels Ratio of sync signal power to the pilot channel power Ratio of paging channel power to pilot channel power
Estimation of signal quality	Errors vector EVM Parameter ρ of signal quality Carrier suppression level
Characteristics of propagation channel	Power delay profile RMS delay spread of propagation channel Multi-path level Level of inter-cell interference
Estimation of reception quality	Ratio of signal/(interference +noise) for pilot signal Bit error probability Packet error probability
Calculation of base station positions	Is fulfilled by amplitude method using results of measurements of signal power in several points of locality
Measurement of field strength in the frequency channel	Is measured with calibrated antennas
Construction of coverage area	Is executed using results of field strength measurement in several points of locality with account to digital map data
Estimation of emission power of the transmitter	Emission power of the sector antenna is estimated at known coordinates of the base station, height of antenna suspension, antenna array pattern, and results of field strength measurements and digital map data

frequency offset is Gaussian and unbiased, and its variance at sufficient number of samples in the correlation interval (which takes place in this case) practically equals to the variance of optimal estimation.

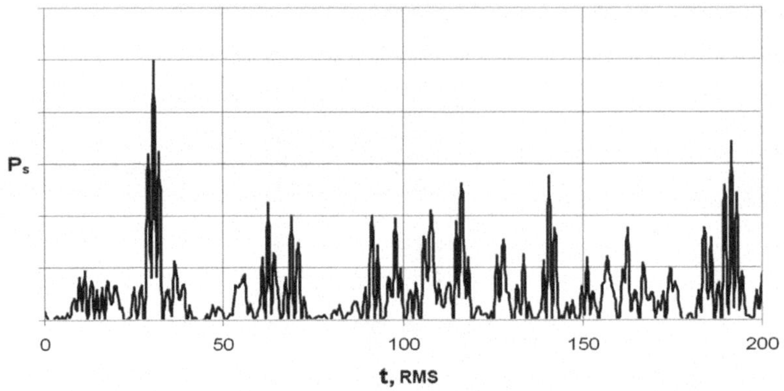

Fig. 9.28 An example of instantaneous power variation of the CDMA signal

The frequency bandwidth of the CDMA signal is defined as the spectral interval within the limits of which 99% of transmitted power falls within the given frequency channel.

The sum signal power is estimated as the mean power of the base station signal in the bandwidth of 1.25 MHz of the CDMA signal. It is known that CDMA signal has a high peak-factor. Therefore, for its estimation, the ratio of the peak power to the mean power of the sum signal must be determined.

Figure 9.28 shows an example of the function of instantaneous power of the sum CDMA signal P_s versus time t. The ratio of the peak power to the mean power for this example is 10.2 dB.

The CDMA signal represents a set of parallel channels expanded by orthogonal Walsh functions. The code area is specific and most important for the CDMA signal. Therefore, its detail analysis must be performed. Powers of all possible channels, which may be present in the CDMA signal under investigation, are estimated and visualized taking into account the multi-path property. At that, in contrast to known analyzers of base stations of cdma2000 networks, Walsh codes of different length are shown on the codogram.

The utilization coefficient of coded channels v represents the ratio of used codes to the total code resource and characterizes the occupancy of the analyzed base station.

The power P of the pilot signal is the most significant characteristic, which defines the coverage area of the base station. The analyzer calculates powers of pilot signals of all detected base stations.

Channel powers of the CDMA signal are different. Ratios of the powers of the sync channel and the paging channel to the power of the pilot signal γ_{SC} and γ_{PC} are estimated.

Figure 9.29 shows an example of the codogram for the signal of the cdma2000 network. Here numbers of the Walsh channel code are on the abscissa axis, and estimations of the channels relative powers are on the ordinate axis. Lines with

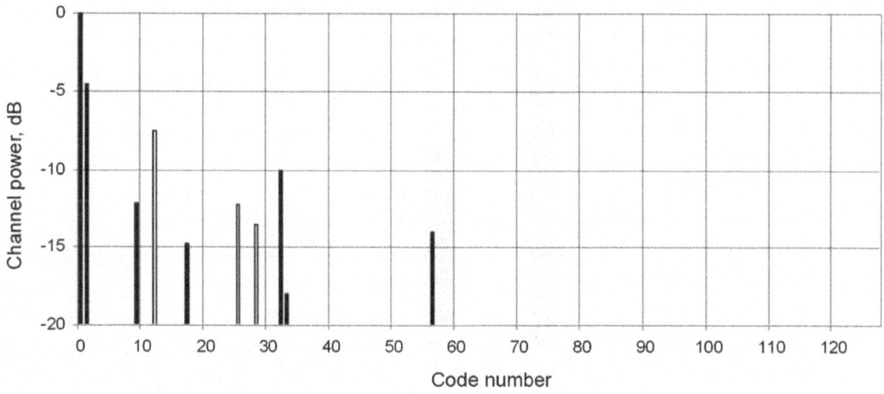

Fig. 9.29 An example of the codogram for the cdma2000 signal

filling correspond to channels with Walsh codes with length of 64 chips; lines without filling correspond to length of 32 chips. The zero Walsh function corresponds to the pilot signal, the 32nd Walsh function with length of 64—to sync channel, the 1st Walsh function with length of 64—to the paging channel. Traffic channels correspond to other Walsh functions with lengths of 32 and 64 chips. Estimation of the utilization coefficient of code channels for this example is 0.203. Estimation of the ratio of the sync channel power to the pilot channel power is −10 dB. Estimation of the ratio of the paging channel power to the pilot channel power is −4.5 dB.

The quality of the base station signal is characterized by several parameters: the Errors Vector Magnitude (EVM) calculated as the RMS value of difference between transmitted and received symbols; the parameter ρ which fully characterizes the signal quality, including synchronization errors and distortions of the signal shape, as well as the level of carrier suppression—the ratio of non-modulated signal power to the modulated signal power.

One of the reasons of reception degradation is time signal dispersion or multi-path property of the propagation channel. Estimation of the power delay profile is performed according to the pilot signal and is defined by the decision function of multi-path signal components search. An example of multi-path profile of the propagation channel is presented in Fig. 9.30.

In addition, auxiliary parameters of the propagation channel are calculated according to the power delay profile: RMS delay spread τ_{RMS}—the square root of the second central moment of the power delay profile and multi-path level—the ratio of signal power of all paths besides maximal to total power of all paths.

The input signal interference can be conditionally divided into intra-cell interference caused by multi-path property of the propagation channel, and inter-cell interference caused by signals of other base stations of the analyzed frequency channel. The last essentially exceeds the level of internal noise of the receiver. The

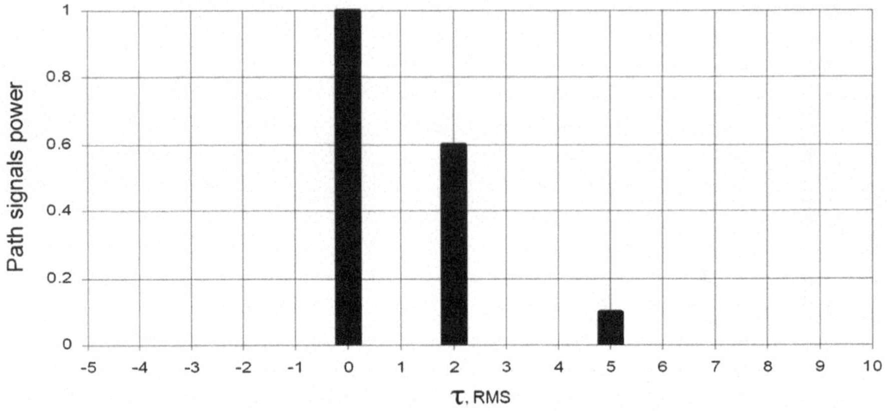

Fig. 9.30 An example of power delay profile of the propagation channel

analyzer estimates level of the inter-cell interference σ^2 using estimations of the mean values of path pilot signal power, path interference, and noise power.

Estimation of reception quality is characterized by bit error rate BER defined as the ratio of unsuccessfully decoded bits to the total number of received bits of successfully decoded packets of the sync channel, and by the packet error rate (PER) calculated as the ratio of unsuccessfully decoded sync channel packets to the total number of received packets. Besides, the analyzer estimates the signal/(interference + noise) ratio (SINR) for the pilot channel taking into account multi-path property of the propagation channel.

Calculation of base station location is provided on the base of the amplitude algorithm of coordinates' determination, which is based on account of source field amplitude dependence on the distance to it. The sample of field amplitude, the analyzer coordinates at the moment of measurements, and the array pattern of the transmitting antenna are the initial data for calculation. For specification of transmitter power, measurements of electromagnetic field in several points are used. The task is solved on the base of discrete search of power values from the range defined by operator, with the aim to search for such a power value, which will ensure maximal coincidence of measurement results with theoretical computations. At that, the locality relief, building up, and vegetation are taken into account. To calculate the base station location, to construct the coverage area, and to specify the transmitter power, the SMO-KN software packet is used [9].

To increase the accuracy of results, the measurement of signal parameters of base stations, such as frequency band, power of the sum signal, EVM, ρ, and level of carrier suppression, should be executed near the base station.

Measurement of parameters related to the propagation channel and the reception quality, such as pilot signal power, power delay profile, RMS delay spread of the propagation channel, signal/(interference + noise) ratio, rates of bit and packet errors, are performed in the arbitrary places of the coverage zone.

9.3.4 An Example of Analyzer Operation

Let us consider two examples of signal analysis of base stations for IS-95B and cdma2000 systems. Measurements were executed in the fixed point of the big city at the 881.25 MHz and 463.975 MHz frequencies, relatively.

The search decision function for different values of the shift index of the pilot PRS PILOT_PN in the cdma2000 network is presented in Fig. 9.31. The horizontal line corresponds to the search threshold. We see that five base stations were detected in the frequency band under investigation. The pilot signal of one of them is powerful enough. Values of the search decision function, at appropriate values of which the shift index exceeded the threshold, characterize the relative signal powers of detected base stations.

The part of the received data and also some measured parameters of detected base stations signals is shown in Tables 9.2 and 9.3. A blank in the tables means the decoding errors at reception of the corresponding message.

It can be seen that three base stations were detected in the IS-95B network, while in the cdma2000—five base stations. At SINR ratio more then −11 dB, for both networks, messages of sync and paging channels of two base stations were successfully received. When the SINR value was approximately −16 dB, the sync channel message was successfully received, while the message of the paging channel was received with error. When the SINR value was approximately −20 dB, messages of both channels were received with error. When signal was weak enough (SINR <−15 dB), some measurements, in particular, of the frequency offset, became unreliable.

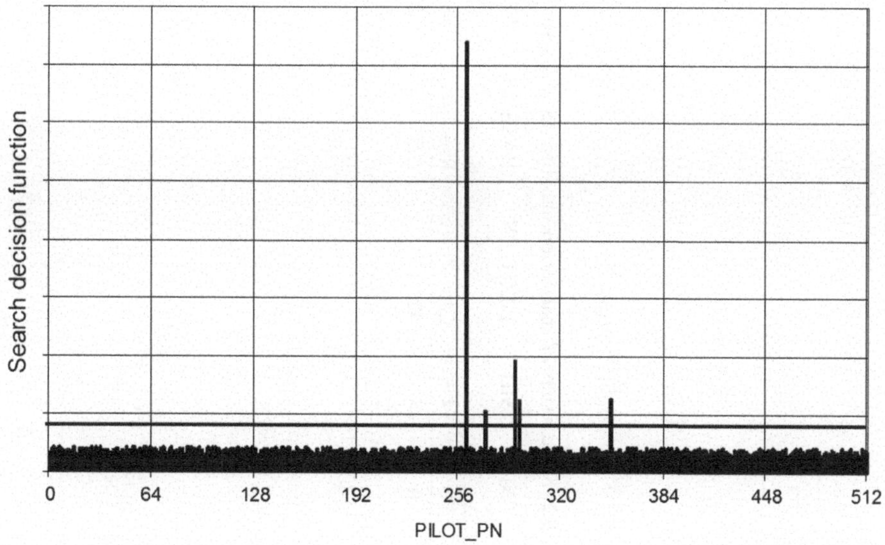

Fig. 9.31 An example of the search decision function

Table 9.2 Some results of analyzer operation for IS-95B system

N o	NID	SID	PILOT_PN	BASE_ID	ΔF, Гц	P, дБ	σ^2, дБ	τ_{RMS},MKC	SINR, дБ	ν	γ_c	$\gamma_п$
1	0	4	240	896	18.63	25.09	33.47	0.00	−8.88	0.27	0.09	0.32
2	0	4	424	275	12.80	23.63	33.92	0.00	−11.09	0.47	0.09	0.30
3	–	–	–	–	−145.00	21.30	34.56	1.85	−20.21	–	0.03	0.09

Table 9.3 Some results of analyzer operation for cdma2000 system

N o	NID	SID	PILOT_PN	BASE_ID	ΔF, Гц	P, дБ	σ^2, дБ	τ_{RMS}, МКС	SINR, дБ	ν	γ_c	$\gamma_п$
1	1	12061	261	936	−19.42	40.79	43.19	0.00	−2.45	0.20	0.10	0.35
2	1	12061	291	4364	−34.02	34.83	45.71	0.00	−11.29	0.81	0.10	0.35
3	1	12061	273	–	−170.00	31.94	45.76	0.00	−16.12	–	0.09	0.25
4	1	12061	351	–	33.47	32.95	45.31	0.00	−15.88	–	0.08	0.22
5	1	12061	294	–	−5.40	32.90	45.45	0.83	−15.99	–	0.07	0.20

9.4 Analyzer of TETRA Base Stations Signals

At present, the Trans-European Trunked RAdio (TETRA) system is one of the most popular trunking communication systems. This trunking system is based on the technical solutions and recommendations of the GSM standard and permits to create digital medium for realization of the complex systems of monitoring and communication. TETRA is the open standard of digital trunking radio communication developed by European Telecommunications Standards Institute (ETSI) for replacement of the old-fashioned standard MPT 1327 [10].

Standard openness represents one of its important advantages, which lies in the fact that equipment of different manufacturers corresponding to the standard is absolutely compatible. The worldwide standard recognition forced the creators to change the abbreviation sense of "TETRA," which now means "TErrestrial Trunked Radio."

Besides the standard function set of the trunking system, the communication TETRA systems essentially increase the data transmission rate, which is a key for development of various applications, such as positioning systems for mobile objects, systems of mobile access to remote databases; and support the unified digital space instead of several uncoordinated systems, which was impossible in analogous systems.

9.4.1 Signals of the TETRA Trunking System of Radio Communication

There are two standards of the system—TETRA PDO for data transmission and TETRA V+D (Voice plus Data). The last standard supports both data transmission and voice exchange; signals of this standard will be described later.

In the TETRA V+D standard, the time division multiple access (TDMA) to the time channels is used—time in four time slots is sequentially assigned for transmission in one frequency band. These four channel intervals (slots) form the TDMA-frames. Offset of adjacent radio channels is 25 kHz as in the common communication systems. The duplex channel offset for reception and transmission is 10 MHz. For systems of TETRA standard, several frequency sub-ranges can be used. In European countries, the ranges of 380–385/390–395 MHz are assigned for security agencies, while for commercial organization the ranges of 410–430/450–470 MHz are adopted. In Asia (first of all, in China), the range of 806–8670 MHz is used for TETRA systems.

Message transmission is provided by multi-frames. One multi-frame consists of 18 simple TDMA frames and has the duration of 1.02 s. Sixty multi-frames form the hyper-frame. The last TDMA frame in the multi-frame is the control frame. The frame duration is 56.67 ms. As has been mentioned above, the TDMA frame consists of four packets (time slots). The packet in the frame corresponds to an

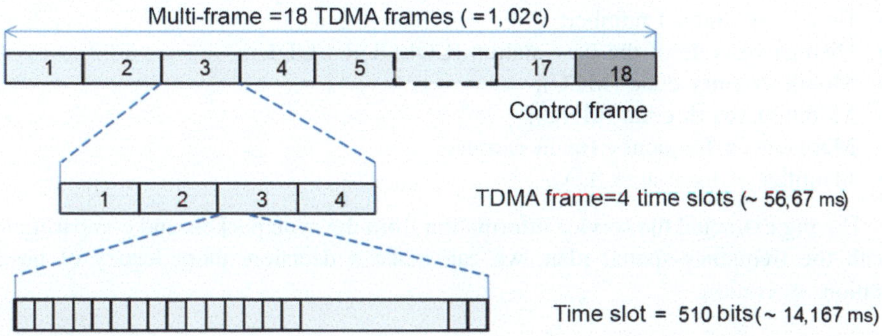

Fig. 9.32 Multi-frame structure

independent channel for data transmission. One packet occupies interval of 14.167 ms and contains 510 bits, which corresponds to 255 modulation symbols. In the middle of each packet, there is the sync sequence for synchronization and training of adaptive channel equalizer in the receiver. The multi-frame structure is shown in Fig. 9.32.

The low-rate voice encoder with the Code Excited Linear Prediction (CELP) algorithm provides the transmission of four voice channels in the band of 25 kHz. The stream speed at encoder output is 4.8 kbit/s. The digital $\pi/4$-DQPSK modulation, which allows decreasing of information stream transmission speed from 36 to 18 kbit/s, is used in this standard.

The complicated time structure of the physical channel, the variety of logical channel types, and great number of operation modes of the system as a whole and its separate elements assume application of different variants of logical channel location at physical level and their dynamic variation during operation. The TETRA standard supposes three states of the physical channel: for control physical channel (CP), traffic physical channel (TP), unallocated physical channel (UP). Depending on the state of the physical channel, the various types of packets and a set of control commands circulate in this channel.

The control channel is divided into main control channel (MCCH) and specific control channel (SCCH). Each base station has one main control channel. In turn, there are five types of control logical channels in this standard: the broadcast control channel (BCCH), the linearization channel (LCH), the signaling channel (SCH), the access assignment channel (AACH), and the stealing channel (STCH). The broadcast control channel is divided into the broadcast network channel (BNCH) and the broadcast sync channel (BSCH).

For base station synchronization with the mobile station and for transmission of control signals, the sync packets are transmitted in the broadcast control channel of the base station. Inside the sync packets, there are data blocks, in which the service information is transmitted including:

- Broadcast channel number;
- Distinctive code of the base station (COLOUR CODE);
- Mobile country code (MCC);
- Mobile network code (MNC);
- Main carrier frequency (main carrier);
- Identifier of local area (LA).

Having extracted the service information from the sync packets and comparing it with the frequency-spatial plan, we can make a decision about legacy of base stations operation.

9.4.2 *The Analyzer of TETRA Base Stations Signals Parameters*

In the previous section, the TETRA standard signal structure was considered, and now, we proceed to examination of the analyzer of these signals.

The analyzer of signal parameters for TETRA base stations is intended for a search for base station signals in the given spectral range, decoding of broadcast information, which is necessary for monitoring of base station parameters correspondence to the frequency-spatial plan.

The analyzer consists of the hardware part—the DRR of the ARGAMAK-family and the software part SMO-BS TETRA, which can be installed on the computer using the unified packet together with other software [1, 2, 7].

Reception of broadcast information assumes the procedures of baseband signal formation with necessary sampling frequency, search for base stations signals, time synchronization as well as direct reception procedure of sync channel messages. The structural scheme of digital processing of TETRA signals is presented in Fig. 9.33.

The digital signal at the intermediate frequency with the sampling frequency $F_s = 6.4$ MHz, from which the baseband signal is formed, is applied to the input of the processing device. For this, positions of frequency channels are calculated, in

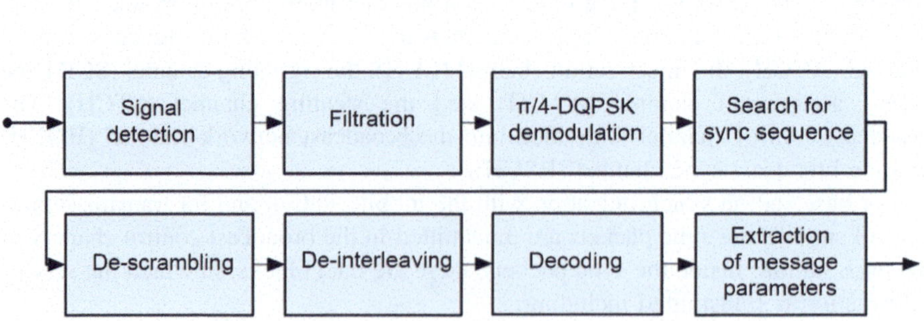

Fig. 9.33 Structural scheme of processing algorithm of TETRA signals

which BCCH can be found. The base station signal frequency is defined by the number of the frequency channel and its level, which is compared with the detection threshold, is estimated. The transfer to the baseband frequency is performed. After frequency transfer, the signal is filtered to suppress higher harmonics and to reduce initial signal bandwidth to the 25-kHz band. As a result, each signal of the analyzed frequency channel proceeds to the $\pi/4$-DQPSK demodulator. At demodulator output, the bit stream appears.

In the obtained stream, the search for sync sequence is performed. After its detection, the frame beginning is calculated, which data pass to the de-scrambler, where they are exposed to reverse scrambling operation. After de-scrambling, the transformed bit stream is exposed to the de-interleaving operation. After de-interleaving, the bit streams pass to the Viterbi decoder, as a result of which the decoded bit stream appears at its output. Then, verification of the message decoding according to check-up bits is executed. After successful check of the checking sum, the information data are extracted from the bit stream.

The SMO-BS TETRA software part of the analyzer is the combined part of the system mathematical support of the base stations automated monitoring complex. This software allows performance of the following functions:

- Detection of broadcasting base stations of TETRA standard and representation of signal spectrum;
- Extraction of service information, which is transmitted by base stations, measurement of signal levels, determination of base station localization, results storage in the database;
- Checking of correspondence of broadcasting base stations to the frequency-spatial plan (FSP);
- Generation of the measurement results report.

The SMO-BS TETRA software has two operation modes: "Monitoring mode" and "FSP control mode."

The "Monitoring" mode is intended for detection of the broadcasting TETRA base stations in the real time, measurement of their parameters, registration of results in the database for further analysis, observation of range scanning results in the deferred mode, determination of base station location according to the saved measurement results in the database, generation of measurement result reports.

The window of the SMO-BS TETRA software is shown in Fig. 9.34 in the mode of base stations monitoring. In the left part of the window, on the "Network" panel, the list of operators and base stations is represented in the form of hierarchical structure ("operator's tree"). In the right part of the window, the spectral diagram and measurement results are shown.

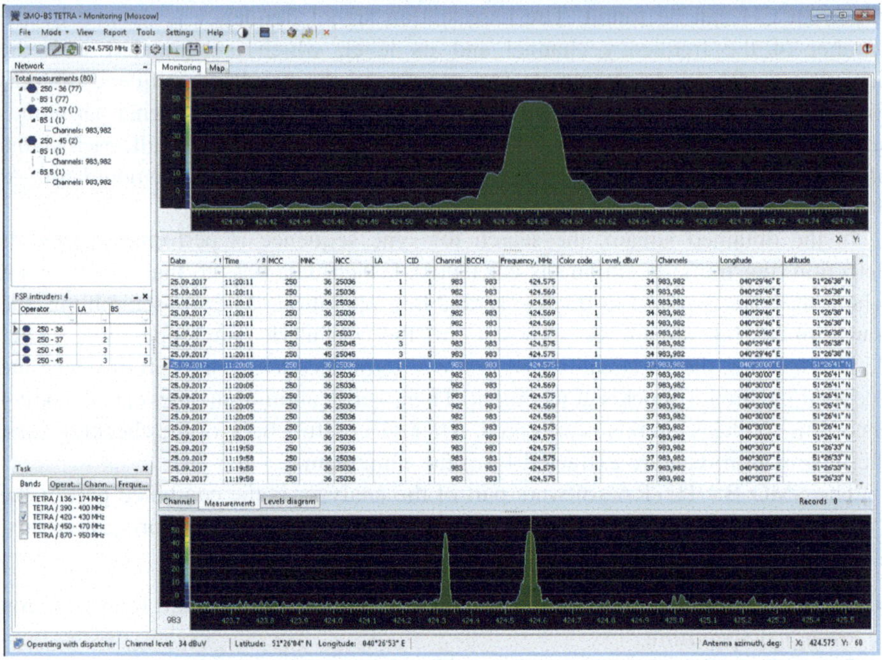

Fig. 9.34 Window of the SMO-BS TETRA software in the "Monitoring" mode

9.5 Analyzer of DECT Wireless Telephony Signals

The Digital European Cordless Telecommunications (DECT) standard for wireless telephony was developed in the beginning of 1990s years by ETSI. DECT is mainly used in systems of micro-cell communication in the medium and large organizations, in the single-cell automatic telephone stations (ATS) for household and small offices. Among single-cell ATSs for household and small offices, the GIGASET equipment by Siemens Co., which consists of one base block for one or two external town lines and from one to eight radio telephone, is the mostly known.

9.5.1 DECT Signals of Wireless Telephony

The DECT radio interface is based on the radio access methodology using the multiple carriers principle, the principle of time division multiple access, and the time division duplex (MC/TDMA/TDD). The frequency range allocated for DECT can be different depending on the country; in Europe, 1880–1900 MHz frequencies are standard.

Duration of TDMA frame in the DECT system is 10 ms, 16 frames are arranged into the multi-frame. The frame itself is divided into 24-time segments with 480 bits (approximately 416.7 μs). The first 12 intervals serve for transmission from the base station, and the second ones serve for transmission from portative terminals. One duplex channel is formed from intervals shifted by 12 slots. At voice transmission, 320 bits are used; the other bits are the sync code (32 bits), the signaling code (48 bits), the code of error protection (16 bits), the check-up symbols (4 bits), and the guard interval (60 bits) [11].

The DECT base station continuously transmits a signal during multi-frame, at least, by one channel in the broadcasting mode, playing a role of the "beacon" for subscribers. The full set of system information is contained in the transmitted signal. The principle of continuous dynamic channel selection, at which each of the subscriber terminals has an access to any of 120 channels, is realized in the system. At connection, the phone selects the channel with the best communication quality, and afterwards, its replacement can happen during the same session. It is provided at the expense of background scanning and estimation of signal level over all frequency–time positions.

The Gaussian minimum shift keying (GMSK) with minimal shift and Gaussian smoothing filter parameter BT = 0.5 is used as the base type of modulation in the DECT standard. The rate of data transmission is 1152 kbit/s. To increase the data transmission rate, the standard anticipates the optional possibility of more complicated modulation schemes (π/2-DBPSK, π/4-DQPSK, π/8-D8PSK, 16-QAM, and 64-QAM). Reception and transmission are executed at the same frequency. The regular time cycles by 24 time-slots repeating each 10 ms and called a frame are arranged at each carrier frequency. The frame consists of 24-time slots, each of which is individually available for DECT devices; i.e., the principle of time division multiple access (TDMA) is realized. Owing to this engineering solution, DECT provides up to 12 simultaneous voice connections per each transceiver, which gives essential cost advantages compared with technologies, which require one transceiver per each active connection.

As we already mentioned, to facilitate implementation of the DECT base standard, the time frame is divided into two halves (TDD—Time Division Duplex); the first 12 time slots are used for base station transmission, and other 12—for subscribers radio blocks transmission as shown in Fig. 9.35.

Each full slot contains 480-bit intervals, which are logically divided into synchronization, control fields, and optional data field. The synchronization field contains the sync sequence, which activates the beginning of information reception. The "control" field contains data about type of the transmitted information, control commands, and the proper check sum to control the reception correctness. The data field contains the subscriber data and can provide voice or data transmission. Five data channels presented in Table 9.4 are arranged in the "control" field.

The DECT base station permanently transmits a signal, at least, in one channel thus acting as a beacon for connection with mobile DECT phones (subscriber radio blocks). Transmission of the base station beacon contains service information about

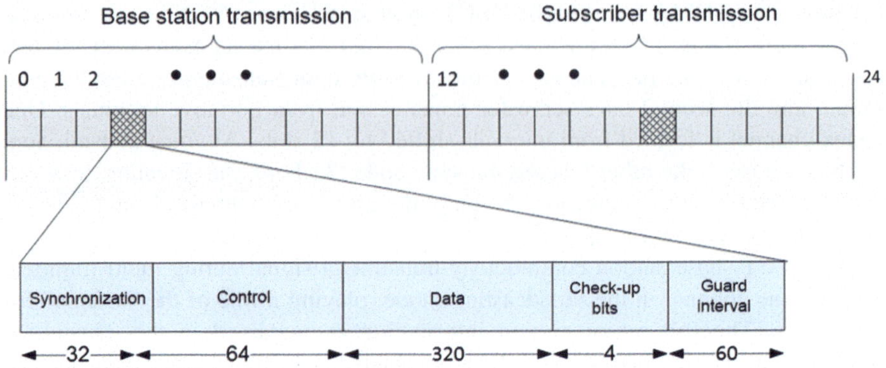

Fig. 9.35 DECT time cycle and the packet structure

Table 9.4 Data channels of the "control" field

Logic channel	Channel function
C	Control of active connection
M	Control of access to the radio channel
N	Control of channel selection of active connection
P	Selective message about call arrived
Q	Broadcast transmission of system information

base station identification, system possibilities and its status, as well as paging information for income communication provision.

DECT base stations can operate both in the single mode (the most typical application of this mode are household wireless phones), and in the structure of the micro-cell network. In this network, the roaming and messages forwarding service transfer is supported at subscriber transition into the activity zone of other base station, which makes DECT standard rather attractive for application in large offices. Possibility of micro-cell DECT network connection with GSM/UMTS networks as well as direct connection between subscriber radio blocks without base station participation is supported by the standard. However, this function is supported by not all subscriber devices.

9.5.2 The Analyzer of DECT Base Stations Signals

The analyzer is intended for identification and parameters measurement of radio signals of the household DECT blocks. The analyzer consists of the hardware part —measuring DRR of the ARGAMAK-family and the software part—the SMO-BS DECT system mathematical support software [7]. The last can be installed on the computer using the unified packet together with other software [1].

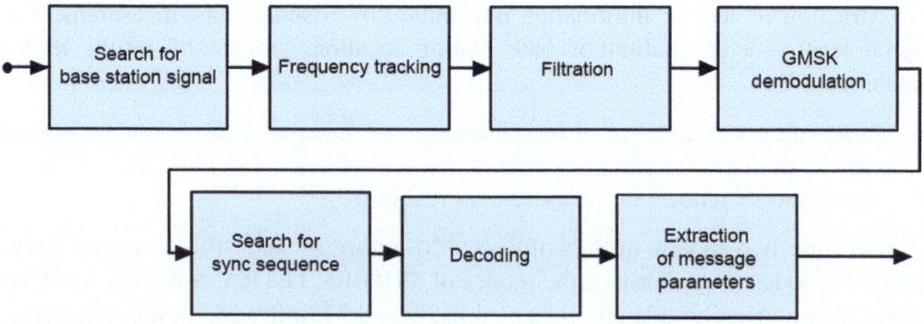

Fig. 9.36 DECT signals processing algorithm

Reception of broadcast information supposes procedures of baseband signal formation with necessary sampling frequency, search for base station signals, time synchronization as well as direct procedure of message reception. The structural diagram of digital processing of DECT signals is presented in Fig. 9.36.

The digital signal at the intermediate frequency with sampling frequency 6.4 MHz, from which the baseband signal is formed, proceeds to the input of processing device. The search for base station signal is carried out according to the energy criterion. Detected signal, after preliminary frequency tuning and filtering, is applied to the demodulator input. The demodulator produces bit stream at the output, in which search for sync sequence is performed. After its detection, the packet of 480 bits is applied to the decoding block and message parameters extraction is performed. In the received packets, the validation of checking sum is provided. In the case of lack of its coincidence, the packet is considered as unreliable and is eliminated from further examination.

Extraction of identification information is carried out from clear broadcast messages, which are transmitted into the "control" field. The analyzer is able to extract from it the following information: network class (single base station, private or public micro-cell system, public access to GSM/UMTS/LTE network, direct connection), Equipment Manufacturer's Code (EMC), DECT network codes (FPN, Fixed Part Number) and base station code in the network (RPN, Radio fixed Part Number).

To receive information about configuration and opportunities of the base station, the service messages of logical channels P and Q are used—the first one from them represents information about available modulation types and current transmitter power, the second—about number of transceivers and available frequency channels.

The SMO-BS DECT software packet, as the SMO-BS TETRA software packet, provides fulfillment of the following functions:

- Detection of broadcasting base stations of the DECT standard, representation of the signal spectrum;

Extraction of service information transmitted by base stations, measurement of signal levels, determination of base station location, storage of results in the database;

- Check of correspondence of broadcasting base stations to the frequency-spatial plan;
- Generation of reports on measurement results.

There are two modes in the software: “Monitoring” and “FSP control,” destination of modes is similar to same modes of SMO-BS TETRA. Software windows in the “Monitoring” mode are presented in Figs. 9.37 and 9.38. In the first figure, the “Measurement” tab sheet is displayed in the software window, in which measurement results of detected base stations are represented. The list of detected base stations is represented to the left in the “operator’s tree.” In the second figure, the results of search for base stations are presented according to the measurement series fulfilled in the motion at analyzer mounting in the motor car.

Operation results of the analyzer are stored in the embedded database of SMO-BS DECT software and can be used in the future for automated monitoring of signal parameters, for comparison with the frequency-spatial plan, creation of coverage areas and for other goals.

Fig. 9.37 Window of the SMO-BS DECT software in the “Monitoring” mode

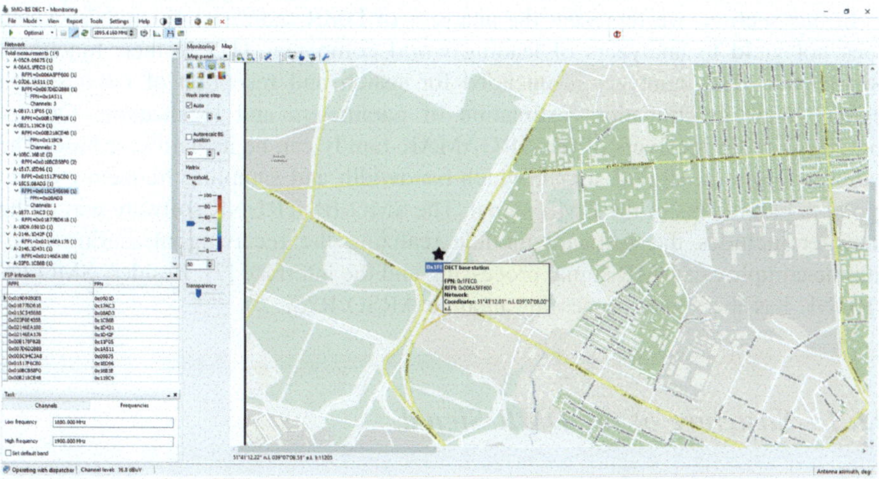

Fig. 9.38 Determination of base station location in the SMO-BS DECT software

9.6 Analyzer of DMR Stations Signals

The Digital Mobile Radio (DMR) standard is the one of the most widespread standards of narrow-band professional radio communication. The first version of this European standard was developed in 2005. Later, this standard was developed and modified.

DMR radio stations are compatible with similar analog radio stations without loss of functionality and have the following advantages compared with them:

- Higher effectiveness of frequency resource utilization;
- Possibility of voice data ciphering;
- Possibility of short text message transmission and other data, in particular, navigation parameters of mobile radio stations;
- Opportunity of Internet connection;
- Enlarged operation period of accumulators, etc.

DMR radio stations are widely distributed in many countries including Russia. Their quantity permanently grows. The Motorola Co. is the largest manufacturer of DMR equipment, which has produced several millions of radio stations.

For planning and exploitation of DMR networks, check of requirements correspondence to the transmitter parameters and check of correspondence to the frequency-spatial plan are necessary to identify functioning radio stations and to analyze their radio signal parameters regularly.

Equipment for signal analysis of DMR radio stations is produced by the leading worldwide manufacturers of radio monitoring equipment: Anritsu (LMR Master S412E), SATCOM Technologies (R8000), Keysight (N7608B).

In this section, we represent the analyzer of DMR radio station signals, which does not yield to analyzers of leading world companies [12] in their functional possibilities. This analyzer is intended for search and reception of radio station signals, their identification, estimation of parameters and localization. One of measuring radio receivers of the ARGAMAK-family can be its hardware base. The receiver forms the digital signal, which bandwidth and sampling frequency correspond to parameters of the DMR signal. The SMO-BS TRUNK software part of the analyzer includes the software module realizing the receiving–measuring algorithmic complex of DMR signal processing and the interface [7]. Besides, SMO-BS TRUNK has the signal processing module ARSO R25.

9.6.1 Structure of the DMR Signal

The DMR standard [13–16] assumes the different communication modes: directly between mobile radio stations (Direct mode) and connection via base stations of the trunking network, in which procedures of channels assignment are centrally controlled by the network.

The bandwidth of the DMR signal is 12.5 kHz, and the multi-station access with TDMA is used. The two-slot time structure is arranged, which allows transmission of voice and service data for two subscribers at one carrier. The slot duration is 30 ms. The data packets of 27.5 ms duration are transmitted in slots. The protective interval between packets is 2.5 ms. The base station signal, where data of the Common Announcement Channel (CACH) are transmitted between packets, forms an exclusion. In the active mode, the base station transmits the signal continuously. An example of the time structure of the DMR signal is shown in Fig. 9.39. Slots of appropriate physical channels are designated by figures "1" and "2."

In each packet, 264 bits are transmitted. From them, 48 central bits serve for sync sequence or service information transmission. The receiver uses the sync sequence for detection of the DMR signal and its frequency–time synchronization. There are 10 variants of the sync sequence, in which information about packet content is indicated. The sync sequence allows determination:

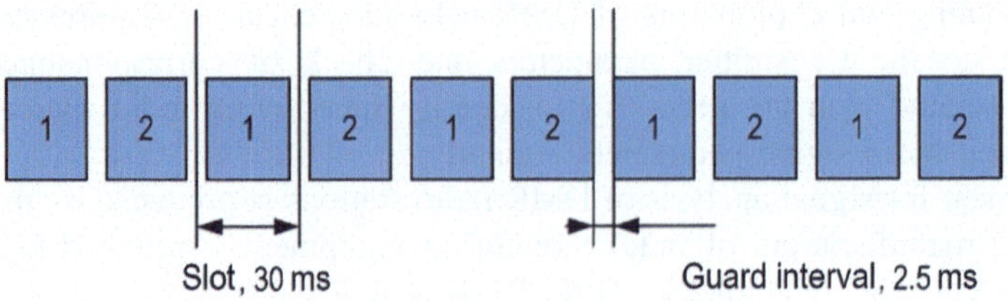

Fig. 9.39 Time structure of the DMR signal

- Whether the signal source operates in the "Direct mode" or via the base station of the trunking network;
- Number of the used slot for the "Direct mode";
- Whether the signal source is the base station or the mobile station in the case of the trunking network;
- Type of the transmitted information in the packet (voice or data).

The structure of the voice packets in DMR is shown in Fig. 9.40. The "Voice" fields in them contain the voice coder frames.

The voice packets are structured in the voice super-frames of 360 ms duration with 6 packets in each (see Fig. 9.41). Packets are designated from "A" to "F." The first packet of each "A" super-frame always contains the sync sequence. In four from five remained packets, the message Link Control (LC) is transmitted in the embedded service information fields.

The coded information of the EMB voice packet indicates whether the packet participates in the transmission of LC message fragment and the number of fragment.

In the LC message, the source address, target address, the address type (individual or group), and other service parameters are transmitted.

On the transmitting side, 72 bits of the LC message are added by 5 check-up bits and are subjected to block turbo-coding, interleaving, and fragmentation into 4 parts of 32 bits. These fragments are represented in the internal service information fields of the appropriated voice packets of the super-frame.

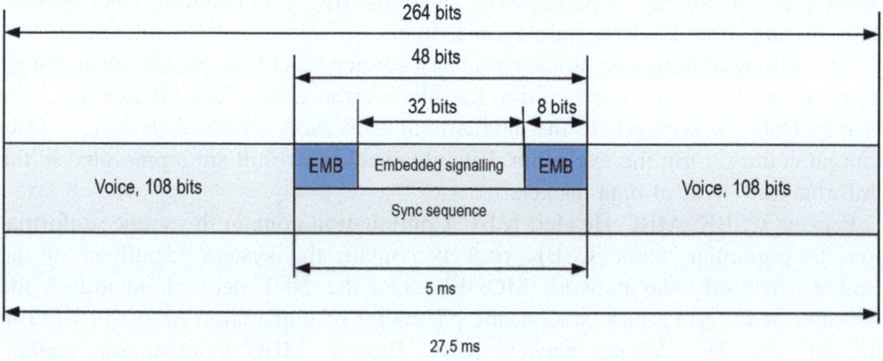

Fig. 9.40 Structure of the voice burst

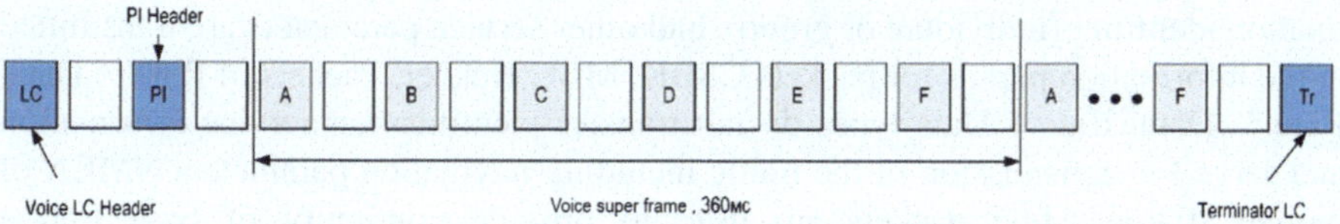

Fig. 9.41 Voice super-frame

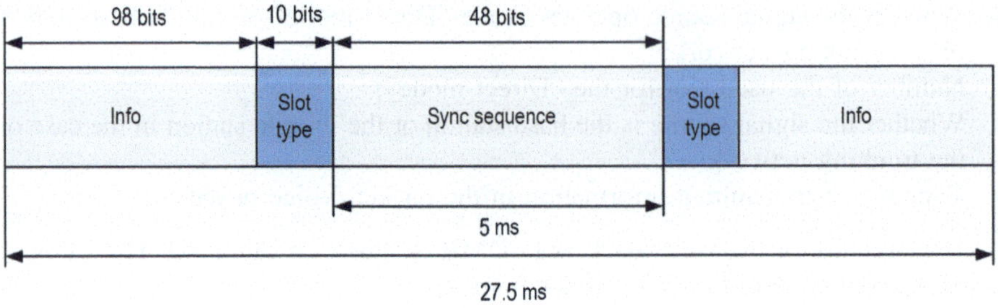

Fig. 9.42 Structure of data packet

The voice transmission may be accompanied by service packets Voice LC Header, PI Header in the beginning of the voice transmission and Terminator LC in its end. The packet Voice LC Header initiates the beginning of the voice transmission, the packet Terminator LC—its ending. In these packets, the LC message is also transmitted. In the case of ciphering mode application, the service packet PI Header is transmitted, which is used for deciphering on the receiving side.

The structure of the data packets in DMR is shown in Fig. 9.42. In the "Information" fields, the informational or service bits are transferred. The sync sequence occupies 48 bits in the packet center. Two fields of 10 bits on the left and on the right of the sync sequence represent the service message of the slot type. Coded service information of slot type indicates the type of data transmitted in the packet. The following variants of the types are possible: PI Header, Voice LC Header, Terminator LC, CSBK, MBC Header, MBC Continuation, Data Header, Rate ½ Data, Rate ¾ Data, Rate 1 Data, Idle.

On the transmitting side, informational or service bits of the packet are added by the check-up bits and are exposed to the block turbo-coding and interleaving. The Rate ¾ Data packets, where the Trellis coding is used, and packets Rate 1 Data without coding, form the exclusion. Bits obtained as a result are represented in the "Information" field of data packets.

Packets CSBK, MBC Header, MBC Continuation contain the service information. In particular, some CSBK packets contain the system identifiers of the trunking network: the network MODEL size, the NET network identifier, the identifier of the SITE base station, the parameter of registration necessity REG in the network. The definite packets MBC Header, MBC Continuation contain information about frequencies of transmission and reception of the base station.

The packet Data Header always precedes to the data packet stream and informs about these data content. Data source and destination identifiers, the type of destination identifier (individual or group), and other service parameters are transmitted in the information part as for packets CSBK, MBC Header. Packets of Rate ½ Data, Rate ¾ Data, Rate 1 Data types do not transfer identification service information and serve for transmission of the traffic including navigation parameters NMEA of mobile stations. Idle packets are used to provide continuity of base station transmission.

Table 9.5 Representation of bits into modulation symbols

Information bits		Symbol	4FSK deviation, kHz
Bit 1	Bit 2		
0	1	3	1.944
0	0	1	0.648
1	0	−1	−0.648
1	1	−3	−1.944

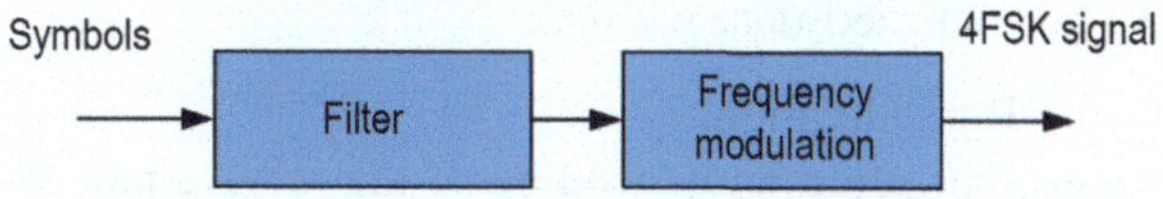

Fig. 9.43 4FSK modulator

In the CACH, the base station can transmit the short LC messages. The CACH signal represents a set of fragments. Each message is represented into four fragments. In addition, there is a block of service information TACT in each fragment. In this block, the fragment number is encoded in the transmitted messages, which is used on the receiving side. In the CACH, the different messages are transmitted, in particular, the message SYS_Parms Short Link Control, in which system identifiers are transmitted.

The DMR signal transmission rate is 4800 bauds, 2 bits are transferred in each symbol. Four-position frequency modulation 4FSK is used. Representation of transmitted bits into symbols is given in Table 9.5.

4FSK modulation is realized in the form of the cascade diagram consisting of the filter of the square root from the raised cosine and the frequency modulator, as shown in Fig. 9.43.

9.6.2 Reception of DMR Signal

To receive DMR digital baseband signal, the received radio signal is amplified, its analog-to-digital transformation is carried out, then the transfer to the baseband, filtering, and oversampling to ensure the correspondence with parameters of the DMR signal is executed.

Processing of the DMR baseband signal assumes the following in connection with each of two channels characterized by its own slot:

- Procedure of detection of sync sequences of packets, time and frequency synchronization, determination of the transmission type;
- For detected voice packets:
  - Demodulation,
  - In the case if the Voice LC Header of the given voice message is not received:

Extraction and decoding of the EMB service part,
Extraction of the service LC message from packets of the voice super-frame of the bit block, its de-interleaving, decoding, validation of decoding,
Determination of transmission confidentiality (ciphered/unciphered),
Determination of source and target addresses and also type of target address (individual/group);

- Recording of voice data;

- For detected data packets:
  - Demodulation,
  - Extraction and decoding of the service part of the Slot Type packet,
  - Determination of data type, statistic calculation of the detected packets data types;
  - For packets of Voice LC Header type:

    Extraction of information part of the packet, de-interleaving, decoding, validation of decoding,
    Determination of transmission confidentiality (ciphered/unciphered),
    Determination of source and target addresses and also type of target address (individual/group);

  - For packets of Data Header type:

    Extraction of information part of the packet, de-interleaving, decoding, validation of decoding,
    Determination of transmission confidentiality (closed ciphered/unciphered),
    Determination of source and destination identifiers and also type of destination identifiers (individual/group),
    Extraction of the DPF field of the packet Data Header information part, check of correspondence to the unified protocol of UDT data transmission,
    Extraction of UDT Format field of the information part of the packet, check of correspondence to NMEA navigation data,
    Extraction of the UAB field of the information packet part, determination of the NMEA data format,
    Demodulation, extraction of the informational part of the next packet/packets (their type must be Rate ½ Data) containing the NMEA data, de-interleaving, decoding, validation of decoding,
    Extraction of parameters of NMEA navigation data;

  - For packets of CSBK type:

    Extraction of information part of the packet, de-interleaving, decoding, validation of decoding,
    Determination of transmission confidentiality (ciphered/unciphered),
    Extraction of LB and CSBKO fields of the information packet CSBK part, the check on correspondence to broadcast notification and presence of system identifiers,

Extraction of system identifiers MODEL, NET/SITE, and REG;

- For packets of MBC Header type:

  Extraction of information part of the packet, de-interleaving, decoding, validation of decoding,
  Determination of transmission confidentiality (ciphered/unciphered),
  Extraction of the CSBKO field of informational part of the MBC Header packet, check of correspondence to broadcast notification,
  Extraction of the Announcement Type field of the informational part of the MBC Header packet, check of correspondence to frequency notification,
  Demodulation, extraction of the informational part of the next packet (its type must be MBC Continuation), de-interleaving, decoding, validation of decoding,
  Extraction of transmission and reception frequencies of base station from the message;

- For base station signal, if system identifiers are not received in the CSBK packets:
  - Demodulation of CACH packet symbols,
  - De-interleaving of each CACH packet,
  - Extraction and decoding of the TACT service part of the CACH packet,
  - Extraction of the CACH service message bits over the LCSS field from TACT,
  - De-interleaving, decoding, validation of demodulation of the CACH message,
  - Extraction of the SLCO field from the CACH message, check of the presence of system identifiers,
  - Extraction of system identifiers MODEL, NET/SITE, and REG.

The sequence of DMR signal processing is presented in Fig. 9.44.

The developed analyzer of DMR stations signal performs the following functions at signal processing at the given frequency in the analyzed time interval:

- Detection of the DMR signal;
- Determination of transmission type (BS sourced voice; BS sourced data; MS sourced voice; MS sourced data; MS sourced standalone RC; TDMA direct mode time slot 1 voice; TDMA direct mode time slot 1 data; TDMA direct mode time slot 2 voice; TDMA direct mode time slot 2 data);
- Determination of the detected packet type;
- Determination of transmission confidentiality: ciphered/unciphered/not defined;
- Calculation of the statistics for transmission type, packet type, transmission confidentiality of the detected packets;
- Estimation of bit error probability for detected transmitting radio stations;
- Reception and record of detected voice data;
- Determination of addresses/identifiers of the source and the target, as well as the target address type (individual/group) of packets of voice and data;

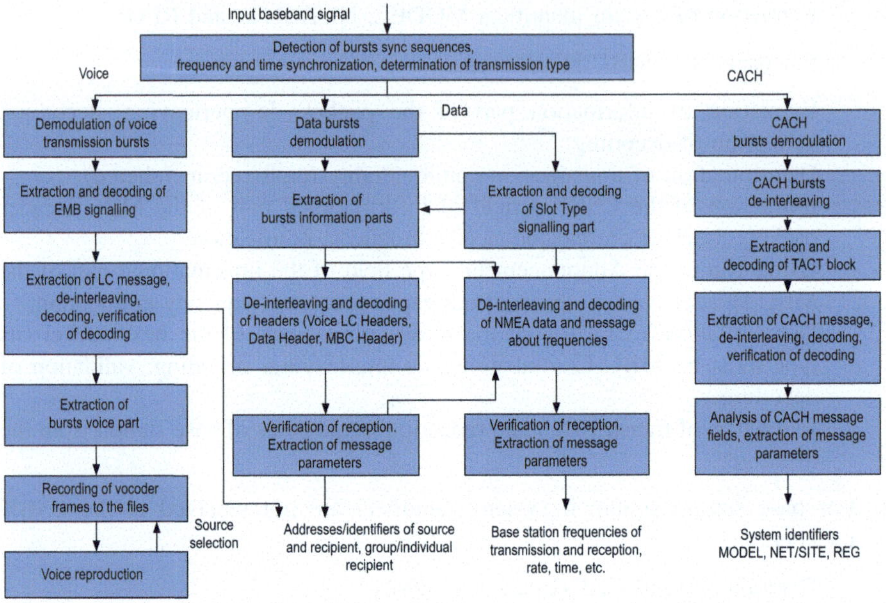

Fig. 9.44 Scheme of DMR signal processing

- Reception of system identifier in the trunking system;
- Search and reception of navigation data NMEA (coordinates, time, rate, etc.) in the trunking system, representation of coordinates on the electronic map of locality;
- Search and reception of messages about base stations frequencies of transmission and reception in the trunking system;
- Construction and analysis of network topology of detected radio stations;
- Storage of analysis results in the database with possibility of further automatic monitoring of situation variation;
- Formation of analysis results in various formats: XML, MS-Word, Excel, flexible choice of outlet information and forms of its representation.

An example of SMO-BS TRUNK window is shown in Fig. 9.45. Signal reception is executed in the fixed point of industrial center. In the left upper part of the window, the tree of detected radio stations, which are grouped according to the standard and the transmission type, is presented. In the left lower part, there is a task panel, in which values or range of analyzed frequencies can be selected. In the lower part of the window, the received signal spectrum in the chosen frequency band is represented. In the window center, there is a table of detected radio stations including time of detection, carrier frequency, signal level, and main transmission parameters. In the upper part of the window, for the chosen radio station, the additional list of received and measured signal parameters is shown.

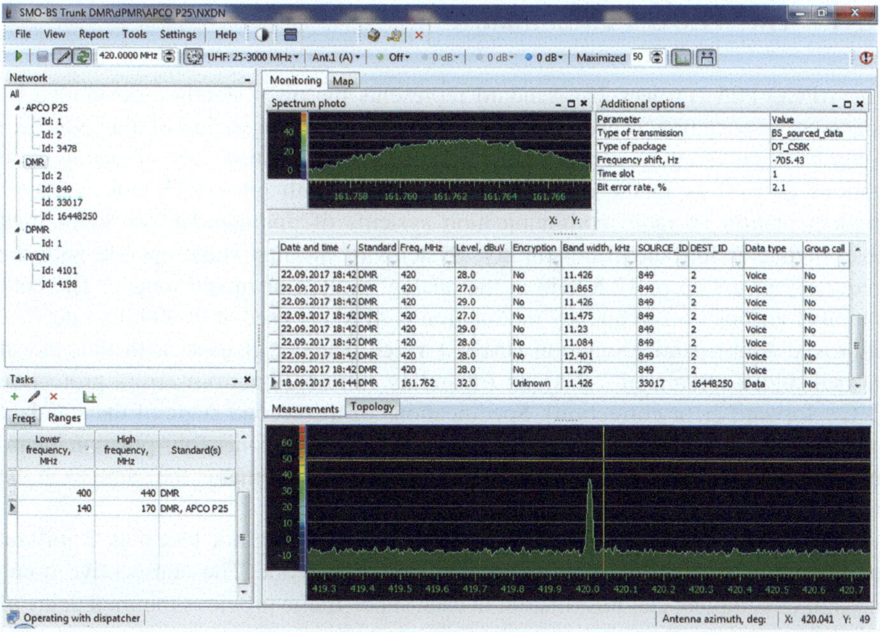

Fig. 9.45 Main window of the SMO-BS TRUNK software

9.7 Analyzer of APCO P25 Signals

The APCO P25 standard has an international importance: representatives of the state organizations and industry from Australia, Canada, and UK are its active participants. China, Denmark, France, Germany, Netherlands, New Zealand, Spain, Taiwan, etc., showed the interest as well. The main goal of the Project 25 is development of new open digital standard of conventional and trunking radio communication for security agencies with the reverse compatibility with the existed analog units.

As we already mentioned, radio stations of APCO P25 standard is widely distributed in many countries including Russia. Their quantity permanently grows. The Motorola Company made serious contribution in formation of this standard by participating in development of this standard. Putting up the huge investments, the Motorola now is able to offer full portfolio of APCO P25 products unified under the ASTRO 25 trademark. Being the worldwide leader, the Motorola has delivered more than 350 radio systems and more than 2 million subscriber terminals ASTRO 25 to their consumers all over the world.

The aim of this section is examination of the signal structure of APCO P25 standard, the procedures of signal processing and functional opportunities of the analyzer [17].

9.7.1 The APCO P25 Standard

The APCO P25 (Project 25) standard represents the open standard developed by consumer association especially for needs of security agencies of the Northern America. Radio stations can operate together with the existing park of analog radio stations as well as in the digital or analog mode with other P25 radio stations. Implementation of radio communication systems of this standard provides high level of interaction and compatibility. Systems of the first stage operate with the frequency grid step of 12.5 kHz in the analog, digital, or mixed modes. The continuous four-position frequency modulation (C4FM) is used at 9600 bit/s rate. The Improved Multi-band Excitation (IMBE) voice decoder is used with the rate of formed digital stream of 7.2 kbit/s (including information from errors protection FEC—forward error correction). Specifications of the second stage of the standard anticipate the more effective method of carrier modulation—four-position phase modulation with phase smoothing. The modulation rate remains the same as at the first stage.

The APCO P25 standard gives a possibility to address not less than 2 million radio stations and 65 thousand of groups in one network. The consecutive transmission of complete call information in each super-frame provides minimal delay of subscriber connection.

Specification of the Project 25 includes determination of radio frequency sub-system and eight interfaces, which are the part of the open architecture of APCO P25 systems:

- Common application interface (CAI);
- Interface with the data transmission terminal;
- Interface with other systems ISSI;
- Interface with the telephone network;
- Interface with the sub-system of network control NMI;
- Interface with networks of data transmission;
- Interface with the console sub-system CSSI;
- Interface with the base station FSI.

The CAI has parameters presented in Table 9.6.

The full standard description contains more than hundred documents. The general structure can be found in [18]. Description of radio interface and used voice coder is contained in [19, 20].

9.7.2 Reception of APCO P25 Signals

Reception of the APCO P25 signal includes the following stages:
Channel detection and baseband signal formation.

Table 9.6 Parameters of the common application interface

	Phase 1	Phase 2
Method of access	FDMA	TDMA
Channel width	12.5 kHz	12.5 kHz (6.25)
Modulation type	C4FM/CQPSK	H-CPM uplink H-DQPSK downlink
Rate of data transmission in the control channel	9.6 kbit/s	9.6 kbit/s
Rate of data transmission in the voice channel	9.6 kbit/s	12.0 kbit/s
Vocoder	IMBE	AMBE

Processing of the baseband signal (packet detection, frequency–time synchronization, demodulation, decoding) and obtaining of bit stream.
Analysis and processing of bit stream—information extraction.
At the stage of channel detection, the wide-band signal processing is used. The total width of frequency ranges recommended for networks APCO P25 (136–174 MHz, 403–512 MHz, 806–870 MHz) is 214 MHz, which imposes the increased requirements to the algorithm performance.

The following algorithm is used as the algorithm of quick channel detection.

1. The dimension n_F of the fast Fourier transform (FFT) is calculated:

$$n_F = 2^{ceil\left(\frac{\ln(2F_S)-\ln(200)}{\ln 2}\right)}, \tag{9.1}$$

   where *ceil* is the operation of up-rounding up to nearest integer, and F_S is the sampling frequency.
2. The amplitude signal spectrum is calculated in the logarithm scale:

$$A = 20\log_{10}(|fft(z * w_F)|), \tag{9.2}$$

   where the operation of FFT is designated as $fft(\cdot)$, z is the sample fragment with the length n_F, $*$ is the operation of component-wise multiplication, and w_F is the Blackman window with n_F length.
   If the length of the analyzed interval exceeds the FFT size, the FFT must be repeated in available intervals and results must be averaged.
3. The correlator response to the channel mask (the Blackman window with n_F width can be used as this mask) is calculated:

$$n_C = 2floor\left(\frac{s \cdot n_F}{2F_S}\right) + 1, \tag{9.3}$$

   where *floor* is the operation of down-rounding down to nearest integer, and $s = 6000$ Hz is the signal spectrum width.

4. Let us numerate samples of the w_c window with numbers from $-(n_c - 1)/2$ to $(n_c - 1)/2$. In this case, in the admissible index values area, the correlator response is calculated according to the formula:

$$c_i = \sum_{k=-(n_c-1)/2}^{(n_c-1)/2} A_{i+k}(w_c)_k, \tag{9.4}$$

5. Analyzing the correlator response c in the area under our interest, we detect the channel according to the following criterion: Simultaneously, the response in c_m channel center is the maximum in the interval $m \pm (n_c - 1)/2$ and exceeds the average response in the analysis interval not less than by given value d. In this case, the following condition is fulfilled:

$$c_m \geq \frac{\left(\sum\limits_{k=-(n_c-1)/2}^{(n_c-1)/4} c_{i+k}\right)}{\left(\frac{n_c-1}{3}\right)} \text{ and } c_m \geq \frac{\left(\sum\limits_{k=+(n_c-1)/4}^{(n_c-1)/2} c_{i+k}\right)}{\left(\frac{n_c-1}{3}\right)}. \tag{9.5}$$

The index of the channel center m is specified using the parabolic approximation of values c_{m-1}, c_m, c_{m+1}. The typical procedure accuracy is about 200–300 Hz depending on the sample length and SNR.

For each detected channel, the standard procedure of baseband signal formation at the frequency F_v, which is multiple to the symbol rate ($F_v = m \cdot 4800$, where m is the integer number), is carried out together with signal filtering in the bandwidth of 8 kHz.

The scheme of baseband signal processing is presented in Fig. 9.46.

As signals of the APCO P25 standard can have two types of modulation, C4FM and CQPSK, the demodulator, which is based on processing of the frequency detector output signal and applicable without additional modification and adjustment to both types of modulation, is used.

Signal detection is executed simultaneously with determination of frequency offset. The fixed preamble (sync sequence) is transmitted in the beginning of the each APCO P25 signal packet (see Fig. 9.47), which permits to use correlation processing for this operation.

In connection with the fact that the APCO P25 signal preamble ($5575F5FF77FF) has various amount of positive and negative offsets, the two responses for each point are calculated for positive and negative preamble parts by the formulas:

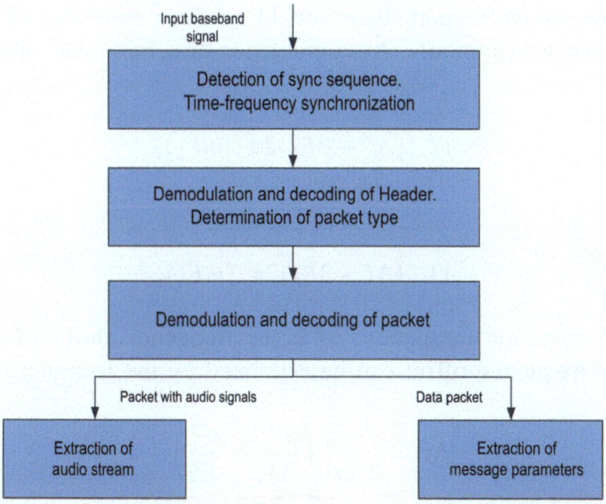

Fig. 9.46 Scheme of APCO P25 signal processing

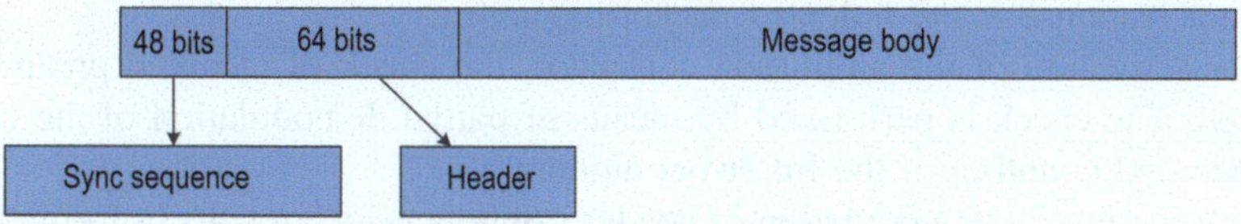

Fig. 9.47 Structure of the signal packet of the APCO P25 standard

$$a_k^+ = \sum_{j=0}^{23} p_j^+ \cdot b_{k+m \cdot j}, \tag{9.6}$$

$$a_k^- = \sum_{j=0}^{23} p_j^- \cdot b_{k+m \cdot j}, \tag{9.7}$$

where $p = [1, 1, 1, 1, 1, -1, 1, 1, -1, -1, 1, 1, -1, -1, -1, -1, 1, -1, 1, -1, -1, -1, -1, -1]$ is the preamble, which is transmitted in the beginning of each packet, p^+ is the preamble consisting of positive elements of the sequence p (negative elements are replaced by zeros), p^- is the preamble consisting of negative elements of the sequence p (positive elements are replaced by zeros).

In the positive preamble part, there are 11 nonzero symbols, while in the negative part—13 nonzero symbols. At coincidence of sample with the preamble, a_k^- will be:

$$13 \cdot (\Delta f - 3F_S)2\pi/(mF_S), \tag{9.8}$$

and a_k^+ will be:

$$11 \cdot (\Delta f + 3F_S)2\pi/(mF_S), \tag{9.9}$$

where F_S is the deviation frequency, Δf is the frequency shift (offset).

The value of frequency offset can be estimated by the formula:

$$\Delta f = \frac{1}{2}\frac{mf_d}{2\pi}\left(\frac{a^+}{11} + \frac{a^-}{13}\right). \tag{9.10}$$

The value of correlation response can be obtained by the formula $c_k = (a^+ + a^-) - 24\Delta f \cdot 2\pi/(mf_d)$ The packet will be considered as detected if the correlator response is more than $0,65 \cdot 72 \cdot 2\pi/(m\,F_S)$. The index of the maximal correlator response is the index of packet beginning.

After activation of the amplitude correlator, the check of received preamble is executed. The check is performed by means of partial demodulation of the testing sequence and counting of the bit errors amount.

The next stage is demodulation of header symbols, which results in formation of rigid decisions of information bits. After demodulation, determination of the received packet type is performed.

Further demodulation and formation of rigid or soft decisions of information bits are carried out by the individual scheme for each packet depending on its type.

9.7.3 Destination and Functional Possibilities of the Analyzer

The hardware–software complex for APCO P25 signal analysis provides:

- Determination of radio electronic system type and signal parameters:
  - Device type (base or subscriber station),
  - Identifier (address),
  - Mean level of the signal,
  - Time of last activity,
  - Frequency offset,
  - Type of transmitted packets,
  - Transmission type,
  - Presence of ciphering,

  - Ciphering type,
  - Bit error rate,
  - Network access code (NAC);

- Analysis of network topology using the intercepted packets;
- Determination of radio equipment location with constructing of the probable localization zone, and also representation of obtained radio monitoring results on the electronic locality map with coordinates affixment by the GLONASS/GPS data;
- Possibility of sorting and selecting of the presented data by the different signs (including locality zone, current activity, date/time, belonging to the given network);
- Storage of analysis results in the database with possibility of further automatic monitoring of situation variation.

The analyzer consists of the hardware part on the base of ARGAMAK-family equipment and the software part. The last can be installed on the computer using the unified packet together with other software, for instance, executing the base stations signals analysis of different cellular communication systems.

When DRR is used, the hardware of the complex provides the baseband signal formation at the required sampling frequency, which essentially increases the analyzer performance. At signal presence, the software performs its further processing and extraction of broadcast information.

The external view of the main window of the SMO-BS TRUNK software is presented in Fig. 9.48 [7].

In the left part of the window, on the "Network" panel, the list of sub-networks and their clients presented in the form of hierarchical structure is displayed. At signal analysis, the station appears in the "Network" panel in the case only if we are able to receive station identifiers. During data exchange between stations, identifiers are transmitted at operation start. The following parameters are registered for each station:

- Data type (data, voice data);
- Station identifier.

For each analysis result, the spectrum photograph is saved for the 20-kHz bandwidth and the central frequency, at which the signal is detected.

For tracking convenience of signal-level variation for the specific identifier, there is the time-level diagram. The signal-level variation for the chosen identifier is graphically represented in this diagram.

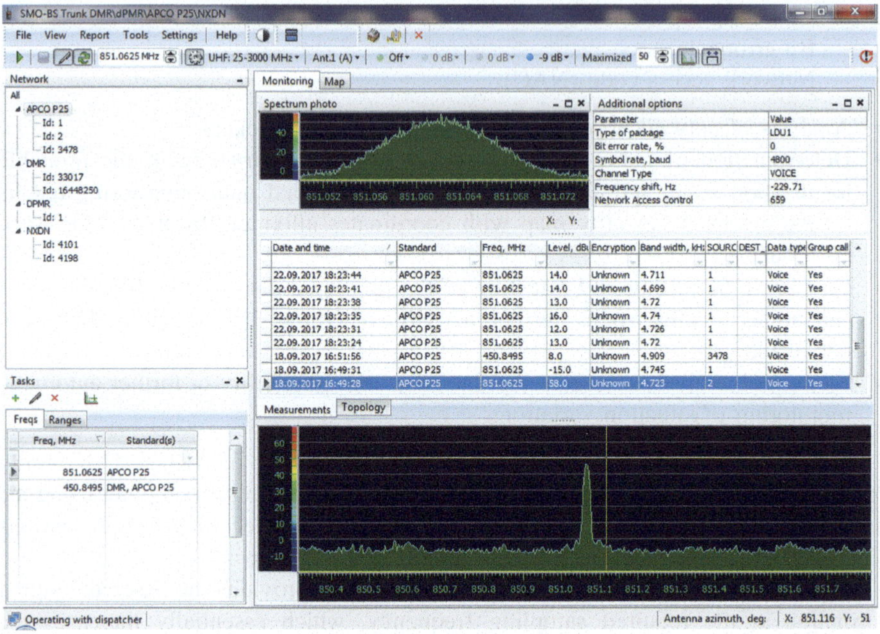

Fig. 9.48 Main window of the SMO-BS TRUNK software

9.8 Analyzer of Wi-Fi Networks and Access Points Signals

The list of existing equipment for parameters analysis of Wi-Fi networks can be divided into several types. The full-functional analyzers of spectrum and signals from the leading world manufacturers of radio monitoring equipment (Rohde & Schwarz, Keysight, Tektronix), which provide fast digital spectral analysis and optional determination of radio signal parameters of different standards, including the standard family 802.11, can be related to the first type. However, as a rule, the similar devices do not suppose monitoring of the frequency-spatial plan, construction of the network topology and estimation of some other parameters, which are significant for radio monitoring agencies. Moreover, the basing technology of analyzer construction, as a rule, is oriented on implementation of universal functional opportunities. Blocks of preliminary signal selection are usually absent in the basing analyzer integration; therefore, its utilization is difficult for solution of radio monitoring tasks.

The equipment of the second type represents the special adapters intended for monitoring of broadcasting radio systems. In contrast to spectrum analyzers, adapters have the narrower list of measuring radio signal parameters, use the non-calibrated antenna systems, and operate only in licensed frequency ranges 2412–2484 MHz, and 5170–5320 MHz [21], which restricts their application in the radio monitoring equipment.

Analyzers of wide-band wireless networks constructed on the base of universal measuring digital radio receivers can be related to the third equipment type. DRRs must have parameters sufficient for providing measurements in the wireless networks, including high frequency stability of the reference oscillator, wide pass-band, small errors of the amplitude–frequency response, digital output of the IF signal. The functionality of similar analyzers is defined by parameters of the measuring DRR and algorithmic possibilities of the software. The cost of wireless network analyzers on the base of digital measuring receivers is commensurable with the cost of the full-functional spectrum and signal analyzers, but its advantage is the possibility of software usage for solution of other radio monitoring tasks not related to measurement of wireless network parameters.

The goal of this section is consideration of construction peculiarities of the analyzer of wireless radio access Wi-Fi signals, which is built on the base of DRR of the ARGAMAK-family, as well as peculiarities of radio signal reception and processing with the aim of estimation and measurement of their parameters [22].

9.8.1 Brief Description of 802.11 Standard Family

Wi-Fi networks operate at frequencies 2.4 GHz (802.11 b/g/n) or 5 GHz (802.11 a/n). From the point of view of network organization, this standard envisages the usage of Ad Hoc and Infrastructure modes [21]. In the Ad Hoc mode, the network nodes are connected directly with each other, while in the Infrastructure mode the node connection is provided via the access point. For access to the data transmission medium, the joint access method with detection of the carrier and prevention of collisions is used.

The set of standards 802.11 defines a set of implementation technologies of physical level (PHY), among which we can single out the following [23]:

- Level of PHY of the standard 802.11b with spectrum spreading by the method of direct sequence;
- Level PHY of the standard 802.11a using the technology of orthogonal frequency division multiplexing (OFDM).

The 802.11b standard supports four data transmission rates: 1, 2, 5.5, and 11 Mbit/s. At low rate, for spectrum spreading, the Barker codes are used, while at high rate, the 8-chips complementary codes (CCK) are used. To encode the information bits, either binary or quadrature relative phase modulations are used.

The 802.11a IEEE standard uses the OFDM technology. The reverse Fourier transform with the window of 64 frequency sub-channels is used for channels division. At that, each orthogonal sub-carrier has the width of 312.5 kHz at total spectrum width of 20 MHz. At low transmission rate, the binary and quadrature phase modulations BPSK and QPSK are used for modulation of sub-carrier frequencies. The BPSK modulation is used for data transmission at 6 and 9 Mbit/s

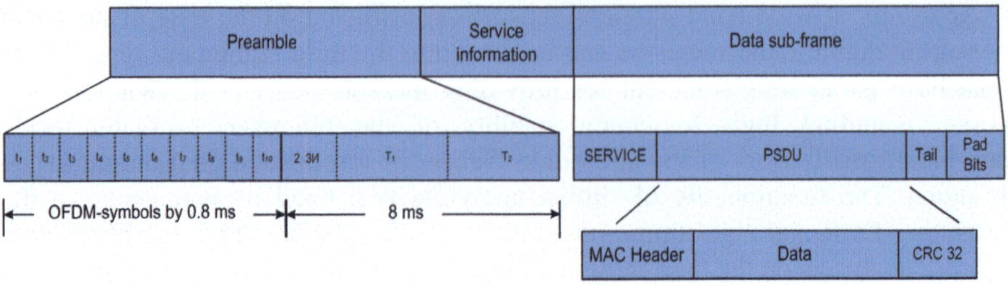

Fig. 9.49 Frame structure of the 802.11a standard

rates, while the QPSK modulation is used at rates of 12 and 18 Mbit/s. To transmit at higher rates (up to 54 Mbit/s), the quadrature amplitude modulations 16-QAM and 64-QAM are used.

The 802.11g standard is the further development of IEEE 802.11b specification and provides the data transmission in the same frequency range of 2.4 GHz. Owing to application of the OFDM technology, this standard supposes growth of the data transmission rate up to 54 Mbit/s.

The OFDM technology is used in the 802.11n standard as well, which provides data transmission rate up to 600 Mbit/c. Essential rate increasing is connected with utilization of the MIMO technology and with increased up to 40 MHz operation bandwidth of these standard devices.

In spite of significant differences of existing standards, the frame structure of physical level for them is similar in type: the header consisting of preamble and service information is added to MAC-frames with informational data. As an example, Fig. 9.49 shows the frame structure for the 802.11a standard [21].

Each frame is begun from the preamble of 16 μ s duration. The preamble consists of 10 short learning symbols (SLS) of 0.8 μ s duration each, and of 2 long learning symbols (LLS) of 3.2 μ s duration each. Learning OFDM symbols are intended for signal detection, frequency–time synchronization and estimation of the frequency response of the spatial communication channel. After preamble, the service information is transmitted containing data about used transmission rate, modulation type, length of the message transmitted in this frame, etc.

Then, the data sub-frame follows, which, in turn, consists of four fields: SERVICE, PSDU, TAIL, Pad Bits. The PSDU field (information field) contains directly the transmitted data. Its duration is variable and may achieve 3 ms. In the beginning of the information block, the header of the MAC level is transmitted: the information about packet type (control, service, informational), hardware addresses of the transmitter and the receiver, network identifier. It should be noted that necessary (from the radio monitoring point of view) broadcast information is available not in all packet types [21]. The 32-bits sequence for control sum verification is situated in the end of the information block.

In the 802.11b standard, two types of preambles of different length of 144 and 72 μ s are used. Both preambles are transmitted at minimal rate of 1 Mbit/s with usage of binary relative phase modulation and Barker codes for spectrum spreading.

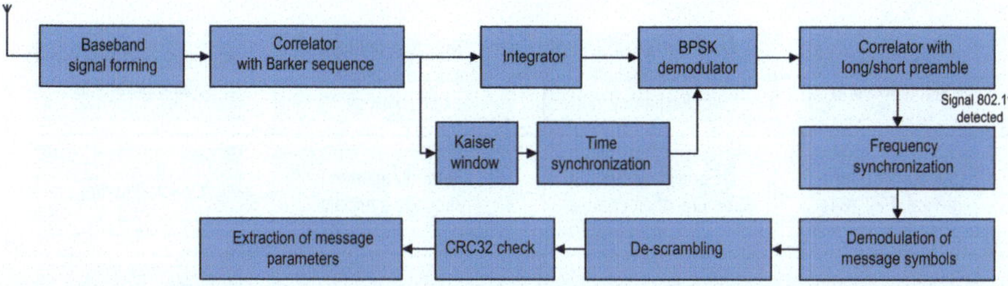

Fig. 9.50 Structural scheme of the processing algorithm for signals of the 802.11b standard

9.8.2 Reception of the Signals of 802.11a/b/g/n Standards

Reception of Wi-Fi signals and extraction of necessary broadcast information supposes the procedure of baseband signal formation with necessary sampling frequency, packet detection, frequency–time synchronization, obtaining the bit stream, as well as extraction of necessary broadcast information. Because of different technologies of implementation of physical level, the algorithms of signal processing for 802.11b and 802.11a/g/n standards have significant differences.

The structural scheme of signal processing for the 802.11b standard is presented in Fig. 9.50.

Detection of the 802.11b signal is based on revealing of the Barker sequence, demodulation of message symbols and further check of demodulated symbols on correspondence to known signal preambles. At first stage, correlation of the input signal with the known Barker sequence is calculated:

$$U_n = \sum_{s=0}^{S-1} b_s^* \cdot z_{s+n}, \tag{9.11}$$

where z_n is the sequence of input signal samples, and b_s is the known Barker sequence of S length.

The signal from the correlator output is multiplied by the Kaiser's window function: According to obtained maximal signal values, the continuous monitoring of the preamble symbol time of arrival is performed for their further demodulation. After demodulation of the preamble symbols, the coincidence of demodulation results with known preambles of the 802.11b signal is estimated, after that the decision is made about signal presence/absence. At packet detection, the phase incursion is measured in the preamble interval, and frequency synchronization is executed. On the base of service information from the message header about modulation type, the information symbol demodulation, de-scrambling, and the CRC verification are provided, after that the broadcast message parameters (necessary for radio monitoring) are extracted.

The structural scheme of signal processing of the 802.11a/g/n standard is presented in Fig. 9.51.

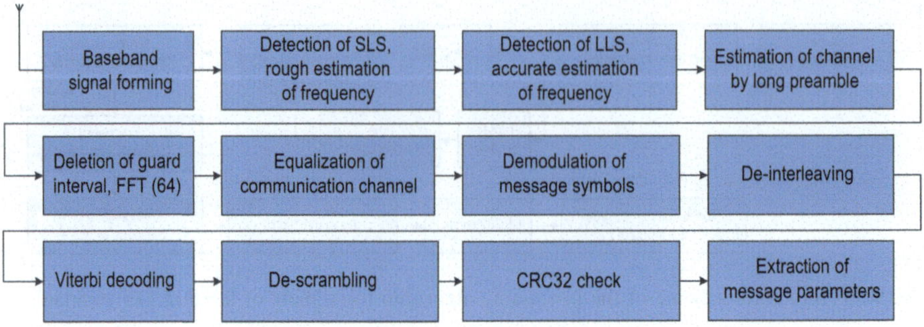

Fig. 9.51 Structural scheme of the processing algorithm of 802.11a/g/n standards signal

Signal detection is executed in two stages: At the first step, SLS is detected and rough frequency–time adjustment is performed, after that the LLS search and accurate frequency–time synchronization are fulfilled. The decision about presence of short preamble is made by means of comparison of resulting statistics m_n with the threshold h_1 of the first stage:

$$m_n = \left| \sum_{l=0}^{L-2} \frac{\sum_{j=0}^{J-1} z_{n+j+l\cdot J} \cdot z^*_{n+j+(l+1)J}}{0.5 \cdot \sum_{j=0}^{J-1} \left|z_{n+j+l\cdot J}\right|^2 + 0.5 \cdot \sum_{j=0}^{J-1} \left|z_{n+j+(l+1)\cdot J}\right|^2} \right|^2 <> h_1, \qquad (9.12)$$

where z_n is the sequence of complex samples of the input signal, $L = 10$ is the quantity of SLS symbols, $J = 16$ is the number of samples in one short OFDM symbol of the learning sequence.

The rough estimation of frequency mismatch between the transmitter and the receiver is

$$\Delta f = \frac{1}{2\pi \cdot \tau_0} \cdot \arg \sum_{l=1}^{L-2} \sum_{j=0}^{J-1} z_{\theta+j+l\cdot J} \cdot z^*_{\theta+j+(l+1)J}, \qquad (9.13)$$

where τ_0 is duration of one symbol of the short learning sequence, «arg» is a phase of complex number within the limits $\pm\pi$, $\theta = \arg\max_n m_n$ is estimation of SLS time position.

In the case of SLS detection (exceeding the threshold h_1 by resulting statistics m_n), the long preamble symbol detection is executed with compensation of rough estimation of Δf frequency mismatch. The decision function of the second stage is compared with the h_2 threshold and the packet is considered as detected in the case of threshold exceeding:

$$U_n = \frac{\left|\sum_{s=0}^{S-1} a_s^* \cdot z_{s+n} \cdot \exp\left(-j \cdot 2\pi\Delta f \cdot \frac{(i+n)}{F_s}\right)\right|^2}{\sum_{s=0}^{S-1} |z_{n+s}|^2} <> h_2, \tag{9.14}$$

where a_s is known LLS, S is the number of samples in one OFDM symbol of LLS, $n = \tau_1 \ldots \tau_1 + 3S$ is the search interval of decision function values, and τ_1 is the estimation of the time position of LLS beginning obtained on the base of the decision function maximum m_n at the first stage of detection.

In the case of packet detection, the final estimation of frequency mismatch is defined as follows:

$$\Delta f = \Delta f + \frac{1}{2\pi\tau_1} \cdot \arg\left(U_{\max 1} \cdot U_{\max 2}^*\right), \tag{9.15}$$

where τ_1 is duration of the LLS OFDM symbol, $U_{\max 1}$ is the function U_n value corresponding to the first maximum, and $U_{\max 2}$ is the function U_n value corresponding to the second maximum. The exact estimation of the time position of packet beginning is defined according to the time position of the $U_{\max 1}$ first maximum.

The signal time samples are corrected taking into account the final estimation of the frequency mismatch (9.15):

$$U_n = U_n \cdot \exp\left(j \cdot 2\pi\Delta f \cdot \frac{(i+n)}{F_s}\right). \tag{9.16}$$

At that, the process of frequency–time synchronization is considered as complete. The initial channel estimation $\dot{H}_{k,0}$ is performed according to LLS:

$$\dot{H}_{k,0} = \frac{1}{2} \cdot \frac{\dot{Y}_{k,0} + \dot{Y}_{k,1}}{L_k}, \tag{9.17}$$

where k is the sub-carrier number, L_k are known values of LLS sub-carriers, $\dot{Y}_{k,r}$ is the result of FFT, which is executed for each block corresponding to the useful part of the OFDM symbol, $r = 0..R-1$ is the number of the OFDM symbol (numbers 0 and 1 corresponding to the LLS symbols), R is the total number of OFDM symbols in the message.

The channel estimation is used for alignment of OFDM symbol sub-carriers:

$$\dot{Y}eq_{k,r} = \frac{\dot{Y}_{k,r}}{\dot{H}_{k,r}}. \tag{9.18}$$

Aligned sub-carriers proceed to the demodulation block, which forms soft solutions of coded bits. For instance, for determination of QPSK soft solutions, the following variables are calculated:

$$\begin{aligned} \Omega_i^x(k,r) &= \left[\mathrm{Re}\left(Yeq_{k,r}\right) - \mathrm{Re}(z)_i/\gamma\right]^2,\ i = 0\ldots\sqrt{M}-1, \\ \Omega_i^y(k,r) &= \left[\mathrm{Im}\left(Yeq_{k,r}\right) - \mathrm{Im}(z)_i/\gamma\right]^2,\ i = 0\ldots\sqrt{M}-1 \end{aligned} \tag{9.19}$$

where $\mathrm{Re}(z)_i$ are values of modulation points along the x axis from the left to the right, $\mathrm{Im}(z)_i$ are values of modulation points along the y axis from bottom to up. For QPSK, M = 4, $\gamma = \sqrt{2}$. Similarly, soft solutions are defined for other modulation types: BPSK, 16-QAM, 64-QAM.

After demodulation of each OFDM symbols, the soft solution stream passes to the block of de-interleaving, Viterbi decoding, and de-scrambling. In the case of decoding results coincidence with the control sum, the message is considered as true decoded and it is sent to the block of message type determination and extraction of necessary broadcast information.

9.8.3 Destination and Functional Possibilities of the Wi-Fi Networks and Access Points Analyzer

The hardware–software complex of signal analysis for wireless communication systems SMO-BS Wi-Fi provides:

- Detection of wireless networks of wide-band data transmission of IEEE 802.11 a/b/g/n standards with the bandwidth of 20 MHz in the frequency ranges of 2.412–2.484 GHz and 5.17–5.905 GHz at using of the external adapter;
- Detection of wireless networks of wide-band data transmission of IEEE 802.11 a/b/g/n standards in the whole operating frequency range in the bands of 5, 10, 20 MHz for the panoramic receiver ARGAMAK-IS and 5, 10, 20, 40 MHz for double-channel radio receiver ARC-D11;
- Determination of radio equipment parameters included in the structure of wireless networks of wide-band data transmission: sub-system identifiers (SSID) of wireless networks; the hardware (MAC) addresses of network equipment; numbers of used channels and frequency nominal values; signal levels; types of devices;
- Analysis of network topology according to the intercepted packets: automatic determination of connected device quantity, their physical (MAC) addresses and type of connection;
- Recognition and identification of signals of legally acting radio electronic units of wireless networks of wide-band data transmission and illegally acting transmitters according to the results of comparison with the database;
- Determination of radio equipment localization for wireless networks of wide-band data transmission, with construction of the zone of probable radio

equipment localization and representation of obtained radio monitoring results on the electronic map of locality with coordinates affixment according to GPS/GLONASS data;
- Possibility of sorting and selection of obtained data by different signs (including zone in the locality, current activity, belonging to the given network, presence/absence of the license, etc.);
- Storage of analysis results in the database with possibility of further automatic monitoring of the situation variation.

The analyzer consists of the hardware part—the receiver ARGAMAK-IS or ARC-D11, or external adapter shown in Fig. 9.52, and the software part. The last can be installed using the unified packet with other software, for instance, performing the signal analysis of base station signals of different cellular communication systems. When using of the external adapter, the analyzed operation is possible without digital receiver, however, scanning is fulfilled only at the standard frequencies assigned for Wi-Fi and signals in the band of 20/22 MHz are exposed to detection. At utilization of ARGAMAK-IS or ARC-D11 receivers, the search for Wi-Fi signals is possible at arbitrary frequencies in the range from 9 kHz to 8 GHz, as well as search for 802.11a/g/n standards signals additionally with the bandwidth of 5, 10, and 40 MHz.

At utilization of DRR, the hardware part of the complex provides the baseband signal formation at the required sampling frequency, and also Wi-Fi signal detection at the monitored frequency, which essentially increase the analyzer performance. At signal presence, the system mathematical software provides its further processing and extraction of broadcast information.

External view of the main window of the SMO-BS Wi-Fi software is presented in Fig. 9.53 [7].

In the left part of the window, on the "Detected (I)BSS" panel, the list of sub-networks and their clients is represented, which is shown in the form of the hierarchical structure. The following parameters are registered for each (I)BSS:

- Network type (Ad hoc, Infrastructure);
- Service Set IDentifier (SSID);
- Basic Service Set IDentifier (BSSID);
- Hardware addresses of sub-network clients.

Fig. 9.52 External Wi-Fi adapter

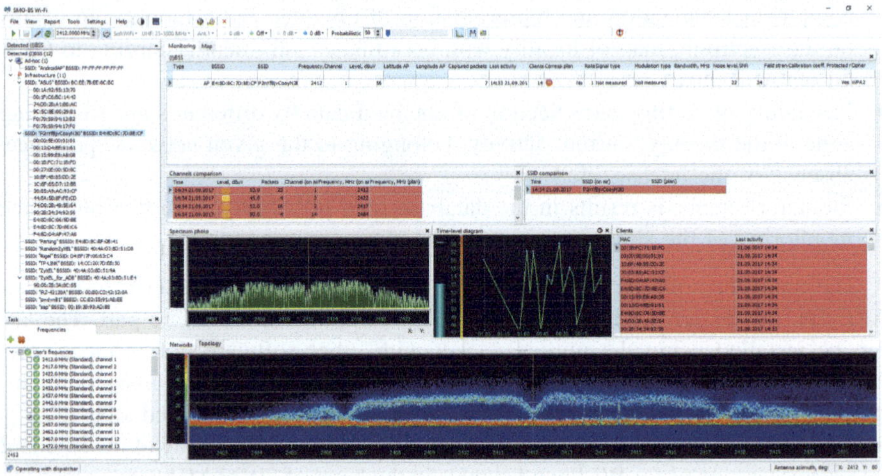

Fig. 9.53 Main window of the SMO-BS Wi-Fi software

In the right part of the window, there are “Monitoring” and “Map” tab sheets. In the upper part of the “Monitoring” tab sheet, there is a table of measurements accumulated during scanning. There are also two tables of frequency-spatial plan (FSP) monitoring: “Channel comparison” and “SSID comparison,” which verify the channel/SSID in ether and in the plan. In the lower part of the tab sheet, the graphical panel with photograph of the signal spectrum in the bandwidth of 20 MHz is represented at the moment of measurement execution. In the “Time-level diagram” panel, the signal-level graph for chosen node is represented in the “Detected (I)BSS” tree. The “Clients” tab sheet contains information (MAC-address and time of the last activity) about clients connected to the network.

In the “Topology” tab sheet, shown in Fig. 9.54, the arrangement of detected sub-networks is graphically represented. Sub-networks of “Infrastructure” type are represented in the form of a “star” diagram, in the center of which there is the access point, and beams outgoing from it are clients connected to the given sub-network. Sub-networks of “Ad hoc” type are represented as “ring” diagrams, in which all clients are distributed at the circumference, and each of them are connected consequently by the line with the following. Clients are presented in the form of rectangles, in which MAC-address and SSID are represented.

At operation of the hardware–software complex on the base of the mobile radio monitoring station and at measurement of the signal level in different locality points, it becomes possible to calculate access point location. For this, for each access point, the own operation zone is automatically built. Then, starting from signal powers obtained from the access point in different points of measurement, the estimation of localization probability in the zone is performed. Obtained estimation is visualized in the form of the probability matrix on the map as shown in Fig. 9.55.

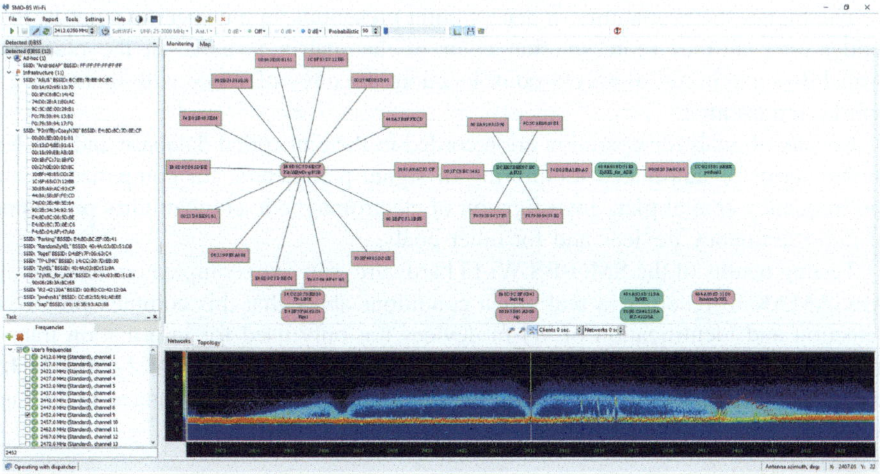

Fig. 9.54 Representation of network topology on the analysis results

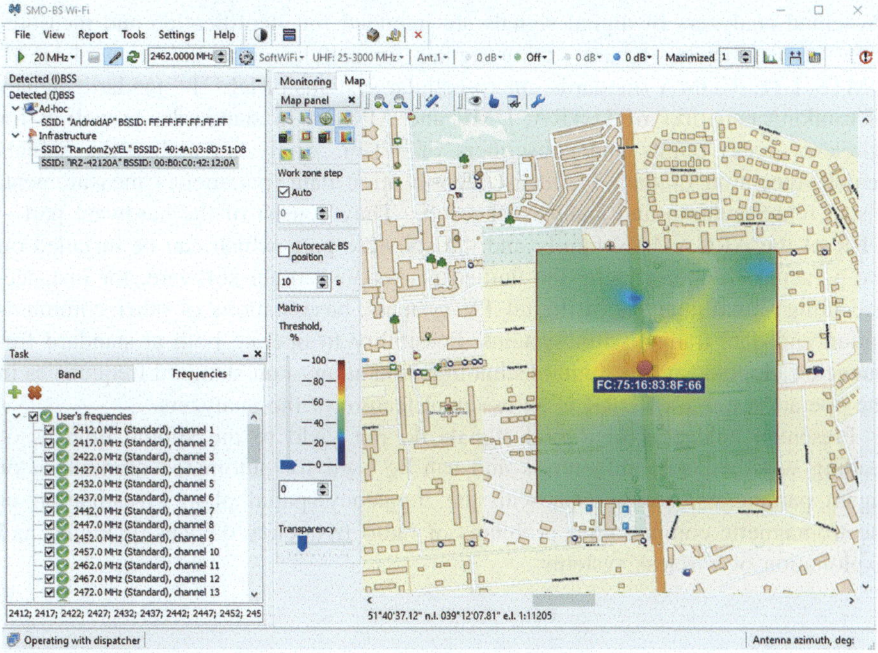

Fig. 9.55 Determination of Wi-Fi access point localization

The higher the probability of access point presenting in the specific point is, the redder area shift of representation color of the matrix is used. In the place of probability maximum of access point location, the circumference with its identifier will be represented.

Results of analyzer operation are recorded in the embedded database and can be further used for automated monitoring of signal parameters, for comparison with the frequency-spatial plan, investigation of electromagnetic compatibility problems of radio frequency devices and for other goals.

Testing results of the SMO-BS Wi-Fi hardware–software complex on the base of ARGAMAK-IS receiver in real urban conditions show that this complex provides detection and identification of Wi-Fi devices not only used for information transmission in the bandwidth of 20/22 MHz and frequency ranges of 2412–2484 MHz and 5170–5320 MHz, but operating in non-licensed frequency ranges including with bandwidths of 5 and 10 MHz.

9.9 Summary

Presented analyzers of digital signals are intended for identification and measurement of parameters of base station signals for cellular communication of GSM, UMTS, LTE, CDMA standards, micro-cellular systems of DECT standard, signals of trunking networks of TETRA, DMR and APCO P25 standards, as well as for wireless access points and subscribers of Wi-Fi. Analyzers realize the signals search, reception of their broadcast identification data, parameters measurement, visual representation and analysis of results. They consist of the hardware part—DRR of the ARGAMAK-family and the software part, which can be installed on the personal computer using the unified packet with other software, for instance, executing signal analysis of digital TV systems, base stations of other communication and data transmission systems. Possibility to operate both at standard frequencies adopted for this digital technology and at any non-standard frequencies in the operation range of DRR is an essential feature of the analyzers.

Presented analyzers of digital signals do not yield to the similar devices of leading worldwide manufacturers and can be used for automated monitoring of signal parameters, comparison with the frequency-spatial plan, investigation of electromagnetic compatibility problems of radio frequency devices, planning and exploitation of wireless systems.

References

1. The catalogue 2017 of IRCOS JSC. http://www.ircos.ru/zip/cat2017en.pdf. Accessed 28 Nov 2017
2. Rembosky AM, Ashikhmin AV, Kozmin BA (2015) Radiomonitoring: problems, methods, means (in Russian). In: Rembosky AM (ed) 4th edn. Hot Line-Telecom Publ., Moscow, 640 p

3. Alexeev PA, Ashikhmin AV, Bespalov OV, Kaioukov IV, Kozmin VA, Manelis VB (2016) GSM, UMTS, LTE base station signal analyzer (in Russian). Spetstehnika i Svyaz (4):50–59
4. Haine G (1999) GSM Networks: Protocols, Terminology and Implementation. Artech House, London, p 447
5. Castro JP (2001) The UMTS network and radio access technology: air interface techniques for future mobile systems. Wiley, 354 pp
6. Sesia S, Toufik I, Baker M (2009) LTE—The UMTS long term evolution, From theory to practice. Wiley, 611 p
7. SMO-BS software packages for wireless data communication system analysis. http://www.ircos.ru/en/sw_bs.html. Accessed 28 Nov 2017
8. Kayukov IV, Manelis VB (2006) Comparative analysis of different methods of signal frequency estimation (in Russian). Radioelectron Commun Syst 49(7):42–56
9. SMO-KN navigation and cartography software package. http://www.ircos.ru/en/sw_kn.html. Accessed 30 Nov 2017
10. ETSI EN 300 392-2 V2.3.2. Terrestial Trunked Radio (TETRA); Voice plus Data; Part 2: Air Interface. European Telecommunications Standards Institute, 2001
11. Prosch R, Daskalaki-Prosh A (2015) Technical handbook for radio monitoring VHF/UHF. Edition 2015. Books on Demand GmbH, Norderstedt, Germany, 437 p
12. Bocharov DN, Kayukov IV, Kozmin VA, Manelis VB (2016) DMR radio station signal analyzer. (in Russian). Spetstehnika i Svyaz (4):106–110
13. ETSI TS 102 361-1 Electromagnetic compatibility and Radio spectrum Matters (ERM); Digital Mobile Radio (DMR) Systems; Part 1: DMR Air Interface (AI) protocol V2.3.1 (2013–07)
14. ETSI TS 102 361-2 Electromagnetic compatibility and Radio spectrum Matters (ERM); Digital Mobile Radio (DMR) Systems; Part 2: DMR voice and generic services and facilities V2.2.1 (2013–07)
15. ETSI TS 102 361-3 Electromagnetic compatibility and Radio spectrum Matters (ERM); Digital Mobile Radio (DMR) Systems; Part 3: DMR data protocol V1.2.1 (2013–07)
16. ETSI TS 102 361-4 Electromagnetic compatibility and Radio spectrum Matters (ERM); Digital Mobile Radio (DMR) Systems; Part 4: DMR trunking protocol V1.6.1 (2014–06)
17. Ashikhmin AV, Bocharov DN, Kozmin VA, Kryzhko IB, Sladkih VA (2016) APCO P25 radio station signal analyzer (in Russian). Spetstehnika i Svyaz (4):111–114
18. Project 25 Document Suite Reference (2010)
19. Tia Standard. Project 25 FDMAFDMA—Common Air Interface. ANSI/TIA-102.BAAA (2003)
20. Tia Standard. Project 25 Vocoder Description. ANSI/TIA-102.BABA (2003)
21. Standard for Information technology—Telecommunications and information exchange between systems—Local and metropolitan area networks—Specific requirements. Part 11: Wireless LAN Medium Access Control (MAC) and Physical Layer (PHY) Specifications. IEEE Std 802.11, 2007, 1184 p
22. Alexeev PA, Kozmin VA, Kryzhko IB, Sladkikh VA (2016) Determination of the wi-fi network and access points parameters (in Russian). Spetstehnika i Svyaz (4):29–36
23. Roshan P, Leary J (2004) 802.11 Wireless LAN Fundamentals. Cisco Press

Conclusion

In this book, the problems of the structure and functioning of geographically distributed automated radio monitoring systems are considered, which were developed, manufactured, and delivered by the Russian Company IRCOS, including hardware for radio monitoring, software, and engineering infrastructure.

It is shown that delivered automated systems are based on the unified hardware–software platform, in which the technical radio monitoring equipment are included: stationary, mobile, and manpack radio receivers and radio direction-finders on the base of digital receivers of the ARGAMAK-family, as well as the SMO-ARMADA software. The client/server architecture consisting of the "thin" clients, the application, and data servers has general properties providing the effectiveness of automated system utilization independently on the system scale and purposes. The following can be attributed to them:

- Hierarchical scaling architecture, application of Web geo-information technologies;
- Possibility to use radio monitoring equipment of various manufacturers on the base of the open unified control protocol;
- Simplicity of interaction with external information systems;
- Fulfillment of radio monitoring tasks in the automatic (planned, background) and manual (operation) modes;
- Eventual mechanism of task, diagnostic and self-diagnostic performing.

At the same time, the composition, configuration, functions of the particular automated system are defined by purposes and tasks being arisen in front of the user of this system. In this book, examples of the specific automated system construction are given, which are intended for application in civil and military agencies: ARMADA, AREAL, and ASU RCHS UNIVERSIADA-2013 systems.

Technical equipment for radio monitoring on the base of radio receivers of the ARGAMAK-family included in the structure of automated systems provide fulfillment of high-precision measurements, spectral and vector analysis of signals, analysis of service identifiers and telecommunication system parameters of signals of standards such as APCO P25, DMR, GSM, UMTS, LTE. Synchronization from global navigation satellite systems (GLONASS) and GPS made possible the

A. M. Rembovsky et al., *Radio Monitoring*, Signals and Communication Technology, https://doi.org/10.1007/978-3-319-74277-9

implementation of the time difference of arrival method for estimation of radio emission source coordinates, which does not require goniometric direction-finder's utilization.

All automated systems have central database for radio monitoring, in which results from the local databases of measuring and direction-finding stations are accumulated. In turn, the accounting and reference data, the planned and operative tasks are transmitted to the radio monitoring stations. The control of the radio monitoring equipment is performed from any automated workstation if its operator has the appropriate access rights.

Automated systems give a possibility of operation in operative and planned modes. In operative mode, the remote control of equipment is performed by the system personnel in the real time. At operation in the planned mode, the set of measurements, the time period, in which these measurements should be done, as well as the list of equipment, which should perform the specified measurements, are specified. Then, the measurements are performed in the automatic mode.

Modern solutions used at the development of the software cause the following positive features of automated systems under consideration:

- Cross-platform property—a possibility of functioning under Windows and Linux operation systems;
- Scalability—a possibility of system deployment both at the separate computer and in the scale of the large country;
- Fault tolerance—a preservation of efficiency under conditions of the partial equipment fault or quality degradation of communication channels up to its loss, determination, and prediction of malfunction appearance on the base of continuous testing of equipment and communication channels' condition;
- Openness—a possibility to exchange radio monitoring data with other information systems, as well as application of technical radio monitoring equipment from various manufacturers;
- Simplicity of the software upgrade, which is performed on the central system server, and then, the upgrade is fulfilled automatically on the secondary servers;
- Separate data utilization—each system operator is associated with the specified agency (or several agencies), which eliminates a possibility of the unauthorized access to data from another agency sections.

The operation effectiveness of the automated radio monitoring system depends not only on perfection of the radio monitoring equipment but also on features of auxiliary sub-systems forming the engineering infrastructure, in which control centers and points, premises, transport basis of radio monitoring stations, system of data transmission are included.

All interaction of the units of the automated system is provided within the limits of the data transmission system, based on protocols with data protection against the unauthorized access. At that, to endure the reliability, scalability, and high performance, the multi-level network is used.

The key components, which connect the network nodes, are communication channels, which may be wired and wireless. Wireless technologies are used in the case where application of wired technologies is difficult or impossible according to technical reasons, for instance, at connection of the mobile radio monitoring stations, or with the aim of arrangement of reserve channels.

ARMADA and ASU RCHS UNIVERSIADA-2013 systems presented as the examples have the hierarchic structure, at the head of which there is the control center. When ARMADA system is used on the country scale, here are additional control points in it, which decrease the response time of the radio frequency service on the local conditions, for instance, appearance of spurious emissions of unauthorized radio frequency equipment, interference arising.

In the automated ARMADA system, the fixed, mobile, portable, and manpack radio monitoring equipment find their application. Their composition and quantity depend on many factors: locality character, type of building up, concentration and types of acting radio frequency equipment, etc. Permissible system cost is also of no small importance factor. To improve economic efficiency, the organization can put the two-stage radio monitoring into the service. Facts of violation can be revealed by comparatively cheap equipment, violation registration and protocols generation can be executed on the base of measurements made by high-accurate measuring equipment in this case. Metrological characteristics of the used equipment satisfy to norms and regulations accepted in Russian Federation and ITU recommendations.

The automated system ASU RCHS—UNIVERSIADA 2013 is a variety of the ARMADA system; it is intended for radio monitoring execution at large international sport actions. The peculiarity of the system is a presence of additional functions, including management of request passage for utilization of radio frequency equipment, bringing in by delegations, their testing and marking. ASU RCHS—UNIVERSIADA 2013 during preparation and conduction of summer Students Games in Kazan-city provided remote control of territory-distributed fixed, mobile, and manpack radio monitoring equipment, testing and marking of radio frequency equipment, interaction with external informational structures. The system enabled effective control of the personnel, coordinated tasks formulation, monitoring of their execution and making necessary decisions in the real time.

A great attention in the book is paid to the task of revelation of unauthorized radio emissions in the objects under monitoring, to features of the search and localization of its sources. Integration of radio monitoring equipment into the automated AREAL system permits to solve this task for twenty-four hours a day in the automated mode informing the system personnel about situations appearance, which require the urgent consideration. The centralized control of radio monitoring equipment enables minimization of automated workstations' number, unification approaches to database conducting according to detected radio signals. Several typical variants of the automated AREAL system structure, which depend on the search and localization conditions of unauthorized emission sources, are presented.

A variant of automated AREAL-1 system is used at operation along the lengthy boundary of the territory under monitoring. The system consists of stationary and mobile radio direction-finders. The AREAL-2 systems are used at single or periodic radio monitoring in the zone under monitoring and AREAL-3 provides enhanced study of RES in the restricted and adjacent territories. It also has mobile, manpack, and portable radio monitoring equipment in its structure. For providing the extended studying of the radio electronic medium and permanent monitoring on the protected territory with fixed boundaries, the AREAL-4 system is used, which consists of the complexes of technical analysis of radio signals. For revelation of unauthorized radio emission sources inside premises and transport facilities, the automated AREAL-5 system is used, in the structure of which the following is included: the distributed ARC-D13R or ARC-D15R system for remote radio monitoring, manpack complexes ARC-D11, manual radio direction-finders ARC-RP3M.

In this book, the constructive features, structural diagrams, technical characteristics, and principles of functioning of used digital equipment for radio monitoring, such as measuring receiver ARGAMAK-IS, RF sensor ARGAMAK-RS, measuring stations for radio monitoring and direction-finding ARCHA-INM, ARCHA-IT, automated radio direction-finders ARTIKUL-S and ARTIKUL-M, handheld radio direction-finder ARC-RP3M, and other equipment, are considered in detail. It is shown that the small weight, dimensions, and the consumed power of the ARGAMAK-family receiver allow its integration directly into the antenna system of the radio monitoring equipment. Such a solution gives a series of advantages compared to traditional receiver accommodation outside the antenna system, which requires lengthy high-frequency feeder line. We can single out the following advantages:

- Preservation of the dynamic range of radio monitoring equipment at the level of the used receiver;
- Reduction or complete elimination of the harmful influence of the antenna effect onto reception and direction-finding of the useful signal;
- Simplification of calibration of measuring systems' sections.

We pay the essential attention in this book to problems concerning the development of mathematical methods and algorithms of digital signal processing in tasks of radio monitoring, including sampling of received signals and formation of quadratic components, complex filtering, variation of sampling frequency, detection and direction-finding of signals, correction of the receiver paths, measuring of radio signals and interference parameters, as well as algorithms of identification and localization of emission sources including the address direction-finding of digital signals, which occupy the same frequency band.

The general approach to solution of the problem of radio emission source localization on the base of the main types of radio engineering measurements is given: amplitude, goniometric, time difference of arrival, frequency, and temporal. The choice of the specific scheme of radio emission source localization depends on the type and parameters of the radio signal, which source must be localized, and the technical equipment of radio monitoring services. Independently on the measurement type, the sequence of steps to solve the localization problem remains the following:

- Formation of the mathematical model of measurements;
- Choice and minimization of the functional;
- Formation of the initial approximation;
- Linearization of the initial task—its reduction to the task "in the small";
- Solution of the problem "in the small" with estimation of obtained solution accuracy.

The algorithms of service parameters' analysis of radio signals are described in this book including signals for digital TV, cellular and trunking networks, networks of wireless access. The implementation of developed algorithms is shown in the form of appropriate modes of software providing the search for digital signals, extraction of their broadcasting identification data, measurement of parameters, visual mapping, and results analysis.

The authors will be grateful to readers for notes and comments concerning the book content, which can be directed in mail: Russian Federation, 129626, Moscow, PO Box 30, "IRCOS" or by e-mail: info@ircos.ru.

Index

A. M. Rembovsky et al., *Radio Monitoring*, Signals and Communication Technology, https://doi.org/10.1007/978-3-319-74277-9

C

D

E

I

J

L

M

N

The manufacturer's authorised representative in the EU is Springer
Nature Customer Service Centre GmbH, Europaplatz 3, 69115 Heidelberg,
Germany. If you have any concerns regarding our products, please
contact ProductSafety@springernature.com

Printed and bound by CPI Group (UK) Ltd, Croydon, CR0 4YY
05/10/2026
02228144-0017